www.KnowledgePublications.com

How to Save Energy and Cut Costs in Existing Industrial and Commercial Buildings

An Energy Conservation Manual

Fred S. Dubin
Harold L. Mindell
Selwyn Bloome

Library of Congress Catalog Card Number: 76-40570
ISBN: 0-8155-0638-4
Printed in the United States

Published in the United States of America by
Noyes Data Corporation
Noyes Building, Park Ridge, New Jersey 07656

www.KnowledgePublications.com

FOREWORD

The gap between national energy supply and demand and the resulting escalation of operating costs we are now experiencing cannot be reduced or eliminated without a concerted and continuous effort to reduce the national energy consumption through the application of conservation practices. Contrary to common belief energy conservation does not mean sacrificing our living standards or living in a bleak utilitarian society. It can be accomplished without discomfort or sacrifice merely by the more efficient utilization of our nonrenewable energy resources.

Buildings consume for heating, air conditioning, lighting, and power more than 33% of all energy used in the United States, the equivalent of 10 million barrels of oil per day at a time when our oil imports are approximately six million barrels per day. Reducing energy used in all buildings by 30% without impairing the indoor environment is a realistic possibility and is the equivalent of cutting our present oil imports by half.

Most of the buildings now in use were designed and constructed when fuels and electric power were readily available and inexpensive and the need for energy conservation was not recognized. The structures and their mechanical and electrical systems were designed to minimize initial costs, not energy usage. Buildings are generally overheated in the winter, overcooled in the summer, overlighted, overventilated year-round, and not operated efficiently. Each year they consume increasing amounts of energy because systems and building components deteriorate as maintenance and service becomes more costly and neglected.

Energy conservation programs which have been undertaken within the past three years in many thousands of existing commercial, institutional and residential buildings have already resulted in a reduction of their annual fuel and electricity consumption of 20-50% and indicate the range of potential savings for almost all buildings now in use.

This book is based on the reports prepared for the Federal Energy Administration by Dubin-Mindell-Bloome Associates. The results reported are compiled in this energy conservation manual (ECM).

Part 1 of this manual (ECM-1) is directed primarily to owners, occupants, and operators of buildings. It includes a wide range of opportunities and options to save energy and operating costs through proper operation and maintenance. It also includes minor modifications to the building and mechanical and electrical systems which can be implemented promptly with little if any investment costs. The measures contained in ECM-1 would result in energy and operating cost savings of 15 to 30% based on present fuel costs.

Part 2 of this manual (ECM-2) is intended for engineers, architects, and skilled building operators who are responsible for analyzing, devising, and implementing comprehensive energy conservation programs which involve additional and more complex measures than those included in ECM-1. ECM-2 includes many energy conservation measures which can result in further energy savings of 15 to 25% with an investment cost that can be recovered within 10 years through lower operating expense.

The urgency of the energy crisis has necessitated the immediate development of these guidelines. They should be used now to conserve energy, reduce operating costs and alleviate hardships as fuel and power supplies become less available.

ACKNOWLEDGMENTS

Many individuals participated in the research for the preparation of these manuals—Barry D. Symmonds, the project manager, Selwyn Bloome, P.E., the principal advisor, Robert Sparkes, P.E., Larry Lawrence, P.E., Lou Mutch, P.E., Philip Fine, David Hartman, Karl Christensen, Jerry Martinchek, Eve McDougal, Bernice Beharry and Steve Selkowitz were the persons most deeply involved in assisting Fred S. Dubin, P.E., the managing partner for this project.

Thomas Ware and David Hattis of Building Technology, Inc., Charles McClure of Charles J.R. McClure, P.E., Associates and Raymond Firmin of G.A. Hanscomb Partnership provided valuable assistance on special problems.

The full cooperation of the Federal Energy Administration is acknowledged, especially Dr. Maxine Savitz, Steve Cavros, Robert Jones, and David Rosoff.

A number of federal agencies, especially personnel from the National Bureau of Standards and the General Services Administration, were particularly helpful in reviewing criteria and offering critical comments during the preparation of the manuals.

Members from ASHRAE, the Energy Committee of the ACEC, the A.I.A., J.C. Blake of Charles E. Smith Companies, Building Owners and Managers Association, Commercial Refrigerator Manufacturer's Association, Ross & Baruzzini, consulting engineers, George Clark of Sylvania Corp., and many other professional engineers and architects gave unstintingly of their time to provide information helpful in preparing these guidelines.

Many manufacturers, too numerous to name, were extremely helpful in providing specific information on the performance of mechanical and electrical equipment.

Fred S. Dubin, P.E.
Harold L. Mindell, P.E.
Selwyn Bloome, P.E.
Dubin-Mindell-Bloome Associates
New York, New York

In order to keep the price of this large and extensive book to a reasonable level, it has been reproduced by photo-offset directly from the original typewritten report. Certain pages are blank, just as in the original report.

TABLE OF CONTENTS

PART 1

BUILDING OWNERS AND OPERATORS MANUAL

ECM-1

Introduction and Scope

ECM-1

INTRODUCTION AND SCOPE

Conserving energy means reducing the amount of fuel and electricity which your building or space uses every month. Reducing fuel and electrical consumption saves money. Cost savings made now will be increasingly greater in the future as raw fuel and electricity prices continue to rise.

Fuel and electricity consumption cost savings are important, but by no means the only reasons for you to conserve energy:

- Energy Conservation can extend the useful life of existing equipment and eliminate the need for early replacement costs.
- Energy Conservation can increase the reserve capacity of the existing central plant systems and meet future building extensions without installing extra boilers, chillers or transformers.
- Energy Conservation reduces the likelihood of shutdown or curtailment of operations due to fuel or power shortages-inevitable if demand continues to outstrip supply.
- Energy Conservation reduces airborne pollution resulting from combustion of oil, gas, or coal and may save installation costs of pollution control equipment.
- Energy Conservation means reduction of waste permitting fuel economy without reduction of health and comfort standards, or curtailment of building services and function.
- Energy management, a form of conservation reduces peak electric loads and electric power demand charges while easing the burden on existing power generating and distribution systems.

In addition to the immediate advantages of Energy Conservation for you, there are advantages to the nation as a whole.

- Energy Conservation conserves natural resources.
- Energy Conservation can enhance economic opportunity where materials and labor are required to improve building thermal characteristics.
- Energy Conservation reduces the need for oil imports and the dependency upon external sources for the internal economic well being and security of this country.
- Energy Conservation combats inflation.

Purpose of this Manual

ECM-1 has been prepared to help you, a building owner, manager, operator or occupant, to conserve and manage energy usage now, without the necessity of investing a significant amount of money to do so. In order to conserve energy, you must understand all of the following factors which are included in ECM-1:

- The basis principles of Energy Conservation and the inter-relationships between specific climatic conditions, building structure, building use, and each of the mechanical and electrical systems.

- The systems, in order of priority, which use the most energy and provide the greatest potential for conservation.
- Methods to identify your building by type, size, function, construction materials and features, mechanical and electrical systems, operation, geographic location (climatic influences) and present energy usage.
- Suggested standards for indoor environmental conditions during occupied and unoccupied periods for spaces with differing functions within the building.
- The opportunities for Energy Conservation with options and guidelines; and the charts, graphs and tables to enable you to quantify the potential savings in fuel and electricity for your own building.

ECM-1 is not strictly a design manual, nor a cost estimating manual, but rather a set of guidelines and procedures to reduce energy usage by reducing waste, and to save energy through more effective operation of the building and its mechanical and electrical systems.

The material, labor and energy costs which are noted or used in the examples throughout the manual are cited to help you gain a perspective of the relative costs and benefits for specific energy conservation measures. The costs are based on specific selected conditions and may vary widely even within the same city. The costs are indications of order of magnitude only. DO NOT ADOPT THEM AS EXACT QUANTITIES.

ECM-1 and ECM-2 were prepared specifically for conservation of energy in existing office buildings, retail stores, and buildings which are generally occupied only a few hours a week such as religious buildings. Many of the examples and analyses refer to these three types of structures. However, most of the measures and operating procedures are equally applicable to other types of existing and new buildings as well, and can be used for hospitals, schools, universities, libraries and other institutional buildings, houses, apartments, warehouses, and industrial buildings (excluding the industrial processes).

ECM-1 and ECM-2 are summations of energy conservation measures which require neither new inventions, new systems, nor exotic solutions. The tables of values, charts, and graphs which are included are based on the authors' analysis, computations, computer programs, judgement and observations.

The opportunities which are described are not universally applicable. Each building is unique, even though others may be of the same type and general appearance. The manuals suggest ideas for consideration with corresponding costs and benefits, but those which are appropriate for any individual building can be determined only after a thorough analysis of the specific building.

Many of the ideas outlined in these guidelines can be executed directly by the building occupants, owners, or managers without delay or the need for further advice; while other measures, even though they do not entail significant capital costs, may require further analysis by an engineer, architect, utility company or service and maintenance organization. Owners should not hesitate to consult these sources. An engineer can complete a preliminary anlysis of energy options to be considered for further analysis, even for a large building in just a few days time.

ECM-1 provides case histories of energy and cost savings for two buildings which have already instituted energy conservation programs, and presents other examples of the calculated savings in energy and dollars for selected major energy conservation options. These two large buildings were able to save \$758,000 per year and \$434,000 per year respectively, with very little capital costs, by following the recommendations of professional engineers retained by the owners. Small buildings can also show proportionate, but dramatic savings as well.

A partial summary of the procedures we suggest you follow to save energy and operating costs for your building are outlined below. The Federal G.S.A. operates more than 10,000 buildings in the United States and by following many of these procedures, they report an average savings of 30% in energy use last year. (Consult Sections 3 thru 10 for a more comprehensive description of major energy conservation opportunities and detailed procedures and guidelines to implement the opportunities.)

1. Walk through your building. Are there areas that are unoccupied, or which can be vacated by making better use of the remaining areas? If so, turn off air conditioning, lights, ventilation and heating (where freezing is not a hazard) permanently. Isolate these areas from other spaces by doors, walls, or other means. If 10 or 15% of the building can be vacated, energy savings will follow almost in the same proportion.

2. Repair broken windows and leaking pipes or ducts; clean filters, radiators, light bulbs and fixtures; caulk leaks around doors, windows, louvers and openings. In many cases, 5 to 15% energy savings are possible, especially in cold climates where infiltration or cold air increases the heating load and causes your heating system to operate longer hours.

3. Shut off lights where not needed. Post colored signs alongside the switch to remind the occupants to do so.

4. Lower thermostats to 68°F in occupied areas during the heating season, and even lower in less critical areas. Lower the relative humidity settings to 20% in the winter. Raise thermostat settings to 78°F* or higher in the summer if your building is air conditioned, and shut off the air conditioner, fans and pumps at night, weekends, and holidays. Savings of 6% to 15% in energy can be realized simply by resetting the control points.

5. Repair all leaky outdoor air dampers, shut off all ventilation systems when the building is unoccupied. Outdoor air which must be heated or cooled often accounts for as much as 30% of the energy used in many buildings. More than half can be saved by night and weekend shutdown since there are more hours in those periods.

6. Have your oil burner and boiler or furnace checked. Clean soot and scale and adjust the firing rate, draft and combustion. 10 to 15% of the heating bill can be saved in many buildings; the colder the climate, the greater the savings.

7. Replace lamps with more efficient ones giving more lumens per watt; remove lamps in unoccupied spaces and disconnect ballasts. Many areas in the building require less illumination than others,reduce lighting levels in less critical areas by removing lamps and disconnecting ballasts. In schools, office buildings, and retail stores, lighting often accounts for up to 40% of all energy used and the heat from the lights also forms a major part of the air conditioning load. 20% to 40% of the energy used for lighting can be saved in many buildings.

8. Clean your windows to let in more natural light. You may find that doing so will permit turning off some of the electric lights near the windows.

9. Set the aquastat lower on your water heater to save energy. In schools, hospitals and housing, domestic hot water often uses from 25 to 40% of the amount of energy required for space heating even in cold climates.

*Exception where terminal reheat systems are installed.

10. There are dozens of other energy conservation opportunities available with little, if any capital costs required (often labor only), depending upon the building orientation, number of windows, the roof and wall materials, the building location and use, and the characteristics of the heating, lighting and air conditioning systems.

In order to take advantage of the opportunities, detailed in Section 4 , you must first understand the particular characteristics of your building and the influences which affect fuel and electricity usage; then analyze each of the measures independently and in combination with each other to assure maximum energy savings without adversely affecting the operation, aesthetic quality, the health and comfort of the occupants.

It is suggested that you also read ECM-2, for additional energy conservation opportunities. These entail more capital expenditure, but the investment can be recovered, through reduced operating costs in relatively short periods of time. ECM-2 includes the procedures to estimate energy savings resulting from modification to the building structure and which can be used in altering or replacing elements of the mechanical and electrical systems with more efficient components. Instructions for estimating costs, methods of economic analysis based on life-cycle cost considerations and suggested energy management plans are also included in ECM-2.

This page blank

Section 1

Principles of Energy Conservation

SECTION 1

PRINCIPLES OF ENERGY CONSERVATION

Before energy can be conserved, the ways in which it is used must be understood. The building structure - i.e. walls, windows, roof - and the passive components of the mechanical and electrical systems - i.e. ducts, pipes, filters, or lighting fixture louvers - do not directly consume energy but they influence the amount that is finally consumed. The primary energy-conservsion equipments, such as coal, oil or gas burners and boilers/furnaces; refrigeration chillers and compressors; motors; and electric lighting bulbs or tubes, consume energy to supply the "building" load and to compensate for the distribution system load.

The term "building" load, as it is used in this manual, refers to the amount of energy in BTU or KW required to maintain desired indoor space conditions and to operate building equipment if the distribution system and energy-conversion equipment were 100% efficient. The distribution or "parasitic" load is a measure of the energy required to deliver energy from the primary conversion equipment to supply the "building" load. The efficiency of the primary energy-conversion equipment (conversion, since the equipment converts fuel to heat and/or electricity to power or light) ultimately determines the actual amount of energy consumed to supply both loads.

Energy usage, then, depends upon two main factors:

1) The magnitude and duration of the loads.
2) The seasonal efficiency of the primary energy-conversion equipment.

"Building" loads will be reduced if the temperature and relative humidity indoors are maintained at lower levels in the winter and at higher levels in the summer; heat loss, heat gain, and infiltration through the building envelope are decreased; ventilation rates are reduced; domestic hot water temperature and quantity are reduced; the level of illumination by electric lighting is lowered; and the number of hours of operation of elevators, business machines, and cooking equipment is reduced.

The "building" load must be considered on a seasonal basis rather than for peak conditions only. Although two buildings may have the same heating load for any particular hour, one of them may have a considerable higher load, on a seasonal basis, than the other, due to the duration of peak conditions. Or, for instance, a religious building, auditorium or warehouse may have a relatively high energy requirement for lighting during a brief period of time, but if the lights are turned off during unoccupied periods, the annual lighting load may be nominal. Generally, the greater the peak loads for any particular system in an individual building, and the longer their duration, the greater will be the seasonal loads. The distribution loads will be decreased by reducing the amount of power required for pumps and fans, reducing heat loss or gain from ducts and pipes, and by eliminating steam, water and air leaks. Distribution loads are often excessive because systems are designed to operate continuously at the maximum capacity required to meet peak "building" loads, even though these peak loads occur for relatively short periods of time. (usually less than 5% of the year)

Peak efficiency is usually based on a one hour performance. Seasonal efficiency, which reflects the average for the entire season, is a better measure. It is the ratio of useful work (in British Thermal Units - BTU's) performed by the equipment over a period of time, to the BTU value of the fuel or electricity consumed by the equipment over the same period.

Reducing "building" loads, then "distribution" loads, and then improving primary conversion equipment efficiency are most effective when done sequentially, since the latter depends upon the magnitude of the first two and the potential for reducing distribution loads depends upon the magnitude of the building loads and upon the operating conditions of the distribution systems.

Changes implemented for a particular purpose often induce secondary effects which also influence energy usage. For instance, reducing lighting levels and increasing the efficacy of the lighting system also reduces the cooling load. On the other hand, it increases the heating load. Additional heat usually can be supplied more efficiently, however, and at lower energy cost, by the heating system rather than the lighting system. In large offices, schools and stores, where lighting is responsible for a large percentage of the energy which is used for cooling, efforts to reduce energy for lighting are particularly important.

A SUMMARY OF THE PRINCIPLES OF AN ENERGY CONSERVATION PROGRAM

-Identify the specific climatic conditions to which the building is exposed. Make a thorough examination of the mechanical and electrical systems and the building structure and determine the normal use and occupancy of the building to become familiar with existing conditions.

-Identify the magnitude of annual energy usage by quantity of fuel and electricity for each system.

-Set a "goal" or target for reduction in energy usage based upon potential savings for each system.

-Reduce the "building" loads which must be satisfied by the mechanical and electrical systems.

-Reduce thermal losses and the power requirements for the HVAC distribution systems. Improve the seasonal operating efficiency on the basis of the new "building" loads by making adjustments to the distribution systems to reflect the reduced "building" loads.

-Improve the seasonal efficiency of the primary energy-conversion equipment and make adjustments to the systems to reflect the reductions in the "building" and distribution loads.

-Replace worn out and inefficient equipment and systems, which can not be improved, with more efficient ones.

-Maintain records of monthly energy usage,

and continuously monitor the systems to assure that the methods which have been adopted are performing as expected, and are meeting or exceeding the energy conservation goals which have been established. As building use and function change, energy conservation measures must be changed accordingly.

Section 2

Major Energy Conservation Opportunities Priorities, Examples

MAJOR ENERGY CONSERVATION OPPORTUNITIES - PRIORITIES, EXAMPLES

A. BACKGROUND

This Section of the Manual identifies the relative order of energy usage of the environmental systems by building type. It is intended to assist the reader to gain a perspective on the relative importance for each type of the building structural elements and segments of the mechanical and electrical systems which use energy. Before establishing priorities for an energy conservation program which will produce the greatest conservation benefits first, the reader must know the order of magnitude of energy usage in his own building. The case histories of actual buildings, where energy conservation programs have already been started, demonstrate the dramatic energy and cost savings which have been accomplished with minimal capital investment and can be repeated in similar situations.

The major energy conservation opportunities, labeled MECO-1 through 20 and the associated examples of the potential savings in fuel,power and dollars are not all inclusive, but they are a summary of the most important opportunities for energy conservation in many buildings. The savings show in each example were calculated to provide insight into the potential results of conservation efforts under specific conditions. A more comprehensive list of energy conservation opportunities and multiple options for each opportunity are described in Sections 4-10 as guidelines.

Before implementing the options in Sections 4-10, first follow the procedures outlined in Section 3 . Identify the particular building structure, its characteristics, and use patterns. Select the options in Sections 4-10 which are applicable to the building; then determine where possible, the potential savings in energy, for those options by using the charts and graphs in that Section.

B. PRIORITIES

Nationwide, the systems that consume the most energy in order of magnitude are: (1) heating and ventilating;

(2) lighting; (3) air conditioning (cooling) and ventilating; (4) equipment and processes; and (5) domestic hot water.

However, depending upon the climate; the building construction; use and mode of operation; and the type, control and efficiency of the mechanical and electrical equipment; the relative order of magnitude of energy use among the first three systems will change.

The amount of energy required for domestic hot water is significant in hospitals, housing, and athletic or cooking facilities in schools and colleges. In many areas of the country the amount of energy to heat water is second only to space heating in the north, and air conditioning in the south in housing. In hospitals the amount of energy to heat hot water may exceed the amount of energy required for lighting.

Religious buildings and public halls which frequently include meeting rooms, offices, and school facilities, are most likely to consume energy in the same pattern as office buildings in the same geographic location—but in smaller quantities per square foot of floor area.

In those retail stores with high levels of general illumination and display lighting, and/or a large number of commercial refrigeration units, electricity consumes the greatest amount of energy.

In climatic zones with mild winters (below 2500 degree days See Fig.4) the seasonal cooling load may be larger than the seasonal heating load and may even consume more energy depending upon the respective efficiencies of each system. In office buildings, schools and retail stores in this zone, the electrical load for lighting, which is relatively independent of climate, may exceed either heating or cooling. Buildings used for only a few hours per week, however, may consume more energy for heating unless indoor temperatures are set back during unoccupied periods and boiler or furance efficiencies are high.

In cold climates, 6000 degree days and above, heating usually consumes the most energy per year in office buildings and schools with lighting and then cooling next. For retail stores in that zone, the most likely order of energy use is lighting-heating-cooling, or lighting-cooling-heating. Generally, heating consumes the most energy for religious buildings or other buildings used for only a few hours/week, in this zone (in fact, in most climatic zones above 3000 degree days), with lighting and cooling following in that order.

In mid-climates 2500 - 6000 degree days, the order of magnitude of energy by systems largely depends upon the type of mechanical and electrical systems and the characteristics of the building structure in which they are installed. The energy required for industrial buildings exclusive of process loads is generally similar in all zones to commercial buildings.

The following Matrix, Table 1, rates the systems by buildings and climates in the general order of annual energy usage with 1 the greatest and 5 the least. However, each building must be analyzed individually to determine its actual annual usage by system. The procedures for determining these values are described in Section 3.

TABLE 1: COMPARATIVE ENERGY USE BY SYSTEM

		Heating & Vent.	Cooling & Vent.	Lighting	Power & Process	Domestic Hot Water
Schools	A	4	3	1	5	*
	B	1	4	2	5	3
	C	1	4	2	5	3
Colleges	A	5	2	1	4	3
	B	1	3	2	5	4
	C	1	5	2	4	3
Office Bldg.	A	3	1	2	4	5
	B	1	3	2	4	5
	C	1	3	2	4	5
Comm.-ercial Stores	A	3	1	2	4	5
	B	2	3	1	4	5
	C	1	3	2	4	5
Reli-gious Bldgs.	A	3	2	1	4	5
	B	1	3	2	4	5
	C	1	3	2	4	5
Hos-pitals	A	4	1	2	5	3
	B	1	3	4	5	2
	C	1	5	3	4	2

Climatic Zone A: Fewer than 2500 degree days
Climatic Zone B: 2500 - 5500 degree days
Climatic Zone C: 5500 - 9500 degree days

The following Pie Diagrams, (Figures 1 and 2) which were produced by Dubin-Mindell-Bloome Associates and the National Bureau of Standards on the basis of a computer analysis, illustrate the relative magnitude of energy consumption for an office building if erected in a cold climate (Manchester, New Hampshire), and the identical building if located in a warm climate (Orlando, Florida). Heating is the single largest user of energy for the office building in New Hampshire, and the amount used for cooling is relatively small. In Florida the energy used for heating has decreased and cooling requires the greater amount.

There is no general rule to determine which part of a particular system accounts for the most energy use of that system. The burner-boiler seasonal combustion efficiency can vary from close to 78% down to 30%; for buildings in climatic zones above 2500 degree days, improving the efficiency of the combustion device may be the single most effective measure. However, lighting accounts for a tangible percentage of energy used in all climates and the potential for conservation is high. Savings of 25% to 50% of the energy required for lighting are possible with little initial cost.

HVAC systems which mix hot and cold air together, or simultaneously heat and cool a space, are particularly wasteful and offer a high potential for energy conservation. These systems i.e. dual duct, terminal reheat, and multi-zone are described in Sections 4-10 which also include guidelines for modifying operating modes to minimize energy usage.

In all cases, reducing the "building" load will conserve energy, but for some particular buildings, the savings in energy by decreasing the distribution loads and increasing the seasonal efficiency of the primary energy-conversion equipment are even greater.

C. CASE HISTORIES

Complete data is not available for buildings in which energy usage was measured before and after a conservation program was undertaken, and the results recorded and published, however two such case histories for buildings in New York are included here to show the results of a

FIGURES 1 AND 2: ENERGY USED BY IDENTICAL OFFICE BUILDINGS IN WARM AND COLD CLIMATES

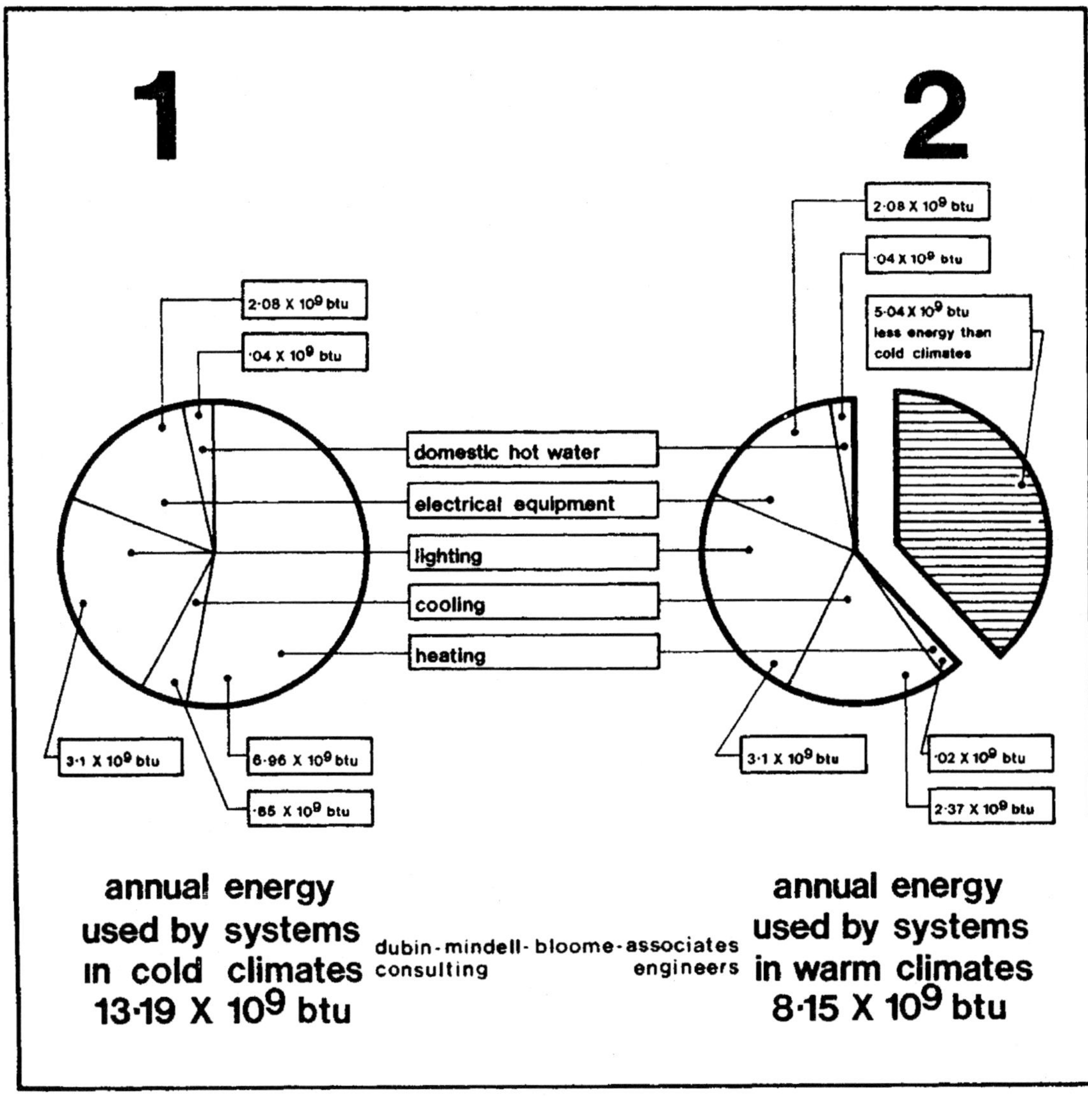

Figures 1 & 2 Engineering Data

Based on the NBSLD program from the National Bureau of Standards which calculated energy consumption for identical buildings in Manchester, New Hampshire and Orlando, Florida.

Gross floor area	- 126,000 $Ft.^2$
Occupants	- 600 people
Operation	- 8 a.m. to 6 p.m. weekdays only
Ventilation	- 15,000 cfm occupied times only
Infiltration	- 4,500 cfm continuous (1/4 air change/hour)
Lighting and power	- 3.4 Watts/$Ft.^2$ occupied times only
Indoor conditions	- Heating 68°F occupied
	58°F unoccupied
	Cooling 78°F occupied only
Outdoor conditions	- Variable - NOAA weathertape

comprehensive energy conservation program only one year after implementation. The energy saved by performing multiple energy conservation options is not always simply the sum of the savings for each individual measure. For instance, reducing the heat loss, or "building" heat load, by a 10% reduction in infiltration, and then increasing seasonal burner-boiler efficiency by 15%, results in a total energy saving for heating of 23 1/2% not 25%. The case histories that follow indicate the net savings achieved through multiple conservation options which were exercised.

For an analysis of the energy savings which are identified for each of a number of individual conservation measures and the net savings due to various combinations refer to Appendix B.

CASE HISTORY 1

A. MAJOR OPERATIONAL CHANGE:

(1) 15°F. NIGHT AND WEEKEND SET BACK. AND INDOOR TEMPERATURES LOWERED DURING OCCUPIED HOURS IN WINTER.

(2) INDOOR TEMPERATURES RAISED TO 78°F. IN SUMMER.

(3) LIGHTING LEVELS REDUCED BY 30%, AND ALL LIGHTS SHUT OFF WHEN NOT NEEDED.

(4) COOLING EQUIPMENT OPERATING HOURS REDUCED BY 20%.

(5) DOMESTIC HOT WATER CIRCULATING PUMPS TURNED OFF AT NIGHT.

B. BUILDING

Office building, New York City
Floor area: 2,000,000 sq. ft. No of stories: 54

C. ANNUAL ENERGY CONSUMPTION BEFORE CONSERVATION PROGRAM:

(1) Purchased steam: 278,464,000 lb. steam/yr. @ $3.20/1000 lbs.*

(2) Electricity: 51,909,000 KWH/yr. @ 4¢/KWH including demand.*

D. ANNUAL ENERGY CONSUMPTION AFTER OPERATIONAL CHANGES:

(1) Steam: 188,017,000 lb. steam, a savings of 32.5%

(2) Electricity: 39,264,000 KWH, a savings of 24.5%.

E. ANNUAL SAVINGS IN DOLLARS;

$758,000

F. COSTS TO IMPLEMENT:

No construction costs. Engineering fees paid are not available.

*Steam and electricity rates as of December 1, 1973.

CASE HISTORY 2

A. MAJOR OPERATIONAL CHANGES:

(1) COMPUTER CONTROL PROGRAM OPTIMIZED TO RESET CHILLED WATER AND TERMINAL REHEAT WATER TEMPERATURES IN ACCORDANCE WTTH LOADS.

(2) OUTDOOR AIR OPERATED ON AN ENTHALPY CYCLE

(3) STEAM CONVERTER REPIPED

(4) OPERATIONAL TIME FOR PUMPS AND FANS REDUCED.

(5) 25% OF FLUORESCENT TUBES REMOVED.

B. BUILDING:

Office building, New York City

Floor area: 800,000 sq. ft. No of stories: 22

Occupants: 6,000/day - 1500 at night (building operated 24 hrs./day - 6 days/week)

C. ANNUAL ENERGY CONSUMPTION BEFORE CONSERVATION:

(1) Purchased steam: 160,000,000 lb. steam/yr (1970) @ $5.30/1000 lbs.*

(2) Electricity: 36,000,000 KWH/yr. @ 5.0¢/KWH including demand.*

D. ANNUAL ENERGY CONSUMPTION AFTER OPERATIONAL CHANGES:

(1) Steam: 97,000,000 lb. steam (1974) a savings of 39%

(2) Electricity: 34,000,000 ** KWH a savings of 6%.

E. ANNUAL SAVINGS IN DOLLARS:

(1) Steam: $334,000

(2) Electricity: $100,000*

F. COSTS TO IMPLEMENT: $12,000 for new steam converter piping (cost to remove bulbs, not available.)

*Steam and electricity rates as of December 30, 1974
**The true savings are unknown, since a substantial amount of new equipment had been installed in 1974.

D. MAJOR ENERGY CONSERVATION OPPORTUNITIES AND EXAMPLES

The following MECO's are only a few of the dozens of opportunities to conserve energy.

There are a large number of options which can be exercised to take advantage of each opportunity. Those which are listed here, in section 2, do not, by any means, comprise a complete listing, but hopefully, they will stimulate the reader to study the wider range of choices to conserve energy, which are described in Sections 4-10.

1. HEATING

The amount of energy or fuel required to heat a building is dependent upon the level of temperature and relative humidity indoors, the amount of ventilation and infiltration air that must be heated, the severity and duration of the outdoor temperature below indoor room conditions, the thermal properties of the building envelope, and the efficiency of the distribution system, burners, boilers and furnaces.

MECO 1 MAINTAIN LOWER INDOOR TEMPERATURES DURING THE HEATING SEASON

Lower indoor temperatures reduce the heating load due to ventilation and infiltration as well as heat loss by conduction through the building envelope.*

Lower the thermostat setting to 68°F or less during occupied hours.**See Section 4, Figure 13, page 103 for "Suggested Heating Season Indoor Temperature Standards."

Lower the thermostat setting to 58°F or less at night, on weekends, and during all other unoccupied periods. See Figure 14, page 105, and example below.

After reducing the heating loads, adjust the distribution system and boiler and burner accordingly to further reduce fuel consumption.**

An Example of Savings by Night Set-Back

1. OPERATIONAL CHANGE: THERMOSTATS LOWERED FROM 68° to 58°F AT NIGHT, WEEKENDS, AND HOLIDAYS.

2. ASSUMPTIONS:

   Office building, Chicago, Illinois (6500 degree days - Fig. 4)

   Floor area: 50,000 Sq. Ft.

   Conditions before operational change: Building occupied 40 hours per week; outside air supplied only during occupied periods; fuel consumption, 64,000 gallons of oil per year, @ 36¢/gal.

3. SAVINGS:

   Energy: 8200 gallons of oil per year; a savings of 12.8%.

   Dollars: $2950/year.

4. IMPLEMENTATION COST; None

* Terminal reheat systems are an exception

** Not included in the example below.

MECO 2 SHUT OFF OUTSIDE AIR DURING ALL UNOCCUPIED PERIODS

Close outside air dampers or shut off outside air fan during unoccupied hours of the heating season, including noon-day periods when buildings may be lightly occupied and periods when areas such as auditoriums, cafeterias, gymnasiums, dormitories and conference rooms are unoccupied.

Reduce or shut off outdoor air entirely in retail stores during periods of the day when occupancy is considerably less than normal.

In religious buildings shut off outdoor air for all days during the week, as well as nights, when the building is unoccupied.

See Section 4 , Figure 16, page 115 for "Suggested Ventilation Standards."

An Example of Savings by Shutting off Ventilation

1. OPERATIONAL CHANGE: SHUT OFF OUTSIDE AIR AT NIGHT AND OTHER UNOCCUPIED PERIODS BY CLOSING OUTSIDE AIR DAMPERS OF AIR HANDLING UNIT.

2. ASSUMPTIONS

   Office building, Minneapolis, Minn. (8400 degree days)

   Floor area: 300,000 Sq. Ft.

   Conditions before operational change: Unoccupied 128 hours per week; outdoor air supplied 24 hours per day @ 30 CFM per person; fuel consumption = 250,000 gallons of oil per year @ 36¢/gal.

3. SAVINGS

   Energy: 68,100 gal. oil per year, a savings of 32.5%

   Dollars: $24,500 per year

4. IMPLEMENTATION COST: None

MECO 3 REDUCE THE QUANTITY OF OUTDOOR AIR FOR VENTILATION DURING OCCUPIED HOURS

1. Reduce the amount of ventilation air during occupied hours by setting outside air dampers and controls. Generally only 5 CFM per person is necessary to maintain proper air quality, but smoking in many areas may raise the average requirements to about 8 CFM per person. See Section 4 , pages 113-118.

An Example of Savings by Reducing Ventilation Rate

1. OPERATIONAL CHANGE: ADJUST DAMPERS TO REDUCE OUTDOOR AIR FROM 22 CFM PER PERSON TO 8 CFM PER PERSON DURING OCCUPIED HOURS.

2. ASSUMPTIONS:

   Office building, Minneapolis Minn. (8400 degree days)

   Floor area: 350,000 Sq. Ft.

   Conditions before operational changes: Occupied 40 hours per week; outdoor air = 30 CFM per person, 2000 occupants = 60,000 CFM; fuel consumption = 450,000 gallons of oil per year @ 36¢ per gal.

3. SAVINGS:

   Energy: 137,104 gallons of oil per year, a savings of 30%.

   Dollars: $4086/per year

4. IMPLEMENTATION COSTS: None

MECO 4 <u>REDUCE THE RATE OF INFILTRATION</u>

Air infiltrates through cracks around doors and windows, through construction joints, and through doors which are frequently opened. Infiltration occurs whether the building is occupied or not. When exhaust fans are operating and outdoor air ventilation is insufficient to provide makeup air, infiltration rates increase.

Reduce air leakage by sealing and caulking leaks around windows and doors.

Seal construction joints.

Reduce exhaust air volume and operating hours of exhaust systems.

See Section 4 pages 113-122 and Figures 17 and 18.

Example Showing Savings Due to Caulking

1. OPERATIONAL CHANGE: CAULK WINDOWS TO REDUCE INFILTRATION RATE FROM 1 AIR CHANGE PER HOUR TO 1/2 AIR CHANGE PER HOUR.

2. ASSUMPTIONS:

   School, Fargo, North Dakota (9000 degree days)

   Floor area: 300,000 Sq. Ft.

   Conditions before operational changes; Occupied 36 hours per week; average indoor temperature during unoccupied periods = 60°F; existing double windows (26,000 Sq. Ft.) = 1000 windows; present fuel consumption = 225,000 gallons per year @ 36¢ per gallon.

3. SAVINGS:

   Energy: 33,300 gallons of oil per year, a savings of 14.8%

   Dollars: $11,988 per year

4. IMPLEMENTATION COST:

   $10,000 for caulking material and labor.

MECO 5 REDUCE FAN HORSEPOWER

The quantity of air (in cubic feet per minute - CFM), circulated for heating can often be reduced in response to reduced heating loads, or upon an analysis which indicates that the current CFM circulated is not required. Air quantities, set for cooling loads, may often be reduced during the heating season when the same system is used for both heating and cooling. Reducing the air quantity for systems operating against high static pressures, with motors 25 horsepower or larger, can result in significant savings during both the heating and cooling season.

See Section 7, page 197 and Figures 23 and 24 for savings resulting from reduced air flow with supply and exhaust fans.

Example of Energy Savings by Reducing Air Flow

1. OPERATIONAL CHANGE: REDUCE SUPPLY AIR QUANTITY BY 20% BY REDUCING FAN SPEED 6%. FAN SPEED IS REDUCED BY CHANGING THE DRIVER PULLEY.

2. ASSUMPTIONS:

   Retail Store, New York, New York

   Floor area: 100,000 Sq. Ft.

   Condition before operational change: Air quantity = 1 1/2 CFM/sq. ft., 1 1/2 CFM x 100,000 sq. ft. = 150,000 CFM @ 6" s.p. Fans operate for 2,500 hours/yr. Energy used 360,000 KWH/yr @ 5¢/KWH.

3. SAVINGS:

   Energy: 118,000 KWH/year, a savings of 33%.

   Dollars: $5940/year.

4. IMPLEMENTATION COST:

   Less than $500. for pulley change and labor.

MECO 6 IMPROVE COMBUSTION AND BOILER EFFICIENCY

The efficiency of the boiler/burner unit or furnace decreases rapidly when combustion is improper, when the combustion surfaces accumulate soot and scale, or when excess combustion air increases the stack temperature. Any percentage increase in seasonal boiler/burner efficiency directly reduces fuel consumption in the same propertion.

Test the combustion efficiency with proper instruments and adjust the firing rate and combustion air rate accordingly.

Adjust the automatic damper to control the draft in accordance with the firing rate.

Remove scale and soot from the boiler.

See Section 4, "Heating", pages 124-125 for detailed guidelines for boilers, furnaces, and burners.

An Example of Energy Savings by Improving Combustion Efficiency

1. OPERATIONAL CHANGE: DESCALE BOILER SURFACES, REMOVE SOOT, ADJUST COMBUSTION EFFICIENCY TO IMPROVE BOILER-BURNER EFFICIENCY BY 10%.

2. ASSUMPTIONS:

   Retail store, New York, New York (4800 degree days)

   Floor area: 100,000 sq. Ft.

   Present operation:

   Fuel consumption = 100,000 gallons of oil per year @ 36¢ per gallon; occupied 72 hours per week.

3. SAVINGS:

   Energy: 10,000 gallons of oil per year, a savings of 10%

   Dollars: $3600 per year.

4. IMPLEMENTATION COSTS:

   $500 per year to service burner and boiler at 4-month intervals during the heating season.

2. COOLING

The amount of energy required to cool a building is dependent upon the level of temperature and relative humidity indoors; the amount of ventilation air and infiltration air that must be cooled and dehumidified; the severity and duration of the outdoor temperature and humidity above indoor room conditions; the thermal properties of the building envelope; the heat gain through walls, roof and windows due to solar radiation; and the magnitude and duration of the internal heat gain due to people, and equipment which emits heat and/or moisture.

MECO 7 OPERATING HOURS

Shut off all refrigeration equipment and auxiliaries, including fans, pumps, cooling towers, and condensers during all unoccupied hours-at night, on holidays and weekends.

Delay the operation of the refrigeration system for one or two hours in the morning and shut off prior to closing time except in the severest hot spells.

See Section 6, "Cooling", page 158.

An Example of Savings Due to Reducing the Number of Hours of Operation of the Cooling System during Occupied Hours

1. OPERATIONAL CHANGE: ENTIRE COOLING SYSTEM OPERATED TWO HOURS LESS PER DAY FOR SIX DAYS PER WEEK.

2. ASSUMPTIONS:

   Retail store (department), New York, New York

   Floor area: 250,000 Sq. Ft.

   Conditions before operational changes:

      Cooling system operated 84 hours per week;
      electric rate = 5¢/KWH

3. SAVINGS:

   Energy: 234,800 KWH per year, a savings of 14%.

   Dollars: $11,700 per year

4. IMPLEMENTATION COST: None

MECO 8 REDUCE THE QUANTITY OF OUTDOOR AIR VENTILATION

Measures to reduce outside air ventilation (and infiltration) to conserve heating energy will also result in decreased summer cooling loads. Where infiltration exceeds 1/2 of an air change per hour for buildings in humid climates, measures to reduce infiltration will provide major energy savings for cooling as well as for heating.

During occupied hours, reduce the amount of air for ventilation as in the measure described in MECO 2.

See Sections 4 and 6, "Heating" and "Cooling", and Figure 21, page 165.

An Example of Savings by Reducing the Amount of Outdoor Air for Ventilation

1. OPERATIONAL CHANGE: ADJUST DAMPERS TO REDUCE OUTDOOR AIR FROM 30 CFM TO 8 CFM PER PERSON DURING OCCUPIED HOURS.

2. ASSUMPTIONS:

   Office building, Miami, Florida

   Floor area: 100,000 Sq. Ft.

   Conditions before operational changes: Occupied 40 hours per week; outdoor ventilation air @ 30 CFM/person x 667 occupants - 20,000 CFM; electric costs = 3.5¢/KWH. Annual energy consumption for chiller = 715,000 KWH/yr. @ 3.5¢/KWH.

3. SAVINGS:

   Energy: 63,000 KWH per year, a savings of 8%.

   Dollars: $2233 per year

4. IMPLEMENTATION COST: none.

MECO 9 USE OUTDOOR AIR FOR FREE COOLING

For the many periods during the year when dry bulb temperature is below the setting for room conditions the use of outdoor air for cooling reduces the hours of operation of the refrigeration system.

Using outdoor air at nighttime to reduce the late afternoon sunloads which are stored in the building mass, and precooling the building, will result in fewer hours of compressor operation on the following day.

Enthalpy control may be even more effective in saving energy in locations where there are fewer than 8,000 wet bulb degree hours.

> On cool days open the damper to circulate 100% outdoor air for sensible cooling. It may be necessary to open the windows slightly to relieve pressure.
>
> If equipped with an enthalpy controller, set it to permit full outdoor air supply when the total heat content of the outdoor air is below room conditions.

See Section 6, "Cooling", page 159.

An Example of Savings by Using Outdoor Air During Occupied Periods in an Economiser Cycle

1. OPERATIONAL CHANGE: OPERATE AN ECONOMISER CYCLE FOR 690 HOURS PER YEAR BY OPENING OUTDOOR AIR DAMPER, CLOSING RETURN AIR DAMPER, AND OPENING A FEW WINDOWS FOR PRESSURE RELIEF.

2. ASSUMPTIONS:

   Office building, Denver, Colorado

   Floor area: 50,000 Sq. Ft.

   Condition before operational changes: occupied 40 hours per week.

   Annual Energy Consumption for Chiller: 245,000 KWH per year @ 3.5¢ per KWH; refrigeration cycle off at night.

3. SAVINGS:

   Energy: 85,000 KWH per year, a savings of 34.7%.

   Dollars: $2975. per year

4. IMPLEMENTATION COST: None.

MECO 10 PERMIT HIGHER INDOOR TEMPERATURES AND RELATIVE HUMIDITY DURING OCCUPIED HOURS.

By allowing higher temperature and humidity conditions in the summer, the cooling load is reduced and chillers or compressors will operate fewer hours and consume less energy per hour of operation. Higher room temperatures will also permit a reduction in supply air quantity with a savings in motor horsepower.

See Section 6, Figure 20, page 163 for "Suggested Cooling Season Indoor Temperature and Humidity Standards."

An Example of Savings by Raising Indoor Temperature and Humidity Levels and Chilled Water Temperatures

1. OPERATIONAL CHANGE: DURING THE COOLING SEASON, RAISE INDOOR TEMPERATURES FROM 72°F to 78°F AND RELATIVE HUMIDITY FROM 50% to 60%. RAISE CHILLED WATER TEMPERATURES FROM 42°F to 46°F.

2. ASSUMPTION:

   Office building, Miami, Florida

   Floor area: 100,000 Sq. Ft. No of Stories: 10

   Conditions before operational changes: Building occupied 40 hours per week; annual energy consumption for chiller - 715,000 KWH @ 3.5¢ per KWH.

3. SAVINGS:

   Energy: 115,000 KWH per year, a savings of 16%

   Dollars: $4025 per year

4. IMPLEMENTATION COST: None, if done manually.

MECO 11 REDUCE THE SOLAR HEAT GAIN THROUGH WINDOWS

The solar heat gain through windows can be a large percentage of the cooling load in office buildings, schools and in small stores where show windows are in direct sunlight.

The greatest amount of solar radiation in the cooling season strikes the west and east glass, and next the southern facade.

> Reduce the solar heat gain through the windows by adjusting existing awning, blinds, or drapes on each window when they are in direct sunlight.
>
> Add reflective solar film to all windows in direct sunlight.

See Section 6, "Cooling", page 173, Figure 22.

Example of Savings by Reducing Solar Heat Gain

1. OPERATIONAL CHANGE: REFLECTIVE SOLAR FILM* ADDED TO EAST, WEST AND SOUTH WINDOWS TO REDUCE SHADE CO-EFFICIENT FROM .9 to .15. AIR AND CHILLED WATER SYSTEMS OPERATION REDUCED ACCORDINGLY.

2. ASSUMPTIONS:

   Office building, Miami, Florida

   Floor area: 100,000 Sq. Ft. No of stories: 10

   Conditions before operational changes: Cooling system in operation 40 hours per week; window area on west, south and east facade = 10,000 sq. ft. existing windows - clear, single glazed. Annual energy consumption for cooling 1,100,000 KWH @ 3.5¢/KWH.

3. SAVINGS:

   Energy: 170,000 KWH per year, a savings of 15%.

   Dollars: $5,900 per year.

4. IMPLEMENTATION COST: At installation cost of $1.25 per Sq. Ft. = $12,500.

*The film may require replacement after 8 - 10 years.

MECO 12 REDUCE SUPPLY AND RETURN AIR FLOW AND CHILLED WATER QUANTITIES

Reducing the cooling loads and analyzing the operation of the existing air conditioning system make possible many opportunities to reduce fluid flow.

While dampering, in the case of air ducts, and valving, in the case of water piping, will reduce flow rate and energy input to motors, greater savings can be gained by reducing motor speed and/or, for piping systems, changing the impellers. For systems with motors drawing 25 brake horsepower (BHP) or more, the yearly savings can be very significant. CAUTION: Don't neglect the smaller motors and systems; their aggregate energy draw may be very high. Follow references in MECO's 6 and 7 for fan and motor modifications during periods when the cooling system is in operation.

MECO 13 CLEAN AND DESCALE CONDENSER TUBES

The efficiency of chillers and refrigeration condensers decreases markedly as scale builds up in the tubes. Check condenser temperatures on a regular basis and descale tubes at least once per year.

See Section 6, page 181.

Example of Savings by Condenser Maintenance

1. OPERATIONAL CHANGE: DESCALE CONDENSER TUBES TO REDUCE FOULING FACTOR TO .0005 OR LESS.

2. ASSUMPTIONS:

   Retail Store (department), Dallas, Texas

   Floor area: 100,000 Sq. Ft.

   Conditions before operational changes: Building occupied 72 hours per week; average fouling factor = .001; annual energy consumption for chillers-650,000 KWH/yr. @ 3.0¢/KWH.

3. SAVINGS:

   Energy: 116,000 KWH per year, a savings of 18%.

   Dollars: $3,480 per year.

4. IMPLEMENTATION COST:

   $1200 per year to maintain the condensers and all air conditioning equipment.

MECO 14 <u>OPERATE AT HIGHER CHILLED WATER AND SUCTION TEMPERATURES TO INCREASE THE EFFICIENCY OF THE REFRIGERATION CHILLERS.</u>

There are extensive periods of time (in some cases the entire year), during which the chilled water temperature can be raised by 4° or 6° or more. Accordingly, the refrigeration compressor or chiller can operate at higher suction temperatures. An increase of 1° in the suction temperature of the chiller, or compressor in direct expansion systems, results in a reduction of 1 1/2 to 2% in the power requirement for the refrigeration unit. The savings in operating costs by increasing chilled water or suction temperatures for systems of 15 H.P. and more are significant.

Raise the chilled water temperature from two to eight degrees higher when cooling load permits.

Raise the chilled water temperature in buildings for those portions of the day that cooling loads are likely to be lower (i.e. during slack business hours in retail stores; noon hours when office buildings are partially vacated; morning, pre-occupancy periods after a night shut down.)

See Section 6, "Cooling" page 178.

<u>An Example of Energy Savings by Raising Chilled Water Temperature</u>

1. OPERATIONAL CHANGE: RAISE CHILLED WATER TEMPERATURE FROM 42° to 48"F. FOR ALL NORMAL OPERATION.

2. ASSUMPTIONS:

   Retail Store (department) Los Angeles, California

   Floor area: 50,000 Sq. Ft.

   Present conditions before operational changes:

   Building occupied 72 hours/wk; cooling system off at night; present energy consumption for refrigeration = 320,000 KWH/yr. @ 3.5¢ KWH.

3. SAVINGS:

   Energy: 32,000 KWH/yr., a savings of 10%

   Dollars: $1120/yr.

4. IMPLEMENTATION COST: None if done manually; nominal cost for controller and sensor for automatic operation.

MECO 15 REDUCE THE CONDENSING TEMPERATURE OF COMMERCIAL REFRIGERATION UNITS TO IMPROVE OPERATING EFFICIENCY

In general, air-cooled condensing units serve a large number of supermarket refrigeration cases. These units are often crowded into storage areas where they are partially blocked, mounted outdoors in direct sunlight, or mounted and neglected on the roof or another remote location.

Refer to Section 8, "Commercial Refrigeration", page 226 for guidelines.

Example

1. OPERATIONAL CHANGE: CONDENSER COILS FOR REFRIGERATION CLEANED TO REDUCE AVERAGE CONDENSING TEMPERATURE FROM 115°F to 95°F.

2. ASSUMPTIONS:

   Supermarket, New York, New York

   Floor area: 25,000 Sq. Ft.

   Conditions before operational change:

   120 H.P. installed capacity; refrigeration units operate an average of 12 hrs./day or 4380 hours/yr. Annual energy consumption for refrigeration = 420,000 KWH/yr. @ 4.5¢/KWH.

3. SAVINGS:

   Energy: 84,000 KWH/yr., a savings of 20%.

   Dollars: $3780/yr.

4. IMPLEMENTATION COST: Negligible

3. LIGHTING

Modifications to the operation of the lighting system and to the system itself provide the greatest opportunity to reduce energy consumption. Simply turning off unnecessary lights, day and night, and making greater use of available daylight for illumination saves energy for both lighting and air conditioning with no added costs. Better cleaning and maintenance practices increase the efficiency of the lighting system and provide the opportunity for lamp replacement (by lamps of lower wattage) or removal, or the switching off of lights, with little or no reduction in illumination levels.

MECO 16 UTILIZE DAYLIGHT TO REDUCE THE LIGHTING LOAD

Windows, properly exploited, can provide a sizeable part of the illumination required in small stores and office buildings during a large portion of the buildings occupied hours.

1. Open drapes and blinds during the day to take advantage of daylight at the perimeters of the building while controlling glare and excessive solar radiation. Drapes should be used to control light greatly in excess of that required. It should be recognized that excess daylight may cause materials to fade and change color.

2. Switch off the lights which are not needed when daylight can supply necessary illumination.

Refer to Section 9, "Lighting" for comprehensive guidelines to maximize the use of daylight for illumination.

Example of Energy Savings by Reducing Lighting Load

1. OPERATIONAL CHANGE: PERIMETER LIGHTS TURNED OFF APPROXIMATELY 50% OF OCCUPIED HOURS.

2. ASSUMPTIONS:

   Office Building, New York, New York

   Floor area: 100,000 Sq. Ft. No of stories: 10

   Glass area: 33% of net wall area

   Available daylight: 50% of occupied hours

   Switching arrangement: Separate switches for perimeter row of lights.

   Electricity used for lighting: 935,000 KWH @ 4.5¢/KWH

3. SAVINGS:

   Energy: 159,000 KWH/yr., a savings of 17%

   Dollars: $7155/yr.

4. IMPLEMENTATION COST: None

   Additional savings in energy for cooling and refrigeration equipment will also result from reduced interior heat gain.

MECO 17 TURN OFF LIGHTS WHEN ENTIRE BUILDING OR PORTIONS ARE UNOCCUPIED

A large percentage of energy used for lighting is wasted when all lights are burning and only a samll portion of the building is in use.

Turn off lights at night.

Turn off lights in auditoriums, conference rooms, cafeterias, computer rooms and other areas when not used during the day.

Schedule cleaning hours during daylight. If this is not entirely possible, illuminate only that portion of the building which is being cleaned at any one time.

Refer to Section 9 "Lighting" for guidelines on the use and operation of lighting systems to conserve energy.

Example of Energy Saved by Turning off Lights

1. OPERATIONAL CHANGE: LIGHTING TURNED ON AT 8:00 AM AND OFF AT 5:00 PM FOR 5 WEEK DAYS. MINIMAL NIGHT LIGHTING DURING UNOCCUPIED HOURS.

2. ASSUMPTIONS:

   Office building, New York, New York

   Floor area: 100,000 sq. ft.

   Conditions before operational changes: Building occupied 40 hrs./wk. Lighting turned on at 7:00 A.M. and off at 7:00 P.M. for 5 week days. Connected lighting load of 4-watts per sq. ft., = 400,000 KWH X 4.5¢/KWH.

3. SAVINGS:

   Energy: 300,000 KWH/yr., a savings of 25%

   Dollars: $13,500/yr.

4. IMPLEMENTATION COST: None.

MECO 18 REDUCE ILLUMINATION LEVELS TO REDUCE LIGHTING LOAD

Lighting levels are frequently higher than necessary for a given task, and can be reduced in many areas of a building during the day to suit the task being performed. The level of illumination necessary over filing cabinets, dead corners, storage areas and some clerical areas, need not be as high as those levels for accounting areas, drafting tables or detail work stations occupied for many hours per day. Uniform lighting in large areas can be unnecessary and wasteful, in sparsely occupied spaces.
Different tasks done in an area may require different light levels; reduce lighting levels when possible.

1. Maintain existing lighting system to improve the footcandles/watt output.

2. Analyze the tasks to be performed and reduce the levels where the tasks are not so critical by:

    (a) Replacing existing lamps with lower wattage, lower output lamps.

    (b) Removing some lamps and ballasts from fixtures.

See Section 9 "Lighting", Figure 26, page 237 for Recommended Lighting guidelines and energy conservation options. See ECM 2 for additional measures involving switching, and the relocation and replacement of fixtures and ballasts.

Example of Saving Energy by Reducing Lighting Level through Removing Lamps and Ballasts

1. OPERATIONAL CHANGE: REMOVE 2 OF THE 4 LAMPS FROM HALF OF THE OVERHEAD 2' x 4', FLUORESCENT FIXTURES AND DISCONNECT THE ASSOCIATED BALLASTS.

2. ASSUMPTIONS:

   Retail store (department), Los Angeles, California

   Floor area: 100,000 Sq. Ft.

   Conditions before operational change: Lights on for 72 hrs. per week; lighting levels average 80 foot candles. Electric use, 1,400,000 KWH/yr. @ 3.5¢/KWH.

MECO 18 REDUCE ILLUMINATION LEVELS TO REDUCE LIGHTING LOAD (Cont'd)

3. SAVINGS:

   Energy: 375,000 KWH/yr., a savings of 28%

   Dollars: $13,000/yr.

4. IMPLEMENTATION COST: $3600 to remove lamps and disconnect ballasts.

5. NEW LIGHTING LEVELS AVERAGE 65 FOOTCANDLES. NOTE THAT THE LEVEL OF ILLUMINATION DECREASED ONLY 16%.

MECO 19 REDUCE THE LEVEL OF ILLUMINATION IN PARKING LOTS TO REDUCE ELECTRIC LOAD

Parking lots require an average of only one Footcandle of even illumination. Overlighting, inefficient light sources, and unnecessarily prolonged periods of operation account for excessive energy usage.

1. Reduce the level of lighting if over one footcandle, while maintaining reasonable uniformity of illumination.

   (a) Replace lamps with more efficient light sources.

   (b) Reduce wattage of lamps or remove unnecessary ones.

Refer to Section 9, "Lighting", page 236 for more comprehensive guidelines.

1. OPERATIONAL CHANGE: REDUCE THE WATTAGE OF THE LAMPS IN A PARKING LOT TO REDUCE LIGHTING LEVELS TO AN AVERAGE OF 1 FOOTCANDLE.

2. ASSUMPTIONS:

   Retail store (2500 car parking lot), Chicago, Illinois

   Conditions before operational changes: Lights operated 4 hrs./day x 315 days/yr. = 1260 hours/yr. to maintain an average of 2 footcandles.

   Electric usage - 256,000 KWH @ 3.5¢/KWH.

3. SAVINGS;

   Energy: 128,000 KWH/yr., a savings of 50%

   Dollars: $4,480/yr. @ 3.5¢/KWH.

4. IMPLEMENTATION COST: None, when lamps are changed at normal relamping periods.

4. DOMESTIC HOT WATER

There are easily-implemented opportunities to conserve energy used to heat hot water at minimal cost by lowering the temperature of hot water at the faucets and reducing the volume which is used.

In general, office buildings use 2 to 3 gallons ot hot water per capita per day, residences about 20 gallons/capita/day and religious buildings and stores less. Hospitals, laundries, cafeterias and restaurant kitchens use considerably more.

Heat lost from storage tanks and circulating pipes is in proportion to the temperature difference between the water and ambient air. Therefore, reducing the maintained temperature of hot water not only reduces the amount of energy required to heat each gallon of water used but also reduces the heat loss from the tank and piping system.

In buildings which have kitchens requiring very hot water for dishwashing, boost the temperature at the equipment, rather than maintaining a high temperature for the entire building hot water system. See ECM- 2 for details.

Reduce the temperature of hot water at faucets to 90°F.

Reduce the consumption of hot water in all buildings by flow restrictors in the piping, self-closing faucets or flow restrictor taps.

Refer to Section 5, "Domestic Hot Water", page 141, for data on usage in buildings and for conservation guidelines.

MECO 20 <u>REDUCE THE QUANTITY AND TEMPERATURE OF DOMESTIC HOT WATER</u>

<u>Example of Savings by Reducing Temperature and Quantity of Hot Water</u>

1. OPERATIONAL CHANGE: INSTALL 1/2 GPM SPRAY NOZZLES ON LAVATORY FAUCETS AND REDUCE TEMPERATURE FROM 135° TO 90°F.

2. ASSUMPTIONS:

   Office building, New York, New York

   Occupancy: 3500 people

   Hot water usage: 2 GPD/capita - 135°F. delivery temperature

   Total consumption: 7,000 gallons/day

   Oil consumption for hot water: 12,000/gallons/yr. @ 36¢/gal.

3. SAVINGS:

   Energy: 6240 gal./oil/yr., a savings of 52%

   Dollars: $2,250 per year

4. IMPLEMENTATION COST:

   $1750 for spray valves

Section 3

Use and Implementation of the Manual

USE AND IMPLEMENTATION OF THE MANUAL

A. BACKGROUND

The opportunities to conserve energy are identified in bold type and numbered ECO-I through ECO-40 in Sections 4-10 of this manual. For each sub-system, i.e. Heating, Cooling, Domestic Hot Water, Lighting, Power, HVAC, a brief discussion describes the conservation concept for each ECO and reference is made to the appropriate Figures which are the Charts, Graphs and Tables for use in quantifying the amount of energy which can be saved for each ECO. The options for consideration are summarized in the guidelines following each section.

Each set of guidelines includes a number of options for consideration, and in many cases, a choice must be made between them. In order to choose the appropriate guidelines and use the Graphs and Charts which are applicable, you must first identify the factors which are particular to your building. Instructions for preparing identification profiles are described in "B" below, which also describes the method of determining the amount of energy your building now uses. Use profile Forms, Figures 3,10,11 and 12, pages 57,73,77,87 to assemble the data you need.

Before assembling your identification and energy use profile, go back and reread the Introduction and Scope, and the Summary of Energy Conservation Principles, Section 1. Then review the major energy conservation opportunities (MECO's) which are summarized in Section 2. The case histories, and examples in Section 2. showing savings are presented to give you an idea of the potential economic benefits as well as energy conservation potential which are possible. The energy savings for each example selected are conservative. More dramatic savings are possible.

Before starting to construct your identification profiles, read all of Sections 4-10 to gain a greater understanding of the general conservation opportunities which are available, but don't try to quantify savings for your building before you construct the four profiles, pages 73-89.*

---

*The implementation costs given are for New York City. Refer to Appendix A to determine costs for other locations.

We strongly suggest that you also should read ECM-2 after you have constructed your profiles to understand the broadest range of energy conservation opportunities which are available. An understanding of the further energy savings which are possible may influence your course of action. For instance, if after consulting Sections 4-10 in ECM-1, you are contemplating caulking windows to reduce heat loss in the winter you may decide, after reading ECM- 2 that you can benefit from lower life cycle costs by investing in storm windows which will also reduce infiltration and conduction losses without the need for caulking. Storm windows will cost more initially, but will result in lower life cycle costs.

The Index at the end of this book (p.719) should be used to quickly find the information on specific subjects or systems which are of interest to you.

B. ENERGY LOADS

1. Building Load

The magnitude of the "building" load and the amount of energy required to maintain the desired indoor temperature and humidity levels is dependent upon (1) the location of the building and climate, (2) the degree of environmental control which is maintained, (3) the number of occupants and period of occupancy, (4) the thermal performance of building structure and (5) the use of the building.

a) Location

The location of the building affects the heating and cooling loads, since the angle and amount of solar radiation which strikes the building exterior surfaces determines the solar heat gain at any period of the year. Solar heat gain may be detrimental in the summer (by increasing the cooling load) and/or beneficial in the winter (by providing heat to the building and reducing the heat load).

The amount of sunshine also determines the amount of available daylight for natural illumination.

The location also determines the climatic conditions to which the building is subjected, and in turn, the peak, and magnitude of the annual heating and cooling loads.

b) Climate

The major climatic conditions which affect the amount of energy used for heating and cooling are: (1) temperature, (2) humidity, (3) wind, (4) the high-low range (diurnal swing) of (1), (2) and (3); (5) the severity of the winter (degree days) and (6) the severity of the summer (cooling wet bulb and dry bulb degree hours)

c) Degree of Environmental Control and Process Loads

The degree and length of time that indoor conditions such as temperature, humidity, amount of ventilation, and quantity and temperature of hot water, influence the size of the "building" load, and the amount of energy required to supply it.

The lighting levels influence the amount of electricity required for illumination when daylight is not available and used.

Energy usage for the operation of elevators, escalators, business machines, commercial refrigeration, cooking equipment, communications and special processes (packaging, etc.) is also dependent upon the period of time they are used. Processes and equipment which emit heat also affect the building's heating and cooling loads.

d) Heat Loss and Heat Gain

The magnitude of the heating and cooling loads due to conduction of heat through the building envelope, and the leakage of outdoor air (infiltration) which must be heated, cooled, and humidified or dehumidified also depends upon the thermal and structural properties of the building as well as on climate, and periods of operation of the heating and cooling equipment to maintain established indoor conditions.

2. Distribution Loads

The energy used in heating, ventilating and air conditioning systems to distribute hot or cold air or water is mainly for motor-driven fans, and hot or chilled water or condenser water motor-driven pumps.

Additional loads are caused by fluid leakage or heat transfer in hot or cold ducts and pipes.

In lighting systems, the fixtures themselves are responsible for reducing the effective lumen output from the lamps, and, along with lamps with low lumen/watt output, power losses from conductors, transformers and switch gear, account for a major amount of the energy used for lighting.

3. Energy-Using Equipment Efficiency

The primary energy conversion units, such as boilers, furnaces, compressors and chillers, which are supplied with energy in the form of fuel or electric power must supply all of the "building" and distribution loads. The amount of energy required to meet the "building" loads and distribution loads is dependent upon the efficiency of the primary energy-conversion equipment.

In order to improve the seasonal efficiency of existing equipment, the characteristics, condition and operating mode of the equipment and systems must first be determined.

C. DEVELOPING THE IDENTIFICATION BUILDING PROFILES - ENERGY CONSUMPTION/PARAMETERS

In order to quantify the total energy savings for your building by using the charts, graphs, and tables in ECM- 1 (and ECM- 2) which provide unit savings, complete the four-part profile which identifies each of the factors listed below which are particular to your building. Enter the information on the appropriate profile form as follows:

1. Location and climate zone profile - Figure 3, page 57.
2. Building type construction, condition and use profile Figure 10, page 73.
3. Mechanical and electrical systems profile - Figure 11, page 77.
4. Audit of present energy use - Figure 12, page 87.

Where the information is required for use with manual ECM- 2 but not necessary for ECM-1, it is asterisked (*) on each profile form.

STEP 1: PREPARE A BUILDING LOCATION AND CLIMATIC ZONE PROFILE

Identify the climatic conditions to which your building is exposed over extended periods of time with the aid of climatic maps, (Figures 4 - 9) on pages 59-69, and enter each value on the appropriate line, Column A, Figure 3. The geographic locations should be plotted on each of the maps and relevant conditions read from the contour lines. If your building lies between contour lines, the values should be interpolated; if in doubt consult your nearest weather station or an enviornmental engineer through the ACEC Chapter in your State.

FIGURE 3: LOCATION AND CLIMATIC ZONE PROFILE FORM

	Column A Value	From
Latitude		Figure 4
Heating Degree Days		Figure 4
Solar Radiation in Langleys		Figure 5
Degree Hours less than 54°F W.B. when D.B. is less that 68°F		Figure 6
Degree Hours greater than 78°F D.B.		Figure 7
Degree hours greater than 66°F W.B.		Figure 8
Degree hours greater than 85°F D.B.		Figure 9

The climatic information shown on the maps Figures 4 - 9 was obtained either from the Climatic Atlas or developed from AFM88/8 (See back of each figure for specific reference)

These maps provide a broad general picture of climatic conditions in any particular location; they cannot however indicate the macro-climate or variation of conditions experienced within local areas. Local knowledge of the climate should be used to modify the general conditions when more precise answers are required. For example, one area may be known to be 10°F colder in winter than the average of surrounding areas and the heating degree days should be modified accordingly.

1. Estimate the Annual Heating Degree Days and Identify Building Location (Figure 4)

Dry bulb degree days influence the annual energy consumption for "building" heating loads. For additional information on degree days in your city, see Reference 11. Degree day information is required for use with Figure 13 (to determine savings due to temperature set-back) and Figure 16.

2. Estimate the Annual Mean Daily Solar Radiation (Figure 5)

Solar radiation, (measured in Langleys or BTU's) influences the annual energy consumption for the "building" cooling load by increasing heat gain by radiation through glazing, and by conduction through opaque roofs and exterior wall surfaces. Solar heat gain reduces the "building" heating load in winter. For additional information on solar radiation effects, refer to Reference 6. Radiation values are useful to determine the benefits of controlling solar radiation to reduce cooling loads, (see Fig. 22).

3. Estimate the Annual Number of Wet Bulb Degree Hours Below 59°F W.B. when the Dry Bulb Degree Hours are Below 68°F. (Figure 6)

This value influences the annual energy consumption for "building" heating load to maintain relative humidity levels in the winter and is needed to determine the energy savings by reducing relative humidity levels using Figure 17.

4. Estimate the Number of Annual Dry Bulb (D.B.) Degree Hours when D.B. Temperatures are Above 78°F.(Figure 7)

This value influences the annual energy consumption for "building" cooling load due to heat gain by conduction through the building envelope, and for "building" cooling load required to lower the dry bulb temperature of outdoor air (for ventilation and infiltration) to room conditions of 78°F. (suggested standards, Figure 20). Use information from Figure 7 to determine the savings due to reduced D.B. degree hours alone by reducing infiltration and ventilation during the cooling season.

5. Estimate the Number of Annual Wet Bulb (W.B.) Degree Hours when the W.B. Temperature is Above 66°F.(Figure 8)

This value influences the annual energy consumption for "building" cooling load to cool and dehumidify outdoor

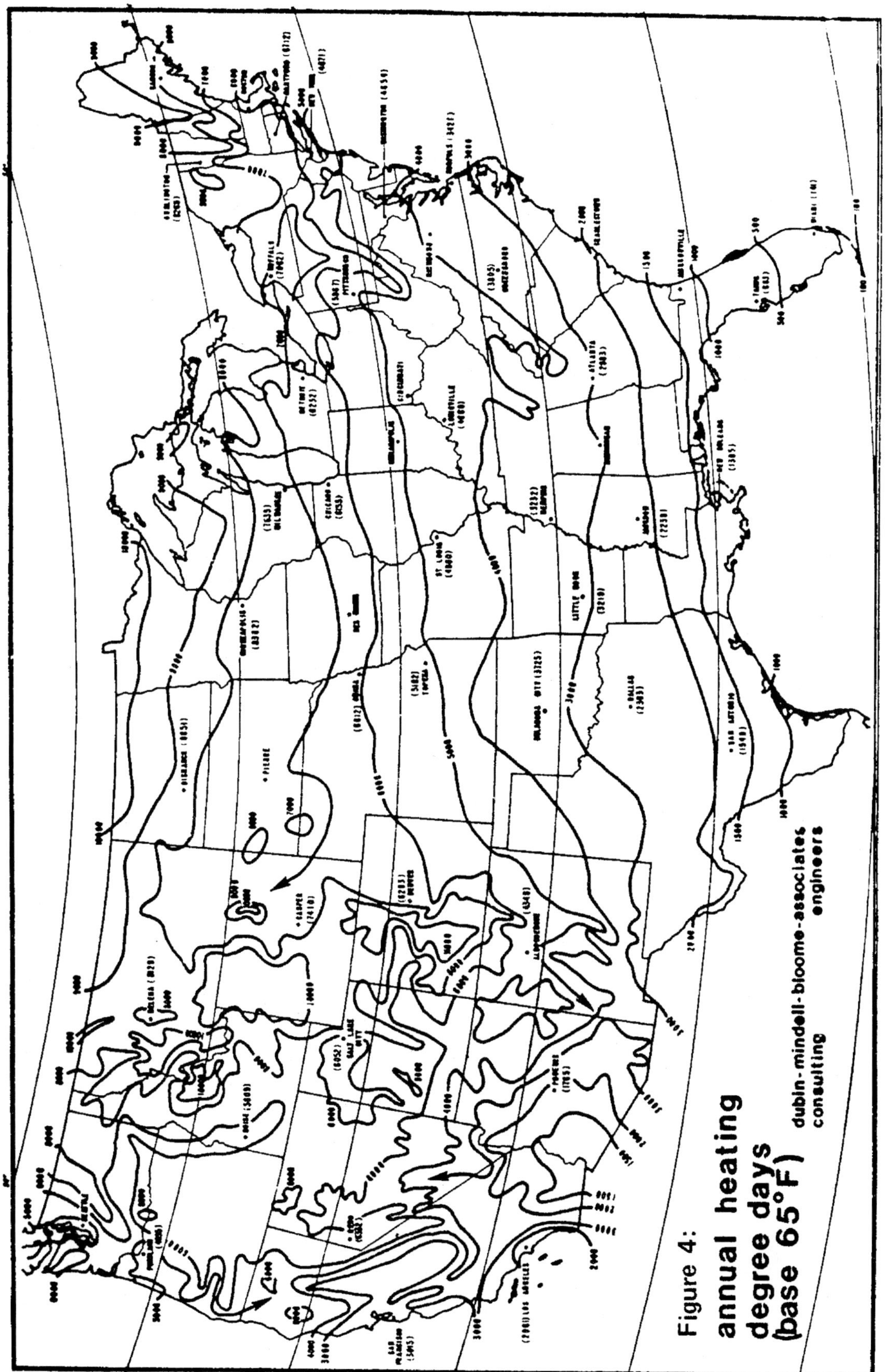

Figure 4: annual heating degree days (base 65°F)

dubin-mindell-bloome-associates consulting engineers

Figure #4 Engineering Data

Source: Climatic Atlas of the United States
U.S. Department of Commerce
June, 1968
Page 36

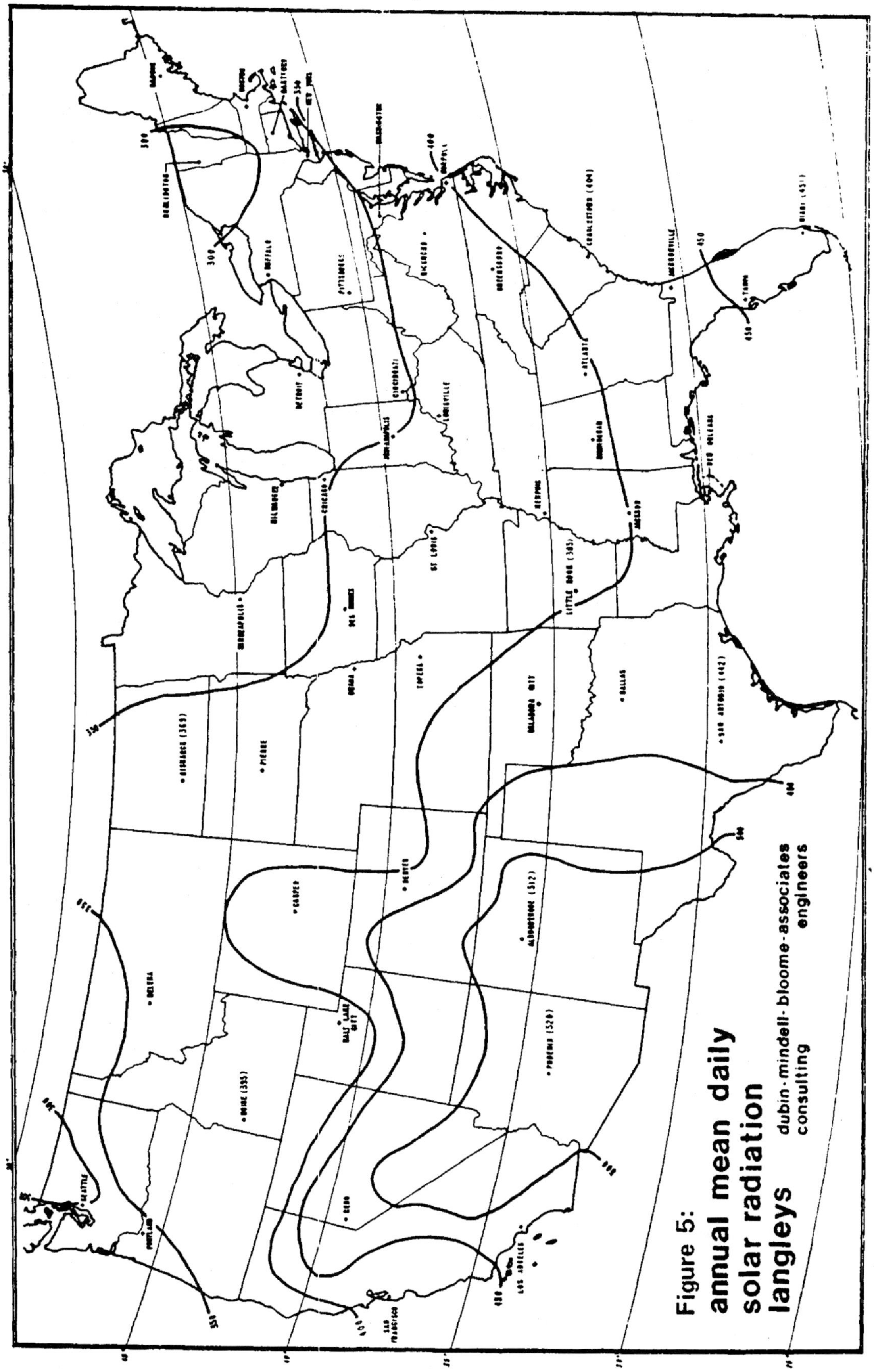

Figure 5: annual mean daily solar radiation langleys

dubin-mindell-bloome-associates consulting engineers

Figure #5 Engineering Data

Source: Climatic Atlas of the United States
U.S. Department of Commerce
June, 1968
Page 70

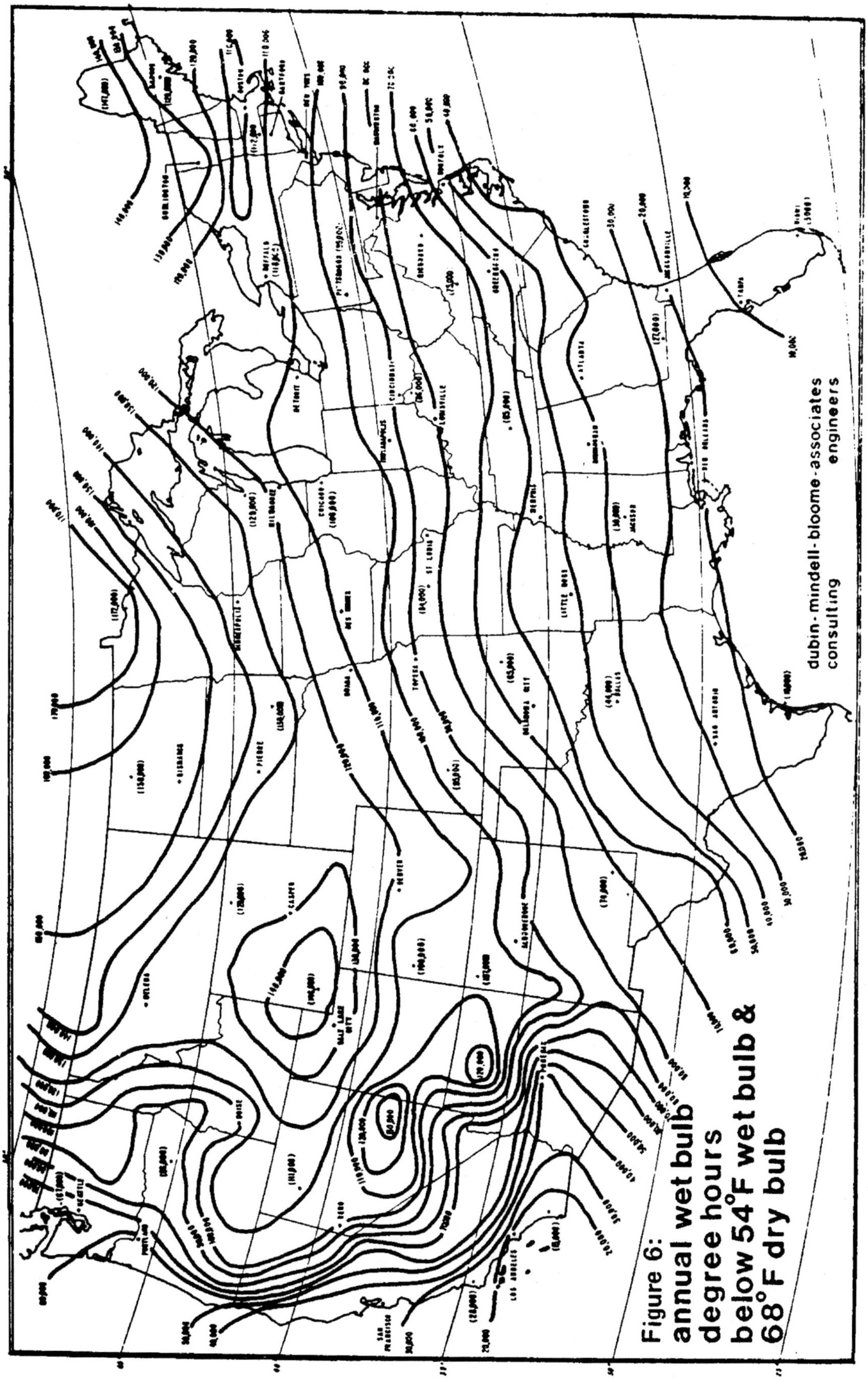

Figure 6:
annual wet bulb degree hours below 54°F wet bulb & 68°F dry bulb

Figure #6 Engineering Data

Source: AFM 88-8
U.S. Government Printing Office
15 June 1967

107 Locations in United States

<u>24 hrs/day October - April</u>

DB less than 68°.

WB taken in 5° increments less than 54°F WB.

Base WB taken at 54°F.

WB temperature difference taken from median of 5° range to 54°, i.e.: for 45 to 49 range, WB temperature difference = 54-47 or 7°F. Temperature difference multiplied by number of hours in each WB temperature range, for October thru April and summed for total. Locations with heating seasons significantly longer than October-April should be analyzed individually for maximum accuracy.

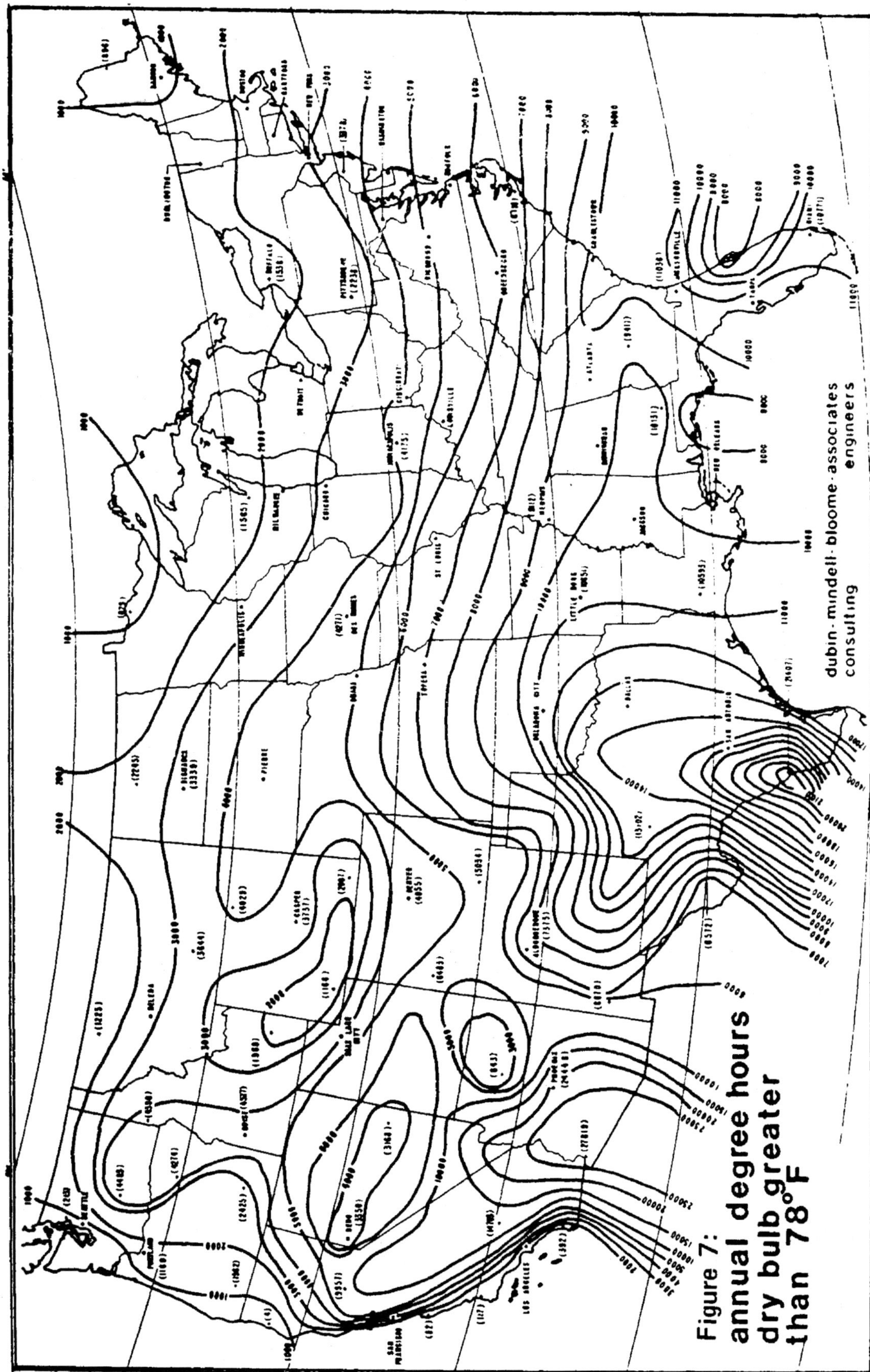

Figure 7:
annual degree hours
dry bulb greater
than 78°F

Figure #7 Engineering Data

Source: AFM 88-8
U.S. Government Printing Office
15 June 1967

107 Locations in U.S.

12 Mos/yr 8 Hrs/Day 0930-1730

DB taken in 5°F increments beginning with 80-84°F.

DB temperature difference taken from median of 5°F range to 78°F, i.e., for 85° to 89°F range, DB Temperature Difference = 87°-78°F = 9°F. Temperature Difference multiplied by No. of hours in each DB Temperature Range and summed for total.

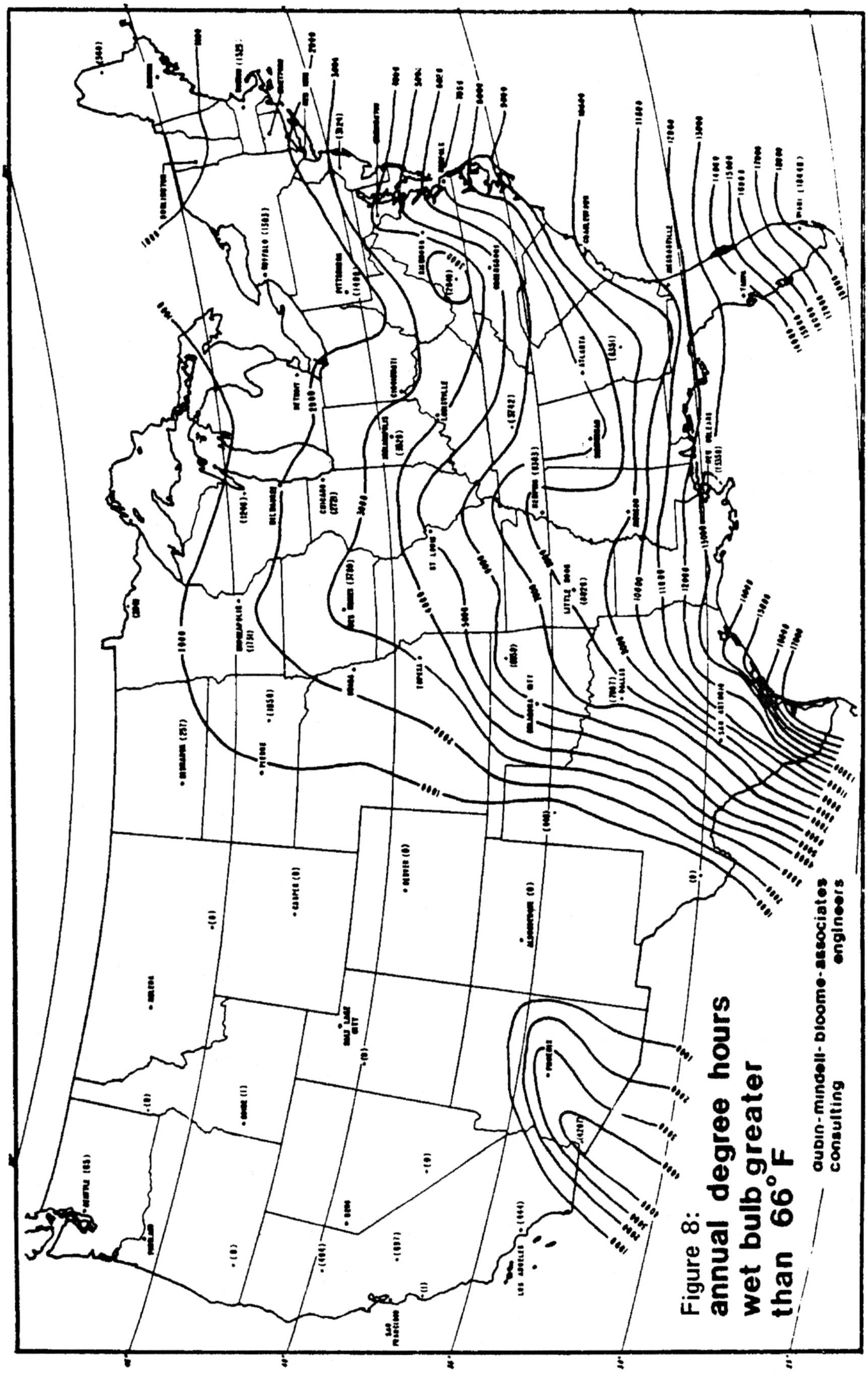

Figure 8: **annual degree hours wet bulb greater than 66°F**

Figure #8 Engineering Data

Source: AFM 88-8
U.S. Government Printing Office
15 June 1967

107 Locations in U.S.

12 Mos/yr 8 hr/day 0930-1730

WB taken in 1°F increments beginning with 67°F

WB temperature difference taken from 66°F

Temperature difference multiplied by No. of hours at each WB temperature and summed for total.

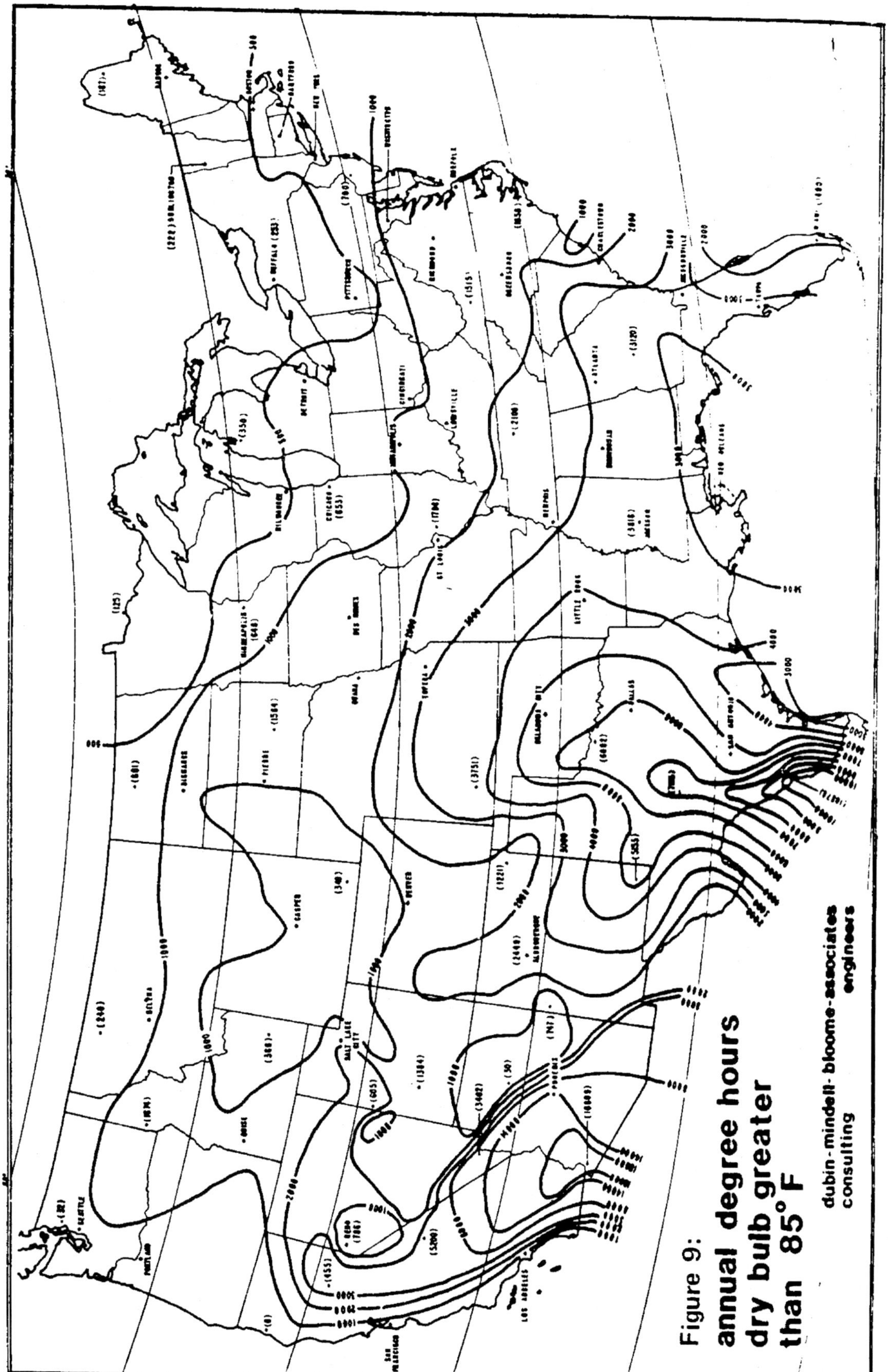

Figure 9: **annual degree hours dry bulb greater than 85°F**

dubin-mindell-bloome-associates consulting engineers

Figure #9 Engineering Data

Source: AFM 88-8
U.S. Government Printing Office
15 June 1967

107 Locations in U.S.

12 Mos/yr 8 Hr/day 0930-1730

DB taken in 5°F increments beginning with 85-89°F

DB temperature difference taken from median of 5°F range to 85°F, i.e., for 90 to 94°F range, DB Temperature Difference = 92-85°F = 7°F

Temperature Difference multiplied by No. of hours in each DB Temperature Range and summed for total.

air to room conditions of 78° D.B. and 55% R.H.(refer to standards, Fig. 20). Use information from Figure 8 to determine total energy savings by reducing outdoor air leakage and ventilation during the cooling season (Figure 21).

6. Estimate the Number of Annual Dry Bulb Degree Hours Above 85°F. (Figure 9)

This value influences the annual energy consumption for "building" cooling load due to outdoor air when enthalpy conditions are low enough to permit full use of outdoor air, but existing cooling coil cannot maintain dry bulb conditions with 100% outdoor air.

STEP 2: PREPARE A BUILDING TYPE, CONSTRUCTION, AND USE PROFILE

Prepare a profile using the form on page 71 (Figure 10) to record the information. Use existing construction or as-built drawings and plans, if available, for building configuration, construction details, and dimensions. Verify the information on the site since changes may have been made to the building after the plans were completed. If plans do not exist, obtain all data by inspection of the building. In particular, note and record the physical condition of the building elements, especially apparent deficiencies, i.e. leaks, missing insulation, leaky windows, etc. Where there are a typical conditions which are not accounted for in Figure 10, i.e. auditoriums, gymnasiums or cafeterias used intermittently in office buildings, or store room areas in supermarkets, it is important to list areas, lighting system details, temperature levels, control systems, and the number of occupants and occupied periods so that these areas can be addressed separately and so that the amount of energy currently used, (and then conserved after implementing the individual guidelines in Sections 4-10) can be determined.

In listing the number of occupants for office buildings, record the average number of employees plus the estimated number of visitors; for retail stores, record the number of employees plus the number of customers; and for religious buildings record the number of people present at one time for various periods during the day and the week. Energy can be saved by reducing ventilation rates when occupancy is reduced even for portions of the day, and temperature can be reset during unoccupied periods for sections of the building.

This page blank

FIGURE 10: BUILDING TYPE, CONSTRUCTION AND USE PROFILE

Configuration and Construction:

Line No. (Circle Appropriate items and fill in blank)

1. Primary Bldg. Use:

2. Length, Feet and Orientation ______Ft. N. W. E. S.

3. Width, Feet and Orientation ______Ft. N. W. E. S.

4. Number of floors: ______

*5. Height from floor to floor ______

*6. Height from floor to ceiling______

7. Floor area, gross sq. ft. : Lines 2 x 3 x 4 = ______sq.ft.

8. Window Glazing: single, double, clear, reflective

9. Window Type: Fixed sash, double hung, casement

10. Window Condition: Loose fitting, medium, tight

11. Windows: Number area, orientation: North - No.______ gross area:______

12. " " " West - No.______ " ______

13. " " " East - No.______ " ______

14. " " " South - No.______ " ______

15. Door types and numbers:
1-single;* 2-vestibule;* 3-revolving

North - No.______ Type______

16. Door Types: East - No.______ Type______

17. Door Types: West - No.______ Type______

18. Door Types: South - No.______ Type______

Fig. 10 (Cont'd.)

Line No. (Circle appropriate item and fill in blank)

*19. Gross wall area and orientation: North: Lines 2Nx3Nx4x5=______sq.ft.

*20. Gross wall area and orientation: West : Lines 2Wx3Wx4x5=______sq.ft.

*21. Gross wall area and orientation: East : Lines 2Ex3Ex4x5=______sq.ft.

*22. Gross wall area and orientation: South: Lines 2Sx3Sx4x5=______sq.ft.

*23. Net wall area and orientation: North: Lines 19-11(area)_____sq.ft.

*24. Net wall area and orientation: West : Lines 20-12 " _____sq.ft.

*25. Net wall area and orientation: East : Lines 21-13 " _____sq.ft.

*26. Net wall area and orientation: South: Lines 22-14 " _____sq.ft.

*27. Exterior opaque wall construction: Circle Type:

1-frame; 2-curtain wall; 3-solid masonry; 4-brick & masonry;

5-masonry cavity

*28. Exterior opaque wall insulation: Material:_____________

Thickness:________

*29. Roof construction: Circle Types: 1-masonry; 2-wood;

3-metal; 4-flat; 5-sloped; 6-pitched; 7-light; 8-dark

*30. Roof insulation: Type:________Thickness:________

"U"Value:____________

*Information required for ECM-2 only. All other items required for ECM-1 and ECM- 2.

*31. Floor: Circle Type: 1-slab on grade; 2-over heated space;

3-over unheated space; 4-wood; 5-concrete; 6-other

32. 1) Number of working hours/wk. 2) Number of Occupants ____

(a) For offices, employees & visitors; stores, employees & customers; religious buildings, occupants.

33. Number of custodial hours per week__________;

After dark summer________; after dark winter________;

Saturdays________; Sundays___________;

34. Temperature and relative humidity inside conditions:

	Season	Temperature	Relative Humidity
34a.	If Heated-Winter	Occupied Hours_____°F	__________%RH
34b.		Unoccupied Hours_____°F	__________%RH
34c.	If Air Conditioned - Summer	Occupied Hours_____°F	__________%RH
34d.		Unoccupied Hours_____°F	__________%RH

35. Ventilation: Outside air:

35a. During occupied hours on/off Amount in total CFM______

35b. CFM/person: Line 35a ÷ Line 32 (2) = __________________

35c. During unoccupied hours on/off, Amount in total CFM___

STEP 3: PREPARE AN ELECTRICAL AND MECHANICAL SYSTEMS PROFILE

Prepare a profile of the lighting, power, heating, ventilating, air conditioning, and domestic hot water systems using Fig. 11, page 77 to record the information.

The original building and engineering plans, and as-built plans and specifications should be used with caution since the mechanical and electrical systems in older buildings have frequently undergone changes and/or the plans may

*Information required for ECM- 2 only. All other items required for ECM-1 and ECM-2 .

This page blank

have been incomplete to begin with. During your on-site survey, note and record a typical conditions such as lighting fixture patterns, or special requirements for portions of the building which are not accounted for in Figure 11. Small variations from the norm will not materially affect the conservation program.

Except for those buildings which employ full time operating and maintenance personnel, it may be difficult or impossible for the owners or operators of the buildings to complete the building and electrical and mechanical profiles without outside help to determine the type and physical characteristics or operating efficiencies of many of the mechanical and electrical systems. Seek assistance from your heating service oil burner (and/or boiler) maintenance company, which may also assist you in identifying other features and conditions of the heating and ventilating systems.

Companies which specialize in balancing HVAC systems are available for preliminary consulation to determine existing conditions. The local utility company can provide help in identifying electrical equipment and systems and information on motor efficiencies.

Where additional information is needed, your State Chapter of the American Consulting Engineers Council can recommend qualified consulting engineers whose services are available to make the preliminary assessments of your existing systems.

FIGURE 11: ELECTRICAL AND MECHANICAL SYSTEMS PROFILE

Line No. (Circle appropriate item and fill in blank)

1. Electric lighting system

Lighting fixtures in primary spaces such as office areas, halls of worship, store sales areas.

(1) Incandescent; (2) Fluorescent: (3) Other note________

(4) # of fluorescent fixtures____;(5) # of lamps per fixture_____;(6) Wattage per lamp______;(7) Total wattage of all fluorescent fixtures_______;(8) Total wattage of all incandescent lamps____(9) Total wattage of incandescent and fluorescent lamps_________.

Fig. 11 (Cont'd)

Line No. (Circle appropriate item and fill in blank)

2. Lighting fixtures in secondary spaces, such as corridors, toilet rooms, storage rooms.

   (1) Incandescent; (2) Fluorescent; (3) Other note______

   (4) # of fluorescent fixtures_____; (5) # of lamps per fixture_______; (6) Wattage per lamp__________;

   (7) Total wattage of all fluorescent fixtures________;

   (8) Total wattage of all incandescent lamps__________;

   (9) Total wattage of incandescent and fluorescent lamps____________.

3. Total installed wattage: Lines 1 (9) + 2 (9)= _________

4. Average installed watts/sq.ft. Lines 3÷7 (Fig.10)______

*5. Type lighting fixtures: (1) Pendant mounted; (2) Surface mounted; (3) Recessed; (4) Wall mounted; (5) Luminous ceiling; (6) Cove mounted; (7) Exterior lighting on walls; (8) Exterior lighting on standards.

6. Total wattage of exterior lighting for: (1) Security____ (2) Parking lots and drive______.

7. Area of parking lots______length x_______width=_____Sq.ft.

8. Parking lot lighting in watts/sq.ft., Lines 6 (2) ÷7 sq. ft. _______w/sq.ft.

9. Hours/wk. parking lot lighting is in operation__________

*Information required for ECM-2 only. All other items required for ECM-1 and ECM-2.

VERTICAL TRANSPORTATION

10. Escalators: Number________Operation hours/day___________

11. Elevators : Number________*Type: Gear, Gearless, Hydraulic Operation Hours/day________*Total connected H.P.__________.

DOMESTIC HOT WATER SYSTEMS

12. Method of generation and storage: Separate water heater; (1) Oil; (2) Gas; (3) Electric; (4) Coal; (5) Tankless heater on space heating boiler; (6) Tank heater on space heating boiler; (7) Storage tank size if any _____gals.; (8)*Tank insulation thickness_______Type________(9) Aquastat setting________°F.

13. Estimated annual usage:

    (1) Office Bldgs: Fig.10 Line 32(2)x750 _______gal/yr.

    (2) Restaurants: Meals served/yr.x3 gal/meal= _____gal/yr.

    (3) Religious Bldgs: Line 32(2)x50 gal/yr=_________gal/yr.

    (Does not include special cooking facilities)

    (4) Stores: Fig.10 Line 32(2)x number of days=_____gal/yr.

    (5) For residential buildings - 7200 gal/capita/yr.

    (6) For Schools - 50 gal/capita/week

    (7) For Hospitals - varies with type

*Information required for ECM-2 only. All other items required for ECM-1 and ECM-2.

HEATING AND AIR CONDITIONING SYSTEMS

14. Boilers or furnace type for space heating (Circle items)

    (1) Hot water; (2) Low pressure steam; (3) High pressure steam; (4) Fire tube; (5) Water tube; (6) Cast iron; (7) Steel; (8) Gravity hot air; (9) Forced warm air.

15. a) Boiler or furnace rating_________ $BTU \times 10^3$/hr. or _______Boiler H.P.

    b) Present measured peak load combustion efficiency______%.

16. Compressors and chillers:

    (1) Number________

    (2) Rating of each in tons of refrigeration___________

    (3) Total tons of refrigeration (1) x (2) = __________

    (4) If electric drive, total motor horsepower ________H.P.

    *(5) If absorption units, total peak steam consumption_________H.P.

17. If central air conditioning systems, indicate: (1) Cooling tower motor sizes total_______H.P.; (2) Air cooled condenser motor sizes total_________H.P. Condenser pumps No_______ Total_______H.P.

18. If room air conditioners or through-the-wall units: Indicate (1) total number_______(2) Horsepower _______/unit. (3) Total connected Horsepower (1) x (2) = ____________.

*Information required for ECM-2 only. All other items required for ECM-1 and ECM-2.

Fig. 11 (Cont'd).

Line No. (Circle appropriate item and fill in blank)

19. If commercial refrigeration, indicate: (1) Number of cold cases or refrigerators______; (2) Number of condensing units________ (3) Total connected horsepower of condensing units________H.P.

HVAC SYSTEMS

Check the systems and fill in appropriate information:

20. All air HVAC systems: Check types-fill in blanks.

(1) Single zone______ a) Number of air handling units____
b) Total Horsepower ______________
c) Total CFM/air handling unit____

(2) Terminal reheat___ a) Number of air handling units____
b) Total Horsepower ______________
c) Static pressure______________
d) Number of reheat boxes_________
e) Type reheat Coil: 1. hot water 2. electric 3. steam
*f) CFM/air handling unit__________

(3) Variable Volume__ a) Number of air handling units____
b) Total horsepower ______________
c) Dump type system______________
d) Vaned inlet__________________
e) CFM/air handling unit___________

*Information required for ECM-2 only. All other items required for ECM-1 and ECM-2.

(4) Induction________ a) Number of air handling units_____

b) Total horsepower ________________

c) Static pressure__________________

d) Number of terminal units_________

*e) CFM/air handling unit___________

(5) Dual duct________ a) Number of air handling units_____

b) Total horsepower ________________

c) Static pressure__________________

d) Number of terminal units_________

e) CFM/air handling unit____________

(6) Multi-zone units a) Number of air handling units_____

b) Total horsepower ________________

c) Static pressure__________________

d) Number of terminal units_________

e) CFM/air handling unit____________

(7) Forced warm air furnaces No.___________

a) Total horse power of blowers_____

b) CFM/furnace______________________

21. Water-air systems

(1) 2 Pipe fan coil__ a) Number of units__________________

b) Total connected horsepower_______

(2) 4 Pipe fan coil__ a) Number of units__________________

b) Total connected horsepower_______

(3) Unitary Heat Pumps_______ a) Number of units__________________

b) Total connected horsepower_______

*Information required for ECM-1 only. All other items required for ECM-1 and ECM-2.

22. Pumps

(1) Chilled water pumps_____________

a) Number of units_______________

b) Total connected horsepower_____

(2) Condenser water pumps___________

a) Number of units _______________

b) Total connected horsepower_____

(3) Boiler feed pumps_______________

a) Number of units_______________

b) Total connected horsepower_____

(4) Hot water pumps for space heating_________________

a) Number of units _______________

b) Total connected horsepower_____

(5) Recirculating pumps for domestic hot water_________

a) Number of units _______________

b) Total connected horsepower_____

23. (1) Outside air fans_________________________________

a) Number of units_______________

b) Total connected horsepower_____

c) CFM/fan unit_________________

(2) Supply air fans(Check the number and total H.P. for all)

a) Number of backward curved multivane fans____HP___

b) Number of forward curved multivane fans_____HP___

c) Number of axial fans______________________HP___

d) Number of propeller fans__________________HP___

e) CFM/fan unit__________________________________

23. (3) Exhaust air fans______________________

a) Number of backward curved multivane fans____HP___

b) Number of forward curved multivane fans_____HP___

c) Number of axial fans____________________HP___

d) Number of propeller fans_________________HP___

e) CFM/fan____________________________________

24. Check if installed:

(1) Fin tube radiators______ (5)Supply and return ducts___

(2) Cast iron radiators_____ (6)Outside air dampers_______

(3) Radiant heating coils___ (7)Steam piping_____________

(4) Hot water piping________ (8)Exhaust duct work_________

STEP 4; PREPARE AN ENERGY CONSUMPTION AUDIT

Prepare a profile of your existing consumption of energy and record the data in the form Figure 12, page 87.

In some buildings (usually only a nominal number of large buildings) there are BTU or energy consumption meters which record quantities of fuel and power used. However, in most cases, these do not exist and the yearly energy used must be ascertained from the fuel and utility bills. To start with, determine the gross number of BTU's which your building uses per year per square foot of gross floor area. This is relatively easy to do and will provide the first bit of information to allow you to compare the energy your building uses with others as a starting point to establish a target, a goal for reduction in energy use. The ultimate savings in energy can be expressed as a percentage of your current consumption. You will be considering energy conservation measures by individual systems and subsystems. It is important to break down the total amount of energy used by the heating system as a whole, and by, or due to each of its sub-components. Energy for cooling, lighting and power must also be broken down into sub-system use.

1. Gather from your monthly utility and fuel suppliers' bills the annual usage of energy in gallons of oil, cubic feet of gas, pounds of propane, tons of coal and kilowatt hours of electricity. Record the gross yearly quantity of

fuel and power in profile Figure 12, Column A, page 87. Convert the units of fuel and electricity to equivalent BTU's by multiplying the quantities of fuel and power by the appropriate conversion factor in Column B. Record the product in Column C.

Conversion factors are based on the number of BTU's in a gallon of oil, cubic foot of gas, ton of coal and kilowatt hours of electricity. Total the gross number of BTU's and enter the result on line 7, of Figure 12. Determine the BTU's in thousands used per year, per square foot of floor space by dividing the gross quantity of BTU's in thousands (line 7, Figure 12) by the gross area in square feet (line 7, Figure 4) Many office buildings use from 80,000 to 500,000 BTU's/sq.ft./yr.

Stores consume energy at a rate within about the same range and religious buildings at womewhat less than the lower end of the range. Hospitals generally use more energy/sq.ft. and housing less. Schools use less energy/sq.ft. than office buildings. New office buildings are now being designed and constructed in an energy conservation mode to consume as little as 55,000 BTU's/sq.ft./yr. or less. A realistic energy budget goal for existing buildings mav be as low as 75,000 BTU's/sq.ft./yr. for office buildings and stores' 60,000 BTU's/sq.ft./yr. for schools, and 35,000 BTU's/sq.ft./yr for religious buildings. See how close to these figures you can approach or if you can better them. Your existing consumption will give you a starting point. The difference between that figure and the target energv budget, will provide an order of the magnitude of savings to strive for.

Next break down the average annual BTU consumption by system, using Fig. 12, Sections 2,3,4 and 5 to record the data. The duration and severity of winter varies from year to year, and affects the quantity of heat required. To obtain the most accurate results, five or more years of energy consumption should be averaged to determine fuel consumption for a typical year for heating. If records are only available for the previous year's fuel consumption, these can be corrected to a typical year by dividing the actual fuel consumption by heating degree days actually experienced for that year, then multiplying by the average yearly degree days for the building's geographic location, as shown in Figure 4. The heating degree days for the year corresponding to actual fuel consumption can be obtained from the weather bureau, or a local fuel supplier.

EXAMPLE

$$\frac{\text{1973 fuel consumption}}{\text{1973 degree days}} \times \text{average yearly degree days} = \text{average yearly fuel consumption}$$

Fuel bills often do not differentiate between the end use for heating or other purposes, and an adjustment must be made. If oil, gas, or coal is the primary fuel, and is used for both heating and domestic hot water, the usage should be broken down between the two. The space heating load occurs in the winter, but the domestic hot water load is continuous for the whole year at a rate that can be assumed constant. To determine the amount of the monthly fuel bills that can be attributed only to heating, select one average winter month's consumption, subtract one average summer month's consumption and multiply the answer by the total number of heating months. The difference between total fuel used and heating fuel use will be the domestic hot water energy consumption.

If the building is heated by electricity and the total electrical usage of the building is metered and billed in a lump sum, the bill will include energy for heating, lighting, and power. To arrive at the amount of electricity used for heating only, it is necessary to assess the quantity used for lighting and power and subtract this from the total billing. In small buildings, a quick assessment of the electricity usage for lighting can be made by counting the number of lighting fixtures and multiplying the wattage of each lamp and the average number of hours that these are switched on during the heating season. This will give the total number of watt-hours consumption that can be attributed to lighting. Divide watt-hrs. by 1000 to get kilowatt-hrs. Similarly, a survey can be made of all electrical motors that are in use during the heating season and their nominal horsepower rating (multiplied by .800) to determine the approximate amount of electricity in KWH used for each hour of running. (This formula assumes an efficiency of 93% for electric motors). The KWH should then be multiplied by the number of hours of operation during the heating season to determine the total kilowatt hours that can be attributed to power. The sum of the kilowatt hours assessed for lighting and power should then be subtracted from the total power consumed by the building for the heating season, to determine the amount used for heating. In large, complex buildings where simultaneous heating and cooling are likely to occur, you should seek professional help to prepare a more accurate analysis of energy flow, if your maintenance staff is unable to do so. To determine the energy used for lighting and for power for the entire year, the same method of determining energy use in the heating season, described above, can be used for a 12-month period.

To determine the amount of energy used for air conditioning, estimate the energy for fans and pumps as outlined above. For electric driven refrigeration units the KWH can be estimated by deducting the energy used for lighting and other motors from the June, July, August, and September electric utility bills.

FIGURE 12: ENERGY USE AUDIT

1. Gross Annual Fuel and Energy Consumption

Line No.

	A	B Conversion Factor		C Thousands of BTU's/yr.
		x 138 (1)	=	
		x 146 (2)	=	
1. Oil - gallons				
		x 1.0 (3)	=	
2. Gas - Cubic Feet		x 0.8 (4)	=	
3. Coal - Short tons		x 26000	=	
4. Steam-Pounds x 10^3		x 900	=	
5. Propane Gas - lbs.		x 21.5	=	
6. Electricity-KW.Hrs.		x 3.413	=	
7. Total BTU's x 10^3/yr.	....			
8. BTU's x 10^3/Yr/Per Square Foot of Floor Area				

(Line 7 ÷ Figure 4, Line 7)

Use for (1) No.2 Oil; (2) No. 6 Oil; (3) Natural Gas; (4) Mfg. Gas

2. Annual Fuel and Energy Consumption for Heating

Line No.

	A	B Conversion Factor	C Thousands of BTU's/yr.
		x 138 (1)	=
9. Oil - gallons	______	x 146 (2)	= ______
		x 1.0 (3)	= ______
10. Gas - Cubic Feet	______	x 0.8 (4)	= ______
11. Coal - Short tons	______	x 26000	= ______
12. Steam - Pounds $x\ 10^3$	______	x 900	= ______
13. Propane Gas - lbs.	______	x 21.5	= ______
14. Electricity-Kw.Hrs.	______	x 3.413	= ______
15. Total BTU's x	....		= ______

16. BTU's x 10^3/Yr Per Square Foot of Floor Area ______
(Line 15 Line 7)

3. Annual Fuel and Energy Consumption for Domestic Hot Water

Line No.

	A	B Conversion Factor	C Thousands of BTU's/Yr.
17. Oil - Gallons		x 138 (1)	= ______
	______	x 146 (2)	= ______
18. Gas - Cubic Feet	______	x 1.0 (3)	= ______
		x 0.8 (4)	=

3. Annual Fuel and Energy Consumption for Domestic Hot Water (Cont'd)

Line No.	A	B Conversion Factor	C Thousands of BTU's/Yr.
19. Coal - Short Tons		x 26000 =	
20. Steam-Pounds x 10^3		x 900 =	
21. Propane Gas - lbs.		x 21.5 =	
22. Electricity-Kw.Hrs		x 3.413 =	
23. Total BTU's/Yr. x 10^3..................			
24. BTU's x 10^3Yr/Per Square Foot of Floor Area (Line 23 ÷ Figure 4, Line 7)			

4. Annual Fuel and/or Energy Consumption for Cooling (Compressors and Chillers)

Line No.	A	B Conversion Factor	C Thousands of BTU's/Yr.
a) if absorption cooling			
		x 138 (1) =	
25. Oil - Gallons		x 146 (2) =	
		x 1.0 (3) =	
26. Gas - Cubic Feet		x 0.8 (4) =	
27. Coal - Short Tons		x 26000 =	
28. Steam-Pounds x 10^3		x 900 =	
29. Propane Gas - lbs		x 21.5 =	
30. Total BTU's/yr x 10^3..................			
31. BTU's x 10^3/Yr Per Square Foot of Floor Area (Line 30 ÷ Fig. 4 Line 7)			
b) If Electric Cooling			
32. Electricity-KWH		x 3.413 =	
33. BTU's x 10^3/Yr Per Square Foot of Floor Area (Line 32 ÷ Fig. 4 Line 7)			

5. Estimated Annual Energy Consumption for Interior Lighting:

Line No.	A	B Conversion Factor	C Thousands of BTU's/Yr
34. KWH	______	x 3.413 =	______

Fig. 10 Line 3 x Fig. 10 Line 33 (1)

35. BTU's x 10^3/Yr/Per Square Foot of Floor Area______
(Fig. 10 Line 35 Col. C ÷ Fig. 4 Line 7)

6. Estimated Annual Electrical Energy Consumption for all Motors and Machines If Building and Hot Water Are Not Electrically Heated: (1)

36. Total Kw. Hrs. ______Less Kw. Hrs. Lighting_____=____Kw.Hrs.
(Line 22, Col. A)

37. Kw. Hrs./Yr/Sq.ft. Floor area = ______________(1)
(Line 37 Col. C ÷ Fig.4 Line 7)

38. BTU's x 10^3/Yr/Sq.ft. floor area = (Line 37) x3.431______(2)

(1) and (2) If building heat and hot water are electrically heated, deduct the Kw.Hrs./Yr/per sq. ft. and BTU's/Yr/Sq.ft. for heating and hot water. (Lines 37 and 38)

D. PROCEDURES FOR IMPLEMENTING DETAILED CONSERVATION OPPORTUNITIES

Each of the following steps, 5-11, are abstracts of more complete options detailed in Sections 4-10, and are listed here to denote the general type of action to be taken. The order of steps may be interchanged for many items depending on the conditions in your building and your resources (labor, mostly) to pursue each suggestion.

STEP 5: REDUCE NEGLIGENT WASTE

Using the guidelines of Sections 4-10 for each system as a checklist, identify those items in your building that can be cleaned, repaired, or serviced to improve their performance and take the appropriate corrective action. Record the action and date performed in a permanent conservation record book which you should set up for your energy conservation program.

STEP 6: SET UP SYSTEM AND MAINTAIN RECORDS

Using your energy conservation record book, which should include all identification profiles and initial energy audit, maintain a running diary of all actions which you take to implement the ECO's appropriate to your building. Provide space for recording future fuel and energy bills, converted to BTU's, to ascertain the effectiveness of your energy conservation program. You will not be able to measure your accomplishments by comparing the utility bills only, due to rising costs; the numbers in BTU's are needed. (But remember, if you hadn't undertaken this program, the bills would have been even higher.) In your record book, outline the additional measures you plan to take and don't delay implementing them. As you accomplish each change, record the date and check off the item. Record the estimated time and expense for in-house labor or contract labor for each ECO you propose to initiate; after completion of the work, record the actual costs. CAUTION: Do not get bogged down in paper work. If you will be delayed by setting up and maintaining a record system, initiate immediate action for steps 5 and 7.

STEP 7: CHANGE THE ENVIRONMENTAL CONDITIONS

Review the way you operate your building. Close off unused spaces and cut off all services to them. Turn off unnecessary lights. Reduce the indoor temperatures during the heating season by resetting thermostats and adjust thermostats to maintain higher temperatures in the summer (See Section 4 for special instructions for terminal reheat systems and core areas of large buildings.) Turn off the air conditioning system and all auxiliary fans and pumps and condenser or cooling towers at night and on weekends. Reduce the quantity of domestic hot water and reset aquastats to maintain lower temperatures, prepare memos to occupants telling them what you are doing. Tag light switches and other control devices to make it easy to identify them; they will get more attention when highly visible. Operate drapes, venetian blinds, windows to maximize solar heat gain in winter, to make maximum use of daylight all year round, and to minimize solar heat gain in summer.

Select the guidelines form each portion of Section 4 that list the options for changing the enviornmental conditions and inaugurate them without delay.

When you find that dampers or other controls are not installed to permit you to take full advantage of any guideline, consult ECM 2 for further instructions.

STEP 8: REDUCE ENERGY CONSUMPTION FOR LIGHTING

Clean walls and other interior surfaces to increase the effectiveness of the lighting system. Clean the lamps, tubes, louvres and fixtures; next, replace lamps with ones of lower wattage, or with lamps or tubes that provide more lumens per watt. Remove lamps from fixtures where the tasks do not require the current illumination levels. Follow guidelines from Section 9, "Lighting" for the measures and methods to follow. Where extra switches would permit a savings in energy without sacrificing visual performance consult ECM 2 for specific instructions for adding them.

STEP 9: REDUCE HEAT LOSS AND HEAT GAIN.

Reduce the "building" heating and cooling loads by reducing the air leakage into the building and heat transmission through the windows. Caulking cracks, weatherstripping, and adding storm windows and doors for small buildings in climatic zones of 5000 or more degree days will always result in a short payback period in fuel savings for each dollar invested. (Refer to Sections 4 & 6 "Heating" and "Cooling"). ECM 2 provides more data on window treatment and insulation for walls and roof to reduce heat loss and heat gain for each typical climatic zone.

STEP 10: REDUCE THE DISTRIBUTION LOADS.

In addition to repairing leaks and insulation, and cleaning filters in air systems to reduce losses, the piping, pumping, duct, and fan systems should be analyzed to determine the potential for reducing energy consumption due to these systems. Bear in mind, though, that the equipment may not have been operating even in accordance with present loads, so there may be a opportunity for conservation by adjusting primary equipment for present loads as well as for new load reductions. Before making any adjustments to motor speeds, dampers, and major controls a survey of the systems which they serve, and the actual conditions of operation and load should be made by competent personnel who understand the equipment and systems. Avoid do-it-yourself adjustments to these systems, since it is possible to seriously disrupt operation and jeopardize equipment if improper adjustments are made. For their guidance, we suggest that you provide ECM-1 and ECM-2 to your operation and maintenance personnel and to outside service organizations which you retain for maintenance and repairs. If you seek professional advice, make sure that architects and engineers whom you consult are familiar with these manuals.

STEP 11: IMPROVE THE EFFICIENCY OF THE PRIMARY ENERGY CONVERSION EQUIPMENT

Use qualified service personnel to measure combustion efficiencies, and the performance of compressors, chillers, cooling towers, condenser pumps and motors and to service and adjust them to operate in accordance with the new loads (which will have been reduced by actions taken in previous steps.) Equipment should be adjusted for "seasonal" efficiency, not simply peak efficiencies. Use the appropriate instructions and guidelines in Sections 4-10 of ECM-1.

ECM- 2 should be consulted for specific measures which entail more than a nominal cost to improve operating efficiencies. It provides guidelines for major alteration or replacements. If systems and equipment are obsolete or beyond significant improvement, replace them.

This page blank

Section 4

Energy Conservation Opportunities Heating and Ventilation

HEATING AND VENTILATION

A. BACKGROUND

The yearly consumption of energy for heating and ventilating can be reduced in three major categories:

(1) Reduce the "building" heat load
(2) Reduce the distribution system load
(3) Increase the efficiency of the primary energy-conversion equipment.

Categories 1 and 3 are addressed in this section, and category 2 in Section 7, "Distribution and HVAC Systems".

B. "BUILDING" HEATING LOADS

One of the factors which determine the "building" heat load is the average difference between indoor and outdoor temperatures -- the larger the temperature difference, the greater the load. Reduce the load by maintaining lower indoor temperatures for as long a period as possible during the heating season. Lowering the temperature to 68°F. or less in the major areas of the building when it is occupied, and maintaining even lower temperatures in less critical areas will conserve energy. The amount of heat produced internally by lights, people and business machines (internal heat gains) and the amount of sunlight impinging upon the structure and transmitted through window panes and doors also reduces "building" heating load.

Realize greater savings by reducing the indoor temperature at night and during weekends. Outdoor temperatures are generally colder at night, and neither solar heat gain nor internal heat gains help at night to offset heat loss.

The amount of outdoor air which is introduced into the building for ventilation or which infiltrates through the building envelope also contributes to the heating load. This air must be heated and humidified to meet indoor conditions. Its ultimate effect on the load depends, as with other heating loads, upon the difference in indoor and outdoor temperatures and the quantity of outdoor air itself. In cold climates, shutting off ventilation air at night -- when no ventilation is required for physiological reasons -- may result in the single largest savings of energy.

"Building" heating load also depends upon the amount of moisture maintained in the building. The measure of the moisture content of the air is relative humidity (R.H.). When cold outdoor air enters the building -- as ventilation or infiltration -- the relative humidity of the interior of the building drops and additional moisture, in vaporized form, must be added. Vaporization requires energy in the form of heat. The greater the volume of outdoor air, the greater is the heat demand to supply humidification. Lowering the level of humidification conserves energy.

A building is more comfortable at lower temperatures if the relative humidity is maintained at a level within a range of from 20 to 40%. In a "tight" building with little air leakage and a small amount of ventilation, the energy saved by lowering the temperature will exceed the energy required to maintain humidity levels higher than 20%. Without using a computer for a seasonal analysis, the calculation to determine this trade-off is complicated. ECM 2 describes a method for making such an analysis. Since there is little medical knowledge available to support the contention that relative humidities higher than 20% are more beneficial to health, the recommendation here is to maintain a maximum of 20% R.H. during the daytime in occupied areas of the building and to add no humidification at night.

Finally, the building envelope -- the roof, exterior opaque walls, windows and doors, all of which are subjected to the outdoor climate -- influences the size of the "building" heating load. The effect of infiltration through the envelope is discussed above. In addition, heat is transmitted through the building envelope by conduction in accordance with the temperature difference between indoors and outdoors and the resistance to heat transfer offered by each of the building components.

The rate of heat transfer through the envelope is expressed as a "U" value -- BTU's /hour/sq. ft. of surface per degree of temperature difference between indoors and outdoors. The thin layer of air surrounding the exterior surface of the envelope adds to the insulating value of the wall or roof material. A lower U value means greater resistance to the transmission of heat and saves energy by reducing heat loss from the building.

Single panes of glass in still air (less than 15 m.p.h.) have a U value of 1.13. Double glazing reduces the U value to about .55. The U values of walls and roofs vary from .4 down to .06 depending upon the structural materials and the thickness of any insulation which has been added. Effective measures to reduce the U value and heating load include adding a storm sash to existing windows, replacing the windows with double glazing (or triple glazing in severe climates), and adding insulation to the interior or exterior surfaces of roofs and exterior walls.

Wind destroys the air film around exterior surfaces and causes the U value of the surfaces to increase. Heat loss, especially through window panes and uninsulated walls, increases accordingly. Use shutters, screens, trees, or other shielding devices to reduce wind velocity on the windows to limit heat losses.

C. DISTRIBUTION SYSTEM LOADS

A distribution system is necessary to supply the "building" heating load. It may carry hot water or steam directly from a boiler to radiators or to fan coil units in the spaces to be heated, or to air handling units which transfer the heat from the steam or hot water piping by means of coils located in the unit. Air warmed by air handling units is forced through ducts to registers or diffusers in the conditioned spaces. Return air, drawn through ducts back to the air handling units, is cleansed, along with fresh air also drawn into the unit, by air filters. The unit, when fitted with cooling coils, employs the same blower and duct system for air conditioning. Gravity hot air furnaces deliver hot air, without a blower, directly to the space through a short duct connection. Forced warm air furnaces are equipped with a blower which delivers warm air to the spaces and return air (through ducts similar to those of the air handling units) to the furnace.

The distribution loads, often called "parasitic" loads since they do not contribute directly to the comfort and requirements of the building occupants, include heat losses from piping and duct work, and electric power to drive fans or pumps against the resistance of the duct or piping system. Air, water or steam leaks from these systems, torn and missing insulation, and broken or ill-fitting windows are flagrant examples of negligent waste -- they increase the load without performing useful work. Distribution loads are discussed extensively in Section 7, "Distribution and HVAC Systems".

D. PRIMARY ENERGY-CONVERSION EQUIPMENT

Ultimately, energy, in the form of oil, coal, or gas is consumed by primary equipment and converted, by way of combustion, into heat in a boiler or furnace. The potential energy in the fuel is not fully realized, because it is difficult to achieve and sustain the correct mix, for combustion of air and fuel. Additional losses, thru the breeching or smoke pipe up the chimney, and heat radiation from the boiler surfaces or furnace jacket further reduce the useful heat output of the unit.

Although the efficiency of a burner-furnace unit at any instant may fall just short of 90% for those fueled by oil and gas (and somewhat lower for those fueled by coal), the seasonal efficiency is generally lower by 10% to 30%, as a result of stack losses during and between on-off firing periods. The amount of heat lost is a function of the amount of excess air required for complete combustion, the amount of draft at maximum and part loads, the amount of air leakage into the combustion chamber, the ability of the burner to modulate in accordance with varying "building" loads, the quality of the fuel, and the amount of soot and scale accumulation (reducing heat transfer) on the combustion surfaces.

Proper adjustment of burners, and maintenance of boilers and furnaces improve efficiency -- and reduce fuel consumption accordingly. Many units originally designed for coal, however, have been fitted with gas or oil burners and cannot be made to operate as efficiently as better quality units designed specifically for oil or gas.

After reducing the "building" distribution load, further improve the efficiency of the burner-boiler or furnace units by adjusting the firing rate to accommodate the new loads. Lower the temperatures of the water or air delivered from the unit. Radiation and convection losses from the unit will be reduced with lowered temperature and steam pressures. If, however, it becomes necessary to circulate a greater quantity of air or water to meet room loads, distribution loads may actually increase. To analyze this trade-off thoroughly, seek professional advice. In general, operate the primary equipment at lower temperatures just sufficient to meet room loads.

ECM 2 describes methods of recovering waste heat from boilers, furnaces, and exhaust air to conserve energy.

E. ENERGY CONSERVATION OPPORTUNITIES

ECO 1 SET BACK INDOOR TEMPERATURES DURING UNOCCUPIED PERIODS

Energy expended to heat buildings to comfort conditions when they are unoccupied (which, actually, is most of the time) is wasted. Save energy by setting back the temperature level at these times. The savings which can result vary with the length of time and the number of degrees that temperatures are set back. The percentage savings will be greater in warmer climates, but the gross energy saved will be greater in cold climates.

In areas where it is not necessary to maintain high temperatures during occupied periods, i.e. corridors and lobbies, maintain even lower temperatures than for the other spaces. See Figure 13, column B, for suggested winter night setback temperatures. Implement setback by resetting thermostats manually (if automatic setback control has not been installed), or adjusting controls to suggested temperatures (if clock, day-nite, or other automatic reset controls are available). Climate, type of system, and building construction will determine the length of the startup period required to attain daytime temperature levels. Experiment to decide upon the optimal setback temperature and startup time for any particular building. If, in extremely cold weather, experience indicates that the heating system does not raise the temperature sufficiently by the time the building opens for the day, set temperatures back to a level higher than those recommended here for those periods of time only.

This page blank

FIGURE 13: SUGGESTED HEATING SEASON INDOOR TEMPERATURES

	A Dry Bulb °F. occupied hours maximum	B Dry Bulb °F. unoccupied hours (set-back)
L. OFFICE BUILDINGS, RESIDENCIES, SCHOOLS		
Offices, school rooms, residential spaces	68°	55°
Corridors	62°	52°
Dead Storage Closets	50°	50°
Cafeterias	68°	50°
Mechanical Equipment Rooms	55°	50°
Occupied Storage Areas, Gymnasiums	55°	50°
Auditoriums	68°	50°
Computer Rooms	65°	As required
Lobbies	65°	50°
Doctor Offices	68°	58°
Toilet Rooms	65°	55°
Garages	Do not heat	Do not heat
2. RETAIL STORES		
Department Stores	65°	55°
Supermarkets	60°	50°
Drug Stores	65°	55°
Meat Markets	60°	50°

FIGURE 13, CONTINUED

SUGGESTED HEATING SEASON INDOOR TEMPERATURES

	A Dry Bulb °F, occupied hours maximum	B Dry Bulb °F unoccupied hours (set-back)	
2. RETAIL STORES (Cont'd)			
Apparel (except dressing rooms)	65°	55°	
Jewelry, hardware, etc.	65°	55°	
Warehouses	55°	50°	
Docks and platforms	Do not heat	Do not heat	
3. RELIGIOUS BUILDINGS*			
		24Hrs. or less	Greater than 24 Hrs.**
Meeting Rooms	68°	55°	50°
Halls of Worship	65°	55°	50°
All other spaces	As noted for office buildings	50°	40°

Use Figure 14 to determine the actual savings in fuel for an average winter for any particular building. From Figure 3, the climate profile, select the degree days for the location and from Figure 12, line 16, find the number of BTU's per square foot per year now consumed for heating.

Enter the graph at the appropriate present heating energy consumption and degree day axes, intersect with the proper setback line, and follow the example line to determine the savings in BTU's per square foot per year. Multiply this value by the gross square foot floor area to give the total yearly savings in BTU's that can be expected for the entire building.

*and other spaces used for only a few hours per week.
**when outdoor temperatures are above 40°F.

FIGURE 14: ENERGY SAVED BY NIGHT SETBACK

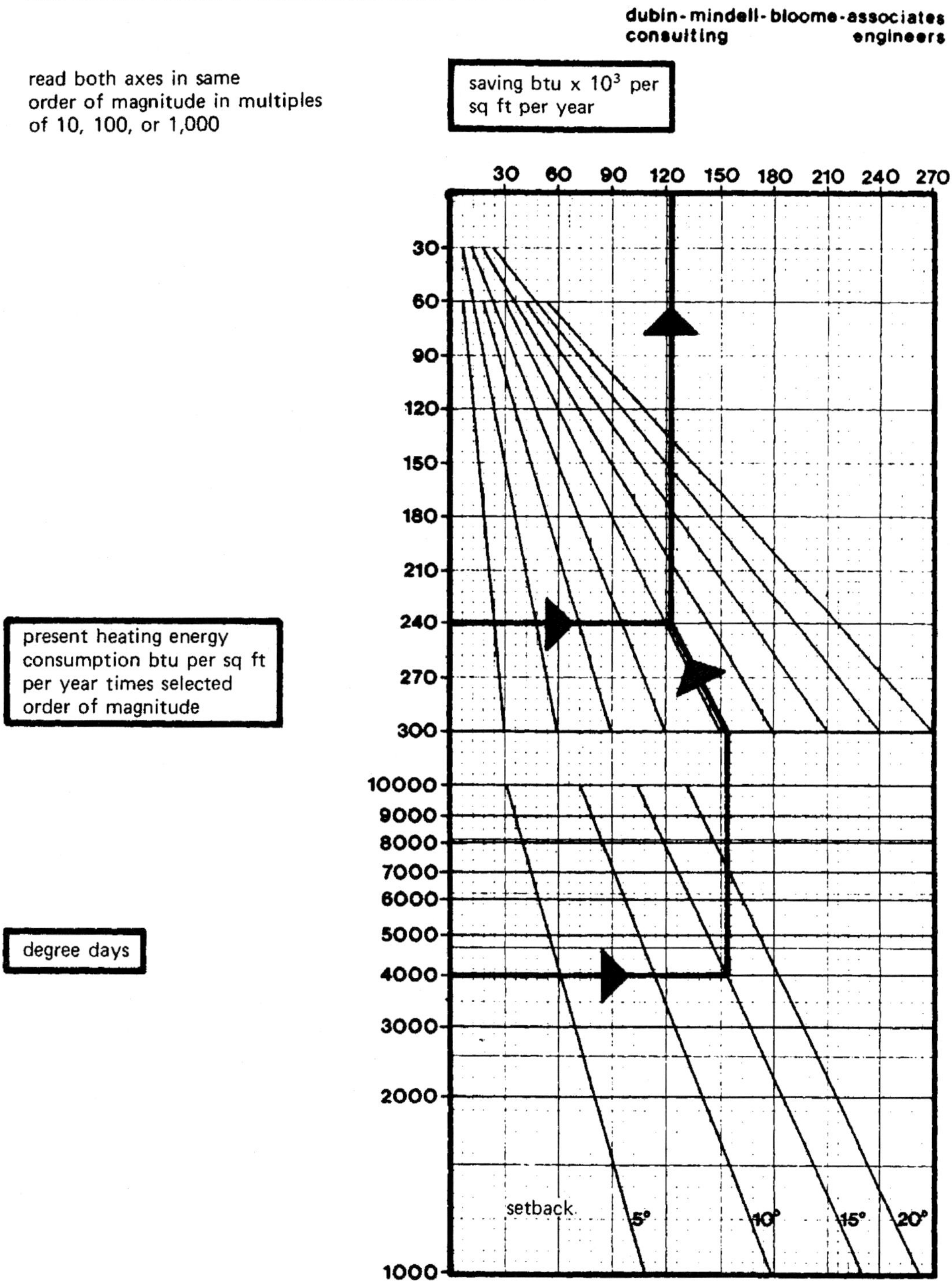

FIGURE #14 Engineering Data

References: AFM 88-8
U.S. Government Printing Office
15 June 1967

Climatic Atlas of the United States
U.S. Dept. of Commerce
June 1968

Five locations with heating degree day totals ranging from 1400-8400 were analyzed regarding time temperature distribution below 68°F DB. The percentage temperature distribution for each of the 5°F ranges which includes the setback from 68°F shown in the lower half of the Figure related to the total number of degree hours below 68°F was determined for 24 hr/day, 365 days/yr. This percentage was then plotted against annual degree days and expanded to cover the entire range of degree days and is shown as the lower half of the Figure.

The upper half of the Figure represents the range of heating energy consumed/sq.ft. for various buildings over the range of 1000 to 10,000 degree days. The extreme right hand line in the upper half represents 90% of the present consumption when projected vertically. Analysis of energy usage for heating by various buildings in several locations showed that it can be safely assumed that approximately 10% of the total heating energy consumption is during occupied hours. Therefore, savings by night setback are applicable to only 90% of the total heating energy consumption. The remainder of the upper half of the Figure simply proportions the energy saved based on the point of entry from the lower section.

Convert this figure to a quantity of fuel by dividing by an appropriate conversion factor. To convert to gallons of #2 oil, divide by 138,000 (the number of BTU's in a gallon); for #6 oil, by 146,000 (BTU's per gallon); for natural gas, by 1,000 (BTU's per cubic foot); for manufactured gas, by 800 (BTU's per cubic foot); and for tons of coal, by $26X10^6$ (BTU's per ton). Multiply this quantity by the unit cost for any type of energy used to calculate cost savings.

Example:

Find the yearly energy saved by lowering the temperature 10°F. during unoccupied hours for the following building:

Type	Office, 40 hrs. occupancy/wk.
Location	Minneapolis, Minn.
Floor Area	100,000 sq.ft.
Fuel	Light Oil (138,000 BTU/gal.)
Present Heating Consumption	76,000 BTU/sq.ft.-yr.
Heating Degree Days from Climatic Zone Profile	8,400

Enter Figure 14 at 8,400 degree days and 76,000 BTU/sq.ft.-year present consumption. Intersect with the 10° set-back line, and follow the example line to determine a savings of 21×10^3 BTU/sq.ft.-year.

Savings: Convert savings in BTU's to gallons at 65% seasonal efficiency

$$\frac{21{,}000 \times 100{,}000 \text{ sq.ft.}}{138{,}000 \times 0.65} = 23{,}411 \text{ gallon}$$

Assuming an average fuel cost of \$.36 per gallon, then the savings in dollars per year is 0.36 x 23,411 or \$8,428.

Results: Energy Saved 23,411 Gallons per year
Dollars Saved \$8428.

ECO 2 REDUCE INDOOR TEMPERATURES DURING OCCUPIED PERIODS

See Figure 13, column A, for recommended heating season indoor temperatures. Maintaining lower indoor temperatures during occupied periods conserves energy, although savings are not as great as those for unoccupied hours. Reduce even lower the temperatures in less critical areas, such as corridors and lobbies, to realize significant savings. The amount of energy conserved will be greatest in buildings which normally have longer periods of occupancy (stores as opposed to schools), and in buildings or sections of buildings with the least internal heat gain and solar radiation to help with the heating load.

ECO 3 <u>AVOID RADIATION EFFECTS TO COLD SURFACES</u>

In cold climates the temperature of an interior surface of an exterior wall or window is considerably lower than room temperature, and people located near to the surface radiate heat to it. Even if the room temperature is 70° or 75°F., these occupants will feel cold, particularly if they are located near windows where the radiation effect is most severe. Often, in order to keep warm, they request that room thermostats be set higher. Overheating of the interior of these particular rooms results and heat loss and energy consumption increase accordingly. A few simple remedies will both save energy and enhance the comfort of the occupants.

ECO 4 <u>REDUCE RELATIVE HUMIDITY LEVELS</u>

Humidification systems vaporize water into the dry ventilating air to increase it's moisture content and achieve the desired R.H. within the building. This humid-ification process requires a heat input of approximately 1,000 BTU's to vaporize each pound of water. The amount of moisture (water vapor) required to maintain any desired level of relative humidity is proportional to the amount of outdoor air which enters the building and it's dryness and the natural moisture input by the building occupants.

Humidification systems while not universally used are often installed to maintain the comfort and health of occupants, and to preserve materials, and prevent drying and cracking of wood, furniture and building contents. Maintain humidification at the level required for occupants where preservation of materials is not a factor. Do not humidify during unoccupied periods. Winter relative humidity in cold climates drops to 5 or 10% in buildings without humidifiers. In the absence of sufficient evidence to support the contention that higher levels are more comfortable or promote health, it is suggested that 20% relative humidity be maintained in all spaces occupied more than 4 hours per day. Shut off the humidifiers completely in all areas at night and during other unoccupied periods. If complaints of dryness and discomfort result, raise the humidity levels in 5% increments until the appropriate level for each area of the building is determined.

FIGURE 15: YEARLY ENERGY USED PER 1,000 CFM TO MAINTAIN VARIOUS HUMIDITY CONDITIONS

dubin-mindell-bloome-associates
consulting engineers

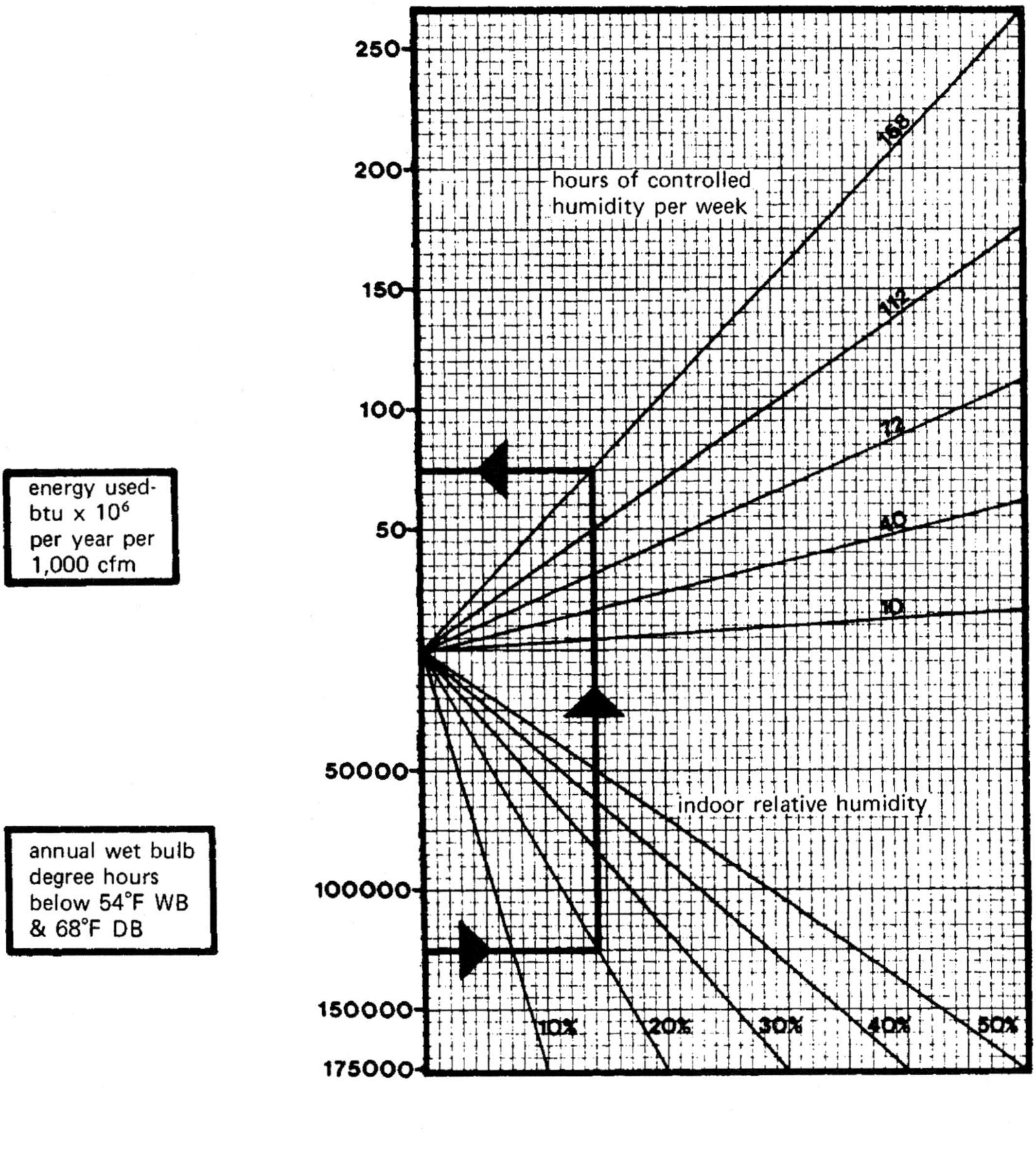

FIGURE #15 Engineering Data

Data from Figure #6.

WB degree hours based on 24 hours/day, October - April.

Base indoor condition for figure is DB=68°F, WB=54°F, RH=40%.

Energy used is a function of the WB degree hours below the base conditions, the RH maintained and the number of hours of controlled humidity. The figure expresses the energy used per 1000 cfm of air conditioned or humidified.

An analysis of the total heat content of air in the range under consideration indicates an average total heat variation of 0.522 BTU/lb. for each degree WB change. Utilizing the specific heat of air, this can be further broken down to 0.24 BTU/lb. sensible heat and 0.282 BTU/lb. latent heat. One thousand cfm is equal to 4286 lb./hr. and since we are concerned with latent heat only, each degree F WB hour is equal to 4286 x 0.282 or 1208 BTU. Further investigation of the relationship between WB temperature, DB temperature, and total heat shows that latent heat varies directly with RH at constant DB temperature. The lower section of the figure shows this proportional relationship around the base of 40% RH. The upper section proportions the hours of system operation with 168 hr./wk. being 100%.

If higher relative humidities are maintained to reduce static electricity and eliminate shocks, (30% RH will usually achieve this) raise the humidity levels only for those periods when the shocks are a serious problem. Refer to Figure 15 and the following example to determine the energy required for humidification and the potential savings through reduced R.H. levels.

Example:

Find the yearly energy saved by lowering winter relative humidity from 50% to 30%.

Type	Retail Department Store, 72 hrs. occupancy/week
Location	Chicago, Illinois
Fuel	Light Oil (138,000 BTU/gal.)
Outdoor Air Rate	5,000 cfm
Annual Wet Bulb Degree Hours Below 54°F. WB and 68°F.	106,000

Enter Figure 15 at 106,000 degree hours, Follow the example line intersecting with 50% R.H. and 72-hour lines and read yearly energy used of 68×10^6 BTU-yr./1000 CFM.

Re-enter Figure 15 intersecting with the 30% R.H. line and read yearly energy used of 40×10^6 BTU-year/1000 CFM.

Savings: The energy saved equals 68 - 40 or 28×10^6 BTU-Year/1000 CFM and for 5000 CFM, the total is 5 times 28×10^6 or 140×10^6 BTU/year.

Fuel Saved at a seasonal efficiency of 60% =

$$\frac{140 \times 10^6}{138{,}000 \times 0.6} = 1690 \text{ Gallons}$$

At \$.36 per gallon, the savings is 0.36 x 1690 or per year.

Results: Energy Saved 1690 gallons per year
Dollars Saved \$608 per year

Note: For any building, the CFM is a combination of ventilation rate and infiltration converted to CFM. A building with 100,000 sq.ft. of floor area could have as much as 15,000 CFM or more of infiltration.

To calculate the BTU's per square foot floor area that are used for humidification at any desired level, multiply the energy per 1,000 CFM times the total CFM of outdoor air divided by 1,000 and divide the answer by the gross number of square feet of building area. (Figure 10, line 7)

ECO 5 SHUT DOWN VENTILATION SYSTEM DURING UNOCCUPIED HOURS

Ventilation is responsible for a large percentage of the "building" heating load. Cold outdoor air, introduced for ventilation, must be heated to the point that it meets the indoor temperature. The load which this imposes on the heating system is directly proportional to the indoor temperature and the quantity of air introduced for ventilation; the yearly energy used to heat this air is a function of heating degree days.

Many buildings are ventilated at a rate far in excess of that necessary to maintain comfort, dilute odors, or meet code requirements and/or they are operating ventilation systems 24 hours per day even when the building is unoccupied or lightly occupied. These buildings contribute to a gross waste of heating energy.

Use Figure 17 to calculate the amount of energy required to heat outdoor air for any selected period of time, and also to determine the yearly savings in energy per 1000 CFM by shutting off the outdoor air during unoccupied periods.

Ventilation may be provided through a fresh air intake duct to an air handling unit; through outside air intakes of fan coil, window, or through-the-wall units; through intake ducts to rooftop or package heating and cooling units; or, by a separate outdoor air fan. Examine the building systems carefully to determine how ventilation is being supplied, and what control or damper devices are available to reduce and shut off the supply of outdoor air.

A method for reducing outdoor air quantities appears in the example for ECO 6. Use it with Figure 17 to determine the savings for complete shut down of ventilation air. The savings which result from shutting off or reducing ventilation air during the cooling season are described in Section 6, "Cooling."

ECO 6 REDUCE VENTILATION RATES DURING OCCUPIED PERIODS

The building owner or operator can perform most of the procedures for shutting off ventilation at night or on weekends, but professional advice and help may be necessary to reduce the ventilation rate in all or portions of the building during occupied periods.

First establish the ventilation rate by measuring the volume or air at the outdoor air intakes of the ventilation system. Then determine the local code requirements and compare them with the measured ventilation rate to find the magnitude of possible savings. If the code requirements exceed "Recommended Ventilation Standards", Figure 16, apply for a code variance. Although code figures are often used as a design criteria, with dampers and ducts installed accordingly, they do not necessarily reflect actual building requirements for the number of people occupying spaces for specific periods of time; they often exceed real requirements.

If ventilation is supplied to a building to provide makeup air for toilet, kitchens and other exhuast air quantities, analyze the exhuast systems and operate them only as needed. Savings of fan horsepower for exhaust fans and supply fans and heating energy to temper the supply air will result. Cooling loads will be similarly reduced in the summer.

If supply fans also handle outdoor air for ventilation, increased return air quantities may compensate for the reduction in ventilation air quantities and preclude fan horsepower savings. Reductions in the "building" heating load, however, will result in significant savings.

When two or more spaces with different requirements are served by one system, it is not always possible to control the separate ventilation rates. Not all of the spaces, however, need be supplied at the rate required for the most critical area. Refer to Figure 16 and calculate the maximum amount of outdoor air by the sum of the individual requirements. This procedure, generally, will provide satisfactory conditions for all spaces.

Where separate ventilation systems are installed, group occupants by type of activity, when possible, to reduce the total ventilation rate and conserve energy.

Outdoor air for economiser cooling is discussed separately under Section 6, "Cooling". Air used for that purpose is not strictly speaking "ventilation".

Use the method outlined in the following example to determine the savings due to reduced ventilation rates during occupied periods.

Example:

From the Identification Profile:

Building Type	Offices
Building Floor Area	100,000 Sq. Ft.
Building Location	Minneapolis, Minnesota
Heating Degree Days	8400
Occupancy	667 people
Indoor Temperature	68°F.
Occupied time	40 hours/week

Present measured ventilation rate - 20,000 CFM

This represents $\frac{20,000}{667}$ = 30 CFM/person

Reduce the rate of ventilation to an average of 8 CFM/person to serve a mixed smoker/non-smoker population. New ventilation rate = 667 x 8 ≈ 5300 CFM Ventilation reduction = 20,000 - 5300 = 14,700 CFM

Determine from Figure 17 that in an area of 8400 degree days, each 1000 CFM for 40 hours/week requires 50×10^6 BTU/year energy for heating. Therefore, heating energy saved by reducing ventilation = $14.7 \times 50 \times 10^6 = 735 \times 10^6$ BTU/year.

Fuel saved if the oil-fired system has an efficiency of 60%, and if using #2 oil at 138,000 BTU/gallon =

$$\frac{735 \times 10^6}{138,000 \times 0.6} = 8877 \text{ gallons/year}$$

@ $.36/gallon $3195/Year

FIGURE 16: VENTILATION RECOMMENDATIONS

1. **Office Buildings**

Work Space	5 CFM/person
Heavy Smoking Areas	15 CFM/person
Lounges	5 CFM/person
Cafeteria	5 CFM/person
Conference Rooms	15 CFM/person
Doctor Offices	5 CFM/person
Toilet Rooms	10 air changes/hour
Lobbies	0
Unoccupied Spaces	0

2. **Retail Stores**

Trade Areas	6 CFM/customer
Street level with heavy use (less than 5,000 sq.ft. with single or double outside door)	0
Unoccupied Spaces	0

3. **Religious Buildings**

Halls of Worship	5 CFM/person
Meeting Rooms	10 CFM/person
Unoccupied Spaces	0

ECO 7 REDUCE RATE OF INFILTRATION

Outdoor air infiltrates a building through cracks and openings around windows and doors, through construction joints between individual panels in a panel wall construction, and through porous building materials of the exterior walls, roofs, and floors (over unheated spaces).

Infiltration increases with wind velocity and penetrates the windward side of the building -- usually the north or western exposures in cold climates. However, in high winds, a negative pressure is often created on the lee side, which if the north and western exposures are windowless and/or tight, may induce air into the building through openings in other exposures and through open doors and passageways.

In tall buildings stack action due to the difference between indoor and outdoor temperatures induces air leakage through cracks and openings. Stack effect is always a potential problem for vertical spaces -- service shafts, elevator shafts, and staircases. The density difference between warm air in the shaft and the cold outdoor air induces air to leak into the bottom of the shaft and out of the top.

Infiltration is also induced into the building to replace exhaust air unless mechanical inlet ventilation balances the exhaust. Follow the suggestions for reducing exhaust air quantities and periods of operation, described in ECO 6 to reduce infiltration rates from this cause.

Infiltration, which often accounts for a major portion of heating load, cannot, like ventilation systems, be turned off at night or during weekends (although it may be decreased if exhaust fans are shut down). It can, however, be reduced at all times. Particularly effective measures include caulking cracks around window and door frames and weatherstripping windows and doors. Weatherstripping doors costs about $50 per door or $75 per double door. To weatherstrip metal frame doors with aluminum and rubber costs about twice as much. Weatherstripping costs for windows depend upon type and number, and range between $25 to $50 per window. To rake out old caulking and recaulk around window edges costs about $15 per window (or about $25 per 30 square feet window). Figure 18 reveals infiltration rates for various types of windows. In most climates the average

FIGURE 17: YEARLY ENERGY USED PER 1,000 CFM OUTDOOR AIR

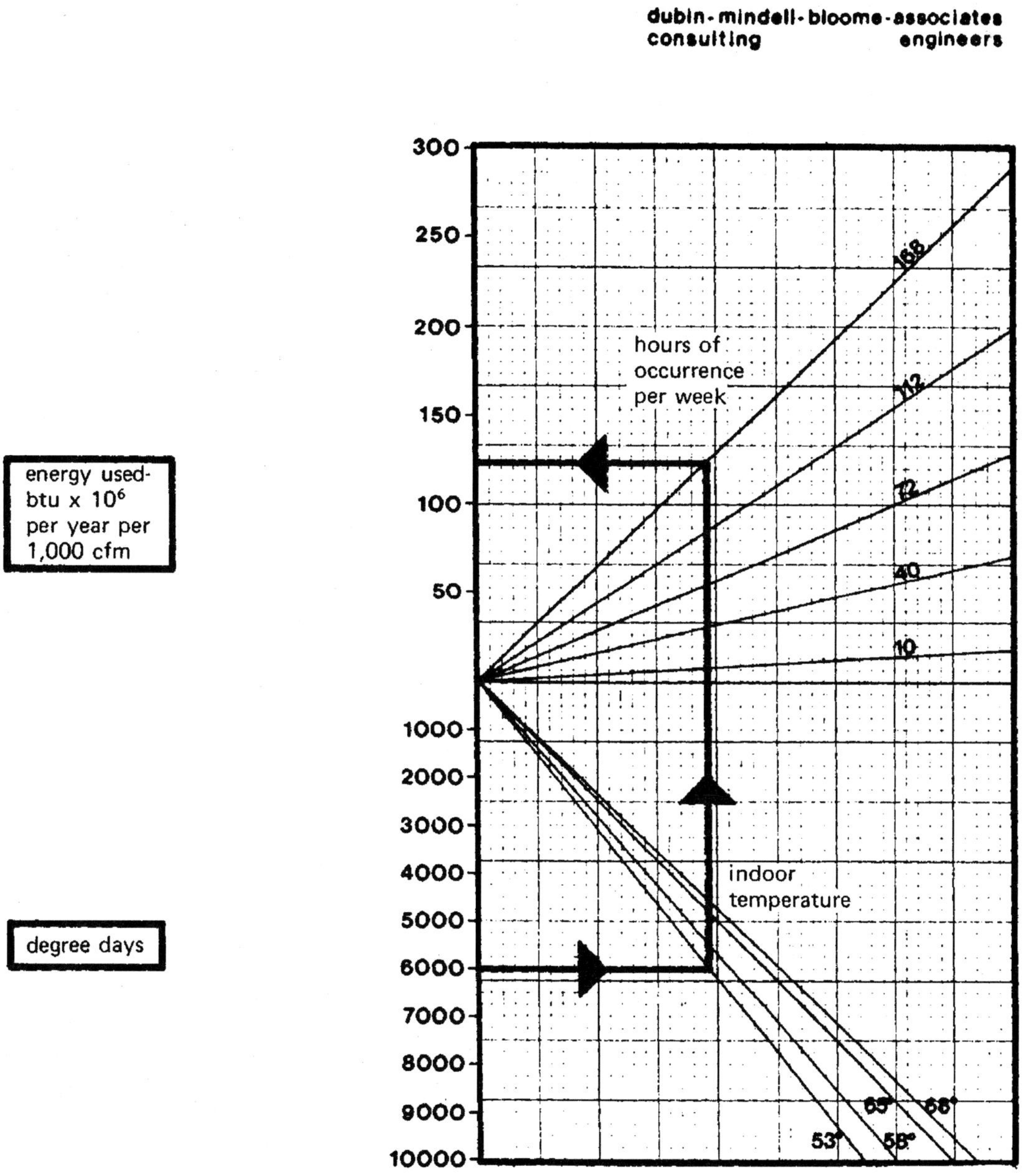

FIGURE # 17 Engineering Data

Data from Figure 4

Energy used is a function of the number of degree days, indoor temperature and the number of hours that temperature is maintained and is expressed as the energy used per 1000 cfm of air conditoned.

The energy used per year was determined as follows:

BTU/yr. = (1000 cfm) (Degree Days/yr.) (24 hr./day 1.08)*

Since degree days are base 65°F, the other temperatures in the lower section of the figure are directly proportional to the 65°F line. The upper section proportions the hours of system operation with 168 hr./week being 100%.

*1.08 is a factor which incorporates specific heat, specific volume, and time.

FIGURE 18: RATE OF INFILTRATION THROUGH WINDOW FRAMES

dubin-mindell-bloome-associates
consulting engineers

key to window infiltration chart
(leakage between sash & frame)

	type	material	weatherstripped?	fit
1	all hinged	wood	yes	avg
		metal	yes	avg
2	all hinged	wood	no	avg
		metal	no	avg
	dbl hung	steel	no	avg
3	all	wood	yes	loose
	dbl hung	steel	yes	avg
4	casement	steel	no	avg
5	all hinged	wood	no	loose

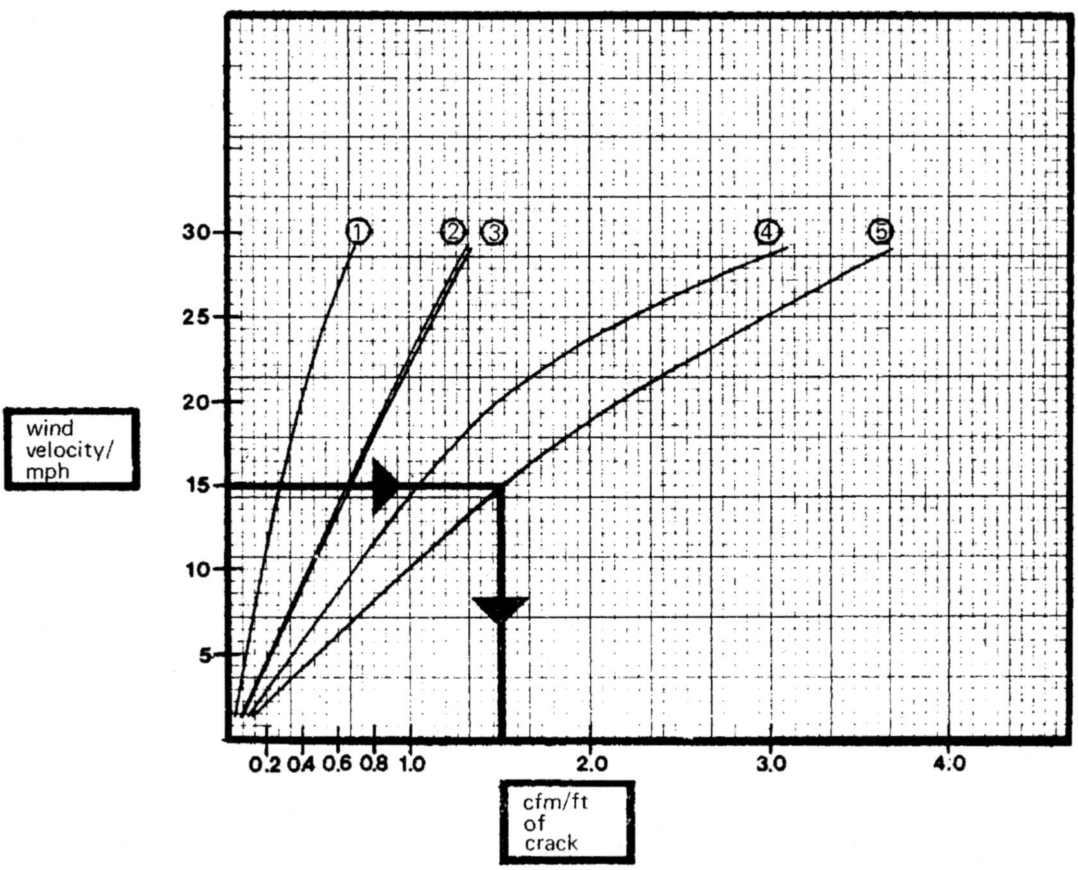

FIGURE #18 Engineering Data

Source of Data: ASHRAE Handbook of Fundamentals, 1972, Pgs. 333, 337,338

wind velocity in the winter is 10 to 15 miles per hour. In some locations it is considerably higher. Check your local weather bureau for relevant information, or draw on individual experience. Wind patterns and velocity vary widely, even on the same street in a city.

The installation of revolving doors and vestibules is an effective measure against air leakage through entrances. Refer to ECM 2 for information on potential energy savings with revolving doors and vestibules as well as additional measures to reduce the effects of stack action.

For porous block walls having leaky mortar joints, point up from the exterior, or paint the wall with epoxy to seal it against air leakage and moisture. A 4 mm coat of epoxy resin costs about 35¢ per square foot for material and labor.*

Use Figure 17, to determine the energy required to heat infiltration air.

To estimate the amount of infiltration occurring in a building because of window leakage, determine, first, the total amount of crack area. Use the following procedure:

<u>Example:</u>

From the Identification Profile:

Building Type	Offices
Building Size	100' x 50' 4 floors
Building Location	Topeka, Kansas
Heating Degree Days	5,000
Indoor Temperature	68°F.
Wind Speed	15 mph
Wind Direction	NW
North Windows	Dimensions: 5'x3'; total no. 28
West Windows	Dimensions; 5'x3'; total no. 56

Window type: loose fitting, double hung wood sash

Determine the crack length/window
Each window perimeter = 5+5+3+3+3 (including the 3'
= 19 ft. horizontal crack between the upper and lower sash)

*The cost of epoxy resin may escalate at a rate higher than that for most costs in this manual.

Total crack length for north and west windows = 19 x 84 windows = 1596 ft.
Determine from Figure 18 the rate of infiltration per foot of crack = 1.5 CFM
Total infiltration due to window cracks = 159x1.5 = 2394 CFM
Determine from Figure 18 the rate of infiltration per foot of crack if windows are weatherstripped = 0.25 CFM/ft.
Therefore, total infiltration = 1596 x 0.25 = 399 CFM
Reduction in infiltration = 2394 - 399 - 1995 CFM

Determine seasonal savings in energy for heating.
From Figure 17, each 1000 CFM required 130 x 10^6 BTU/year
Therefore, reducing infiltration by 1995 CFM would result in a savings 130 x 10^6 x 1.995 = 259.1 x 10^6 BTU/year.

Savings due to weatherstrip = 228.1 x 10^6 BTU/year
Fuel saved, if the heating system uses #2 oil (138,000 BTU's/gallon) and has a seasonal efficiency of 60% =

$$\frac{259 \times 10^6}{138{,}000 \times 0.6} = 3128 \text{ gallons}$$

@ $.36/gallon $1126/year

The cost of weatherstripping 84 windows of 15 sq.ft. each in New York City is $25/window. Adjustment for Topeka (use Denver from Appendix A) = 0.81 x $25 = $20.25 each window.

Therefore, total cost for 84 windows = 84 x 20.25 = $1701

In summary,

$1701 spent on weatherstripping windows will realize a savings in fuel cost of $1126 per year. Capital pay-back will be about 1 1/2 years making the investment worthwhile. If fuel costs continue to escalate, the pay-back period will be even shorter. Because weather-stripping the windows results in additional savings during the cooling season (even though wind velocities and infiltration rates are less), the total yearly savings in energy will be greater than that shown.

ECO 8 INCREASE THE SOLAR HEAT GAIN INTO THE BUILDING

Although solar heat gain adds to the cooling load, it can be very helpful in reducing the heating load. Solar radiation impinging upon an opaque building envelope raises the surface temperature and reduces conduction losses. In an existing building it is difficult to increase this effect. However, if the walls are not heavy construction and are not well-insulated, consider refinishing the exterior south wall in particular, and the east and west walls next, in that order, to increase solar radiation effects. See ECM 2 for details.

Other measures to permit available sunshine to enter the building through windows and glass doors can be implemented readily and the energy conservation benefits, especially in the northern latitudes, are considerable. The amount of sunlight that penetrates the windows depends upon the number of panes of glass, the area of the windows, the orientation of the windows, the type and cleanliness of the glass, the type of solar control device, the latitude where the building is located, and the percentage of sunshine at the location. Charts which address all the variables and permit calculations for exact benefits of increasing the solar heat gain in the winter are impossible within the scope of this manual. A general rule is possible, however. In Minneapolis, there is approximately 25% more sunshine on the south glass and 15% more on the east and west glass than there is on the north face; in Florida, the amount of sunshine striking all facades is nearly equal. Between these two locations, due to sun angles at different latitudes, the percentages of sunlight change linearly with latitude.

The heat contributed by solar radiation through windows can save about 3/4 of a gallon of oil per square foot of south facing glass in Minnesota, and about 1/4 of a gallon of oil per year for east or west facing glass in Miami.

About 10% less sunlight penetrates double glazing than single glazing. However, double glazing reduces the heat load due to conduction, and the benefits from this more than offset the loss of solar radiation.

ECO 9 REDUCE HEAT TRANSMISSION DUE TO CONDUCTIVE LOSSES THROUGH THE BUILDING ENVELOPE

Heat loss through the envelope depends upon the temperature difference between indoors and out; the mode of operation of the heating system; and the mass, color, and insulating value of the exterior walls, roof, windows and floor (over unheated spaces). Wind impingement on the exterior surfaces increases heat loss by transmission. Solar radiation compensates for it by raising the temperature of the exterior surfaces. A single layer of glass transmits 1.1 BTU/square foot of surface per degree of temperature difference (td); double glass transmits about .55 BTU, an uninsulated frame wall about .3 BTU, a 12" masonry and 4" brick wall about .25 BTU's and insulated walls of various types and thicknesses down to .027 BTU's per square foot per degree td.

Insulation and double glazing are discussed in ECM 2 . For buildings in climates of 4500 degree days or more, adding a storm sash to windows and adding 6" of fiberglass insulation (to the underside of the roof, or over the ceiling of uninsulated buildings) will have a payback period less than 8 years with oil prices at 36¢ per gallon. Contact at least 2 local contractors for prices before awarding a contract. ECM 2 contains data on the relative effectiveness of double glazing and insulation in various climatic zones in the United States, as well as cost data for various measures to reduce heat loss.

The installation of storm windows reduces infiltration rates as well. Therefore, adding storm windows, may preclude weatherstripping and caulking loose fitting windows. (Removing a single pane of glass and replacing it with a pane of double glazing reduces conduction losses but not infiltration).

For small buildings with few windows, a temporary storm sash, non-operable, can be erectdd quickly with clear heavy duty plastic sheet nailed in place and caulked for tightness at about $4 or $5 per window.

ECO 10 IMPROVE THE BURNER-BOILER/FURNACE SEASONAL EFFICIENCY

Although there are many types and sizes of boilers, furnaces and burners in use today, all have certain common characteristics, and similar techniques can be used to improve their efficiency and conserve energy.

In theory, combustion of oil, gas, or coal requires a given fuel/oxygen ratio for complete burning and maximum efficiency. In practice, air (mixture of oxygen and nitrogen) is used to provide the necessary oxygen for burning and must be supplied in excess of the theoretical requirements to insure complete combustion. The quantity of air that just gives complete combustion is the optimum amount. Any reduction in the optimum air quantity prevents complete combustion and wastes energy -- the maximum heat value of the fuel cannot be released. Any increase over the optimum air quantity for combustion will reduce not the efficiency of combustion itself, but rather the rate of heat transfer to the boiler or furnace; an increase in the stack temperatures (heat lost out the chimney) will also occur. It is important, therefore, that air into the combustion chamber be controlled to achieve the most favorable fuel/air ratio for any given burning rate. Take measures to prevent any other uncontrolled source of air entry through leaks into the combustion chamber.

Clean and adjust burners each year and monitor them periodically during the year to assure optimal combustion efficiency (adjust them at these times as necessary). In large installations make combustion control and monitoring a daily procedure. Use orsat apparatus to check the CO_2 content of the flue gas, and thermometers to check stack temperatures. Taken together, the amount of CO_2 and the stack temperature constitute a measure of the combustion efficiency.

After reducing the "building" and distribution heating load, clean and correct dirty oil nozzles, oversized or undersized nozzles, fouled gas parts, and improperly sized combustion chambers. Reduce nozzle sizes and modify combustion chambers for proper combustion.

The condition of the heat transfer surface directly affects heat transfer from the combustion chamber and/or hot gases. Keep the fire-side of the heat transfer surface clean and free from soot or other deposits and the air and water-side clean and free of scale deposits. Remove deposits by scraping where they are accessible, by chemical treatment, or by a combination. In the case of steam boilers, once the water-side of the boiler is clean, institute correct water treatment and blow down to maintain optimum heat transfer conditions.

Boilers and furnaces may achieve relatively high instantaneous full-load efficiencies (80 - 87%), but because they are operated most of the time at part load, they have lower seasonal efficiencies. Generally, measures which increase the full-load instantaneous will also increase the seasonal efficiency. However, when the peak loads are of very short duration, it may be advantageous to tune the boiler for maximum efficiency at part-load conditions in order to gain greater seasonal efficiency.

Flue gas temperatures from boilers range typically from 250°F. for gas-fired low temperature boilers up to 600°F or more for oil and coal-fired high temperature boilers. Furnaces operate in the same range. Hot flue gas contains useful heat which can be reclaimed for space or air heating, or to preheat feed water, by the installation of heat transfer coils or heat pipes in the breeching. (This measure requires capital expenditure and is dealt with in ECM 2).

GUIDELINES TO REDUCE ENERGY USED IN HEATING

SETBACK TEMPERATURES DURING UNOCCUPIED PERIODS

-Shut off radiators or registers in vestibules and lobbies.

-Reduce the hours of occupancy to the greatest extent possible during periods of severely cold weather.

-Adjust automatic timers or add time clocks to automatically set-back temperature for night and weekend operation.

-When buildings are used after hours for meetings, conferences, cleaning or scattered activities, for instance, reduce the number of spaces occupied and, to the extent possible, consolidate them in the same section of the building. Reduce the temperature and turn off humidifiers in all other parts of the building.

-When there is no danger of freezing, turn off radiators or supply registers in areas that do not have a separate thermostat. Open them when building is occupied.

-Check relevant codes requiring protection for plumbing, fire sprinkler systems, and standpipes before implementing extreme temperature reduction. Public utilities may set temperature requirements for their equipment rooms.

SET BACK TEMPERATURES DURING OCCUPIED HOURS

-Stores are commonly overheated and uncomfortable for transient patrons who are wearing outdoor clothing. To conserve energy, reduce store temperatures to a level that is comfortable for heavily dressed patrons and encourage staff to dress more warmly or provide local heaters.

-In all buildings encourage occupants to wear heavier clothing so that they are comfortable at lower indoor temperatures.

-Some buildings contain large heated areas such as storage spaces that are only occupied by one or two people. In such areas, reduce the temperature to a low level just sufficient to prevent damage to other systems (freezing sprinklers, etc.) and provide local radiant heaters for one or two occupants.

-Corridors and stairwells are unoccupied areas, used only by people who are physically active in moving from one heated space to another. Providing that the temperature does not fall below 55°F., turn off heating in these areas. Keep closed all doors between unheated corridors and heated spaces.

-Some areas of the building require no heating -- spaces which are heated by adjacent areas or which receive solar heat through windows. If thermostats are unavailable in these areas, shut off radiators, registers, fan coil units. or any other terminal heating devices until the temperature levels suggested in Figure 13 can be maintained.

-Refer to applicable codes. Minimum temperature requirements are specified in OSHA and other occupational regulations to assure employee working conditions.

AVOID RADIATION EFFECTS OF COLD SURFACES

-When the winter sun is not shining on the windows, draw drapes or close venetian blinds to reduce radiation of heat from occupants to cold surfaces. (The cost of adding venetian blinds ranges from $1.00 to $1.60 per square foot.)

-Encourage those working near the exterior walls and windows to wear heavier clothing in winter.

-Close windows tightly in the winter.

-Reduce drafts and increase comfort by caulking and weatherstripping windows to reduce heat loss and to reduce the need to overheat the entire space in order to achieve comfort near the exterior walls. (See ECO-7, page 116).

-Have occupants sit together in interior spaces away from cold walls in sparsely occupied spaces.

-Rearrange desks and task surfaces away from cold exterior surfaces.

-Other methods to reduce transmission to cold exterior surfaces are described under ECO-8 page 122.

REDUCE WINTER HUMIDIFICATION

-Turn off all humidifiers at night and during unoccupied cycles.

-Reduce the amount of infiltration and outdoor air ventilation.

-Reduce or eliminate any introduction of moisture for humidification in corridors, store-rooms, equipment rooms, lounges, lobbies, laundries,supermarkets, stores with high density occupancy, religious buildings, cafeterias, kitchens, etc.

-Use waste steam condensate for winter humidification. Refer to ECM 2 for details.

-Whenever condensation is running freely on the inside of window surfaces, shut off humidifier. Excess moisture will damage structure.

-When humidifier is maintained to eliminate static electricity, shut off humidifier when shocks are not a problem.

-If humidifiers are located in the return air or outdoor air mixing box in the air handling section of the system, relocate them to the hot duct section.

-In pan type humidifiers adjust float or control to eliminate overflow onto hot furnace sections.

-Whenever moisture is condensing on duct work or humidifier casing section, insulate the casing or duct work and/or reduce flow rate.

SHUT OFF VENTILATION AIR DURING UNOCCUPIED PERIODS

-If there are one or more separate fresh air supply fans, deenergize them; use an automatic timer if one is available; or switch off manually. Where an automatic damper closes when the fan shuts down, check it to make sure it closes tightly. Repair it with felt edges if it by-passes air.

-If air handling units which supply warm (or cold) air to spaces are equipped with fresh air inlet duct and damper, close damper as described above. If no damper exists, install a felt-lined damper with remote or local control device and timer control.

-If windows are used to provide ventilation, close them at night.

-If exhaust fan hoods serving kitchens, bakeries, cafeterias, and snack bars are interlocked with outside air fans or dampers, to sure to shut down exhaust system when not needed.

-Close outdoor air dampers for first hours of occupancy when outdoor air has to be heated or cooled.

-Provide ventilation in accordance with occupancy and not on a continuous basis. In many buildings heavy occupancy can be monitored or occurs at regular intervals, so that ventilation can be shut off for certain periods during the day. In a department store designed for peak loads of 1,000 customers, ventilation can usually be shut off for those slack sales periods when only 10% or 20% of the peak occupancy occurs.

-If during the heating season outdoor air temperature in the morning is above desired room conditions, use it for heating by opening damper.

-Generally the amount of outdoor air quantities for window units, through-the-wall units, and fan coil units is fixed. When these units are used to maintain nighttime or unoccupied cycle room temperatures, the outdoor air intakes are generally open. If dampers are available, close them when outdoor air is not needed. Where infiltration is sufficient for daytime ventilation requirements, block off the outdoor air damper completely for some or all of the units.

REDUCE VENTILATION RATES DURING OCCUPIED PERIODS

- Close outdoor air damper, if provided, until desired volume reduction is achieved.

- If damper is not provided, either fit one or blank off part of the outdoor air intake.

- Take particular care in setting outdoor air quantities where window air conditioners, through-the-wall units, and fan coil units and unit ventilators with fresh air intakes are used. Generally, the amount of outdoor air is fixed but can be reset.

- Use an anemometer at the outdoor inlet or louver to measure the amount of outdoor air introduced into the ventilation system. The outdoor air damper setting often differs from the damper indicator setting. Also, a hinged volume damper blade does not proportion air directly with its opening. A blade with 10% opening permits much more than 10% flow.

- Cut off direct outdoor air supply to toilet rooms and other "contaminated" or potentially odorous areas. Permit air from "clean" areas to migrate to the "dirty" areas through door grilles or under-cut doors. This is effective if toilet areas have a mechanical exhaust system, or even if ventilated by open windows (especially those on the lee side of the building).

- Use odor absorbing materials in special areas rather than providing outdoor air for dilution.

- Supply ventilation air to parking garages according to levels indicated by a CO_2 monitoring system.

- Provide baffles so that the wind does not blow directly into an outdoor air intake.

- Operate exhaust systems intermittently throughout the day. Turn them off when possible at times when they are not needed. Operate them at other times such as noon, coffee breaks, heavy cooking times.

- Shut off exhaust hoods in kitchens, snack bars and cafeterias when cooking or baking operations are completed.

REDUCE VENTILATION RATES DURING OCCUPIED PERIODS (Cont'd)

-Many hoods exhuast much more air than necessary. Reduce the quantity of exhaust air from hoods in kitchens and similar operations by closing off a portion of the hood, changing hood type from canopy to high velocity and slowing down exhaust fans to the point just necessary to satisfy exhaust requirements.

-If outdoor temperature is below 45°F., don't operate the ventilation system any longer than needed. Noticeable odors are good indicators of the need to operate the fans or open outdoor dampers. Flushing out the building until odors disappear will usually be satisfactory.

-Concentrate smoking areas together so that one ventilation system can serve them. Adjust outdoor air to serve those areas and reduce all other outdoor air to systems.

-Use window ventilation where possible. This will reduce the power required for mechanical exhaust systems and also the air intake which normally balances the exhaust. Shut off supply air system when window ventilation, alone, is adequate.

-In high ceilinged buildings (religious buildings, auditoriums, etc.) used for short periods of intermittent operation, outdoor air is frequently unnecessary. For longer periods of light occupancy, outdoor air is still unnecessary. A warehouse for example, may have less than one person per 1,000 cu.ft. for less than four hours and no need for outside air ventilation.

-Check codes, where applicable, for specification of minimum exhaust CFM and for rules on recirculation or filtering of air.

IMPROVE BURNER-BOILER EFFICIENCY

The following suggestions are recommended operating and maintenance items to increase boiler, furnace, and heating system efficiency for hot water systems, high pressure steam systems (about 15 psi), low pressure steam systems (less than 15 psi), forced warm air systems, and gravity hot air systems. Review the list and select those items which are applicable to any particular system.

None of these measures can be quantified, in this manual, to show energy savings -- any increase in efficiency depends on the equipment status before and after the maintenance item has been carried out. However, each item will result in greater efficiency and consequent energy savings. Many building owners report fuel savings of 20 to 30%.

-Clean and scrape fire-sides to remove soot and scale.

-Clean water-sides, remove built-up scale.

-Scrape scale from steam drum

-Clean air-sides, remove soot, and scrape scale in forced warm air and hot air furnaces.

-Maintain water level or pressure to radiators or coils on the highest level of the building.

-Insulate units which are in unheated spaces, on roofs, or in air-conditioned spaces. Repair insulation where it is in need. (If the boiler or furnace casing is 10-15% warmer than room temperature, radiation loss could be 10% or more of the capacity of the unit.)

-Check for and seal air leaks between sections of cast iron boilers to improve combustion efficiency.

-If the combustion efficiency is at a maximum but stack temperatures are still too high (over 450°F.), install baffles or turbulators to improve heat transfer. Consult your boiler manufacturer.

-Seal all air leaks into combustion chamber, especially around doors, frames, and inspection ports.

-Maintain the lowest possible steam pressure suitable for supplying radiation or coils.

-Vary the steam pressure in accordance with the space heating or process demands. Steam pressures can be reduced most of the year. Standby losses are reduced

IMPROVE BURNER-BOILER EFFICIENCY (Cont'd)

as pressures are reduced.

-Maintain the lowest possible hot water temperature which will meet space or domestic hot water needs.

-In the absence of indoor-outdoor modulating controls, raise or lower operating temperature (for hot water systems) to conform to indoor-outdoor conditions. For example, operate a boiler at 120°F. with outdoor temperature at 60°, and raise the level to 160° when it is 20° outdoors.

-Clean filters regularly in gravity and forced warm air units to reduce the operating time of the furnace.

-Reduce firing rate and/or enlarge return air opening if hot air temperature of a gravity furnace is over 150°F. at full load. (Refer to ECM 2 for adding a blower).

-Shut down hot air furnaces completely when building is not occupied and there is no danger of freezing.

-Set operating aquastats on steam and hot water boilers to 100°F. during shut-down periods.

-Schedule boiler blowdown on an as-needed basis rather than on a fixed timetable. Smaller and more frequent blowdown quantities are preferable to larger quantities and less frequent blowdown.

Note: Be sure that boiler blowdown procedures adhere to specifications outlined by the manufacturer, the National Board of Boiler and Pressure Vessel Inspectors (of Columbus, Ohio), and local codes. With few exceptions it is illegal, and in all cases undesirable, to discharge boiler blowdown directly to a sanitary sewer.

-Use warm exhaust air from adjacent areas, or from the ceiling of the boiler rooms, to preheat combustion air.

-Use chemical fuel additives to reduce the flashpoint temperature of fuel oil, especially #4 and #6 oils. Proper chemical treatment will reduce soot deposit on #2 oil systems also.

-Interlock combustion air intake with burner operations; maintain prepurge and postpurge as required for some burners.

IMPROVE BURNER-BOILER EFFICIENCY (Cont'd)

-Seal all air leaks into natural draft chimneys, especially where flue pipe enters the wall.

-Repair or rebuild oil burner combustion chambers to the correct size for providing optimum efficiency at 90% of the full load firing rate. Construct chambers with bricks of the refactory type, not common bricks. Incorrect matching of burner and combustion chamber and broken brickwork can result in losses of from 10 to 20%.

-Turn off gas pilots for furnaces, boilers, and space heaters during the non-heating months and during long unoccupied periods.

-Provide an automatic draft damper control to reduce the heat loss through the breeching (smoke pipe) when the gas or oil burner is not in operation. Adjust draft-control with combustion testing equipment to match the firing rate.

-Adjust oil burner efficiencies to achieve proper stack temperature, CO_2 and excess air settings. Adjust setting to provide a maximum of 400° - 500° of stack temperature and a minimum of 10% CO_2 at full load conditions. Excess air through a boiler can waste 10 to 30% of the fuel. Accurate testing is essential for the correct burner adjustment to attain maximum efficiency. Use appropriate instruments and institute combustion testing as part of a planned general maintenance program. (A simple combustion testing kit can be purchased for approximately $100).

-Adjusting the firing rate of gas or oil burners at too high a rate will cause short cycling and excessive fuel consumption. Too low a rate will require constant operation and inadequate heat will be delivered to the spaces. If the boiler is oversized, adjust the firing rate to the building load, not the boiler.

-If there is more than one boiler, operate one only up to its maximum load before bringing other boilers on the line. It is inefficient to operate two or more boilers at very low capacity to carry part loads.

-Details or reclaim heat systems and other measures which require capital investment but can show a quick payback period are described in ECM 2.

REDUCE INFILTRATION

-Operate the exhaust systems as outlined in the guidelines for ECO 5 and 6.

-Inspect the building exterior and interior surfaces and caulk all cracks that allow outdoor air to penetrate the building skin.

-Caulk around all pipes, louvers, or other openings which penetrate the building skin.

-Repair broken or cracked windows. Reglaze with standard or tempered glass of proper thickness, as per building code requirements, or with wire glass if fire rating is required.

-As a temporary measure cover windows with 4 mil plastic sheets and extend the covering over the frame. Hold plastic sheet in place with continuous nailing strip.

-Weatherstrip exterior doors and windows in climates with more than 2000 degree days.

-Cover porous exterior wall and roof materials with epoxy resin.

-Cover window air conditioners with plastic covers in the winter (when not used for heating).

-Install automatic closers on exterior doors. (Cost ranges from $60 to $90 per door depending on whether closers are installed by maintenance staff or outside labor)

To reduce infiltration due to stack effect:

-Reduce temperature in stairwells. (Protect piping from freezing)

-Seal elevator shafts top and bottom and insure that machine room penthouse door is weatherstripped and closed.

-Seal vertical service shafts at top and bottom and, in tall buildings, at every sixth floor.

-Weatherstrip and close doors in basement and roof equipment rooms where these are connected by a vertical shaft which serves the building.

-Check building code for venting requirements and check fire resistance ratings of materials used. Skylights or smoke relief vents may be required.

INCREASE BENEFICIAL SOLAR HEAT GAIN INTO THE BUILDING

-Clean windows to permit maximum sunlight transmission.

-Operate drapes and blinds to permit sunlight (when available) to enter windows during the winter; move desks or work stations out of the direct path of sunlight to avoid occupant discomfort.

-A percentage of direct solar radiation is stored in the structure and furnishings where it will help to offset the heat load at night. Permit the space temperature to rise so that excess heat can be stored in the structure and be available for heating at night or cloudy periods. Even on cloudy days diffuse radiation is considerable; allow it to be transmitted into the occupied spaces.

Note: treat skylights and display windows in the same manner as windows.

-If windows are not fitted with blinds, drapes, or shutters, consider installing them to control the rate of heat flow into and out of the building.

-During heavily clouded weather, and at night, reduce the heat loss through the window by drawing shades and drapes or closing shutters where fitted.

-If direct sunlight or excessive window brightness causes glare, add a light transluscent drape which cuts glare but permits solar heat to enter.

-Readjust blinds during the day if, during particular times of the year, overheating occurs.

-Before installing shades or blinds, check fire code for prohibitions against certain types of materials.

-Where possible, treat the site to increase useful solar heat gain in the winter.

-Trim all foliage shading the southern, eastern or western face of the building in winter. Reduce any evergreen foliage grossly blocking the winter sun.

-Where possible, remove shading devices and any other objects casting shadows on the building surfaces during winter.

This page blank

Section 5

Energy Conservation Opportunities Domestic Hot Water

DOMESTIC HOT WATER

A. BACKGROUND

The amount of energy consumed in heating hot water is about 4% of the annual energy used in most large commercial buildings. In smaller commercial buildings, the percentage is smaller. However, in facilities which include restaurants, cafeterias and especially laudromats, the percentage of energy for hot water compared to other systems will be greater.

If domestic hot water is heated by the same boiler which heats the building, and if the load is only 10 or 20% of the total boiler load in those months when the building is heated, the energy used in the fall and in summer months for domestic hot water may be considerably higher than in the winter as the boiler will be operating at low part load efficiency. To determine the amount of energy used for domestic hot water follow the method described in Section 3 , Figure 12.

The opportunities to conserve energy for heating domestic hot water can be summarized as follows:

Reduce the load
- -decrease the quantity of domestic hot water used
- -lower the temperature of the domestic hot water

Reduce the system losses

- -repair leaks and insulate piping and tanks
- -reduce recirculating pump operating time

Increase the efficiency of the domestic hot water generator

B. EXISTING CONDITIONS

1. Average Usage

TABLE 1

Office Buildings

(Without kitchen or cafeteria services) . . .	2 to 3 gallons per capita per day for hand washing and minor cleaning (based on an average permanent occupancy which includes daily visitors)

Department Stores

(Without kitchen and cafeteria services) . . .	1 gallon per customer per day

Kitchen and Cafeterias for hand washing

Dishwashing, rinsing and hand washing	3.0 gallons per meal plus 3 gallons/employee/day

Schools

Boarding	25 gallons per capita per day
Day	3 gallons per capita per day (Does not include cafeteria or athletic facilities)

Apartments

High rental	30 gallons per capita per day
Low rental	20 gallons per capita per day

Hospitals

Medical	30 gallons per capita per day
Surgical	50 gallons per capita per day
Maternity	50 gallons per capita per day
Mental	25 gallons per capita per day
Hotels	30 gallons per capita per day

2. Average Temperatures

The usual temperature at which hot water is supplied - from 120°F. to 150°F. - is too hot to use directly and must be mixed with cold water at the tap. For dishwashing and sterilization the delivery temperature is generally 160°F. or higher. Often hot water supplied to all faucets is at temperature required for the kitchen. Frequently, the hot water, generated and stored in tanks at 150° to 160°F., loses heat by conduction and radiation from the tank and piping, even before the delivery at wasteful temperatures.

When hot water is supplied by a tankless heater, it is within 5° or 6°F. of the boiler water temperature maintained to heat the building. A mixing valve is often used to control the delivery temperature, but frequently the temperature at which it is set is excessive. If the tankless heater, or tank heater, is installed inside the boiler, the losses from the domestic heater may be considerable.

3. Methods of Generation and Storage

a) By a tankless heater from a hot water boiler used to heat the building, or by a below-the-water line tankless heater on a steam heating boiler.

b) By a tank heater and storage tank combination which is either a hot water or steam-heating boiler. The tank heater may be integral with the storage tank, or separately mounted and connected to the boiler and tank by piping.

c) By a separate oil, gas, coal or electric domestic hot water heater with integral storage tank.

d) By separate electric booster heaters without storage tanks.

4. Distribution

Hot water is distributed either by gravity circulation or by a recirculating hot water pump through separate piping to the fixtures. The recirculating hot water pump delivers hot water instantly at the faucets and reduces the total quantity of water used by saving the cold water which is usually drawn upon first opening the faucet. However, because the pump requires electrical power for operation, and because its piping system must always be filled with hot water and experience heat loss, the use of the recirculating pump could be energy-wasteful in systems where all faucets are close to the tank.

C. ENERGY CONSERVATION OPPORTUNITIES

ECO 11 REDUCE THE TEMPERATURE OF DOMESTIC HOT WATER SUPPLIED TO TAPS

Lowering the temperature of the hot water reduces both the "building" domestic hot water load, as well as the distribution load. The building load for hot water heating is expressed by the following formula:

> Yearly BTU's = Q x Td_B, where Q = Quantity of domestic hot water used per year in pounds, and Td_B = Magnitude of the difference, in °F., between the temperature of cold water entering the heater, and the temperature of the hot water at the faucets.

The parasitic load is determined similarly, except:

> Yearly BTU's = Q x Td_p, where td_p= Magnitude of the difference, in °F., between the generation temperature and the temperature of the water at the taps.

Total load, then, is calculated as follows:

$$\text{Yearly BTU's} = (QxTd_B) + (QxTd_p).$$

Or, because:

$$Td = Td_B + Td_p$$

(that is, the difference between the temperature of the water as it enters the heater and the generation temperature), it is calculated more simply as follows:

$$\text{Yearly BTU's} = QxTd$$

Figure 19 indicates energy used for domestic hot water at various generation temperatures and usage rates. An incoming water temperature of 50°F. and 251 days of occupancy per year are assumed.

FIGURE 19: SAVINGS FOR REDUCTION OF FAUCET FLOW RATE AND WATER TEMPERATURE

dubin-mindell-bloome-associates consulting engineers

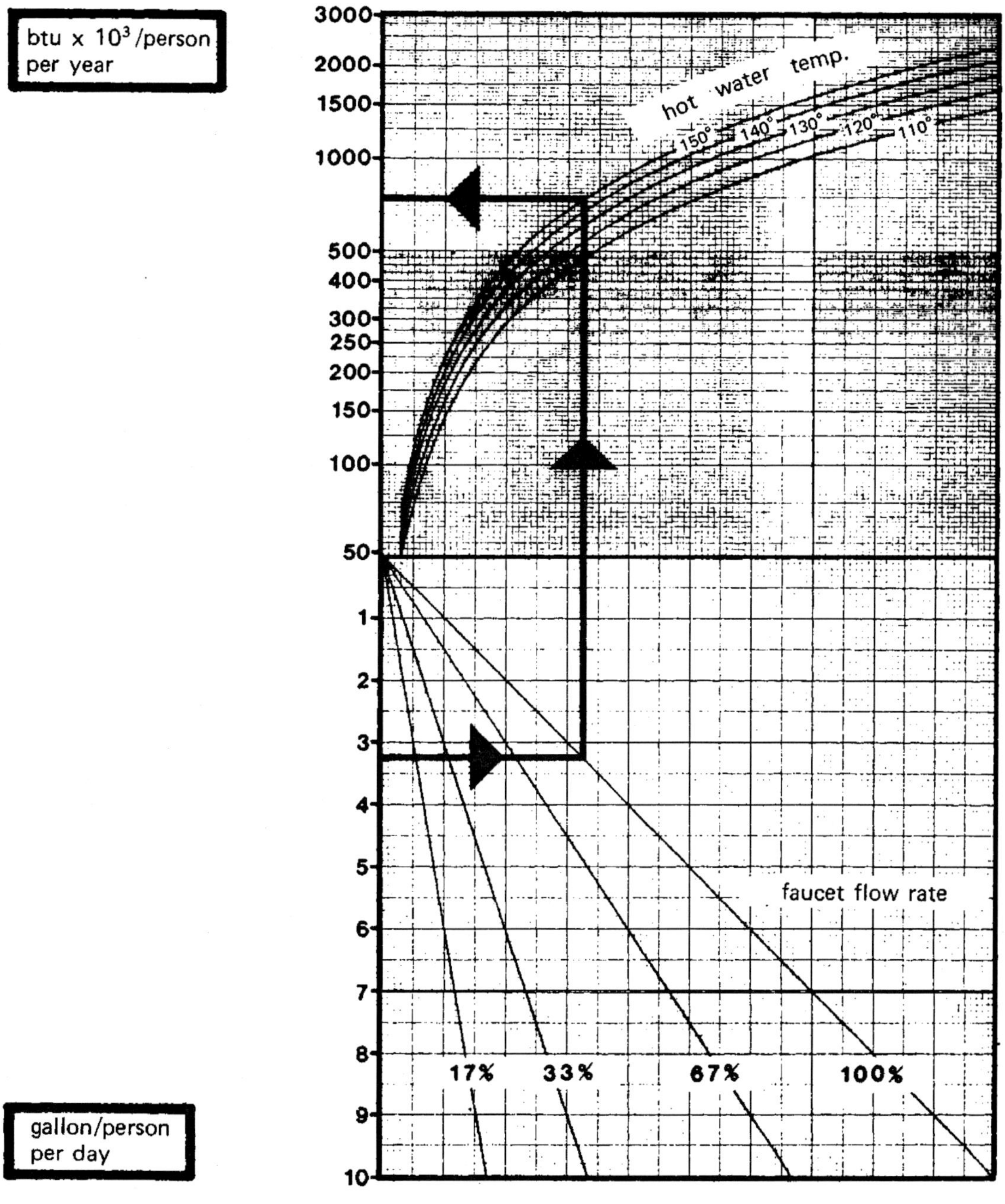

Figure #19 Engineering Data

Method: Straight heat transfer calculations

Assumption: Users open faucet for a set amount of time regardless of flow rate.
i.e. Washing hands is based on the time it takes rather than the water quantity.

The actual amount of energy required to supply the total load depends upon the seasonal efficiency of the heater, "E", which varies with the type of heater and the fuel used. On a seasonal basis, the following are average efficiencies:

a) Oil-fired heating boilers used year round, but with domestic hot water as the only summer load = .45.

b) Oil-fired heating boilers used year round with absorption cooling in the summer = .7.

c) Gas-fired heating boilers used year round, but with domestic hot water as the only summer load = .50.

d) Gas-fired heating boilers used year round with absorption cooling in the summer = .75.

e) Separate oil-fired hot water heaters = .70.

f) Separate gas-fired hot water heaters = .75.

g) Separate electric water heaters = .95.

h) Separate coal fired water heaters = .45.

To determine actual energy consumption, divide the value obtained from Figure 19* by the appropriate efficiency or use the following formula:

$$\text{Yearly BTU's} = \frac{QxTd}{E}$$

Example:

An office building has 500 occupants, each of whom uses 3 gallons of hot water per day for 250 days each year. The temperature of the water as it enters the heater is 60°F. (an average for the year) and it must be heated to 150°F. in order to compensate for a 20°F. drop during storage and distribution, and still be delivered, at the tap, at 130°F. Hot water is generated by an oil-fired heating boiler, used year round with domestic hot water as the only summer load. The fuel is #2 oil, which contains 138,000 BTU's to a gallon.

* If incoming temperature differs from 50°F., adjust valve before dividing. If incoming temperature is 60°F., for instance, at a generation temperature of 150°F., multiply value by $\frac{150-60}{150-50}$.

Building load

Q = 500 occpt x 3 gal./day/occpt x 250 day/yr. =
375,000 gal./yr.
1 gal. = 8.3 lbs., therefore
Q = 390,000 gal./day x 8.3 lbs./gal - 3,112,500 lbs/yr.
Td_B = 130°F. - 60°F. = 70°F.
Yearly BTU's - 3,112,500 lbs. x 70°F. - 217,875,000

Parasitic load

Td_P = 150°F. - 130°F. = 20°F.
Yearly BTU's = 3,112,500 lbs. x 20°F. - 62,250,000

Total load

Yearly BTU's = 217,875,000 + 62,250,000 = 280,125,000

Total energy used

E = .45

$$\text{total BTU's} = \frac{280{,}125{,}000}{.45} = 622{,}500{,}000$$

Total fuel consumption

$$\text{Yearly gallons} = \frac{622{,}500{,}000}{138{,}000 \text{ gal.}} \text{ BTU's} = 4{,}511$$

To calculate the amount of fuel needed at a reduced delivery temperature, 90°F., for example, perform the following procedure:

$$4{,}511 \text{ gal.} \times \frac{90°\text{F.}}{130°\text{F.}} = 3{,}123 \text{ gal.}$$

This is actually a conservative figure, as the total savings in heating, storing, and distributing the water would include reduced storage and distribution losses as well.

Table 2 indicates the yearly energy loss in BTU's for various sizes of tanks, located in a space with an ambient temperature of 65°F., and with fiberglass insulation.

TABLE 2

Insulation Thickness	Tank size in gallons	BTU's in millions/year. Lost at Hot Water Temperatures of: 100°F	120°F	160°F
1"	50	1.9	3.0	5.2
	100	3.0	4.7	8.2
2"	250	3.1	4.9	8.4
3"	500	3.1	4.9	8.4
	1,000	5.2	8.2	14.1

Costs for insulating hot (or cold) water tanks with 3# density fiberglass - foil scrim craft facing, finished with pre-sized glass cloth jacket - are as follows:

Material Thickness	Cost/sq. ft. of Surface Area
1"	$2.60
1-1/2"	$2.70
2"	$2.95
3"	$3.60

ECO 12 REDUCE THE QUANTITY OF DOMESTIC HOT WATER USED

A primary benefit of reducing the quantity of hot water used is that energy consumption will be decreased to the same extent as with an equal percentage reduction in temperature.

A secondary benefit is the reduction in raw source energy which occurs because less water needs to be treated in the water supply treatment and sewage treatment plants, whether on-site or off-site. For municipal facilities, the diminished energy requirements will result in lower operating costs than otherwise possible, which in turn will mean that less taxes will be needed to support the facility. In areas where there is a charge based on total water consumption flowing into the sewer, the reduction in consumption of water will result in direct savings, as well. Water consumption can be lowered to 1-1/4 or 1-1/2 gallons per person per day in office buildings today without inconvenience to the occupants. Additional opportunities to reduce water consumption are summarized in the guidelines.

Example:

An office building has 500 occupants, each of whom uses 3.5 gallons of hot water per day for 250 days each year. The water, as it enters the heater, is at 40°F., and it is heated to 150°F. The separate gas-fired heater has an efficiency of 0.75.

Enter Figure 19 at 3.5 gal./person per day. Follow the example line intersecting with the 100% flow rate and 150°F. temperature lines and read yearly energy used of 800 x 10^3 Btu per person per year.

Re-enter Figure 19 intersecting with the 33% and 120°F. lines and read yearly energy used of 190 x 10^3 Btu/person/yr.

Savings: The energy saved equals 800-190 or 610 x 10^3 Btu/person/year and for 500 people, the total is 5 x 610 x 10^3 or 305 x 10^6 Btu/year.

$$\frac{305 \times 10^6}{138{,}000 \times 0.75} = 2{,}947 \text{ gal.}$$

Convert to Cost: At $0.36/gal., the savings is 0.36 x 2,947 or $1,061 per year.

Results: Energy Saved - 2,947 gallons/year
Dollars Saved - $1,061/year

ECO 13 <u>IMPROVE THE EFFICIENCY OF THE STORAGE AND DISTRIBUTION SYSTEMS</u>

Repair and replace all torn insulation to reduce heat loss. All measures to improve the efficiency of both space heating and domestic hot water heaters are noted in Section 4 , "Heating", and the measures to reduce heat loss from piping are detailed in Section 4 , "HVAC Systems and Distribution".

ECO 14 <u>GENERATE HOT WATER MORE EFFICIENTLY</u>

All of the measures for improving combustion units for space heating apply equally well to hot water heaters. Keep in mind, however, that when more than one heater, be it boiler or hot water heater, is installed on a project, it is more efficient to operate one for the total load if it can carry it, rather than to operate all boilers at partial loads.

The greater opportunities for conserving energy by improving the efficiency of the hot water generator, after normal service operations and minor modifications have improved the existing equipment to the extent possible, will require major modifications or the replacement of equipment.

ECM- 2 details these opportunities which include the following:

- -Provide a separate hot water heater for summer or year round use.
- -Heat hot water by use of rejected heat from refrigeration system condenser water or hot gas heat exchangers.
- -Heat hot water with recovered energy from incinerators, heat pipes, hot water heat exchangers, or heat pumps.
- -Replace resistance electric hot water heaters with gas or oil heaters or heat pumps.
- -Add separate booster heater for kitchen or laundry service.
- -Heat hot water with condensate return to steam operated systems.
- -Heat hot water with solar water heaters.

GUIDELINES TO REDUCE ENERGY USED FOR DOMESTIC HOT WATER

REDUCE DOMESTIC HOT WATER TEMPERATURE:

- -Where possible, use cold water only for hand washing in lavatories when cold water temperature is 75°F. or above. This is most readily accepted in retail stores, religious buildings, owner-occupied small office buildings and in washrooms used primarily by the public on an infrequent basis.
- -Where tenants insist upon hot water for hand washing, heat tap water to 90°F.
- -Do not maintain an entire hot water system at the same temperature required for the most critical use.

Do not heat water for hand washing, rinsing or cleaning to the same temperature required for dishwashing sterilization.

-If the space heating boiler is also used to supply domestic hot water, lower the aquastat setting in the summer time to 100°F. The same setting should be used for storage tank temperature control, summer and winter.

-Where higher temperatures are required, at a dishwasher, for example, a small gas or electric booster used only as needed saves more energy than a large storage tank, piping

and distribution system which heats all domestic hot water in the building to the most critical temperature.
-Use cold water detergents for laundries and laudromats and set water temperature to 65°F. -70°F.
-Refer to Health or Food Handling Codes, if applicable, for minimum temperature specifications.

REDUCE HOT WATER CONSUMPTION:

-Insert orifices in the hot water pipes to reduce flows.
-Install spray type faucets that use only 1/4 gallons per minute (gpm) instead of 2 or 3 gpm, at a cost of about $50 a unit.
-Install self-closing faucets on hot water taps.
-In buildings with cooking facilities that are used only periodically, such as meeting rooms in religious buildings, shut off the hot water heating system, including gas pilots where installed, when the facilities are not in use.
-Re-examine the need to heat entire tank of water when only a small quantity or no hot water is needed.
-Simplify menus to reduce the need for large pots and pans that require large amounts of hot water for cleaning. Where practical use short dishwashing cycles and fill machine fully before use.
-Reduce the number of meals served and/or serve more cold meals to reduce the hot water requirements for dishwashing.
-In areas where water pressure is higher than a normal 40 to 50 lbs., restrict the amount of water that flows from the tap by installing pressure reducing valves on the main service. Do not reduce pressure below that required for fire protection or for maintaining adequate pressure on the top floor for flushing.
-Refer to Plumbing Codes, if applicable, for hot water supply requirements.

REDUCE SYSTEM LOSSES

-Repair insulation of hot water piping and tanks or install it where missing (unless piping and tanks are located in areas which require space heating or aire air-cooled).
-Where forced circulation of hot water is used, shut off the pump when the building is unoccupied; when hot water usage is light consider using gravity circulation without the pump.
-Flush water heater during seasonal maintenance of heating systems.
-Repair leaky faucets.
-Repack pump packing glands of recirculation hot water heaters to reduce leaking of hot water.
-For boilers with immersion tankless domestic hot water coils, make sure boiler water covers coils.

Section 6

Energy Conservation Opportunities Cooling and Ventilation

COOLING AND VENTILATION

A. BACKGROUND

The annual consumption of energy for cooling and ventilation can be reduced by applying the same three major categories as for heating and ventilation:

1. Reduce the "building"* cooling load.

2. Reduce the distribution system load.

3. Increase the efficiency or performance of the primary energy conversion equipment.

Categories 1 and 3 are addressed in this section and Category 2 in Section 7, "HVAC Systems and Distribution".

B. THE "BUILDING" COOLING LOAD

Two components - sensible or dry heat, and latent heat, a function of the heat content of the moisture in the air - determine the "building" cooling load. To maintain comfort, the dry bulb temperature which measures sensible heat, and the wet bulb temperature (or the relative humidity), which measures latent heat, must be controlled. Because, for most comfort cooling installations, air is dehumidified when it is cooled, separate control of the relative humidity is not usually required.

Average temperature difference between indoor and outdoor conditions is one of the factors which defines the sensible heat gain portion of the annual "building" load; reduce this load by maintaining higher indoor temperatures for as long a period as possible during the cooling season, whenever further operation of the mechanical refrigeration system would be required to maintain a lower one. When the sensible heat gain is decreased, reduce the amount of cool supply air accordingly to realize additional energy savings in fan horsepower.

The cooling load, is in part, due to conductive heat gain from outdoors to indoors, through the building envelope.

The average difference in temperature between indoors and the exterior surfaces of the walls and roof depends upon outdoor dry bulb temperatures and the amount of solar radiation impinging upon the outside walls and roof (and warming

*"Building" cooling load is used to identify the loads to maintain interior conditions and should not be confused with "room load" which is standard terminology for building loads not including the ventilation load.

them to a level - termed "sol-air" temperature - which is most often above ambient temperatures). To conserve energy, shut off, during unoccupied periods, the refrigeration system and its auxiliaries (chilled water and condenser water pumps, cooling towers and air cooled condensers) and maintain, during occupied periods, a dry bulb temperature of 78°F. or higher. Maintaining even higher temperatures in less critical spaces will conserve more energy. If the relative humidity in the building is permitted to rise to 55% (the usual level is 45 to 50%), or to fluctuate normally (within the limits set by the refrigeration system's ability to maintain wet bulb temperatures), considerable energy will be conserved. The system requires power to remove moisture which originates from internal loads - people and cooking - and from outdoor air as it infiltrates and/or ventilates the building. See Figure 20 for "Recommended Cooling Season Indoor Temperature and Humidity".

"Heat gain" - a term used to quantify the amount of heat which is added to a space and must be removed to maintain desired space conditions - is an addition to the cooling load. Solar heat gain through windows frequently constitutes a major portion of the sensible heat "building" cooling load. The benefits from using the sun to reduce the heating load in the winter are considerable, but remember to block it out in the summer. Solar control devices which can be adjusted either to accept or block sunlight reduce energy consumption year'round.

Just as ventilation and infiltration add to the heating load in the winter, they also increase the cooling load in the summer. Outside air must be cooled and dehumidified. Maintaining higher indoor temperatures and relative humidity reduce the cooling load contributed by infiltration and ventilation as well as conductive heat gains. However, whenever outdoor air is cool enough to lower the indoor temperatures, it may be advantageous (and energy conserving) to cool or precool the building at night,using an economiser cycle, without mechanical refrigeration. When the wet bulb (W.B.) temperature of the outdoor air is lower than the W.B. temperature of the return air from the interior spaces, an enthalpy controller will open the outside air dampers and close the return air dampers to take in 100% outdoor air; the controller saves power for the refrigeration system. Internal heat gains-heat emitted from electric lighting fixtures, business machines, motors, cooking equipment and people form a large part of the total cooling load.

Reducing the "building" cooling load by a fixed percentage allows further savings of energy since distribution loads can be decreased accordingly. A reduction in the flow of cool air or chilled water to meet the reduced load and an increase in the suction temperatures at which the refrigeration chillers or compressors operate (improving their Coefficient of Performance - C.O.P.) will result in significant savings - less horsepower per ton of refrigeration as well as less tons to be produced.

C. DISTRIBUTION SYSTEM LOADS

The distribution system may carry chilled water to units in the spaces to be conditioned (fan coil, induction, or small air handling units) or to remote air handling units (which supply air, via ducts to local registers and diffusers). In some systems the compressors discharge a vaporized refrigerant to a condenser, when it is condensed to a liquid; then, it is allowed to expand directly in a cooling coil which cools or dehumidifies circulating air. In all cases, air may be circulated from the conditioned space, from outdoors, or from a mixture of both: the air is supplied after it is filtered, cooled and dehumidified in the air handling unit or air-handling section of the air conditioning system.

Piping and duct losses - heat gain and water or air leakage - increase the distribution loads, and as a result, the load on the primary energy conversion equipment. See Section 4 , "HVAC Systems and Distribution" for energy conservation opportunities and guidelines to reduce distribution system losses.

D. PRIMARY ENERGY CONVERSION EQUIPMENT

Energy is used to supply the cooling "building" and distribution loads. In the form of heat, energy operates absorption refrigeration units, and as electricity, it operates reciprocating, centrifugal or screw type compressors and/or chillers. The refrigeration equipment produces either chilled water or, in the case of direct expansion refrigeration units with air handling units, cooled and dehumidified air. (Note: Refrigeration equipment is often referred to as the "high" side of the refrigeration cycle.) The air-cooled or water-cooled condensers of the refrigeration equipment control the condensing temperature of the system - the lower the condensing temperature, the more efficient the refrigeration system. The manner and conditions under which air cooled condensers or evaporative condensers with cooling towers are operated exert a significant influence on the energy required to run the refrigeration units. The efficiency of the refrigeration system is expressed as a coefficient of performance (C.O.P.).

The capacity of the system is measured in tons of refrigeration; one ton will produce a cooling effect equal to 12,000 Btu's/hr. Seasonal C.O.P. is the ratio of the tons of refrigeration produced, expressed in Btu's, to the energy required to operate the equipment, also in Btu's. If one kilowatt hour of electricity (equivalent to 3,412 Btu's) is required to produce one ton of refrigeration (12,000 Btu's) the C.O.P. is $\frac{12,000}{3,412} = 3.52$. To improve the C.O.P. - reduce energy usage - lower the condensing temperatures and/or raise the suction temperatures. The opportunities to do either depend upon load reduction and upon the operation and maintenance of the equipment, outlined in ECO's 20 and 21. The opportunities to improve C.O.P. by reducing distribution loads are outlined in Section 7, "HVAC Systems and Distribution".

E. ENERGY CONSERVATION OPPORTUNITIES

ECO-15 TURN OFF COOLING SYSTEM DURING UNOCCUPIED HOURS

Turning off the entire air conditioning system, at night and during days when the building is unoccupied, generally will save energy. Manually shut down cooling or condenser fans, chilled water and condenser pumps, and supply and exhaust fans as well as chillers or compressors. If the cooling tower fans and pumps are interlocked with the compressors, they will shut off automatically when the compressors are shut off. Shut down boilers which supply steam or hot water to absorption units; do not maintain boiler water temperature or steam pressure when absorption units are inoperative. If, after shut down, the air conditioning system is not capable of achieving and maintaining temperature and humidity conditions in hot spells, install a control to activate the equipment a few hours before occupancy instead of operating the system all night or throughout the entire weekend.

In geopraphic areas such as parts of Arizona or California with a large diurnal swing (large temperature difference within 24 hours), night operation of the refrigeration system for periods of the summer may be economical. If climatic conditions at night permit lower condensing temperatures, perform an analysis to indicate whether energy can be saved by operating the refrigeration system at night to precool the building or to store chilled water in the piping system.

An engineering and economic analysis, comparing refrigeration operation and economiser cooling is necessary to determine the mode of operation which will result in minimum energy consumption.

ECO-16 USE OUTDOOR AIR FOR COOLING

A. Night and Unoccupied Periods

When nighttime outdoor temperatures are below indoor temperatures by 5° or more, shutting off the refrigeration system and using outdoor air for night cooling will save energy in most areas of the country. The opportunities to use outdoor air for cooling during unoccupied periods depend upon the length of time that the outdoor air temperature is below 73°F. In most areas there will be no advantage to bringing in outdoor air above that temperature; heat from the fan motors will raise the air temperature 2 to 3°; and the power required to drive the fans will outweigh any savings due to precooling and reduced operation of the refrigeration system. The temperature at which outdoor air cooling is advantegeous may be even lower in large buildings which have high velocity systems, as the power requirements for fan motors in these systems are higher.

In buildings which have operable windows, however, night cooling, even at higher outdoor air temperatures, is worthwhile, as fan motors need not be operated.

If the dry bulb temperature of the outdoor air is above indoor temperature, the wet bulb temperature of the outdoor air can be higher than the room wet bulb temperature. There may be periods when air at higher wet bulb temperatures, introduced directly into spaces that have been cooled during the day, will cause condensation on walls or furnishings. Maintaining higher space temperature conditions, 78°F. and above, will minimize the amount of time that this could occur and all but eliminate the constraint in most areas of the country. Even in humid climatic zones, wet bulb temperatures above 78°F. occur for relatively few hours at night; the need to limit economiser operation exists only during those hours.

Where there are no automatic controls to operate an economiser cycle, operate dampers and fans manually, or consider installing controls.

To fully utilize outdoor air for cooling, it may be necessary to install return air and outdoor air dampers, and provide a means of relieving air pressure. Some possible options include partially opening some windows, operating an exhaust system, or installing some propeller type exhaust fans in the wall. Refer to ECM-2 for more details on economiser or enthalpy control.

B. Occupied Periods

During occupied periods, the opportunities to use outdoor air for cooling depend not only on outdoor dry bulb temperature, but also upon the wet bulb temperature; if outdoor air is brought into the building above 66° F.W.B. (the equivalent to the wet bulb temperature when the spaces are maintained at 78°D.B. and 55% R.H., it adds to the cooling load. The wet bulb temperature is a measure of the total heat content of the air. If the wet bulb temperature is lower than 66°F., outdoor air can be introduced through the cooling coil, with the refrigeration system in operation in any quantity. At less than 66°F W.B. outdoor air, in fact, will reduce the cooling load. For many existing systems, however, (even when the temperature of outside air is less than 66°F. W.B.) the cooling coils are not designed to be able to handle outdoor air at temperatures above 85°F. D.B. and still maintain room conditions at 78°F. D.B. - this, despite the fact that outdoor air at less than 66° W.B. actually has a lower total heat content than room air. Figure 9 indicates for areas all over the country, the number of wet bulb degree hours below 66° W.B. when the dry bulb temperature is less than 85°F. in the summer. If a building is in a location where there are 3,000 or more such degree hours, an enthalpy controller will be a good investment. Denver, Colorado, on a seasonal basis, has low wet bulb temperatures, and it would appear that it would be more economical to utilize 100% outdoor air through the air conditioning coils in the daytime, rather than to recirculate air. Summer outdoor dry bulb temperatures in Denver, however, often rise above 90°F. If the existing coils have not been selected to reduce the temperature of outdoor air at those times, they will be incapable of handling the conduction, solar and internal heat gains which occur. To conserve energy in this type of a climate, open the outdoor air damper fully except when indoor dry bulb temperatures cannot be maintained. Operating systems manually to reflect temporary conditions and conserve energy is sometimes difficult but the effort yields significant savings. Automatic controls are available to optimize the operation of most systems and to meet varying and selective conditions. Refer to ECM- 2 for details.

ECO - 17 INCREASE INDOOR TEMPERATURE AND RELATIVE HUMIDITY LEVELS DURING OCCUPIED HOURS*

The air conditioning systems in many buildings were designed to maintain 72° to 75°F. D.B. and 50% R.H. during peak loads in the cooling season. They are operated to maintain those levels at peak conditions, and to achieve even lower levels during the part load conditions which occur most of the time.

Maintain dry bulb temperature and relative humidity for various spaces at the levels suggested in Figure 20, "Recommended Cooling Season Indoor Temperature and Humidity" to reduce the cooling load and energy and operating costs. Realize even greater savings, without serious discomfort, by maintaining higher levels in areas which are only occupied for short periods of time.

If unoccupied periods in these areas exceed 60% of the day, shut off the cooling unit or close registers, grilles and diffusers; allow temperature and humidity to rise when the areas are unoccupied. (Auditoriums, corridors, cafeterias and conference rooms are all spaces frequently unoccupied for most of the day.)

Increasing the indoor temperature and humidity levels from 74°D.B. and 50% R.H. to 78° D.B. and 55% R.H. will save approximately 13% of the energy required for cooling. The exact amount depends upon the amount of ventilation air and infiltration which enters the building; the conduction losses, solar heat gain, and internal loads of the building; and the type of air conditioning system it has.

Use the multipliers in the following table to determine the amount of energy saved (in Btu's per hour) per 1,000 CFM of outdoor air (ventilation plus infiltration) by raising the dry bulb temperatures at constant relative humidities. (All other factors are considered constant, and since the effect of raising the dry bulb temperature on conductive heat gain is actually negligible, it has been disregarded here.)

*With terminal reheat systems in operation, the indoor space conditions should be maintained at lower levels to reduce the amount of reheat and save energy. If cooling energy is not required to reduce the temperatures, maintain 74°F. D.B. instead of 78°F. See Section 7, "HVAC Systems and Distribution" for discussion of terminal reheat systems and guidelines for operation.

TABLE 1

Relative Humidity	50%	60%	70%
Dry Bulb Temperature			
72°F.	0	0	0
73°F.	2,700	2,433	3,000
74°F.	2,657	2,400	3,257
75°F.	3,000	2,572	3,000
76°F.	3,000	2,572	3,000
77°F.	3,000	2,572	3,429
78°F.	3,000	2,572	3,429

Example:

Project the savings achieved, in cooling outdoor air, by raising the indoor dry bulb temperature from 72° F. to 78°F.

Type	- Office, 40 hours occupancy
Annual WB degree hours above 66°F. WB	- 8,000
Total outdoor air including infiltration	- 10,000 CFM
Relative Humidity	- 50%

From Table 1, in the 50% RH column, determine that raising the DB temperature from 72° to 73° saves 2,700 Btu's/hour/1,000 CFM, from 73° to 74°, 2,657; and from 74° to 75° 75° to 76° to 76° to 77°, and 77° to 78°, 3,000 Btu's/hour/ 1,000 CFM. To calculate savings over total temperature change, add the 6 figures together:

2,700 + 2,657 + 3,000 + 3,000 + 3,000 + 3,000 = 17,357 Btu's/hour/1,000 CFM

For an outdoor air rate of 10,000 CFM, savings will be:

10 x 17,357 = 173,570 Btu's/hour

For 40 hours of cooling per week, and a cooling season of 20 weeks, yearly savings will be:

173,570 x 40 x 20 = 128 x 10^6 Btu's/hour

FIGURE 20: SUGGESTED INDOOR TEMPERATURE AND HUMIDITY LEVELS IN THE COOLING SEASON

I. Commercial Buildings — Occupied Periods

	Dry Bulb Temperature *	Minimum Relative Humidity
Offices	78°	55%
Corridors	Uncontrolled	Uncontrolled
Cafeterias	75°	55%
Auditoriums	78°	50%
Computer Rooms	75°	As needed
Lobbies	82°	60%
Doctor Offices	78°	55%
Toilet Rooms	80°	
Storage, Equipment Rooms	Uncontrolled	
Garages	Do Not Cool or Dehumidify.	

II. Retail Stores — Occupied Periods

	Dry Bulb Temperature	Relative Humidity
Department Stores	80°	55%
Supermarkets	78°	55%
Drug Stores	80°	55%
Meat Markets	78°	55%
Apparel	80°	55%
Jewelry	80°	55%
Garages	Do Not Cool.	

* Except where terminal reheat systems are used.

Savings: Energy saved for a mechanical refrigeration system with a C.O.P. of 2.5 will be:

$$\frac{139 \times 10^6}{2.5} = 56 \times 10^6$$

$$\frac{56 \times 10^6 \text{ Btu's/year}}{3{,}413 \text{ Btu's/KWH}} = 16{,}408 \text{ KWH/year}$$

Energy saved for an absorption refrigeration system with a C.O.P. of 0.68 will be:

$$\frac{139 \times 10^6}{0.68} = 204 \times 10^6 \text{ Btu/year}$$

If this energy would have been supplied by a steam boiler with a seasonal efficiency of 60%, $\frac{204 \times 10^6 \text{ Btu's}}{0.6 \times 138{,}000 \text{ Btu's/gallon}} = 2{,}463$ gallons

Assuming a cost for electricity of \$0.4/KWH, savings for the mechanical system will equal:
.04 x 16,408 = \$656/year
At an oil cost of \$0.36 per gallon, savings for the absorption system will equal:
.36 x 2.463 = \$887

If the relative humidity in the building is allowed to rise from 50% to 70%, the savings are calculated as follows:

Enter Figure 21 at 8,000 WB degree hours. Follow the example line with the 50% RH line and the 40 hour line and read yearly energy used at 22.5 x 10^6 Btu/year/1,000 CFM.

Re-enter Figure 21 intersecting with the 70% RH line and read yearly energy used at 16 x 10° Btus/year/1,000 CFM/

Savings: The energy saved equals 22.5 - 16 or 6.5 x 10^6 Btu's/year/1,000 CFM. For 10,000 CFM, the total reductions in energy input are 7,618 KWH/year and \$305 for a mechanical refrigeration system and 1,160 gallons of oil/year and \$415 for an absorption refrigeration system.

Results: Energy Saved - 7,618 KWH/year or 1,160 gallon/yr.
Dollars Saved - \$305 or \$415

FIGURE 21: YEARLY ENERGY USED PER 1,000 CFM TO MAINTAIN VARIOUS HUMIDITY CONDITIONS

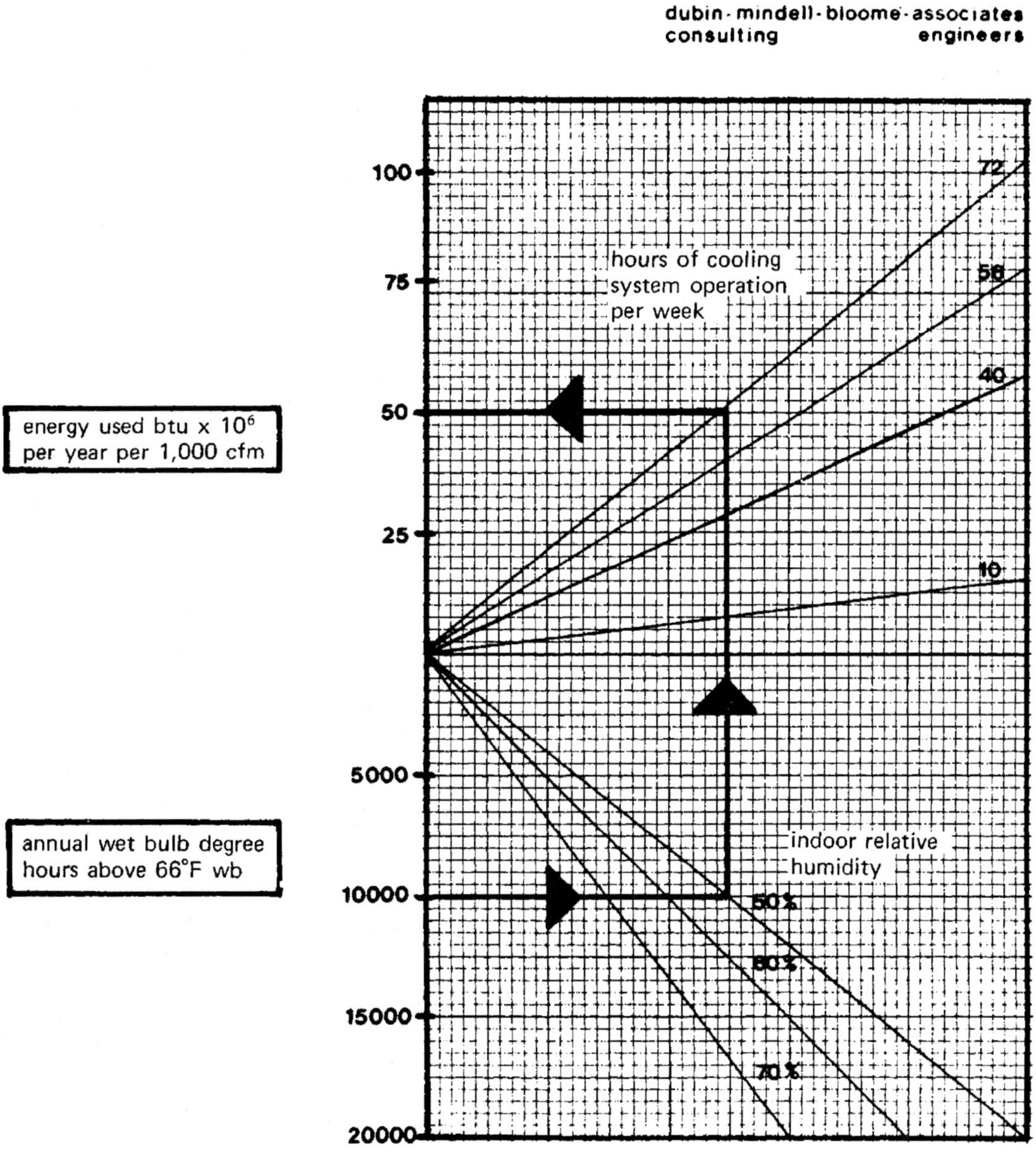

Figure #21 Engineering Data

Weather data from Figure #8

WB degree hours based on 12 Mos/yr, 8 Hr/Day

Energy used is a function of the WB degree hours above the base of 66°F, the RH maintained the No. of hours of controlled humidity. The base RH is 50% which is approximately 78°F DB, 66° WB. The figure expresses the energy used per 1000 CFM of air conditioned or dehumidified. An analysis of the total heat content of air in the range under consideration indicates an average total heat variation of 0.93 Btu/lb for each degree WB change at constant DB temperature and that the total heat varies nearly directly with RH. One thousand CFM is equal to 4286 lb/hr so each degree F WB hour is equal to 4286 x 0.93 or 3986 Btu. The lower section of the figure shows the direct relationship from the base of 50% RH and the upper section proportions the hours of system operation with 56 Hr/Wk being 100%.

ECO-18 REDUCE VENTILATION RATES DURING OCCUPIED PERIODS*

Except where outdoor air can be used productively for economiser cooling (and the temperature and enthalpy are suitable as described in ECO-15 above), cooling and dehumidifying outdoor air consumes large quantities of energy. Any measure that reduces the quantity of ventilation to a minimum and is compatible with physiological needs, will result in substantial energy and dollar savings. Refer to Figure 16, "Suggested Ventilation Standards" for both summer and winter periods. The suggestions for ventilation discussed in Section , "Heating and Ventilating" apply, generally to the cooling season, as well. Some codes permit further reductions in ventilation rates if the air is cooled. During the cooling season, the enthalpy difference between the outdoor and indoor air determines the quantity of energy required to cool and dehumidify the ventilating air to indoor conditions. As enthalpy, or total heat, is very closely approximated by wet bulb temperature, the difference in wet bulb temperature between outdoor air and room conditions can be used to determine the energy required for cooling and dehumidification.

To determine the approximate yearly energy savings resulting from a reduction of outdoor air, first determine how much air is being introduced into the building, and second, determine the minimum quantity that legally, must be introduced into the building. The difference determines the potential reduction of ventilation air and energy savings. For the cooling season, use Figure 21 to determine the energy required to cool and dehumidify a given amount of ventilation air. The following example indicates the method to be used to determine the savings.

Example:

Building Type	-	Office
Location	-	Miami, Florida
Occupied Hours/Week	-	40
Number of Occupants	-	667
Indoor Relative Humidity	-	50%
Annual et Bulb Degree Hours Above 66°F	-	18,500

*Also read "Heating and Ventilating", ECO-6, guidelines.

Determine energy saved by reducing outdoor air from 30 CFM/person to 8 CFM/person (mixed population of smokers and non-smokers).

Determine total CFM reduction:

30 - 8 = 22 CFM/person x 667 people = 14,674 CFM. Enter Figure 21 at 18,500 degree hours, intersecting with the 50% RH and 40 hour lines, follow the example line and read yearly energy used of 52.4×10^6 Btu/year/1,000 CFM.

Savings: The total energy saved is 14.5 times 52.5×10^6 Btu/year. If cooling is by a mechanical refrigeration system with a C.O.P. of 2.5, the reduction in energy input is $\frac{771 \times 10^6}{2.5} = 308 \times 10^6$ Btu/year

$\frac{308 \times 10^6}{3{,}413} = 90{,}243$ KWH/year

If an absorption machine is used, the reduction in energy input is $\frac{771 \times 10^6}{0.68} = 1{,}134 \times 10^6$ Btu/year

and if this energy is supplied by a steam boiler with a seasonal efficiency of 65%, the gallons of oil saved is

$\frac{1{,}134 \times 10^6}{0.65 \times 138{,}000} = 12{,}642$ gallons/year

Convert to Cost: At $0.04/KWH, the savings for the mechanical refrigeration system are 0.04 x 90, 243 = $3,609/yr. At $0.36/gal. the savings for the absorption system are 0.36 x 12,642 = $4,551/year.

Results: Energy Saved - 90,243 KWH/year or 12,642 gal/yr.
Dollars Saved - $3,609/year or $4,551/year

If, in the previous example the refrigeration system had been operating nights and weekends with the 30 CFM/person ventilation rate, the savings due only to shutting off the outdoor air during unoccupied periods and reducing it from 30 CFM/person to 8 CFM/person for 40 hours per week would be about $4,460 per year with electric refrigeration and $7,000 per year with absorption refrigeration. These figures do not include the additional savings (in power) which would accrue from not operating the refrigeration system and its auxiliaries during unoccupied periods. If a refrigeration system is operated at night for precooling, or to handle late evening events, shut off the outdoor air intake.

ECO-19 <u>REDUCE INFILTRATION RATES</u>

Leakage or infiltration of air into the building, as described in Section 4., "Heating and Ventilating", adds, to the same extent as an equal amount of ventilation air, a load on the "building" cooling system. Generally, because wind velocities are lower in the summer, the infiltration rates are somewhat lower as well. They may still, however, be a significant factor in the magnitude of the cooling load. Infiltration imposes an added load on the distribution system. Unlike ventilation, which is not a "space" or "room" load, infiltration is an additional room load, and the sensible heat component of infiltration increases the amount of air which must be circulated (as the amount of air circulated in a cooling system depends only on the magnitude of the room sensible heat load). Because infiltration occurs for 24 hours per day, it creates a load during the night which must be removed the next day by the refrigeration equipment. In retail stores, the greatest amount of infiltration occurs through door openings. Refer to EMC- 2 for information on infiltration rates through door openings and the installation of revolving doors as a measure to reduce it.

Determine infiltration rates through windows by using Figure 18, and calculate the cooling load for any given amount of infiltration converted to CFM by using Figure 21.

From all causes, the infiltration rate for most buildings is between 1/2 and 1 1/2 air changes per hour, depending upon the condition of the doors and windows, length of cracks, height of the building, and the number of door openings. (See Section 4, "Heating and Ventilation", ECO-5). To convert air changes per hour to CFM, multiply the building volume (area per floor, times the number of floors, times the ceiling height of one floor) by the number of air changes per hour, and divide by 60. For example a building of 10 stories with 10,000 square feet of floor area per story, a floor to ceiling height of 10' and 1 1/2 air changes per hour would have:

$$\frac{10 \times 10 \times 10{,}000 \times 1.5}{60} = 25{,}000 \text{ CFM}$$

Use Figure 21 to determine the cooling load due to the air quantity in CFM for any location.

Weather stripping and caulking (usually done to reduce infiltration rates in the winter) will reduce the summer cooling load. To determine the savings in energy and dollars during the cooling season for reduced infiltration due to window leakage, use the method indicated in the following example.

Example: Determine the crack area first. Use the identification profiles to gather pertinent data.

Building Type	-	Offices
Building Size	-	100' x 50', 4 floors

Building Location	-	Topeka, Kansas
Wet Bulb Degree Hours Greater than 66°F.	-	3,000
Wind Speed	-	8 mph
Wind Direction	-	SW
South Windows	-	5' x 3'; 28 total
West Windows	-	5' x 3'; 56 total

Window Type: double hung, loose fitting.

Determine the crack length/window
Each Window perimeter = 5+5+3+3+3(center joint)=19 feet
Total crack length - 19 x 84 windows = 1,596 feet

Determine from Figure 18, the rate of infiltration:
Per foot of crack = .9 CFM/ft.
Total infiltration = 1,596 x .9 = 1,436 CFM

Determine from Figure 18, the new rate of infiltration per foot of crack if windows are weather stripped = 0.1 CFM/ft.
Therefore total infiltration = 1,596 x 0.1 = 159.6 CFM
Infiltration Reduction = 1,436 - 160 = 1,276 CFM

Determine additional cooling energy required:
Enter Figure 21 at 3,000 degree hours and intersecting with the 60% RH and 40 hour lines, follow the example line and read yearly energy used of 7.5×10^6 Btu/year/ 1,000 CFM.

Savings: The total energy saved is $1,276/1000 \times 7.5 \times 10^6$ 9.6×10^6 Btu/year. For mechanical refrigeration system, the reduction in energy input is $\frac{9.6 \times 10^6}{2.5 \times 3,413}$ = 1,125 KWH/
For absorption system, the reduction in energy input is $\frac{9.6 \times 10^6}{0.68}$ = 14.1×10^6 Btu/year and,
For steam boiler supplying the steam and gallons of oil saved is $\frac{14.1 \times 10^6}{0.65 \times 138,000}$ = 161 gallons/year

Convert to cost: At $0.04 KWH, the savings for the mechanical refrigeration system are $0.04 x 1,125 = $45/years
At $0.36/gallon, the savings for absorption system are 0.36 x 161 - $58/year.

ECO-20 REDUCE SOLAR HEAT GAINS

Solar heat gain is a major contributor to the cooling load in all parts of the country where the ratio of window to wall area (on the east, west, and south facades) is 20% or more with clear unshaded glass, and 50% or more with any type of glass exposed to direct sunlight. Solar radiation on the roof (which receives the most in the summer) and on the east, west, and south walls increases the outside surface temperature and heats the surfaces. The increase in heat transmission through the surfaces which then occurs is in excess of conduction which would have resulted from temperature difference between indoors and outdoors alone.

Solar radiation on skylights and windows is transmitted through the glass almost instantaneously. Annually, almost 80 percent of all the solar radiation striking a vertical single sheet of clear glass surface is transmitted through it. It constitutes an addition to the cooling load, but a reduction of the heating and electric loads (if lights are turned off when daylight is available). Conservation measures that reduce solar heat gain include shading the interior or exterior of the glass and treating it to increase its reflective properties. Evaluate devices that prevent solar radiation from entering the building in hot weather, against the need for solar heat in cold seasons.

Solar shading devices on the exterior of the building are more effective than inside blinds, drapes, or reflective coatings or heat absorbing or reflective tinted glass. Refer to Figure 22 for the relative effectiveness of various shading devices. Achieve external solar control through the use of trees that shade the building surfaces, horizontal exterior louvers, or eyebrows on the southern exposure and vertical and horizontal fins or extensions to shade the east and west facing glass surfaces. Install thin sun screens on the exterior face of the window to reduce solar heat gain by 50 to 75%. Sun screens cause the loss of some natural lighting and the blocking of some of the useful sun rays in the winter; however, sun screens on windows that are subjected to winter winds (usually from the west) reduce heat loss considerably in the winter as well as heat gain in the summer. Sun screens are commercially available for about $3.50/per square foot of glass surface.

Use internal shades, venetian blinds, curtains or sun screens (similar to the exterior type which can be removed in the winter) to reduce solar heat gain in the summer and still permit direct radiation to enter in winter. To install venetian blinds costs about $1.60/sq.ft. in small quantities down to about $1.00/sq.ft. for 400 windows, or more.

Painting the exterior surface of the exterior walls with white coatings to increase reflectivity is not so effective as increasing the emmissivity of the exterior roof surface. Using light colored coatings or cooling the roof with a roof spray are two options. However, measures to increase emissivity and reduce cooling loads for exterior surfaces which have a mass of 50 to 100 lbs/sq.ft., and/or which are insulated to a U factor of .1 or lower are ineffective. It is also difficult to maintain the light color on roofs in urban areas.

ECO-21 REDUCE INTERNAL HEAT GAIN

Internal heat gain (from lights, business machines, computers, motors, occupants, and in some buildings, ovens, ranges and refrigeration units) make up a major portion of the cooling load in many buildings. Except in small restaurants which can have major heat gains from cooking equipment, heat gain from lighting is the most significant.

Guidelines to reduce illumination levels, in the many buildings which are overlighted, to more reasonable levels are described in Section 9, "Lighting". The reduction of lighting is a simple and effective way to reduce the cooling load. The following example indicates the potential savings for many buildings, from decreasing cooling load by reducing lighting.

Example:

First determine the approximate length of the cooling season based on past experience.

From the Identification Profile:

Building Type	-	Office
Building Location	-	New York
Building Size	-	50,000 sq. ft.
Lighting Intensity	-	4.5 watts/sq.ft.

The building has a total of 1,400 fixtures with 4-40 watt lamps in each. If 2 lamps are removed from 75% of the fixtures, then the reduction in watts = 1,400 x .75 x 2 x 40 = 84,000 watts or 84KW. The present total wattage = 225,000 and the new total = 225,000 - 84,000 = 141,000 watts or $\frac{141,000}{50,000}$ = 2.8 watt/sq.ft.

The heat gain from lights is reduced by 84KW for each hour of operation during the cooling season, or 84 x 3,413 = 286,692 Btu/hr.

FIGURE 22: SHADING COEFFICIENTS

GLASS	
1/8" Clear Double Strength	1.00
1/4" Clear Plate	0.93 - 0.95
1/4" Heat Absorbing Plate	0.65 - 0.70
1/4" Reflective Plate	0.23 - 0.56
1/4" Laminated Reflective	0.28 - 0.42
1" Clear Insulating Plate	0.80 - 0.83
1" Heat Absorbing Insulating Plate	0.43 - 0.45
1" Reflective Insulating Plate	0.13 - 0.31

SHADING DEVICE	WITH 1/4" CLEAR PLATE GLASS	WITH 1" CLEAR INSULATING GLASS
Venetian Blinds - Light Colored, Fully Closed	0.55	0.51
Roller Shade - Light Colored, Translucent, Fully Drawn	0.39	0.37
Drapes - Semi-Open Weave, Average Fabric Transmittance and Reflectance, Fully Closed	0.55	0.48
Reflective Polyester Film	0.24	0.20
Louvered Sun Screens - 23 Louvers/In.	0.15 - 0.35	0.10 - 0.29
- 17 Louvers/In.	0.18 - 0.51	0.12 - 0.45

If the cooling season is 100 days long, and occupancy is 8 hours per day, the yearly reduction in heat gain = 286,692 x 100 x 8 = 229 x 10^6 Btu/year. With a mechanical refrigeration system, the reduction in energy input = $\frac{229 \times 10^6}{2.5 \text{ (Average C.O.P.)}}$ = 91.6 x 10^6 Btu/yr or $\frac{91.6 \times 10^6}{3413}$ = 26,838 KW/yr

At 5 cents/KW, a savings of $0.05 x 26,838 = $1,342 for refrigeration (plus approximately $600/yr additional savings in fan and motor horsepower for the auxiliaries). If an absorption refrigeration system were used, the reduction in energy input = $\frac{229 \times 10^6}{.60}$ = 382 x 10^6 Btu/yr.

Assuming that heat is supplied to the absorption unit by steam boiler with seasonal efficiency of 65%, then the gallons of #2 oil saved = $\frac{382 \times 10^6}{.65 \times 138{,}000}$ = 4,459 gallons.

At $0.36 per gallon, the savings would be .36 x 4,259 = $1,533/yr

It is very important to keep in mind that a reduction in lighting intensity not only reduces the cooling load, but also reduces the building electrical consumption throughout the year.* In this case, the reduction amounts to 84 KW x 2,100 hrs/yr (occupied time) = 176,400 KWH at $0.05 KWH = $8,820.

Whenever motor operated equipment which is located in air conditioned spaces is operated for fewer hours, the cooling load is reduced by about .8 KW, or about 2,500 Btu's per horsepower hour. In office buildings, equipment consumes an average of 0.75 watts /sq. ft. of floor area. For a building of 50,000 sq. ft. calculate savings from a reduction in operating time from 8 to 7 hours per day as follows: 50,000 x 3/4 x 1 hr. = 37,000 watt hours/day. Multiply by a conversion factor to get equivalent Btu's: 37,500 x 3.4 = 127,500 Btu's per day. This is equivalent to about 10 ton hours, or 50 tons hours per week. If the air conditioning system uses 1 1/4 KW/ton, the savings for a 30 week cooling season would be 30 x 50 x 1.25 = 1,875 KW hours. In New York electricity is now about 5 cents/KW, so the savings there would be $94/yr. This may seem small compared to the savings which can be accomplished by reducing the lighting intensity, but the results of many small measures add up to significant totals. Calculate the savings in cooling energy for any building in the same manner.

*By re-arranging tasks and removing fluorescent tubes from non critical areas, it is usually possible to maintain the same footcandles on critical tasks. See Section 9, "Lighting", ECO's 31 and 34.

ECO-22 REDUCE CONDUCTIVE HEAT GAIN TRANSMISSION THROUGH THE BUILDING ENVELOPE

The conduction of heat from outdoors to indoors is only a significant portion of the cooling load when the building is one or two story with black uninsulated roof and has low internal heat gains. The difference between outdoor and indoor dry bulb temperatures is smaller during the cooling season (especially if indoor conditions are maintained at 78° or higher).

Insulation, added to reduce heat loss in the winter, will have a small added benefit in the summer. ECM- 2 includes a section on insulation and storm sash, roof sprays, and reflective roof and wall surfaces showing the relative benefits of each for both heating and cooling. Roof insulation will cost from $2.00 to $3.00 per sq. ft. depending upon type and construction details.

ECO-23 IMPROVE THE CHILLER AND COMPRESSOR PERFORMANCE

A building may have one or more types of chillers and/or compressors installed that may operate at the same time or sequentially. All have some common features, but there are differences, expecially between electric compression refrigeration systems and absorption chillers, which make each type unique. Identify the systems in the building and enter the information in the proper spaces, Figure 11. The characteristics and services of each type are listed below to assist in the identification process. Since most of the measures to improve chiller and compressor performance should be done by competent service or engineering personnel, seek their help at this time if your own maintenance department is not able to analyze and modify the equipment operation and maintenance.

The performance, or seasonal efficiency, of an existing refrigeration system can be increased by 1) changing the mode of operation and the operating conditions to conform more closely to the part load conditions which are the rule, rather than the exception, in virtually all buildings 2) improving maintenance and service procedures.

1. Types of Mechanical Compression Refrigeration Systems

Mechanical compression refrigeration systems all have compressors to raise the gas refrigerant pressure and temperature, condensers to reject the heat of compression and change the state of the refrigerant from a gas to a liquid, and evaporators to absorb heat from the refrigerant to chill water or air.

a. Compressors may be:

   Reciprocating
   Centrifugal
   Positive displacement screw
   Fully or partially sealed hermetic compressors

b. Compressors may be driven by:

   Electric motors
   Diesel or gas engines
   Steam Turbines

c. Condensers may be:

   Water-cooled shell and tube
   Water-cooled evaporator type
   Air-cooled

d. Evaporators may be:

   Shell and tube water chillers
   Direct expansion coils in HVAC duct systems or air handling units.

If the present refrigeration system includes steam turbine-driven or gas or diesel engine-driven mechanical chillers, consider reclaiming waste heat from the exhaust steam or the hot exhaust gases and using this as the source of energy for an absorption machine or to heat domestic hot water. Because this suggestion involves capital expenditure, it will be dealt with more fully in ECM- 2.

Unitary air conditioning systems (such as window air conditioners, through-the-wall units, packaged heat pumps, and 3 to 150 ton self-contained packaged air conditioners) have electric-driven compressors; the smaller sizes are equipped with reciprocating or hermetically sealed compressors, and the larger sizes with centrifugal or positive screw displacement type. The incorporation of compressor, condenser, evaporator (cooling coils), filters and supply fans in one insulated casing, characterizes all of these units.

The same types of components are often used in "built-up" systems. Each component or group of components may be installed either in one package or in locations which are remote from each other but inter-piped to form an operating unit.

Larger "central-station" systems include reciprocating compressors (of up to 150 tons capacity) and/or centrifugal compressors (from 100 to 8,000 tons in size), screw type compressors, and, as later described, absorption chillers. Each has separate air or water-cooled condensers, and separate air or water-cooled condensers, and separate air handling unit or fan coil units with filters and cooling coils. They serve one area directly, or multiple areas through a duct system. Any of the three types of drives (listed in b above) may be used with all compressors and chillers except the absorption type.

Large chillers with double bundle condensers and/or evaporators are large heat-pump type installations, The C.O.P. of the refrigerating machine can be measured as:

$$\frac{\text{Heat absorbed in evaporator}}{\text{(Heat rejected in condenser-Heat absorbed in evaporator)}}$$

Typical values for C.O.P. range from 2 to 5 at full load. Air-cooled condensers are in the lower range. When refrigeration units are operated as heat pumps, the C.O.P. increases, since the heat rejected from the condenser is put to useful work for heating.

The C.O.P. is related directly to evaporating and condensing temperatures which, for a water chiller with cooling tower, are typically 40°F. and 100°F. respectively. If the evaporating temperature can be raised (by using chilled water at a higher temperature of, for instance, 50°F., instead of 40°F) and/or if condensing temperatures can be reduced, then the C.O.P. will increase and a greater cooling effect will result from the same power input. Consider each individual refrigeration machine separately to determine the extent to which its C.O.P. can be increased; in general, however, raising the chilled water temperature 10° and reducing the condensing water temperature 10° will result in an increase in efficiency of approximately 20 or 25%. At part load conditions, this increase in efficiency is more marked; on a seasonal basis it will have a greater effect in conserving energy than would be indicated by consideration of full load operating conditions only.

2. Characteristics of Absorption Chillers:

Absorption chillers achieve a cooling effect without the use of mechanical compression; instead, they use heat directly as the driving force. They normally include a heat activated generator-absorber, shell and tube condenser, and shell and tube evaporator. Water is used as a refrigerant with an absorbant such as lithium bromide. The C.O.P. of an absorption machine is not as favorable as that of a mechanical chiller and will normally be in the order of .67. Absorption machines are particularly sensitive to condensing temperature and will show a good improvement in C.O.P. at lower condensing temperatures, but take care - if condensing temperatures are too low the absorbant for most existing absorption units will crystallize. If the absorption machine is operated from waste heat, run it (to be more economical) as close as possible to its maximum output and modulate other mechanical refrigeration machines, if installed, according to load. Refer to manufacturers for information on specific machines before implementing a program to lower condensing temperatures.

3. Improving the Efficiency by Raising Evaporator Temperatures

The evaporator temperature is a function of operating mode and maintenance. All measures to reduce the loads, as discussed earlier in this section, provide opportunities to raise the temperature for the peak loads, and present even greater opportunities for part load conditions. Adjustments for load often permit an increase in the supply air temperatures even with a reduction in the total amount of air circulated. This will increase the evaporating temperature and suction temperature of a direct expansion system, and conserve power for the compressor. Raising supply air temperatures with chilled water systems permits higher chilled water temperatures, higher evaporator temperatures, higher suction temperatures, and a reduction in power used by the chiller.

Systems with an inadequate charge of refrigerant and fouled evaporator surfaces have a greatly reduced refrigerating capacity and waste energy.

4. Increasing Efficiency by Lowering Condenser Temperatures

The condenser temperature is a function of the outdoor or ambient D.B. conditions (for operation of the air cooled condensers) and W.B. conditions (for operation of the cooling towers or evaporative condensers) and of the condition and operating mode of the air cooled condensers, cooling towers and shell and tube water cooled condensers.

Cooling towers, air cooled condensers, and evaporative condensers were usually selected initially, to provide a given condensing temperature for maximum expected outdoor conditions; consequently, they will provide lower condensing temperatures when outdoor conditions are below the maximum expected level. Maximum cooling load, however, usually occurs at the same time as maximum outdoor conditions. Rather than allow chiller operation to follow the load, it is often more effective to operate chillers at full load in the morning - when outdoor wet bulb temperatures are low, and low condensing water temperatures can be obtained from the cooling tower.

5. Description of Cooling Tower Operation

Cooling towers lower the temperature of condenser water by direct evaporation of the water to outdoor air. The condenser water is sprayed over a series of baffles or fill, and then drains by gravity into a sump. Outdoor air is drawn through the tower and psses over the fill and is then discharged to the atmosphere. The intimate mixing (though counter flow of air and water) promotes evaporation of the condenser water, increasing the moisture content of the air. Each pound of water evaporated removes 1,000 Btu's of heat from the condenser water system. The rate of evaporation is directly affected by the wet bulb temperature of the incoming air and the condenser water temperature. The difference between these temperatures is known as the "approach" temperature; cooling towers are commonly sized for 10°F. approach (i.e. if the design outdoor wet bulb is 75°F., the lowest temperature condenser water that can be obtained from the cooling tower at its full rating will be 85°F.). Any reduction in condenser water flow rate or air flow rate through the tower, or any fouling or blocking in the fill will reduce the tower's effectiveness and increase the approach temperature, thus increasing the condenser water temperature, which in turn lowers the chiller efficiency.

Because water is constantly being evaporated in the tower, total dissolved solids in the condenser water system increase and promote scaling at the spray nozzles and on the baffles and/or fill. Scale formation on the spary nozzles will not only reduce the quantity of water flow, but will also inhibit the fine atomized spray necessary for evaporation. Clean spray nozzles carefully to remove all scale and dirt. Remove scale formation on the fill either manually (by chipping away) or chemically. Correct rates of blow down will hold total dissolved solids in the condenser water system to a tolerable level and correct water treatment will prevent scaling both in the tower and in the refrigeration machine.

If the cooling tower is located in an area that experiences strong sunshine, there is a danger that rapid algae growth will clog spray nozzles and coat the fill, reducing the tower's efficiency. If the cooling tower is contaminated with algae or bacterial slime, have it thoroughly cleaned with chlorine and flushed through to remove all deposits; then institute periodic treatments with algicides. To inhibit the growth of algae, it is helpful to shade the cooling tower from direct sunlight.

In addition to increasing the efficiency of the compressor or chiller system by improving the performance of the cooling towers and air cooled condensers, it is also possible to reduce the energy consumption of the tower itself. The opportunities are listed in the summary of guidelines later in this section.

If the existing cooling towers are at or near the end of their useful life, consider replacing them with larger or more efficient cooling towers that will give lower approach temperatures. As this involves capital expenditure, it will be dealt with fully in ECM- 2.

6. Description of Air-Cooled Condensers

Air-cooled condensers discharge heat to a flow of air through a finned coil containing the hot refrigerant gas. the rate of heat rejection is directly affected by the dry bulb temperature of the air stream and by the efficiency of the heat transfer surface. In geographic locations that experience long periods of high dry bulb temperatures, the efficiency of an air-cooled condenser can be increased by using the cool exhaust air from the building as a source of cooling air for the condenser.

Because it forms part of the refrigerant system, the air-cooled condenser (with its connecting pipework) imposes a resistance to refrigerant flow and, therefore, increases the pressure at which the compressor must operate. Condensers are frequently installed in locations remote from the refrigeration compressor, but often they can be easily relocated to reduce the length of connecting pipework. Any reduction in pipework length will decrease the loss, and increase the efficiency of the refrigeration machines.

Air-cooled condensers, serving process refrigeration equipment, such as food display cases (See Section 8 "Commercial Refrigeration") are frequently located with the refrigeration compressors in a store room where heat builds up and reduces the efficiency of condensing units.

7. Maintenance Procedures:

Each individual piece of equipment that comprises a chiller system has built inefficiencies. If they are minimized, the total efficiency of the unit will increase. Leaks from the refrigerant high pressure side of the system will reduce the refrigerant charge and, hence, the refrigeration effect that can be obtained for a given power input. Leaks on the low pressure refrigerant side (if the pressure is sub-atmospheric) will allow the entry of air into the refrigerant system. Air is comprised of non-condensable gases and will reduce both the rate of heat transfer of the condenser and evaporator and, again, the refrigeration effect available for a given power input. Check the refrigerant system first, therefore, to ensure that it contains a full charge of refrigerant and that all non-condensable gases are removed. Then check the system for leaks, using a Halide lamp or other similar equipment. Common sources of leaks include shaft seals, inlet guide vane seals, valves and pipe fittings.

Compressor prime movers and drive trains, if poorly maintained, can absorb as much as 15% of the total energy input into the compressor. Examine speed reducing gear boxes for quality and quantity of lubricating oil, gear backlash and wear, thrust bearing condition and main bearing condition. Check "V" belt drives for correct tension. Replace frayed belts or belts with a frayed driving surface. Where multiple belts are used, all belts should be replaced at the same time, or different tensions will result, causing loss of transmission efficiency.

Maintain water-cooled shell and tube condensers to give the greatest rate of heat transfer. Heat transfer is inhibited if the tubes become fouled with scale or bacterial slime. Reducing the condenser fouling factor from .002 to .0005 will result in approximately a 10% increase in efficiency. Where it is accessible, remove scale from the tube surface by chipping it, by chemical means, or by a combination of both. Once the tubes are clean and free of scale, enforce a policy of correct water treatment. The appropriate type of water treatment depends largely on the quality of water, which varies in different geopraphic locations. Remove bacterial slime from the condensers by flushing and chemical treatment and prevent its reappearance by periodic shock treatments with bactericides and algicides such as chlorine. Select the frequency and type of treatment to suit local conditions. By maintaining the heat transfer surfaces in shell and tube condensers, a supermarket in Atlanta, Georgia, of 50,000 sq.ft., equipped with a 200 ton centrifugal electrical chiller could reduce the annual power requirements from 600,000 to 540,000 KWH. The maintenance program would entail cleaning the surface twice yearly to reduce the average fouling factor to .0005. At 4 cents/KW, a savings of 10% in annual energy usage saves $2400 per year.

Water-cooled shell and tube evaporators are not as prone to scale or bacterial fouling as condensers, but they should be inspected and, if necessary, cleaned. If the fouling factor of a water-cooled evaporator decreases from .002 to .0005 the efficiency of the machine will increase by approximately 14%.

Direct expansion evaporator coils, installed in duct systems, quickly become fouled with dust, particularly if the filtration systems are ineffective. Inspect these coils on a periodic basis and clean them with steam or compressed air jets.

GUIDELINES TO REDUCE ENERGY USED FOR COOLING

USE OUTDOOR AIR FOR COOLING

-Use outdoor air for economiser cooling whenever the enthalpy is lower than room conditions during occupied periods, and whenever the dry bulb temperature is 5°F lower than indoor design conditions during unoccupied periods. (If the dry bulb temperature is only 2 to 3°' lower than room temperature, heat from absorbed fan horsepower will eliminate the value of outdoor air cooling.)

-Use operable windows, without fan operation, for outdoor air cooling. Outdoor temperature during unoccupied periods, should be below room D.B. and during occupied periods below room D.B. and W.B. conditions. (Unfavorable acoustics and air quality may preclude implementation of this option.)

-If cool outdoor air is available, consider cooling and building well below normal during the night and early morning hours preceding any day that is expected to be extremely hot.

-Whenever the volume of outdoor air is increased for economiser cooling, and if there is no exhaust system which can handle an equal quantity of air, provide pressure relief.

REDUCE VENTILATION RATE DURING THE COOLING SEASON

-Refer to guidelines to reduce ventilation rates during the heating season.

-When the enthalpy of the outdoor air, on a seasonal basis, is lower than room conditions, outdoor air dampers should be fully opened. Close them only at times when the enthalpy of the outdoor air is higher, or dry bulb temperature of outdoor air is above 85°F.

-Check local and state codes to determine if ventilation rates can be legally lowered when spaces are air conditioned.

-Do not reduce ventilation rate when outdoor air can be used for economiser or enthalpy cooling.

-Refer to ECM- 2 for energy recovery "heat" exchangers for latent and sensible heat exchange between exhaust air and outdoor air for ventilation.

REDUCE INFILTRATION RATES DURING THE COOLING SEASON

-Before investing any money to reduce infiltration rates, perform an analysis for both heating and cooling.

-Whenever infiltration rates are reduced for winter conditions, additional benefits accrue from reducing the cooling load in the summer.

-In areas of the country which experience fewer than 10,000 wet bulb degree hours, energy conserved by weatherstripping windows and doors and/or caulking window frames does not generally justify the cost. In climatic zones which experience a greater number of wet bulb degree hours, make an engineering analysis of weatherstripping and caulking to reduce cooling loads.

-The rate of infiltration in stores through door openings may be considerably higher than leakage through windows or cracks. Refer to ECM- 2 for infiltration rates through doors and an analysis for vestibules and revolving doors in all climates.

-Refer in Section 4, "Heating" to "A Summary of Guidelines for Reducing Infiltration Rates" for additional guidelines.

CONTROL SOLAR HEAT GAIN

-In hot weather, adjust existing blinds, drapes, shutters or other shading devices on windows to prevent penetration of solar radiation into the building. (In the case of certain types of shading devices, this modification may conflict with requirements to utilize natural illumination; an engineering analysis of relative energy consumption due to solar radiation loads versus artificial illumination loads may be required).

-Install blinds, drapes, shutters, or other shading devices on the inside of all south, east and west facing windows which are subject to direct sunlight in hot weather and/or exposed to a large expanse of sky. (Fire codes may limit the use of some materials. Where dependent on natural light, do not reduce available light below statutory limits.)

-Use lightweight drapes, with reflective properties for effective solar radiation control. (Again, check fire codes before selecting appropriate material.)

-Use vertical or horizontal reflective blinds. (Vertical blinds or louvers generally are most effective on the west and east sides of a building; and horizontal blinds are most effective on the south side.) PVC vertical louvers cost between $2.90 and $2.30 per square foot.

-Add a reflective film coating to the inside surface of glazed areas on the south, west and east windows. (Where dependent on natural light, refer to occupational regulations or health and safety codes before drastically reducing it.)

Note: Reflective mylar sheets (or rolls) and plexiglass sheets are available from a number of manufacturers at relatively low cost. Not only do they help control solar radiation, but they may also increase the strength and resistance of glazed areas. The thicker materials are good wind insulators - a 1/4 inch sheet of plexiglass is as good as a single pane of glass. Available in different colors, the films, depending on quality and make, allow a maximum transmission of from 55 to 90 Btu's/hr/sq.ft. The maximum heat transmitted in Btu/sq.ft. for a clear single pane of glass is 215 Btu/hr/sq.ft. The films transmit 9 to 33 percent of the visible light spectrum and reflect 5 to 75 percent of the solar radiation which strikes them.

For example, reflective coating on the south, east or west exposure glass that reduced solar radiation by 50 percent would save from 30,000 to 50,000 Btu's.sq.ft. window per season, in hot weather, for energy required for space cooling. (The value is higher in northern latitudes and lower in southern latitudes because of the angle of the sun striking vertical surfaces.) The cooling load can be reduced by about 3 ton hours/sq.ft. of glass per year by proper use of shading devices. In southern climates, the north facing glass can receive a surprising amount of diffuse solar radiation. If heat gain from north windows is excessive, treat them similarly to the other exposures.

-It is not cost effective to add storm sash solely to reduce solar heat gain.

-Skylights on a roof transmit between two and four times as much solar heat in the summer time as an equal area of east or west facing glass. Exterior solar control over skylights is most effective, and permits solar radiation and daylight to enter when desirable. Interior shades with mylar reflective coatings cost about $3 per sq. ft. to install and reduce solar heat gain in the summer by up to 80%. White paint on the exterior of skylights, venetian blinds or sun screens can also be employed to reduce solar heat gain.

-Do not prune trees which shade the building in the summer.

REDUCE INTERNAL HEAT GAIN

-Turn off unnecessary lights and heat producing equipment.

-Reduce lighting levels by removing lamps. See Section 9, "Lighting", ECO's 31 and 35.

-Exhaust the heat from ovens, ranges, and motors directly outdoors when the enthalpy of the outdoor make-up air is lower than the enthalpy of the space.

-Refer to Section 8, "Commercial Refrigeration" for guidelines for supermarkets and other areas with commercial refrigeration.

-See "A Summary of Guidelines to Reduce Usage and Operating Time", in Section 10, "Power".

-Insulate hot surfaces of tanks, piping and ducts which are in air conditioned spaces.

-Disconnect dry type transformers which are in conditioned spaces when there are no operating loads.

REDUCE CONDUCTIVE HEAT GAIN THROUGH THE BUILDING ENVELOPE

-In climates without a significant heating season adding a storm sash or double glazing is not usually economically feasible.

-Do not confuse solar radiation heat gains through windows with conductive gains . Solar radiation control cannot be neglected.

-Make an engineering and economic analysis before insulating walls or roof, installing a roof spray, or treating exterior surfaces to increase emissivity.

-Insulate roofs, or ceilings below roofs, which have a U factor of greater than .15 to improve U factor to .06.

CONSERVE ENERGY BY OPERATION AND MAINTENANCE

1. Increase Evaporator Temperature -

-Raise supply air temperature

-Raise chilled water temperature

-Operate one of multiple compressors and chillers at full load, rather than two or more at part loads.

-Maintain full charge of refrigerant.

-Maintain evaporator heat exchange surfaces in clean condition

-Maintain higher relative humidity levels in air conditioned space.

-See Section 7, "HVAC Systems and Distribution" for specific measures which will result in lower evaporator temperatures and for measures to re-schedule chilled water temperatures.

-Clean all cooling coils, air and liquid sides.

2. <u>Reduce Condensing Temperatures</u> -

-Clean all condenser shells and tubes

-Clean all air cooled condenser coils and fins on a regular basis with compressed air or steam jets.

-Remove obstructions to free air flow into cooling tower and fans.

-Direct cool exhaust air from the building into the air intake of air cooled condensers or cooling towers. Refer to ECM- 2 for details.

-Under light load conditions, when the refrigeration load is small and the ambient wet bulb temperature is likely to be low, the cooling performance of the tower will exceed the needs of the refrigeration machines. Under these conditions the cooling tower fan or fans can be cycled on and off to maintain a desired condenser water temperature, thus saving the horsepower required to drive the fans.

-Adjust air flow and water rates to air cooled condensers and cooling towers to produce the lowest possible condensing temperature. Generally, the savings in refrigeration power will exceed any increase in added power for condenser fans or pumps.

-Use well water, if available, for condenser cooling.

-Shade cooling towers and air cooled condensers from direct sunlight.

-Remove bacterial slime and algae from cooling towers.

-Institute and maintain a continuous water treatment program for cooling towers.

-At partial refrigeration loads, operate cooling towers as natural draft towers with cooling tower fans turned off.

-Clean and descale spray nozzles, and descale fill in cooling towers.

-It is often more effective to operate chillers at full load in the morning when outdoor wet bulb temperatures are low, and low condensing water temperatures can be obtained from the cooling tower. Then, use the machine to sub-cool the building, turn the chiller off when wet bulb temperatures rise and allow the building temperature to drift up. In large buildings, the extensive chilled water piping system provides a degree of thermal storage.

3. Improve the Efficiency by Improved Maintenance Practices -

-Repair refrigerant leaks

-Lubricate speed reducing gear boxes

-Replace worn bearings

-Maintain proper tension on "V" belt drives

-Adjust water flow control valves to maintain lower condensing temperatures.

-If well water is used for condenser cooling, provide proper treatment to prevent scale build-up.

-Clean and replace, as necessary, all strainers to reduce resistance to refrigerant or water flow.

-Select a water treatment system for cooling towers that allows high cycles of concentration (suggested target greater than 10.7) and reduces blow-down quantity.

-Maintain boiler and burner efficiencies where steam or hot water is generated for absorption cooling units. Refer to Section 4 , "Heating and Ventilation"

-Many or most of the options can be implemented by the building maintenance staff. An alternative, however, would be to let a service contract. With an Inspection and Labor type service contract for servicing simple air conditioning and fan coil units on an annual fee basis to cover costs of labor (for inspection, maintenance, breakdown repair) and material (such as filters, oil, grease, etc.), typical costs might be:

A/C Units

System Tonnage	2.5	5	7.5	10	15	20	25	30	40
Price	$ 168	202	253	308	410	495	595	685	865

Fan Coil Units

System Tonnage	1/2	3/4	1	1 1/2	2
Price	$ 25	35	45	70	95

-Confirm adjustments with mechanical and occupational codes.

Section 7

Energy Conservation Opportunities Distribution and HVAC Systems

DISTRIBUTION AND HVAC SYSTEMS

A. BACKGROUND

The "parasitic" distribution loads for individual or various combinations of heating, cooling and ventilation systems depend upon (1) the performance, for air systems, of the terminal room devices, air handling units and duct work systems, which, with the fan and motor performance influence the amount of energy required for air; (2) the terminal devices and piping for hot water systems, which, with pump and motor performance, influence the amount of energy required for hot water circulation and (3) the piping and appurtenances for steam systems, which impose a heating load on the burner-boiler system.

The same system may handle the heating, cooling, and ventilation loads at separate times or concurrently, to serve common or separate areas of a building.

Consider the opportunities for HVAC systems and distribution in the following distinct but related areas:

a) Improve the performance of the terminal devices to reduce their resistance to fluid flow and increase their heat transfer characteristics.

b) Lower the resistance to flow in duct and piping systems to reduce the required horsepower for fans and pumps.

c) Modify the control systems and modes of operation of air handling and piping systems to reduce simultaneous heating and cooling.

d) Decrease fluid leaks and thermal losses from piping, air handling equipment, and other vessels holding hot or cold water, air, or steam.

e) Improve the performance of fans, pumps and motors by maintenance and operating procedures.

f) Reduce the hours of fan and pump operation.

If boilers are used to generate hot water or steam for any HVAC systems in your building, refer to "Heating", ECO 10 for guidelines on the primary energy-conversion equipment and ECOs 26 and 27 below for guidelines on piping, pumps and appurtenances.

Refer to Section 6, "Cooling", ECO 23 for suggestions for compressors, chillers, condensers and cooling towers and for cooling equipment not included in this section, and refer to ECO 25 below for guidelines on duct work and fans.

Even though an individual building does not have all of the systems discussed, read all of the ECOs and guidelines. Many energy conservation opportunities are common to a number of systems, but for simplicity, have been noted only once. For instance, thermal losses through the casings of air handling units, mentioned under "Single Zone Duct Systems", are valid for each type of system in ECO 24.

Before making changes to equipment, control cycles, or air or water flow quantities, and if the building personnel are not experienced or qualified to implement them, consult an air conditioning and heating service maintenance company, or an HVAC engineer to analyze proposed changes and perform the work.

B. ENERGY CONSERVATION OPPORTUNITIES

ECO 24 REDUCE ENERGY CONSUMPTION THROUGH OPERATION AND/OR MAINTENANCE OF SPECIFIC HVAC SYSTEMS.

a) Direct Hot Water or Steam Systems employ direct radiation, fin tube convectors, fan coil units, cabinet heaters, or fan driven horizontal or vertical unit heaters with hot water, steam or electric coils.
Fan coil units, and induction units equipped with heating coils, are used for both heating and cooling.

b) Single Zone Duct Systems deliver warm air from a furnace (see Section 4 , "Heating", ECO 10) or from an air handling unit or section of a package air conditioner. The same system, when equipped with cooling coils, is used to deliver cool and dehumidified air as well as warm air through common duct work.

Control cycles may provide fluids (hot water, steam, refrigerant or chilled water) to the coils continuously, may modulate the flow, or may cycle it on and off. At the same time, the fan may be on continuously, modulating, or cycling on and off in accordance with room temperature demands and fluid flow. The characteristics and performance of each sytem vary widely and preclude a general rule for the optimal control cycle, but any measure which prevents simultaneous heating and cooling, and delivers warm or cool air to spaces in accordance with thermal needs, rather than in bursts of over-heated and over-cooled air, is more efficient and provides more comfort to occupants.

Fan and motor performance are discussed in ECO 25 below and Section 10, "Power", ECO 41, respectively.

c) Terminal Reheat Systems are commonly used and very wasteful. Basically, supply air is cooled to a fixed point (usually 55°F.) The cool air is supplied through ducts to all outlets, individually or in groups, where it is then reheated to a temperature compatible with the demands of that space. Under partial or light loads, most or all of the re-heat coils will still be operating because the supply air is too cold for direct use in certain parts of the building. Only under peak cooling loads do the majority of reheat coils turn off.

To conserve energy, reduce the supply air quantity and reset the temperature to a higher level. Raise the temperature in increments of 3°F. to determine the highest supply temperature which will maintain satisfactory room conditions.

-Terminal reheat systems were usually installed to control dehumidification. Wherever dehumidification can be eliminate and zone control can be satisfactorily maintained (or is not needed), operate the terminal reheat system on a temperature demand cycle only; considerable energy will be saved.

Refer to ECM 2 for further conservation options by major modifications to, or replacement of, terminal reheat systems.

d) Dual-Duct Systems, as the name implies, have two ducts, one trunk to supply warm air and the other to supply cool air to mixing boxes which then deliver air in one duct to the conditioned space. Warm and cool supply air is proportioned in the mixing boxes at the point of use, to provide supply air to the room at the temperature which meets the thermal demand. This type of system, while not usually described as a terminal reheat system, does, in fact, operate essentially on the same principles. Air is cooled down to a fixed level and then reheated to the point of use by being mixed with warm air. It is, inherently, a wasteful system.

e) A Multi Zone Unit, with the exception of new ones which incorporate recent changes, normally consist of a fan which discharges air either through a cooling coil to a cold deck or through a heating coil to a hot deck. Both hot and cold decks are equipped with modulating dampers that automatically control the quantities of hot and cold air to one supply duct serving the space. This system is also a form of reheat; the overheating and overcooling of the supply air carries the penalty of inherent waste of energy.

In many units the dampers, which control the quantity of hot or cool air, leak. As the system compensates for the additional cool or hot air it wastes energy.

f) Induction Systems, normally found in large office buildings, include an air handling unit which supplies cooled or heated primary air at high pressure to individual induction units normally located on the outside wall in each room or zone. The flow of high-pressure air through a nozzle system in the unit induces a flow of room air through a cooling or heating coil unit. The room air then mixes with the primary air to provide supply air at a temperature to meet the heating or cooling load within the space. Typically, the ratio between primary air and induced air is 1:4.

The major causes of high energy use with induction systems are:

1) The horsepower to maintain the high pressure to deliver primary air and induce room air flow through the unit.

2) Simultaneous heating and cooling in a "terminal reheat" type of cycle.

3) Restriction in air passages impeding the flow of room air through the coils.

4) Dirty coils which reduce heat transfer to room air, and also impose an extra load on the primary air system to compensate for the reduced induction effect.

g) Variable Air Volume Systems (VAV) include an air handling unit which typically supplies either heated or cooled air at constant temperatures to variable air volume boxes for each zone or for each individual outlet. Variable air volume boxes regulate the quantity rather than the temperature of the warm or cool air supplied to the space in the heating or cooling season, respectively. Varying air quantity with the appropriate supply temperature is one of the most efficient modes of operation in terms of energy usage. Run the fan at full volume only when all VAV dampers are fully opened.

Fan volume is not controlled in some systems. When full volume is not needed in the conditioned spaces supply air is dumped into a ceiling plenum the air is returned to the supply unit by inlet openings in the return air side of the air handling unit. This "dump" system is necessary because virtually 100% of the supply air passes over direct expansion coils. This VAV system is not as efficient as a fan controlled type but it does save energy by effectively reducing the static pressure of the system.

h) Fan Coil Systems are comprised of one or more fan coil units set up to be two, three or four pipe systems; each unit contains a fan which blows air through heating and/or cooling coils into a single outlet to supply a zone or room. Typically, the heating and cooling coils are controlled to maintain a supply air temperature to meet zone heating or cooling loads. However, in many systems the fan speed can be manually controlled - high - medium - low - to control heating or cooling output. Often fans are operated at low speeds to reduce noise.

For those fan coil systems which have separate coils for heating and for cooling, it is important to prevent simultaneous heating and cooling, and a "terminal reheat" type of operation.

i) Window and Through-the-Wall Air Conditioners have similar self-contained compressor air-cooled condensing units. They may be equipped with electric coils for heating, or, in the case of through-the-wall units, coils supplied with hot water or steam from a central source.

Because both systems are relatively inefficient during the cooling cycle (due to small - usually undersized-condensers), maintenance of coil surfaces, previously discussed for other types of equipment with evaporator and condenser coils, and reduction of operating hours are very important.

Most of the older unitary conditioning units have a low cooling EER (Equivalent Efficiency Rating) - sometimes as low as 5 or 6 Btu's/Watt. If these units are at or near the end of their useful operating life, replace them with new units having an EER rating of 9 or more. New units now in the development stage, will produce an EER of 15 or more. Units with electric resistance heating coils for winter use, consume a large amount of energy for tempering or heating. If the existing units to be replaced are equipped with electric resistance heating coils for winter use, exchange them with air-to-air heat pumps, which provide for 1.75 to 3 times as much heat per KW input as resistance heaters. Outdoor air blowing into the room directly through the coils of the unit or through cracks around the frame of window air conditioners and through-the-wall units, can be a major source of infiltration and heat load in the winter. If not used for heating, remove window units and store them in the winter, or caulk cracks and cover units with fitted enclosures made for that purpose.

j) Unitary Heat Pumps are either of the air-to-air, or water-to-air type. The older air-to-air models have the ame built-in condenser and evaporator inefficiencies as room air conditioners (refer to i) above). The water-to-air heat pump operates at lower condenser temperatures which are more efficient. All guidelines for condensers, compressors, evaporators, and fans apply to both types.

During the cooling season, improve the operating efficiency of air-to-air heat pumps by shielding the condenser air intakes from direct sunlight and directing cool exhaust air from the building to the condenser air inlet.

Where possible, follow the options for pumps, piping, fans and duct systems outlined in ECO's 25, 26 and 27 below for heat pump installations.

ECO-25 REDUCE ENERGY CONSUMPTION FOR FANS

To the extent possible, reduce resistance to air movement throughout the HVAC system, reduce quantity of air supplied to or returned from conditioned spaces, and limit the periods during which fans operate. Ordinarily, the three steps provide for maximum savings when taken sequentially.

Implementing energy conservation methods which reduce the heating and/or cooling loads first will then permit reducing the volume of air handled by the HVAC systems. Whenever the volume of air is reduced, the resistance to air flow in that section of the system is also reduced, since the resistance of any system depends upon the quantity of air circulated in the system as well as the characterisitics of the various portions of the system including:

a) Intake louvers and air outlet duct caps or elbows

b) Heating and cooling coils

c) Filters

d) Ductwork system, balancing dampers, and exhaust hoods

e) Registers, grilles, and diffusers

f) The fan scroll and housing

If, as is often the case, the original HVAC system has been oversized for the contemplated loads (due to a conservative "safety" factor in the design, or a built-in reserve for the future), or if the original loads have been reduced, the opportunities to reduce the air volume will exceed that which would result from a further reduction in "building" loads only.

Reducing the resistance to air flow in an existing system results in an increase of air delivery which, in turn, permits a reduction in fan speed to bring air volume down to the original level. Because for multi-vane centrifugal fans, (the type most commonly used in HVAC systems), the power input (BHP) varies directly with the cube of the speed, any reduction in speed, which is proportional to air volume, results in a very sizable decrease in power input.

The basic fan laws for centrifugal fans follow:

1) the volume varies directly with the speed

2) the pressure varies directly with the square of the speed

3) the power input (BHP) varies directly with the cube of the speed

4) the volume varies directly with the square of the pressure

If it is possible to reduce the volume of air in a system by 10%, the savings in power will be about 27%.

To conserve energy for fan horsepower, first reduce heating and cooling loads, second, reduce the resistance to air flow in each of six areas (a-f) listed above, third, measure the increased volume which results and determine the new air volume required to meet new loads, fourth, reduce fan speed accordingly, and fifth, change the motor if necessary. Use Figures 23 and 24 (for forward-curved and backward-curved multi-vane blowers, respectively) to determine potential savings.

FIGURE 23: YEARLY ENERGY CONSUMED CENTRIFUGAL FANS FORWARD CURVE BLADES

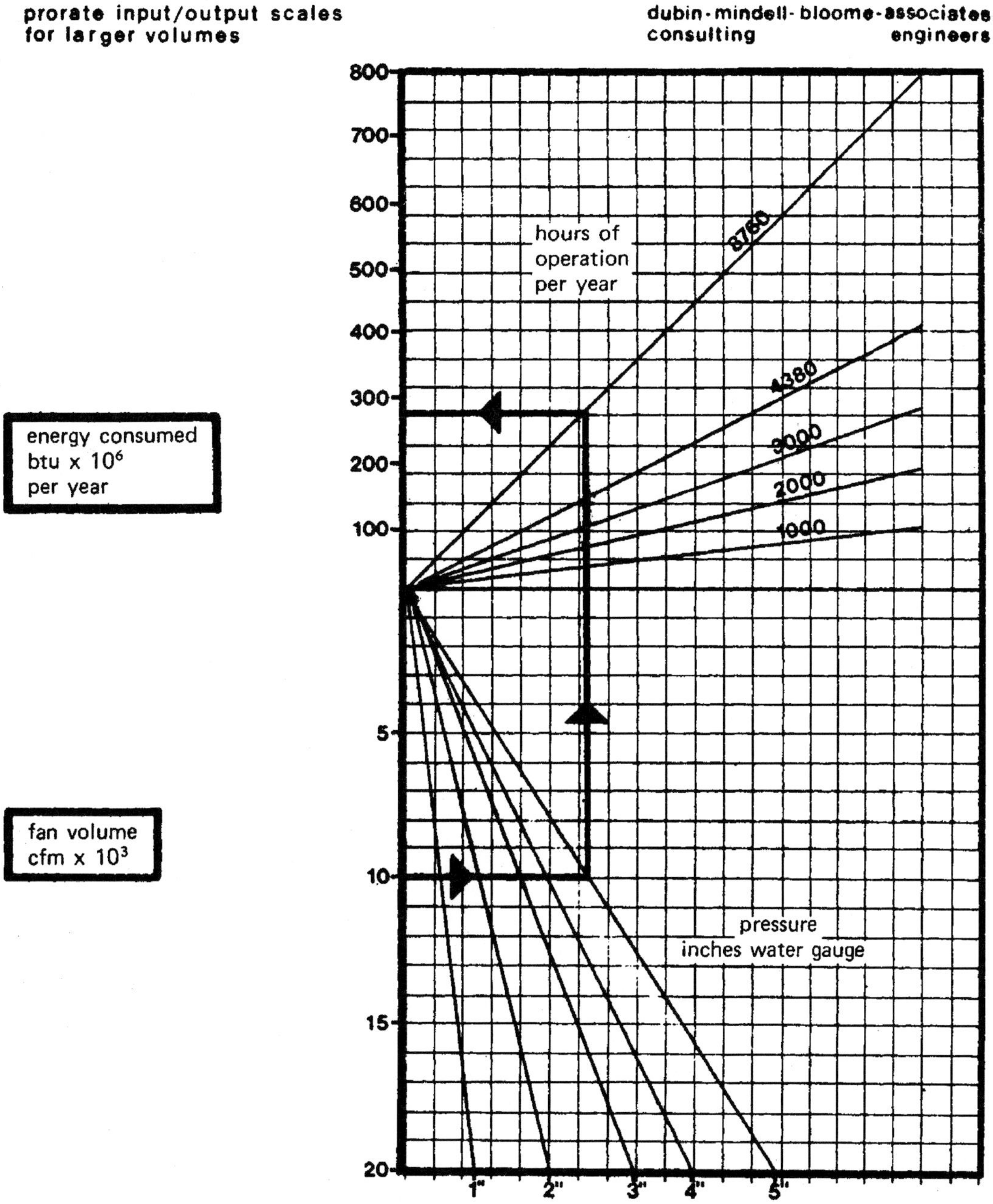

FIGURE #23 Engineering Data

Source: Manufacturers fan capacity tables for forward curve centrifugal fans.

FIGURE 24: YEARLY ENERGY CONSUMED–CENTRIFUGAL FANS, BACKWARD CURVE BLADES

prorate input/output scales
for larger volumes

dubin-mindell-bloome associates
consulting engineers

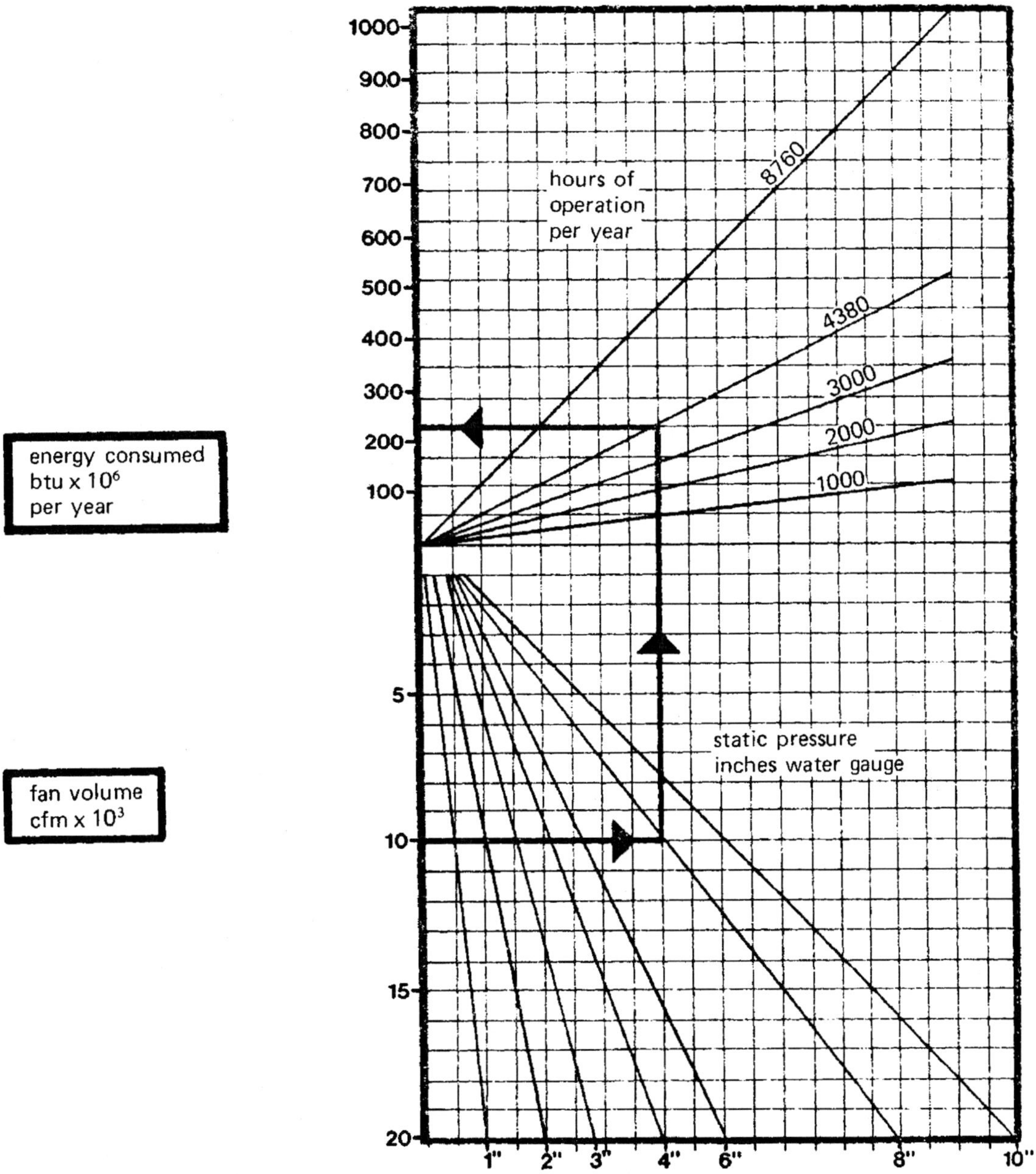

FIGURE #24 Engineering Data

Source: Manufacturers fan capacity tables for backward curve centrifugal fans.

<u>Example</u>:

A building has a heating load of 620,000 Btu/hr. and a fan volume of 15,000 cfm. Energy conservation measures have been implemented to reduce the heating load to 510,000 Btu/hr. If the temperature drop remains the same, the volume of air required to meet the reduced load is 12,340 CFM. In addition, system resistance has been reduced by modification to filters and heating coils.

Initial system characteristics were measured at 15,000 CFM and 4" w.g., point A in the diagram above. The measurements taken after lowering system resistance were 16,620 CFM and 3.8 "w.g., point B on the new system curve.

Now the fan speed is to be reduced to give the new required volume of 12,340 CFM. The new pressure at this volume is measured at 3.27", point C.

Assuming a backward curve fan operating 4,380 hours per year, use Figure 24 to determine energy savings.

Enter Figure 24 with the initial conditions of 15,000 CFM at 4" w.g. and read that yearly energy consumed is 170×10^6 Btu. Re-enter Figure 24 with the final conditions of 12,340 CFM at 3.27" w.g. and read yearly energy consumed at 110×10^6 Btu.

Savings: The energy saved equals 170–110 or 60×10^6 Btu/yr or -

$$\frac{60 \times 10^6}{3{,}413} = 17{,}580 \text{ KWH/year}$$

At \$0.03/KWH, the savings is 0.03 x 17, 580 or \$527 per year.

Results:	Energy Saved	17,580 KWH/year
	Dollard Saved	\$527/year

ECO-26 REDUCE ENERGY CONSUMPTION FOR PUMPS

Pumps used to circulate hot water, chilled water and condenser water are generally of the centrifugal type and are governed by the same basic laws of operation as fans. Any reduction to system resistance will result in energy savings, providing the pump flow rate is held at the same level. With variable speed pumps, adjust for any increase in pump flow rate due to reduction in system resistance by reducing the speed of the pump (as described for fans). With direct drive pumps (for which it is not so easy to reduce the speed), reduce the flow rate by replacing the impeller or reducing its' size.

Balancing valves are normally installed and adjusted toward the closed position to achieve the desired flow in any given circuit. The precise position of the balancing valves is a function of the resistance of the longest or index circuit.

If resistance in this circuit is decreased, all balancing valves can be opened proportionately, reducing the overall resistance to water flow and the pumping energy requirements accordingly. To realize these savings substitute sections of larger size pipes for portions of the index circuit. Boilers, chillers, heating and cooling coils, valves, and strainers all affect resistance to flow, as well. Refer to sections of the manual dealing with these types of equipment for specific suggestions.

In the case of strainers, the quantity of dirt they contain and the condition of the filtration media, often rusty, corroded, and deformed, determine resistance. Inspect these items, and clean or replace them on the basis of a regular maintenance schedule.

ECO-27 REDUCE DISTRIBUTION THERMAL AND PIPING LOSSES

In addition to frictional resistance, energy losses from piping systems occur in the form of heat gain to chilled water and heat loss from hot water or steam systems. Pipe insulation, which normally keeps these losses to a minimum, frequently becomes damaged. Inspect each piping system; where no insulation exists, add it, and where it is damaged, make all necessary repairs. The heat energy loss from bare pipe and insulated pipe can be calculated from Table 1. Determine the annual loss for any system by comparing the losses (per foot of pipe) for bare pipe to those for insulated pipe. Multiply the unit loss, first, by the total number of feet in the system; then, by the difference, in °F., between hot water temperature and ambient air temperature; and finally, by annual hours of operation.

TABLE 1: HEAT LOSS THROUGH PIPES

(Based on an ambient temperature of 68°F and a K for insulation = 0.3)

Pipe Size (In.)		Insulation Thickness						
	Bare Pipe	1/2	3/4	1	1 1/4	1 1/2	1 3/4	2
1/2	0.63	0.163	0.135	0.116	0.105	0.098	0.091	0.086
3/4	0.76	0.191	0.155	0.135	0.120	0.110	0.103	0.096
1	0.93	0.211	0.179	0.153	0.136	0.125	0.115	0.108
1 1/4	1.14	0.263	0.210	0.178	0.158	0.143	0.132	0.122
1 1/2	1.27	0.287	0.232	0.194	0.172	0.154	0.142	0.132
2	1.53	0.345	0.271	0.229	0.198	0.178	0.163	0.151
2 1/4	1.87	0.425	0.325	0.270	0.237	0.210	0.190	0.175
3	2.15	0.487	0.368	0.309	0.251	0.214	0.211	0/195
4	2.65	0.600	0.447	0.375	0.305	0.279	0.252	0.231
5	3.2	0.663	0.500	0.407	0.346	0.305	0.271	0.245
6	3.7	0.852	0.628	0.536	0.432	0.379	0.341	0.305
8	4.75	1.090	0.828	0.650	0.549	0.486	0.433	0.388
10	5.75	1.341	0.990	0.778	0.678	0.580	0.511	0.457
12	6.75	1.550	1.152	0.920	0.802	0.664	0.604	0.541

Estimated costs for insulating hot and cold water pipes with pre-sized fiberglass material with a standard jacket are listed in Table 2.

TABLE 2

Pipe Sizes	Price per L.F. Installed Insulation Thickness: 1"	1 1/2"
	$	$
1/2"	1.35	2.05
3/4"	1.40	2.10
1"	1.45	2.15
1 1/4"	1.50	2.20
1 1/2"	1.55	2.25
2"	1.60	2.35
2 1/2"	1.65	2.45
3"	1.70	2.50
4"	2.00	3.05
5"	2.25	3.15
6"	2.55	3.30
8"	3.15	4.20
10"	3.85	5.00
12"	4.50	5.60
14"	5.20	6.45
16"	6.00	7.20
18"	6.70	7.60
20"	8.25	8.50
24"	9.00	9.70

For costing purposes, add to total lineal feet of piping three (3) lineal feet for each fitting or pair of flanges to be insulated.

Example:

Bare pipe 4", 300' in length
Ambient air temperature: 60°F., hot water temperature: 200°F.
Operating time: 200 days, 24 hours a day
Heat loss from bare pipe: 2.65 Btu/ft/degree temperature difference
Annual heat loss for bare pipe = 300 x 2.65 x (200 - 68) x (200 x 24) = 504 x 10^6 Btu's

Example con't.

If hot water is generated by an oil-fired boiler with 65% seasonal efficiency, the energy used would be $\frac{504 \times 10^6}{.65 \times 138,000}$ = 5,619 gallons
Cost/yr. @ 36¢/gal. = 5,619 x .36 = $2,023/yr.
If 1-1/2" fiberglass insulation were used, the yearly heat loss would be 300 x .305 x (200-68) x (200 x 24)= 57 x 10^6 Btu/yr. The energy used would be Gal. of Oil= $\frac{57 \times 10^6}{.65 \times 138,000}$ = 647 gallons
Cost/yr = 647 gal. x 0.36 = $233
Yearly savings in fuel $2,023 - $233 - $1,790
Initial cost to install 1-1/2" insulation = 300 x $3.05/ft.= $915
Pay back period .51 years

TABLE 3: COSTS OF STEAM LEAKS IN A 125 PSI SYSTEM AT $4/1,000 LBS

Size of Office	Lb. Steam Wasted Per Month	(Per Month) Total Cost	(7 Months) Total Cost
1/2 inch	708,750	$2,040	$14,280
3/4 inch	400,000	1,600	11,200
1/4 inch	178,000	712	5,000
1/8 inch	44,200	176	1,232
1/16 inch	11,200	45	315
1/32 inch	3,000	12	84

For steam at 50 lbs. pressure, the waste is about 75% of the figures given; at 20 lbs. it is about 50%, and at 5 lbs, the loss is about 25%.

Steam traps are installed to remove condensate, air and carbon dioxide from the steam utlizing unit as quickly as they accumulate. With time, internal parts begin to wear and the trap fails to open and close properly. Among other undesirable effects on system performance and maintenance, malfunctioning steam traps reduce fuel economy and equipment capacity.

FIGURE 25: STEAM LOSS THROUGH LEAKING STEAM TRAPS

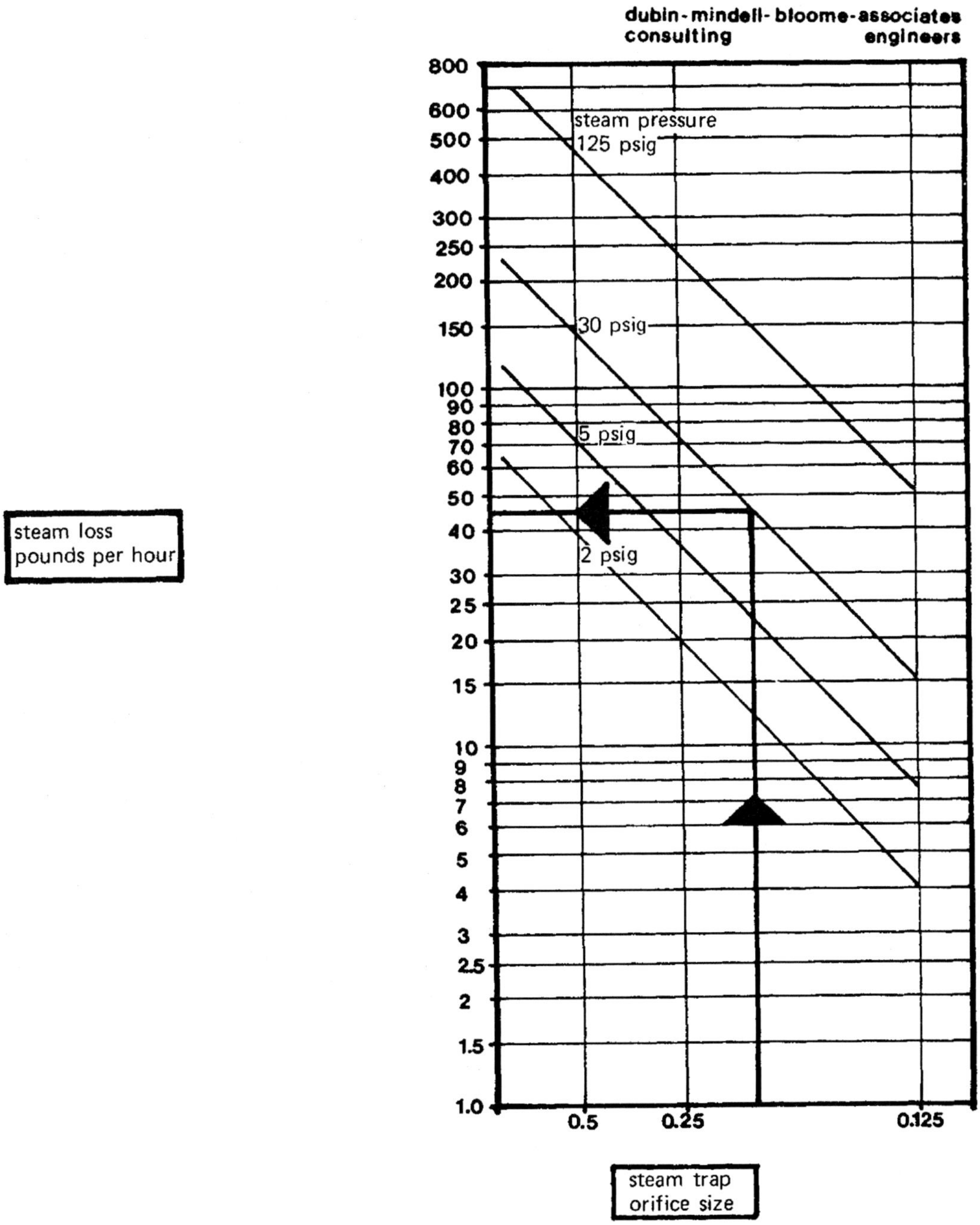

FIGURE #25 Engineering Data

Source: Manufactuers data

The following example illustrates the use of Figure 25 to determine the amount of steam lost through leaking steam traps.

Example:

Assuming a 30 psig system with 0.125 inch orifice traps, the loss from Figure 25 is 15 lbs. of steam per hour per trap.
At 1,000 Btu/lb., and if the steam is being produced by a boiler burning light oil with an efficiency of 70%, the energy input is equal to $\frac{15 \times 1,000}{0.7}$ = 6,987 gal/yr
At $0.36/gal. this amounts to 0.36 x 6,987 = $2,515/yr potential savings if all traps are replaced.

Results:	Energy Saved	6,987 gal/yr
	Dollars Saved	$2,515/yr

GUIDELINES TO REDUCE ENERGY USED FOR DISTRIBUTION AND HVAC SYSTEMS

HEATING SYSTEMS

- Clean the air side of all direct radiators, fin tube convectors and coils to enhance heat transfer.
- Keep radiators free from blockage. A one foot clearance in front of convectors, radiators, or registers is desirable. Heating systems, particularly hot water or electric baseboard radiators and low level warm air supply registers, work more efficiently if they are not blocked by furniture. Keep all books or other impediments from blocking heat or air delivery from the top of horizontal shelves or cabinets which enclose radiators, fan coils, unit ventilators or induction units.
- If radiator is set directly in front of a window where the glass extends below the top of the radiator, or in front of an uninsulated wall, insert a one inch thick fiberglass board panel, with reflective coating on the room side, directly between the radiator and the exterior wall to reduce radiation losses to the outdoors.
- Vent all hot water radiators and convectors to assure that water will completely fill the interior passages.
- Check radiator steam traps to assure that they are passing only condensate, not steam.
- Make sure that all fans, frequently inoperative in unit heaters, fan coil units, and unit ventilators are running normally to increase the heat transfer rate from heating coils.
- Use electric or infrared units as spot heaters for remote areas, a reception desk in a large lobby, for example, rather than operating an inefficient central system for a small area in the building.
- In the public spaces of all buildings such as lobbies, corridors, stairwells, vestibules and lounges - conserve

energy as follows:

> Where heat is provided by a unitary terminal equipment valve, turn off such equipment and remove the handles from control valves. If balancing cocks are included, turn them to the "off" position. In each stairwell of multi-level buildings, shut off all but the unit located at the bottom. Turn off heat in vestibules and foyers.

-Overhead unit heaters should direct heat to floors. Add directional louvers to focus heat to floor or area requiring heat; where possible draw return air from floor.

SINGLE ZONE DUCT SYSTEMS

-Maintain filters to reduce resistance to air flow, permitting a reduction in the time that the blower and the primary heat or cold-producing equipment run to satisfy room loads. Where possible change filters to low resistance type. A V type arrangement, if space permits, will provide more filter surface and less resistance, and thus requires less frequent cleaning than a straight single panel filter.

-Clean coils of lint and dirt to increase heat transfer efficiency in air handling units.

-To reduce heat loss and heat gain, repair insulation where torn, or insulate the exterior surfaces of casings where insulation is not on the interior surface. Units located outdoors on-grade or on the roof, or in non-conditioned spaces should be insulated with the equivalent of 3" of fiberglass.

-The relationships between the supply air temperature and quantity, and the chilled or hot water temperature and quantity, and space loads cannot be oversimplified in an energy use analysis.

The further relationship between fluid temperatures and flow and the primary energy-conversion equipment can be understood only after a thorough seasonal analysis of all factors considered together. However, consider the following:

1) Lowering supply air temperatures in the winter conserves energy, reduces "parasitic" loads by reducing thermal losses from the duct and piping systems, and improves the efficiency of the boiler

or furnace. It also reduces or eliminates wide swings in the space temperature because of overheating. However, the equipment may operate for a longer period of time and/or require a larger volume of air and more horsepower, thereby cancelling some or all of the savings in energy gained by reduced thermal losses.

2) Raising supply air temperatures in the summer conserves energy by reducing heat gain in ducts and piping, and by improving the efficiency of the chillers or compressors whenever they are in operation. Over cooling is minimized and wide temperature swings are eliminated but, again, a greater amount of supply air may be required to handle the load, necessitating more fan horsepower and/or longer periods of operation for fans, blowers, chillers, and condensers or cooling towers. In general, however, operation at the higher air temperature will prove to be more efficient.

-Where electric coils are used, operate the coils in stages, rather than all off - all on. Take care that coils are not on when air is not circulating, or they will burn out.

-Where humidity control is not essential, consider using the cooling coil for both heating and cooling by modifying the piping connections and removing the heating coil. The benefits are (1) the elimination of the heating coil reduces total system resistance to air flow and allows a lower fan speed to achieve the same volume through-put of supply air with less fan horsepower; and (2) when used for heating, cooling coils, which typically have a much greater surface area for heat transfer than heating coils, allow considerably lower water temperatures for a given heat output. The reduction in water temperature, in turn, minimizes heat losses from the system and allows more efficient operation of the boilers.

-Maintain the hot water and the chilled water at constant flow, but reset for part load conditions of more that 2 hours duration.
Adjust the temperature to provide the desired supply air temperature (which can vary with small load changes) and reduce the heat gains and heat losses of the piping systems.

-Avoid simultaneous heating and cooling except when required for humidity control in critical areas.
Do not operate a large system to meet the needs of a small area for humidity control.

TERMINAL REHEAT SYSTEMS

-Shut down system completely at night, weekends, and other occupied periods.
-Raise the supply air temperature to allow the use of higher chilled water temperatures which in turn will increase the efficiency of the chiller.
-Reduce supply air quantity.
-Operate on demand schedule, without reheat, whenever reheat has been used for humidity control in comfort applications.
-Operate on demand schedule without reheat when zone control can be sacrified.
-Relax humidity requirements and allow the relative humidity to drift naturally up to 65%. Higher relative humidity levels will reduce the need for low supply air temperatures allowing the suction temperature of the refrigeration unit to rise, decreasing energy required for refrigeration.
-Schedule supply air temperature according to the number of reheat coils in operation. If 90% of the reheat coils are working, the supply air temperature is too low and should be reset to a higher level until the number of reheat coils in operation falls to about 10%.
-Ensure that simultaneous heating and cooling cannot occur by providing interlocks between the heating and cooling control systems.

DUAL-DUCT SYSTEMS:

-Refer to "Terminal Reheat Systems" for general recommendations applicable to dual-duct systems.
-Raise the temperature of the cold duct and lower the temperature of the warm duct.
-Under conditions where there is no cooling load, close off the cold air duct, turn off the cooling systems, and then operate as a single duct system by rescheduling the warm duct temperature according to heating loads only.
-Under conditions where there is no heating load, close off the warm air duct, shut down the heating system and, by rescheduling the supply air temperature according to cooling loads, operate the system with the cool air duct as a single duct system.
Note: Dual duct systems are usually designed so

that either of the ducts can handle 80% of the total air circulated. Reducing the system to single duct operation, therefore, is likely to reduce the total quantity of air circulated and save a corresponding amount of energy. While the reduction may not affect space conditions adversely, it could possibly introduce noise problems requiring adjustment of dampers and/or fan speeds and acoustical treatment of the terminal boxes.

- -Repair leaking dampers in mixing boxes.
- -Do not bleed air into mixing box in the cooling mode in order to control dehumidification.

MULTI-ZONE UNITS:

- -Set controls to reduce the hot deck temperature and increase the cold deck temperature.
- -If not so equipped, provide the heating coils with an automatic control valve that will modulate to reduce the hot duct temperature according to demand. The valve is arranged so that when all hot duct dampers are partially closed, it will progressively reduce the hot duct temperature until one or more zone dampers is fully open.
- -Provide the cooling coils with an automatic control valve that will modulate the cold duct temperature according to demand. When all of the cold deck dampers are partially closed, the valve will raise the cold duct temperature until one or more of the zone dampers is fully opened.
- -Repair leaky valves and dampers.
- -Shut off the fan and all control valves during unoccupied periods in the cooling season.
- -Shut off the cooling valve during unoccupied periods in the heating season.

INDUCTION SYSTEMS

- -Refer to "Terminal Reheat Systems" for general measures, also applicable to induction systems.
- -Reschedule the temperature of the heating water and the cooling water according to the load. If the building has a light cooling load, the chilled water temperature should be raised, if the building has a light heating load the hot water temperature should be lowered. Avoid simultaneous heating and cooling in any one zone.
- -Typically, the primary supply air temperature is maintained at constant levels for cooling and heating which

necessitates reheating or recooling by the individual units to meet fluctuating load conditions. The system operates essentially as a reheat or recool system with the inevitable consequence of wasted energy. To reduce this waste, schedule the temperature of the primary supply air at the air handling unit according to load; i.e. under light heating load conditions, the primary air heating coil should be modulated to reduce the supply air temperature and under light cooling load conditions, the supply air cooling coil should be similarly modulated to increase the supply air temperature. Achieve the modulation preferably by adjusting water temperatures (rather than water volumes) to the central coils. Chillers and boilers will operate closer to their peak efficiency.

-Remove lint screens in induction units and clean coils regularly.

-During unoccupied hours in the heating season, realize a major opportunity for saving fan horsepower by shutting off the primary air supply system and operating the heating coils in the induction units as convectors.

VARIABLE VOLUME SYSTEMS

-When dampers begin to close (indicating that the requirement for supply air has decreased) reduce the fan volume (and the corresponding fan power input) to the point where one or more VAV dampers are fully open again. Adjust fan volume with either inlet vortex dampers or a variable speed driver such as a multi-speed motor or SCR control. If controls of this type do not exist in the building, refer to ECM-2 for installation details.

-Rescedule the supply air temperature to the point where the damper of the variable air volume box serving the zone or room with the most extreme load is fully open. Maintaining constant supply air temperature, even when all VAV dampers are partially closed is wasteful, as it increases the duct heat losses or heat gains and precludes the advantages which accrue from hot or chilled water temperature modifications. (Before reducing fan volume and/or rescheduling supply air temperature, carry out an analysis to determine the optimal levels of adjustment for the individual building.)

-Lower the hot water temperature and raise the chilled

water temperature according to thermal demands.
-Reduce total volume of air handled by system.
-Insure all VAV boxes operate in accordance with room requirements to prevent overheating or overcooling.
-Be sure, when resetting air volumes and where outdoor air is supplied by the same system, that code requirements for outdoor air are still met.

FAN COIL SYSTEMS

-Refer to "Heating Systems" and "Single Zone Duct Systems" for general recommendations applicable to fan coil systems
-Where a large number of fan coils are used in a building, the coils are provided with heating and cooling media of constant temperature, and control is achieved by varying flow rates through the coils. Consider adjusting the heating and cooling systems according to load. For example, when the heating load is light, and most of the heating control valves which vary the flow rate are partially closed, reduce the hot water temperature until one or more valves is fully open.
-In mild winter weather, and at night, shut off fans and permit the coil to operate as a convector. In severe winter weather, when the building is unoccupied but must be heated, or when excessive noise is not a problem, operate fans at high speed.
-Clean filters and coils.
-Close outside air dampers anytime infiltration equals ventilation requirements, or block off permanently.
-Block off inlets where no dampers are installed if infiltration meets ventilation requirements.
-Keep air outlets and inlets free of obstrucions.
-Where fan coil units are not located in conditioned areas, insulate casings to reduce heat loss/gain.

WINDOW AND THROUGH-THE-WALL UNITS

-Clean condenser and evaporator coils and intake louvers.
-Caulk openings between unit and window or wall frames.
-Remove units or cover them with plastic hood during the winter.
-Provide thermostatic control or timers to shut them off automatically.
-Turn off cooling unit and fan when leaving the space.
-Outdoor air inlets to units with resistance heaters are usually fixed, bringing in outdoor air whenever the units are operating. Close these inlets when unit

is not in operation or block off entirely if infiltration meets outdoor air requirements.
-See discussion above and ECM- 2 for replacement of inefficient units.

REDUCE FAN POWER BY REDUCING SYSTEM RESISTANCE

-Most duct systems are provided with balancing dampers which ensure the correct flow of air to each register or diffuser. The method of balancing which probably was used originally was to first determine the longest run, which normally had the highest resistance to air flow, and then to leave the balancing dampers in that run in a wide open position and close all other balancing dampers to achieve the required air flows. Thus, the shorter the run to any particular outlet, the more closed its balancing damper and the greater loss of power across it. Reduce resistance in the longest or index circuit so that all balancing dampers can be open, and total system resistance reduced. Further reduce resistance within all other branches or portions of the system.
-Remove unnecessary dampers or other obstructions from both supply and return air ductwork.
-Eliminate high resistance turns and elbows. One bad fitting may increase the entire system resistance by more than 1/2" S.P.
-If the majority of dampers in an air system are closed or partially closed to "balance" the system, consider reducing the total air volume and opening some of the dampers to save fan horsepower.
-Clean filters, blower fan wheels and fan scroll blades frequently to reduce friction.
-Replace filters which have low efficiency and high resistance with filters offering less air resistance and greater efficiency.
-Clean the air side of all coils in air handling units. Cleaning coils also increases their efficiency so that less energy for operation of the heating or cooling primary equipment is required.
-Seal and caulk leaking joints in ductwork, air handling units, and flexible duct connectors.
-Although air leakage from the ventilating system does not increase the resistance to air flow as such, it does waste

energy in two other ways:

1) It increases the total quantity of air handled by the fan over and above that required to meet the room temperature.
2) It increases the quantity of air that must be cooled or heated resulting in an expenditure of energy over and above that which otherwise would be required to meet desired room conditions.

Air leakage in the longest or most resistant duct run imposes a particularly large resistance load on the entire system since all other runs must be dampered down.

-Where direct drive fans are used and it is difficult or costly to change fan speed or motor, blank off unused portions of exhaust hoods in kitchens and cafeterias to reduce air volume and horsepower.

-Insulate ductwork where heat losses occur. Insulate ductwork used for air conditioning where it passes through non-conditioned spaces and picks up heat.

REDUCE FAN POWER BY REDUCING AIR VOLUME

-Reduce air volume by reducing the speed of rotation -

1) Where motor sheave is adjustable, open the "v" to reduce its effective diameter and adjust motor position or change belts to maintain proper tension.

2) Change motor sheave if no further adjustment is possible.

3) If motors are variable speed, set controller for reduced speed.

-After implementing changes to the system to decrease the required fan horsepower, the existing motor may be too large. If the full load on the motor is less than 60% of the nameplate rating, consider changing the motor to a smaller one. (Further guidelines for motors are included in Section 10, "Power" ECO-41).

-If the fan is direct drive, changing the speed of rotation is expensive. However, electric motors also tend to be oversized and if the system resistance is reduced significantly, the next size smaller motor, running at a slower speed, may well be suitable. In this case, calculate the savings that will accrue due to reduced power requirements to determine whether changing motors to achieve the speed reduction is economical.

-Losses in the drive train and bearings can, if maintenance is poor, amount to as much as 20% of total power input. Examine all bearings for wear and resistance to movement. Excessive belt tension will also impose an added load on the motor. Adjust or replace slipping fan belt drives.

-Operate exhaust systems intermittently to reduce total operating time. In many buildings the toilets and other areas require full exhaust for limited periods only. With proper programming and wiring to integrate the operation of exhaust fans and dampers with light switches, an office building with a system which exhausts 100 CFM through each of 20 toilet rooms, could reduce its exhaust to an average of only 400 CFM throughout the day.

REDUCE POWER FOR PUMPS

-Regularly adjust all pumps to control leakage at the pump packing glands. Curtail excessive waste of water by repacking, not only to conserve water and reduce losses, but also to avoid erosion necessitating costly repairs of the shaft.
-Reduce water flow where the terminal heating device can deliver heating or cooling adequately with the lower rate. For a chilled water system, take care to avoid reducing flow to the point that a lower suction temperature must be used. In general, the energy consumed at that lower suction temperature to reduce chilled water temperature (to maintain temperature or humidity control) would outweigh the energy saved by reducing the power for pumping.
-Globe valves can be used to restrict flow and save some energy, but since most hydronic systems are equipped with direct drive pumps, it is not possible to reduce the pump speed when the system friction is reduced by opening valves or removing resistances; use globe valves to restrict flow and save some energy. It is often possible to replace the pump impeller to save motor horsepower. Refer to pump maufacturers information for instructions.
-Refer to ECM-2 for changes to variable speed pumps, modular pumping and impellers.
-Check the power input to the pump motor against the nameplate rating. If the motor is drawing less than 60% of the rated input, analyze the potential energy and cost savings with a smaller motor and pump against the initial cost of replacement. See ECM-2 for details.

PIPING SYSTEMS

-Repair torn or missing insulation on water and steam piping which pass through unconditioned spaces.
-Insulate piping, valves, fittings and heat exchangers where heat is lost to unconditioned spaces or where heat lost from the pipes adds to the summer cooling load.

-Insulate cold piping used for air conditioning where it passes through uncooled spaces and picks up heat.
-Protect pipe insulation from water and moisture to preserve its thermal resistance qualities.
-Check for high temperature differences in heating and cooling heat exchangers, which may be an indication of air binding, clogged strainers, or excessive scale.
-Repair leaks in and/or replace steam traps. Confirm the location of a leaking trap by testing the temperature of return lines with a surface pyrometer and measuring temperature drop across the suspected trap. Lack of drop indicates steam blow-through. Excessive drop indicates the trap is holding back condensate.
-Chilled water and hot water systems require near complete elimination of air: check vents before reducing water temperature to improve cooling performance or increasing water temperature to meet room heating loads.

This page blank

Section 8

Energy Conservation Opportunities Commercial Refrigeration Systems

COMMERCIAL REFRIGERATION SYSTEMS

A. BACKGROUND

Commercial refrigeration systems account for as much as 50% of the total annual energy consumption in some supermarkets, and a significant amount of the energy consumed in cafeterias and restaurants. The actual energy usage is analogous to usage by air conditioning systems; total quantity depends upon the "produce" load, (the equivalent to the cooling system's "building" load), the distribution or "parasitic" load, and the seasonal efficiency of the primary conversion equipment, which in this case consists of the compressors.

To "produce" load is a function of the mass of the materials to be cooled, the temperature difference between the refrigerated materials and the ambient conditions and the period of time that the conditions are maintained.

The "distribution" load is a function of the performance of the refrigeration distribution system (refrigeration, piping, valves, controls, cold air ducts and diffusers) and the performance (seasonal efficiency) of the refrigerators, cold display cases, and other cold storage boxes.

The performance of the compressors, as noted in Section 6, "Cooling", ECO-22 is contingent upon the condition and operating mode of the evaporator, condensers, motors and drives and the performance of the air-cooled condensers. Review that section of the manual, as many of the energy conservation opportunities and options detailed there are applicable to commercial refrigeration systems as well.

Since commercial refrigeration units must operate at suction (evaporator) temperatures lower than those of evaporators of air conditioners they use more energy per tin of refrigeration produced - the C.O.P.* is lower. Because the commercial refrigeration system operates for the entire year - and conditions must be maintained for 24 hours per day - any measures to increase the C.O.P. will be even more productive per ton of refrigeration than equivalent measures

*Section 6, "Cooling" has an extensive discussion of C.O.P.

for improving the C.O.P. of air conditioning systems. A supermarket, with a 150 ton air conditioning system driven by a 150 H.P. motor, could have 200 H.P. of installed commercial refrigeration consuming from 2 to 6 times as much energy as the air conditioning system.

B. ENERGY CONSERVATION OPPORTUNITIES

ECO-28 REDUCE ENERGY CONSUMPTION FOR COMMERCIAL REFRIGERATION GUIDELINES TO REDUCE ENERGY CONSUMPTION FOR REFRIGERATION:*

- Use night covers over cold products, but be careful to avoid compressor damage or frost build-up on product.
- Do not permit refrigerated products to stand in the aisles, on docks, or anyplace where they will warm up, and create an additional refrigeration load when the products are placed in fixtures. Reduce the volume of items requiring cold storage.
- Avoid setting controls (pressure and temperature) any lower than necessary. Too often, a freezer may be operating at -30°F. air temperature when most often a -10°F. or higher is all that is necessary.
- Keep products below clearly marked load lines. An overloaded display case decreases product quality and increases energy use as much as 10 to 20% for each fixture by:
  a. Increasing the amount of store air mixing with fixture air, thereby increasing load on compressor.

  b. Increasing running time of compressors.

  c. Increasing fixture temperature.

  d. Increasing loss of refrigerated air out of fixture.

  e. Increasing defrost requirements consuming extra energy to clear frost from coils.

---

* Many of the items suggested are from recommended guidelines in "Retail Food Stores Energy Conservation", from the Commercial Refrigerator Manufacturers Association.

-Turn off refrigeration for cutting rooms, prep rooms, and some display fixtures, such as meat cases, when not in use. Put all food products in coolers where possible.
-Where possible, construct a tight partition or hang a heavy drape from roof to floor between sales and storage areas and maintain storage areas at 60°F or lower in winter, to prevent interchange of air.
-Keep return grilles of fixtures clear of stacked products, otherwise refrigerated air will flow into aisles.
-As customers shop from fixtures throughout the day, repack product displays and keep them below load lines.
-With multi-shelf fixtures, follow the recommendations of manufacturers in regard to shelf position and size to prevent increased refrigeration loads.
-Consider reducing - or turning off entirely - the internal shelf lights to reduce both refrigeration requirements and the lighting load.
-Automatically shut off all prep rooms at night and weekends. Set up for automatic startup when required.
-Consider unloading meat and produce display and shut off refrigeration at night and weekends. Set up on time clock to automatically turn off at closing and on in the morning, early enough to allow fixture temperatures to drop to the required level.

IMPROVE PERFORMANCE OF EQUIPMENT THROUGH OPERATION AND MAINTENANCE

Although some of the options listed below involve capital costs, they are included in ECM-1 because the payback periods should not exceed three years for any one option.

-If refrigeration load is decreased by any means, light reduction or case cleaning, for example, recheck temperature and pressure control setting to avoid freezing of products or short cycling of compressors. Enforce closed door rules when case is not in use, even for short periods of time.
-In multi-shelf low temperature equipment, check all fixtures for inoperative fan motors, often unnoticed, which affect efficiency.
-Minimize head pressure, to increase compressor capacities and reduce energy use, by increasing air supply over condenser. Clean condenser coils regularly. Maintain lowest head pressure at which the commercial refrigeration system can operate without

short-cycling or impairing expansion valve and coil efficiency. In cool months, set head pressure with controls:

a. Adjust hold back valves to control condenser surface, to lowest possible head pressure.

b. Set fan cycling with small differential and low cut-in point. Where more than one fan is used, consider cycling one or more fans.

-To prevent pressure drop and loss of compressor capacity avoid the use of suction line controls.
-Insulate suction and liquid line together, except where not recommended (hot gas defrost), to increase system efficiency.
-For systems where product is required at 32°F. or above, use time defrost rather than an added heat source. With existing equipment, disconnect the heater and reset the controls. Energy will be saved because the compressor does not operate and no electricity is used for defrost heaters.
-On forced defrost system, (electric or hot gas) use defrost terminating thermostat on each fixture to avoid over-defrosting individual fixtures and bring compressor back on as soon as all fixtures are satisfied.
-Consider demand defrost for all types of defrost systems. The number of defrosts is normally set up for the most adverse store conditons that may occur in a year. These conditons usually exist for only short periods of the year. Demand defrost compensates for these periods by causing fixtures to defrost only when it is required. Consult the fixture manufacturer before specifying demand defrost.
-Consider separate wiring circuit for anti-condensate heaters. Energy use by these heaters, which operate 24 hours a day, 365 days a year, is high, particularly on glass door freezers. Heaters are only required when humidity is high. Be aware, however, that most existing installations have fan and anti-condensate heaters on same circuit. Do not turn off the fan. Consider eliminating incandescent lighting over top display meat cases. In particular minimize spotlighting.
-Clean display fixture and cooler coil regularly. Be

sure to shut off refrigeration before using water for cleaning.

a. Use pressure spray to clean flues.
b. Remove discharge grilles to thoroughly clean inlet side (unseen from outside). A reduction of air flow can result in as much as a 10°F rise in product temperature and cause more frequent defrost.
c. Drain should be pressure flushed regularly to prevent build up in fixture bottoms.
d. Clean back side or inlet side of cooler units.
e. Check cooler door seals for loss of refrigerated air. Install spring-loaded door closers, reminder sign in plain view, buzzers, lights, etc., to be sure store personnel keep doors closed.

-Check all electrical circuits for power leak to ground. A leak to the ground may be small enough to go undetected for years with substantial accumulated loss of energy.
-Check all systems for correct refrigerant charge to avoid excessive compressor operation. Shortage will usually show up when low ambient air conditions exist.
-Where possible, provide staged cooling controls. Consider replacing one large compressor and coil with two or more circuits. Where possible, consider removal of hot gas bypass capacity control.
-Redirect outlets which are discharging into refrigerated fixtures.

This page blank

Section 9

Energy Conservation Opportunities Lighting

LIGHTING

A. BACKGROUND

Except in buildings that are electrically heated, electric lighting is one of the major uses of electrical energy and accounts for the most significant portion of total energy usage in offices and retail stores. When lights are left on in areas which are unoccupied or unused for lengthy periods of the week - in religious buildings and outdoor parking lots, for example, the inadvertent waste of energy often approaches or exceeds the amount of energy used by other building systems much of the week; the cost of this waste for one year may equal the initial cost of installing automatic controls to eliminate it.

Electric lighting is also the major contributor to internal heat gain in stores and core areas of office buildings. Because it increases the cooling load, electric lighting induces additional expenditure of energy to operate the refrigeration system, including air handling units, chilled water, and cooling tower or condenser fans.

To conserve energy for lighting and air conditioning:

1. Utilize available daylight more effectively for illumination.

2. Improve lighting system efficiency by increasing the light delivered to the task through more frequent fixture maintenance.

3. Switch off lights at night and in unoccupied areas during the daytime.

4. Reduce the levels of illumination at selected task locations, and lower them further between or beyond the tasks.

5. Relamp with more efficient light sources.

6. Modify the existing lighting fixtures and their location in the room to provide a greater amount and better quality of lighting.

7. Reduce the losses in the power distribution system which serves the lighting system. (When ballasts require replacing use high power factor type).

Office buildings and store sales areas are frequently illuminated at levels from 75 to 300 footcandles or more, requiring an average of 3 to 8 or more watts per square foot. In studies and design work for a number of projects (including the General Services Administration (GSA) energy conservation office building in New Hampshire, two Argonne National Laboratories offices, the Atomic Energy Commission office in Illinois and the New York Botanical Gardens office and laboratory in Millbrook, New York) Dubin-Mindell-Bloome Associates has found that a 2 watt per square foot average of good quality lighting is adequate for illumination in these office buildings and should be adequate for all new office buildings. Studies by Ross and Baruzzini, Consulting Engineers, for GSA and FEA confirm that adequate lighting can be provided in office buildings, utilizing only 2 watts per square foot lighting energy usage. Satisfactory illumination level for stores can also be possible within the 2 watt per square foot budget; in religious buildings, the budget can be even lower.

Without major modifications to both the pattern of lighting fixtures and the fixtures themselves,it may be possible to attain the 2 watts per square foot goal in existing offices and stores while maintaining proper lighting. If, however, the guide to illumination levels, Figure 26, is followed, average energy usage for lighting will approach the target.

Lighting levels on tasks, have traditionally been measured by footcandles. A room with a high level of footcandles, however, does not necessarily have a high level of effective illumination. Impediments to visual acuity, veiling reflection and disability glare, are not accounted for in footcandle measurements. The higher footcandle levels and added wattage may contribute ineffective illumination only.

The Illuminating Engineering Society (IES), which in the past has set lighting standards based on footcandles, in 1972 adopted the measure of effective light called Equivalent Sphere Illumination (ESI). ESI is a measure of lighting on tasks that accounts for both quality and quantity in the lighting system. Due to the complexity and time involved in E.S.I. calculations, consider using computer simulation programs now available, thru some fixture manufacturers, to approximate ESI footcandles. This method should be employed for areas with critical tasks. When high E.S.I. lighting valves are provided, they will yield very effective lighting at low energy input. Refer to ECM- 2 for further details.

Measures suggested here can provide for a lighting system with higher ESI footcandles. When there are opportunities to switch off or remove some lamps from fixtures, plan the eventual pattern of lighting fixtures to be that layout which locates the most remaining lamps just beyond the sides of the working edge of desks and task surfaces.

The simplest and most obvious option to save energy is to turn off lights when they are not needed and during unoccupied periods. Examine the building, first, to locate the areas where this can be done with available switches. ECM- 2 suggests the installation of additional switching where necessary to provide for switching of areas, utilization task lighting and more effective use of daylight.

Painting or covering walls, ceilings, and floors with light colored, more reflective finishes, and cleaning interior surfaces improves the effectiveness of any lighting system - daylight, electrical, or combination. Rooms with dark or dirty surfaces and furnishings, which absorb light, require either more lamps or lamps of higher wattage to maintain the desired level of illumination.

As a general rule, for all light sources except low pressure sodium, as the remaining useful life of the lamp decreases, its output of light, measured in lumens, decreases. This decrease is called the lamp lumen depreciation (L.L.D.)

Dirt accumulation on lamps, fixtures and lenses decreases the quantity of light falling on the task. This reduction of available light is called the Luminaire Dirt Depreciation (LDD) and combined together with the lamp lumen depreciation (LLD) comprises the maintenance factor (MF). The maintenance factor is used in lighting calculations to determine how many fixtures and lamps must be installed initially to give the designed illumination when lamp and dirt depreciation has occurred.

The installation will obviously provide more lumens on the task initially than required and if light fall-off can be reduced by improved maintenance, some of the fittings and lamps can be removed without reducing illumination below that required.

The discussions and guidelines of ECO's 29-36 reveal opportunities and options for lighting energy conservation which are in accord with the following principles and practices.

- -Illuminate specific task areas to that level, for each area, which provides adequate visual acuity, to perform the required task satisfactorily and reduce illumination levels in adjacent areas.
- -Provide artificial (electric) lighting only where and when it is required (to perform the task or for safety).

- -Consider all available light sources, including daylight, and select the most efficient that suits the application(highest lumen output per input watt.).
- -Institute a thorough maintenance program to sustain the highest value of light loss factor (L.L.F.)
- -Minimize power losses in ballasts and the distribution system.

B. ENERGY CONSERVATION OPPORTUNITIES

ECO 29 REDUCE ILLUMINATION LEVELS

Conserve energy for lighting by reducing illumination levels where they need not be high and eliminating illumination where it is not needed at all. Consult Figure 26, "Recommended Lighting Levels" for suggested levels in specific areas of the building. If several tasks requiring different levels of illumination occur within the same space, first consider their visual severity and then modify maintenance procedures, redecorate the area, and implement changes to the lighting system while reducing illumination levels to the appropriate level for each task. A uniform modular lighting pattern of general illumination throwing light equally on all areas regardless of task may waste up to 50% of the energy used for lighting in the building. Orient lighting to suit the tasks to be performed.

If one task with a critical lighting requirement is confined to a specific work area - i.e. drafting table, typewriter, desk top - in the midst of a larger work area with less critical requirements, provide a lower general illumination level for the overall area and a portable light at each critical task to raise the level of illumination locally (less than $25/lamp). Use fluorescent portable lamps in preference to incandescent.

In many cases it is less costly to move tasks to suit an existing lighting pattern than to add or rearrange fixtures. If task areas are widely dispersed, more light spills into adjoining areas where it may not be needed. Group tasks requiring similar lighting levels to limit the spill of higher level illumination and to allow lower lighting levels at less critical work areas.

Light levels in standard footcandles can be determined with portable illumination meters such as a photovoltiac cell connected to a meter calibrated in footcandles. The light meter should be accurate to about ± 15 percent over a range of 30 to 500 footcandles and ± 20 percent from 15 to 30 footcandles. The meter should be color corrected (according to the CIE Spectral Luminous Efficiency curve) and cosine corrected. Generally, measurements refer to average maintained horizontal footcandles at the task or in a horizontal plane 30 inches above the floor.

Measurements should be made at many repesentative points between and under fixtures, an average of several readings may be necessary. Daylight should be excluded during illumination-level readings for a true determination of level without light contribution from daylight.

The suggested illumination levels for office buildings, listed in Figure 26, agree closely with new standards recommended by G.S.A. for public office buildings. Keep in mind, however, that even lighting at lower intensities is very wasteful if lamps are burning when not needed.

FIGURE 26: SUGGESTED LIGHTING LEVELS*

With proper attention to quality the following levels should generally be adequate for tasks of good contrast:

Circulation Areas between Work Stations: 20 footcandles.

Background beyond Tasks at Circulation Area: 10 footcandles.

Waiting Rooms and Lounge Areas: 10-15 footcandles.

* Unless otherwise noted, all levels are average.

Conference Tables: 30 ESI footcandles with background lighting 10 footcandles.

Secretarial Desks: 50 ESI footcandles with auxiliary localized (lamp) task lighting directed at paper holder (for typing) as needed. In secretarial pools, 60 ESI footcandles.

Over Open Drawers of Filing Cabinets: 30 footcandles.

Courtrooms and Auditoriums: 30 footcandles.

Kitchens: non-uniform lighting with an average of 50 footcandles.

Cafeterias: 20 footcandles.

Snack Bar: 20 footcandles.

Testing Labs: As required by the task, but background not to exceed 3 to 1 ratio footcandles.

Computer Rooms: As required by the task, (consider 2 levels, 1/2 and full). In computer areas, reduce general overall lighting levels to 30 footcandles and increase task lighting for critical areas for input. Too high a level of general lighting makes it difficult to read the self-illuminated indicators.

Drafting: Full time 80 ESI footcandles at work station, part time 60 footcandles at work stations.

Accounting Offices: 80 ESI footcandles at work stations.

Note: Where applicable, refer to health and safety codes and federal standards (OSHA) for minimum lighting specifications.

The goal of the above standards is to reduce store and office lighting energy usage to less than 2 watts/sq. ft. gross floor area, or 2.5 watts/sq. ft. net area and 1.5 watts/sq. ft. for religious buildings. To determine net area subtract from the gross building floor area, the corridors, storage rooms, lobbies, mechanical equipment rooms, stairwells, toilet rooms, and other unoccupied, or seldom occupied areas.

ECO 30 CONSERVE ENERGY BY TURNING OFF LIGHTS

Utilize Existing Switching

Contrary to common belief, letting a light burn rather than turning it off never saves electrical energy. When electric lighting is not required, switch it off.

Figure 27 indicates - for an office lighting system of 1,000 fluorescent luminaires each with +100 4' 430 MA lamps - energy usage when lights burn only when needed.

ECO 31 IMPROVE THE EFFECTIVENESS OF THE EXISTING LIGHTING INSTALLATION

Increase the effectiveness of the present lighting system, then reduce the number of wattage of lamps required to meet the suggested standards. In order to reduce the number and/or wattage of lamps, consider the following measures to increase the maintained footcandles provided by existing lamps and fixtures.

Improve the maintenance factor by regularly cleaning fixtures and lamps and instituting a relamping program. Clean a fixture, at the least, every time it is relamped. (When relamping and cleaning are combined, the additional labor is very small.) In extremely dirty atmospheres, clean fixtures between lamp replacements, as well. Clean the lenses and interior housings. Use an anti-static compound to reduce electrostatic dirt collection.

The approximate cost of cleaning ceiling - mounted 48" 4-tube fluorescent fixtures - or incandescent fixtures of similar size - varies as follows: (for rough estimate only)

Number of Fixtures:	10	50	100	200
Price	$14.00	$58.00	$115.00	$224.00

A group relamping program will give a higher level of maintained footcandles, because lamp lumens will not depreciate to the low levels apparent at the end of their lives. Refer to Figures 28 and 29 for the effects of lamp and fixture cleaning and improving the maintenance factor.

When there are no critical reading or writing tasks, and where some glare from exposed lamps is acceptable (check with codes, where applicable), remove louvers or lenses from fixtures. Consider this suggestion, for instance, in regard to corridors, storage areas, stores with high ceilings, equipment rooms, and snack bars.

Increase the reflectance of ceilings, walls, and floors by cleaning or by painting with colors of higher reflectances: more illumination on tasks, with the same lumen output, will result. Increasing the reflectivity of the interior surfaces may open the possibility of reducing the number of lamps and luminaires while maintaining recommended lighting levels. Greater reflectances enhance the performance of natural daylight as well as artificial lighting. For example, a 25 x 25-ft. room with a 9-ft. ceiling and a ceiling reflectance of 80%, wall reflectance of 70% and floor reflectance of 30% will provide 35% more footcandles (with no increase in power consumption) than the same room with a ceiling reflectance of 50%, wall reflectances of 30% and floor reflectance of 10%. Caution: high gloss finishes may reduce visual comfort by increasing glare. Discretion should be used when redecorating.

The approximate cost of applying one coat of flat paint on smooth finish ranges between 15 and 20¢ per square foot, depending on total square footage. (for rough estimate only).

ECO 32 USE DAYLIGHT FOR ILLUMINATION

Use windows effectively as a primary source of illumination in perimeter spaces. The amount of available daylight in a building is a function of operating hours, latitude, weather, time of year, air quality, window size and location, shading and glazing details, reflectivity of interior surfaces and furnishings. Control of natural light, for effective use and integration with artificial light, is important, because the amount of natural light available varies. Control may be manual or automatic. Often blinds or drapes, which should be closed only for short periods of excessive sunshine or glare, remain closed all day with the electric lights turned on, when daylight could be used. Emphasis should be placed on proper control of daylight; otherwise the heat gain imposed on the cooling system may outweigh savings in energy for lighting. Indirect daylighting, however, generates less heat per lumen of light than electrical lighting. When equivalent lumens from daylight replace electrical lighting, savings in cooling load will occur.

FIGURE 27: EFFECT OF TURNING OFF UNNECESSARY LIGHTS ON POWER CONSUMPTION

dubin-mindell-bloome-associates
consulting engineers

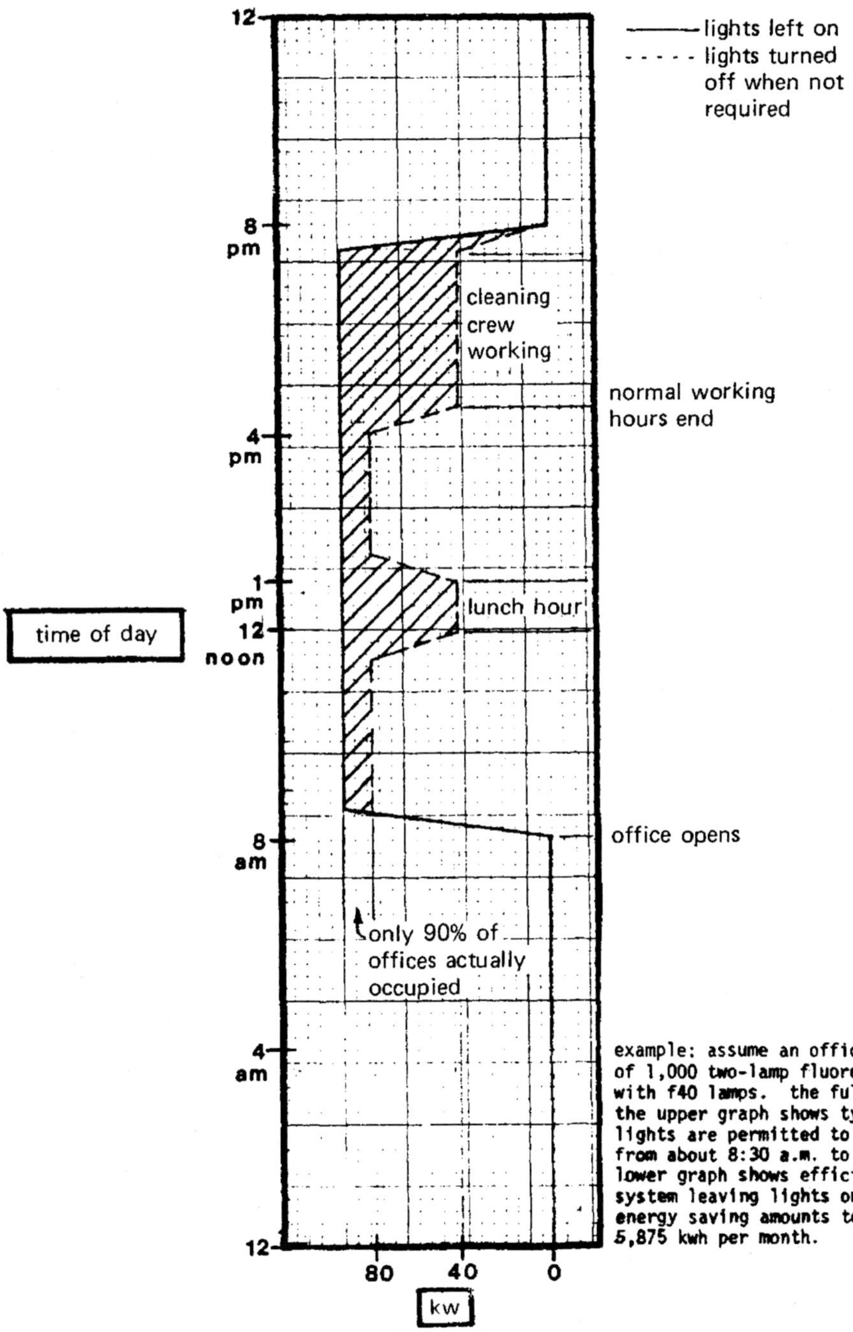

example: assume an office lighting system of 1,000 two-lamp fluorescent luminaires with f40 lamps. the full system uses 92 kw. the upper graph shows typical usage when lights are permitted to burn continuously from about 8:30 a.m. to 7:30 p.m. the lower graph shows efficient use of the same system leaving lights on only when used. energy saving amounts to 262.5 kwh daily or 5,875 kwh per month.

Figure # 27 Engineering Data

Graph developed from theoretical calculations based on typical office building practices.

FIGURE 28: FIXTURE CLEANING CYCLE

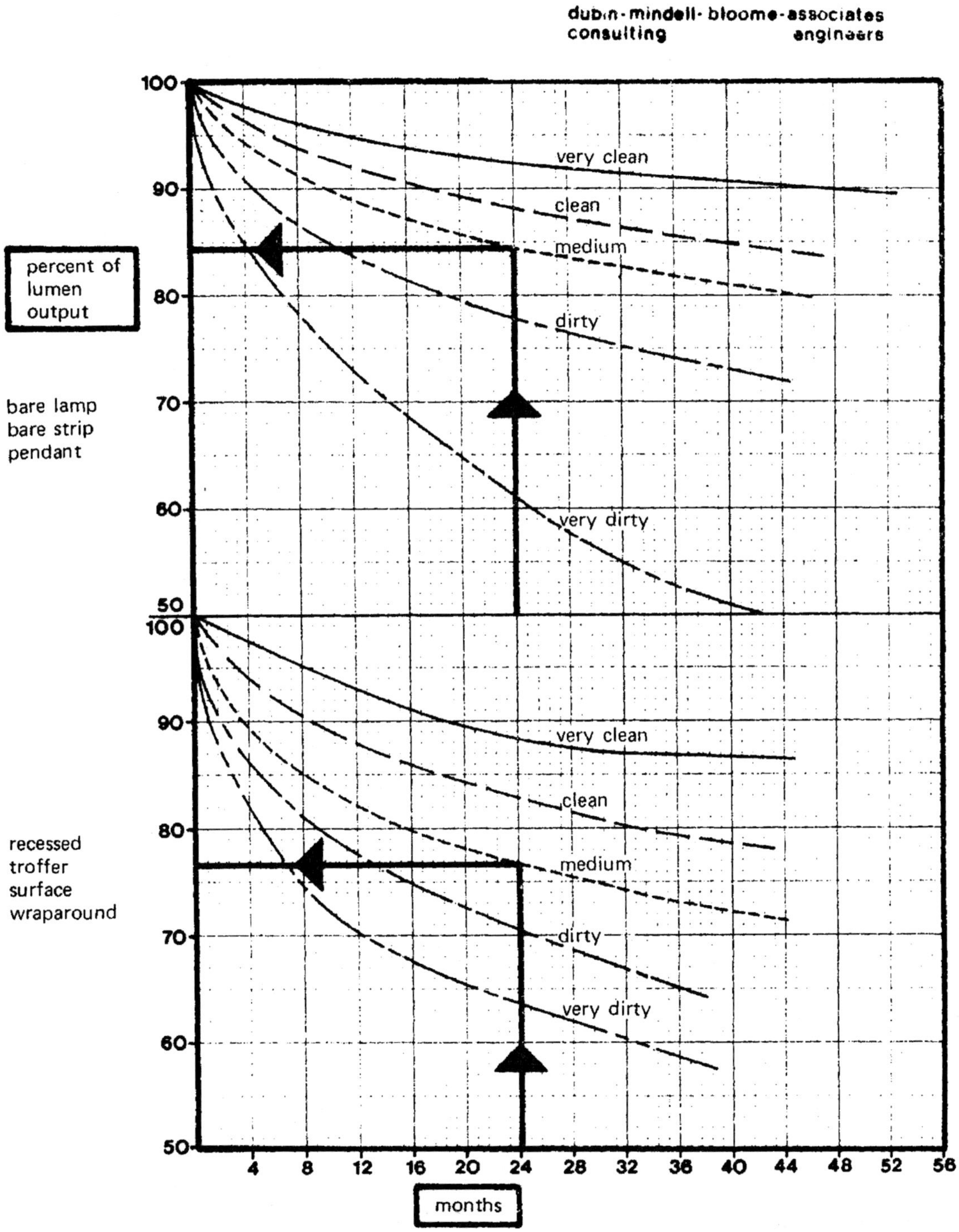

Figure #28 Engineering Data

Reference: I.E.S. Lighting Handbook, 4th edition 1968, page 9-17, figure 9-7 Illuminating Engineering Society 345 East 47th Street, New York, N.Y.

The data is computed from actual measurements of lumen output from lamps and fixtures exposed to various environments for different lenghts of time.

FIGURE 29: EXAMPLE OF MAINTENANCE FACTOR EFFECT ON FOOTCANDLES

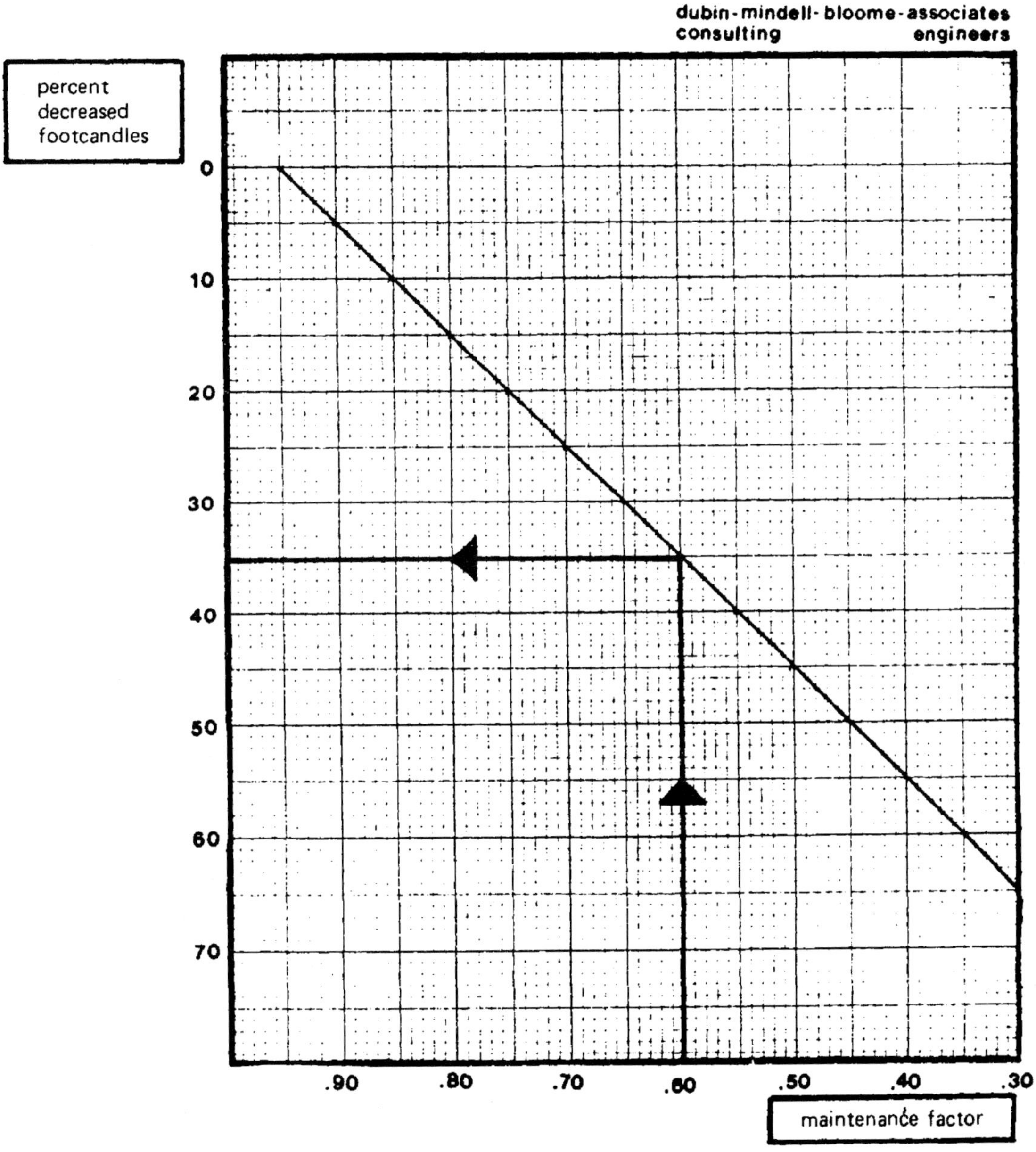

Figure #29 Engineering Data

Example Only - C.U. = Coefficient of Utilization
M.F. = Maintenance Factor

$$\text{footcandles} = \frac{\text{No. lamps (lamp lumens) No. fixtures (C.U.) M.F.}}{\text{area}}$$

$$\text{footcandles} = \text{M.F.}\left[\frac{\text{No. Lamps (lamp lumens) No. fixtures (C.U.)}}{\text{area}}\right]$$

footcandles = M.F. [constant - (for any specific area)]

Therefore, footcandles vary directly with maintenance factor

The luminous efficacy (lumens per watt) of various light sources as compared to daylight are shown below (no allowance for ballast of luminaires):

Low Pressure Sodium	183 lumens/watt
Natural light	120 lumens/watt (varies)
High pressure (HID)	105-120 lumens/watt
Metal halide	85-100 lumens/watt
Fluorescent	67-91 lumens/watt
Mercury vapor	56-63 lumens/watt
Incandescent	17-22 lumens/watt

ECO 33: <u>ADD SWITCHING AND TIMERS TO SHUT OFF LIGHTS</u>

If there are insufficient switches installed for unwanted lights to be turned off, consult an engineer, contractor, or utility company to determine the cost and procedures for installing additional switches. ECM- 2 includes further details.

There are many types of surface mounted flat ribbon conductors available for installation in existing spaces; they can be installed with the minimum dislocation of existing wiring or damage to interior decorations. Locate new switches near doors or where they are most convenient for occupant use: Switches in inconvenient locations will not be used. If switches are group-mounted lable each one to indicate the area it controls.

Provide time switches for areas which are commonly used for short times, and in which lighting is inadvertently but frequently left on. (i.e. reference rooms and stock rooms.) At a pre-determined time after the switch has been turned on, it will automatically shut off; if the area is to used for a long period of time, the switch can be manually overridden.

To ensure the success of any switching program, instruct all occupants in the use of available switching. Ensure that any changes to switching are in compliance with electrical codes.

ECO 34: <u>REMOVE UNNECESSARY LAMPS</u>

Remove unnecessary lamps when those remaining can provide the desired level of illumination. When removing fluorescent or high intensity discharge lamps, also remove the ballast (or disconnect it in place). If it is left connected to the power source it will continue to consume energy even though it serves no useful purpose. When two lamp fluorescent fixtures are mounted in a row, remove lamps in alternate fixtures of the rows (rather than removing one entire row) to derive higher quality lighting. In order to maintain the recommended lighting level it may be necessary, after removing some lamps, to use increased output lamps in place of some of the remaining lamps. Even so, the measure should result in net savings of energy. In many cases it will be possible to relamp the remaining fixtures with more efficient lamps to provide more lumens/watt -- increased illumination levels without increased wattage. Refer to Figure 30 to determine appropriate footcandle output with some common lamp replacements.

Lamp efficiencies vary between lamps consuming the same number of watts, and lamps of the same type. Different colors, shapes, gaseous fills, cathode construction and internal coatings cause those variations. For example, a fluorescent 40-watt, T-12 lamp, rated 430 millamperes, can have a rating from around 53 lumens to around 92 lumens per watt. Or, for example, a natural white color fluorescent lamp at 2100 lumens (53 lumens/watt) provides 1/3 less foot candles than cool white fluorescent lamps at 3200 lumens (80 lumens/watt.) In some cases, it may be desirable to relamp with lower wattage lamps as an alternative to removal of lamps.

An example of energy and money saved by removing two lamps from every other fixture (and disconnecting ballasts) is given below.

Example: <u>BUILDING TYPE - RETAIL STORE</u>

Number and type of fixtures (4-40W) 2' x 4'	375
Floor area	24,000 sq. ft.
Illumination level before change	100 FC
Hours of operation/yr.	3720 hours
Watts saved by the change = $100W \times \frac{375}{2}$ =	18,750 watts
Energy saved/yr. = $\frac{18,750W}{1000} \times 3720$ hrs. =	69,750 kwh

At an electricity cost of 4¢/KWH, dollar savings = 69,750 x .04

= $2790

Additonal savings in energy for cooling, due to reduced heat gain from lighting, will more than compensate for any loss of heat from lighting which occurs in the winter.

FIGURE 30: RESULTANT FOOTCANDLES AND WATTS PER SQUARE FOOT FOR VARIOUS LAMP CHANGES

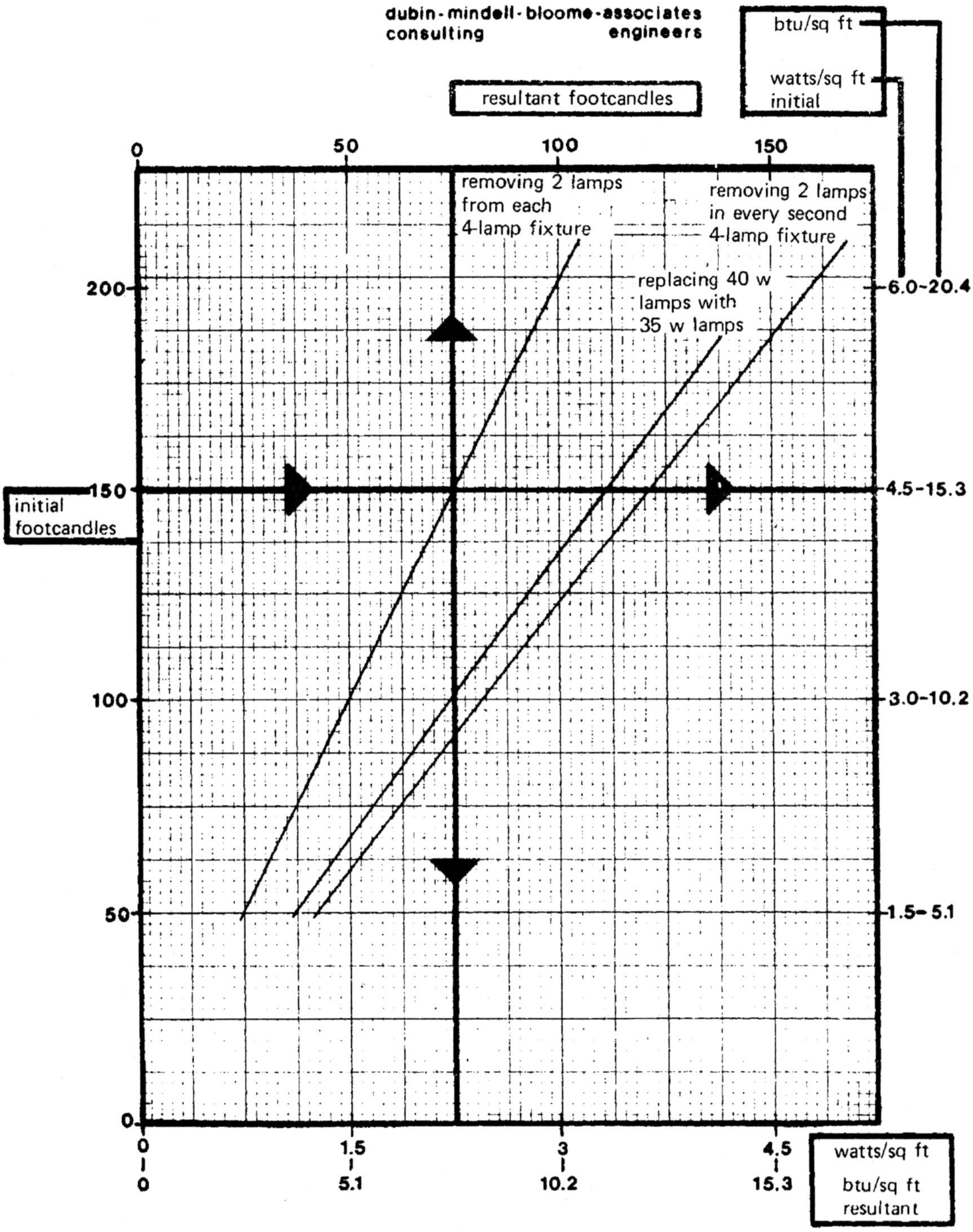

Figure #30 Engineering Data

This graph is derived from calculations based on manufacturers data and should be used only to obtain order-of magnitude savings that may be available.

C.U. = Coefficient of Utilization
M.F. = Maintenance factor

$$\text{footcandles} = \frac{\text{No. lamps (lamp lumens) No. fixtures (C.U.) M.F.}}{\text{area}}$$

$$\text{footcandles} = \text{No. lamps (C.U.)}\left[\frac{\text{(lamp lumens) No. fixtures (M.F.)}}{\text{area}}\right]$$

footcandles = No. lamps (C.U.) [constant - for any specific area]

Therefore, footcandles vary directly with No. lamps & C.U. Coefficient of utilization varies by only about 10%, hence, has been neglected for the approximation presented in this figure.

Since wattage consumption depends on the number of lamps and ballasts used; wattage consumption varies directly with number of lamps used.

ECO 35: REDUCE LIGHTING LOAD BY RELAMPING

Efficiency of lamps is measured in lumens or useable light produced per watt input. Selecting lamps with higher lumens per watt could permit the removal of some lamps, providing the lumens produce the required footcandles. More efficient lamps also impose smaller heat loads on the air conditioning system. In supermarkets and produce markets, reduced lighting reduces loads on commercial refrigeration systems as well. In winter, any heat lost by the reduction in wattage can generally be supplied more efficiently by the heating system itself.

When relamping, use more efficient light sources. Refer to Figure 31 for suggested changes to increase lighting efficiency without modifications to luminaires or ballasts.

The variation in efficiency of lamps (measured in lumens per watt) between incandescent lamps and high pressure sodium vapor lamps, is about 1 to 7 -- an extreme example. Proper application of lamps, however, precludes the higher efficiency lamps in all applications. For example, using a very bright concentrated light source, i.e. a high pressure sodium vapor lamp, in an office with a low ceiling would not provide comfortable lighting. Instead fluorescent lamps with outputs up to about 93 lumens per watt will be both effective and energy conserving. Or, for example, a 20-watt incandescent lamp in a fixture directly attached to a paper holder at a typist's desk could provide 60 footcandles on a manuscript with low ambient background lighting. It would require 100-watts of fluorescent lighting from a ceiling fixture, even though the fluorescent lamp produces more lumens per watt. Esthetics, size, efficiency, color, initial cost and operation and maintenance cost all help to determine choice of lamps. Energy conservation in fighting is not synonymous with sacrificing quality. Quality is not a measure of quantity only; it is obtained by intelligent application of the many interrelated factors. Correctly applied, energy conservation can increase the quality of lighting while reducing operating costs.

Generally, the lamp type recommended for each application in Table 1 is the most favorable for energy conservation.

TABLE 1: ENERGY CONSERVING LAMP APPLICATIONS

Lamp Type	Applications
Incandescent	1. Short time uses such as decorative display lighting or out-of-doors (Christmas trees).
	2. Religious worship halls
	3. In projection lamps for illuminating work closets or other very confined spaces.
	4. Stage spotlighting.
	5. Where small light source are required to light the task.
Fluorescent	1. Offices and other relatively low ceiling applications.
	2. Flashing advertising signs.
	3. Service islands at service stations
	4. Display cases in stores.
	5. Desk lamps.
	6. Classrooms or training centers.
	7. Cafeterias (with color correction).
High Intensity Discharge	1. Stores and other relatively high ceiling applications.
	2. Auditoriums
	3. Outdoor area lighting
	4. Outdoor flood lighting
	5. Outdoor building security lighting
	6. Marking of obstructions

It is also desirable to adopt a group relamping program so that when a majority of the fluorescent tubes reach the point of only 70% of their rated output they will be replaced. Labor costs are generally less for group relamping than for individual relamping throughout the year.

ECO 36: REDUCE POWER FOR LIGHTING BY REPLACING BALLASTS

When the standard ballasts in existing systems fail or must be replaced, substitute high efficiency types.

For instance, in an installation of 1,000 2-lamp 40-watt fluorescent luminaires each ordinary ballast consumes 12 to 16-watts amounting to an annual energy consumption of 24,000 to 32,000 kilowatt hours in buildings operating 2,000 hours/year. On the other hand, an efficient ballast having only 10 watts loss will only use 20,000 kilowatt hours annually, for a savings of 4,000 to 12,000 kilowatt hours a year.

GUIDELINES TO REDUCE ENERGY USED FOR LIGHTING

TURN OFF LIGHTS

- -Mark all ganged switches with identification of the lights controlled. Color code switches and institute a program of use. (i.e. Blue 7 a.m.-6 p.m., Red 9 a.m.-12/1 p.m.-4:30 p.m. etc.)
- -Instruct occupants and maintenance personnel to switch off all lights which are not required, even for portions of the day, including lights for:

Storage rooms
Vending machines (Use vending machine illumination).
Mechanical equipment rooms
Auditoriums, conference rooms and cafeterias (when not used)
Meeting rooms (when not used)
Bulletin boards
Office directories
Unassigned office areas
Any areas where natural light is available
Loading docks.

- -Turn off all lights other than those needed for security when the building is unoccupied.
- -Substitute small table or floor-mounted lamps in lounge areas or waiting rooms and turn off modular ceiling fixtures.
- -For cleaning which must be done at night, turn on lights only in that portion of the building which is being cleaned immediately.
- -Switch off lights in each area when moving to the next.
- -Turn off display case internal lighting (in food cases of a supermarket, for example) when premises are unoccupied.
- -Turn off incandescent lighting over top display of meat cases in a supermarket.
- -Turn off flood lighting which is strictly decorative.
- -In kitchens, avoid keeping infrared food warming lamps on when no food is being kept warm.
- -Turn off lights in areas of religious buildings which are not used during the week.
- -Maintain hazard and egress lighting at all times as required by building and fire codes.

INCREASE THE EFFECTIVENESS OF THE EXISTING LIGHTING SYSTEM

- -Clean and wash walls, ceilings and floors.
- -When repainting, use light colored paint on ceilings, walls and floors but avoid objectionable specular reflections from gloss finishes. (Consult applicable fire codes for flame spread ratings of interior finishes.)

-When recarpeting or retiling, put in lighter colored carpets or tiles. (Consult local fire codes for ratings of materials.)
-When refurnishing, select lighter color furnishings.
-Remove partitions where no longer required. (Identify fire partitions and leave intact, as specified by building code. Also partitions may contain or retard noise deemed hazardous by federal standards(OSHA). If wall outlets with partitions are eliminated, electrical codes may require alternate outlets in floor.)
-Decrease partition heights where possible. (Again, check for compliance with building and fire codes, as well as occupational standards for noise, etc.)
-Lower luminaire closer to the lighting task especially in high bay areas such as storage and equipment rooms. (If alteration of ceiling assembly is involved check for compliance with fire ratings.
-Where possible change the relationships between the light source and task by moving the luminaire or relocating the task so that the light source is to one side of the task rather than directly in front of directly over it.
-Direct security lighting where it is most required, such as at windows and entrances, and reduce it where the security problems are minimal.
-For display or merchandising lighting, establish grouped, highlighted display islands where many products can be lit with the same lighting sources and reduce the total number of display islands.

MAKE BETTER USE OF DAYLIGHT

-Clean windows and skylights.
-Where practical, schedule periods of occupancy, cleaning and meetings to make maximum use of daylight.
-Locate tasks that need the best illumination closest to the windows, with the task-viewing angle parallel to the windows.
-Switch off electric lights in areas when natural light is available.
-Add additional switches to turn off electric lights where luminaire arrangement and circuiting permits. Refer to ECM-II for details on modifications to the switching system.
-Increase the interreflectances of interior surfaces.
-To reduce galre, rearrange work stations so that side wall daylight crosses perpendicular to the lines of vision.

-In the winter, open blinds and drapes even if space mildly overheats.

-In buildings without air conditioning systems, open blind and drapes in summer even if space mildly overheats.
-Block out or paint out surfaces of skylights which receive excess amounts of direct sunlight in the summer; leave other surfaces clear.

REMOVE EXCESS LAMPS

Removing both Lamps of a Two-Lamp Fluorescent Fixture

When removing both lamps, disconnect the black and white ballast power leads from the power supply. In addition to the wattage saved from the lamp removal, there will be savings in ballast wattage as follows:

LAMP TYPE	BALLAST WATTS SAVED
F40	7
Slimline	11-13
High Output (800 mA.)	12
High Output (1,500 mA.)	13.5

Removing Two Lamps in a Four Lamp Fluorescent Fixture or One Lamp in Three Lamp Fixtures

For rapid start 430 milliampere lamps in a four-lamp fixture, remove both inside lamps and disconnect power wiring to their associated ballast. (Savings will be lamp wattage plus ballast wattage). In a 3-lamp fixture remove the middle lamp.

Two lamps in a 2'wide fluorescent fixture is actually more efficient than four lamps in the same fixture. The resultant lighting levels using only two lamps will be greater than half the level if four lamps were used.

Removing One Lamp in a Two Lamp High Intensity Discharge Fixture

Removing a lamp from a two lamp mercury or metal halide ballast will generally cause no adverse effect. However, there will still be a current flow consuming as much as 20 watts in a 400-watt lamp ballast or 50 watts in a 1,000 watt lamp ballast, disconnect the ballast to save energy.

Removing a lamp from a two lamp high pressure sodium ballast for more than a short period of time will damage the starting circuit in the ballast because the circuit will operate continuosly with the lamp removed. A failed lamp can cause the same type of damage. Remove both lamp and ballast.

USE MORE EFFICIENT LAMPS

-Use a single, larger incandescent lamp where possible, rather than two or more smaller lamps. Higher wattage general service incandescent lamps are more efficient than lower wattage lamps. For example, one 100-watt (at 1,750 lumens outout) produces more light than two 60-watt lamps (at 2 x 860 lumens output per lamp = 1,720 lumens).

-Avoid multi-level lamps, where light levels are not often changed. The efficacy of a single lamp is higher per watt than a multi-level lamp. For example, a 50-100-150 watt multi-level lamp operating at 100 watts has an efficacy of 17.1 lumens per watt.

-Avoid replacing incandescent lamps by self-ballasted mercury vapor lamps, as they give less light than the same wattage incandescent.

-Avoid using extended service lamps except in special cases, such as recessed directional lights, where short lamp life is a problem. In those cases use one size smaller extended service lamp. For the same wattage, general service lamps at 750 to 1,000-hour life are more efficient than extended service lamps at 2,500 hours. An extended service lamp (100 watt) producing 14.8 lumens per watt will require 17.5% more power to provide the same lighting as a general service lamp (100 watt) producing 17.5lumens per watt. Most cost studies show extended service lamps are not economical.

-When relamping, replace 40-watt fluorescent lamps with 35-watt lamps to achieve a reduction in lighting level of approximately 18%, while saving 20% in fixture electrical energy.

USE MORE EFFICIENT BALLASTS

-When ballasts burn out, replace with the following:

For fluorescent lights, 430 milliampere, replace low power factor high-watt ballasts with high power factor (90% or more), low-watt ballasts. (Circuits with 50% power factor ballasts have 240% more energy losses in wiring than those with 90% power factor ballasts).

This page blank

FIGURE 31: RELAMPING CHANGES TO INCREASE LIGHTING EFFICIENCY WITHOUT MODIFICATIONS TO LUMINAIRES OR BALLASTS

(All lamps operating at 120 volts)

Incandescent Lamps

Present Light Source						Replacement Light Source				
Watts	Lamp Description	Lamp Volts	Lumens @ Socket Volts	Watts @ Socket Volts	Hrs. Life @ Socket Volts	Watts	Description	Lumens	Hours Life	Reduction in Watts
40	40 A/99	130	323	35	7,000	25	25 A	235	2,500	10
60	60 A/99	130	597	53	7,000	40	40 A	455	1,500	13
						40	40 A/99	420	2,500	13
75	75 A/99	130	770	66	7,000	54	54 A	775	3,500	12
						55	55A	670	2,500	11
						60	60 A	870	1,000	6
						60	60 A/99	775	2,500	6
100	100 A/99	130	1,147	88	7,000	75	75A/99	1,190	750	13
100	100 A21/99	130	1,109	88	7,000	75	75 A/99	1,000	2,500	13
150	150 A23/99	130	1,779	132	7,000	90	90 A	1,290	3,500	42
150	150/99	130	1,771	132	7,000	92	92 A	1,490	2,500	40
						100	100 A	1,750	750	32
						100	100 A/99	2,500	1,490	32
150	150 R/FL	120	1,870	150	2,000	75	75PAR/FL	765	2,000	75
150	150PAR/FL	120	1,740 ea	150 ea	2,000	250	Q 250 Par38	3,220	3,000	50
200	200 A/99	130	2,626	176	7,000	150	150 A	2,880	750	26
200	200/99 IF	130	2,510	176	7,000	150	150 A23/99	2,310	2,500	26
				135		135	135A	2,100	3,500	41
				41		138	138 A	2,300	2,500	38
						160	HSB160/SS/M	2,700	20,000	16

FIGURE 31: (continued)

Incandescent Lamps

Present Light Source						Replacement Light Source				
Watts	Description	Lamp Volts	Lumen @ Socket Volts	Watts @ Socket Volts	Hrs.Life @ Socket Volts	Watts	Description	Lumens	Hrs. Life	Reduction in Watts
60	60/99IF Extended Service	120	740	60	2,500	54	54/99IF Extended Service	645	2,500	6
60	60A19/35 Industrial Service	120	670	60	3,500	54	54A19/35 Industrial Service	590	3,500	6
100	100/99IF Exten-Ded Service	120	1,480	100	2,500	90	90/99IF Extended Service	1230	2,500	10
100	100A21/35 Extended Service	120	1,280	100	3,500	90	90A21/35 Extended Service	1090	3,500	10
150	150A/99IF Extended Service	120	2,350	150	2,500	135	135A/99IF Extended Service	1990	2,500	15
150	150A25/35 Industrial Service	120	2,150	150	3,500	135	135A25/35 Industrial Service	1790	3,500	15
300	300	120	4,900	300	3,000	250	250PS-35 Self-Ballasted Mercury	4800	11,000	50
1000	1000	120	18,300	1000	3,000	750	750R-57 Self-Ballasted Mercury	17650	15,000	250
1500	1500	120	28,400	1500	3,000	1,250	1250 Bt-56 Self Ballasted Mercury	38000	15,000	250

FIGURE 31: (continued)

Incandescent Lamps

Present Light Source						Replacement Light Source				
Watts	Lamp Description	Lamp Volts	Lumens @ Socket Volts	Watts @ Socket Volts	Hrs. Life @ Socket Volts	Watts	Description	Lumens	Hours Life	Reduction in Watts
300	300 M/99 IF	130	3,996	264	7,000	200	200A	4,010	750	64
300	300/99 IF	130	3,996	264	7,000	200	200A/99	3,410	2,500	64
						300	HSB 300/SS/M	7,800	20,000	0
300	300	120	5,820	300	1,000	300	HSB 300/SS	7,800	20,000	0
500	500/99 IF	130	6,984	440	7,000	300	HSB 300/SS	7,800	20,000	140
500	500/99 IF	120	9,070	500	2,500	450	HSB 450/SS	9,500	16,000	50
750	750/99	130	10,934	660	7,000	500	500	10,850	1,000	160
						500	500/99	9,070	2,500	160
						450	HSB 450/SS	9,500	16,000	210
750	750 R 52	120	13,000	750	2,000	750	HSB 750R/120	14,000	16,000	0
1,000	1,000/99	130	15,246	880	7,000	750	750	17,040	1,000	130
						750	750/99	14,200	2,500	130

FIGURE 31: (continued)

Incandescent Lamps								
Present Light Source				Replacement Light Source				
Watts	Description	Beam Candle-power	Hrs. Life	Watts	Descripticn	Beam Candle power	Hrs. Life	Reduction in Watts
100	Reflector Floodlight R-40	800	5,000	75 PAR-38	Projector Floodlight	1,430	5,000	25
150	Reflector Floodlight R-40	1,200	5,000	100PAR-38	Projector Floodlight	2,230	5,000	50
150	Reflector Floodlight R-40	1,200	5,000	75 PAR-38	Projector Floodlight	1,430	5,000	75
150	Reflector Floodlight R-40	1,200	5,000	100 BR-40	Floodlight	1,200	5,000	50
200	Reflector Floodlight R-40	1,600	5,000	150PAR-38	Projector Floodlight	3,450	5,000	50
300	Reflector Floodlight R-40	2,450	5,000	200PAR-38	Projector Floodlight	4,560	5,000	100
500	Reflector Floodlight R-40	3,600	5,000	250PAR-38	Projector Floodlight	5,850	5,000	250

FIGURE 31: (continued)

Incandescent Lamps

Present Light Source				Replacement Light Source				
Watts	Description	Color	Lumens	Watts	Description	Color	Lumens	Reduction in Watts
40	F40T12-48"	CW	3150	34	F40/RS/EW	CW	2800	6
40	F40T12-48"	WW	3200	34	F40/RS/EW	WW	2900	6
75	F96T12Slim line	CW	6300	60	F96T12/EW	CW	5220	15
75	F96T12Slim line	WW	6400	60	F96T12/EW	WW	5340	15

FIGURE 31: (continued)

High Intensity Discharge Lamps

Present Light Source					Replacement Light Source					
Watts	Description	Color	Lumens	Hours Life	Watts	Description	Color	Lumens	Hours Life	Reduction in Watts
	Mercury Vapor					Mercury Vapor				
400	H33CD-400	Clear	21,000	24,000	300	H33CD-300	Clear	14,000	16,000	100
	H33G1-400/DX	Deluxe White	23,000	24,000	300	H33GL-300/DX	Deluxe White	15,700	16,000	100
400	Mercury Vapor		23,000	24,000	360	High pressure sodium		34,200	12,000	40
175	Mercury Vapor		8,500	24,000	150	High pressure sodium		12,000	12,000	25

Section 10

Energy Conservation Opportunities
Power

POWER

A. BACKGROUND

The amount of electric power used within a building - measured in kilowatt hours (kwh) - depends upon (1) the demands of the systems which use power to supply the "building" load (lighting, heating, ventilating, cooling, domestic hot water, commercial refrigeration, elevators, escalators, business machinery, communications systems, cooking, snow melting and other processes), (2) the power losses of the conversion and distribution systems which supply those loads, and (3) the characterisitics of the electric service and distribtuion systems. The opportunities to reduce "building" and "parasitic" distribution loads have been detailed in other sections. As an example, when the quantity of water or air circulated in a HVAC system is reduced, less power is needed to drive fans and pumps and, accordingly, it is possible to reduce annual energy consumption even further. Improper distribution voltage, low power factor, transformers at inefficient loading points, losses from standby transformers, excessive operation of motors and equipment, improper control and low load factors of motors, and excessive voltage drop in undersized conductors or wiring are the major causes of the waste of electric energy. Reducing the kWh of electricity used by the building reduces the consumption of raw source energy (the amount of energy required to generate electricity at the power plant) by a factor of more than three. The boiler and generator losses at the power plant and transmission and distribution losses from the plant to the building account for about 70% of the energy used in generating electricity. Nationally, conserving electrical energy helps to conserve oil, gas, and coal supplies which are consumed in the generation of electricity at fossil-fueled power plants and in the mining and processing of uranium for nuclear plants.

Before changing or adjusting equipment or operating procedures, and if building personnel are inexperienced or not qualified to implement any options which are chosen, consult an electrician or electrical engineer to analyze the proposed changes and perform the work.

B. ENERGY CONSERVATION OPPORTUNITIES

ECO 37 REDUCE OPERATING TIME OF ELEVATORS AND ESCALATORS

The amount of power required annually to operate an elevator is a function of the height of the building, the number of stops, passenger capacity and load factors, and the efficiency of the hoisting mechanism. For example, a 2,500 lb. capacity "local" elevator making 150 stops per car-mile consumes about 5 kwh per car-mile, while an "express" elevator making 75 stops per car-mile consumes about 4 kwh. A 4,500 lb. capacity elevator in a 12-story department store stopping at every floor will use 13 kwh per car-mile. Consumption will vary between elevators of the same capacity depending on the type of hoisting motor and control, whether the elevator is hydraulic, geared or gearless and the kind of service and the amount of load offset by the counterweight. Select speeds that are as slow as possible, while keeping maximum waiting time to no more than 2 minutes. Seek the assistance of the elevator manufacturer or a consulting engineer to study the traffic patterns and to reschedule as necessary. Where multiple elevators are installed reduce the number in service during light traffic periods.

Assuming 35% equivalent full load operation, escalator energy consumption may vary from 1.3 kw per hour for a 32-inch wide model operating at 90 feet per minute with a 14-foot vertical rise, to 3.0 kw per hour for a 48-inch wide model operating at 90 feet per minute with a 25-foot vertical rise. Escalators consume energy whether they are carrying passengers or not.

ECO 38 REDUCE ENERGY CONSUMPTION FOR EQUIPMENT AND MACHINES

Most buildings contain many electrically driven machines which are left switched on and idling but which are used only for short periods of time or when the building is occupied. Make an inventory of all office and business machines, convenience machines etc. and determine their true periods of use. Install time clocks to control machines such as automatic vendors. Encourage typists and clerks to turn off electric typewriters, desk calculators etc. when not actually in use.

ECO 39 REDUCE PEAK LOADS

Utility rate structures are based not only on the building's total usage of electricity, but also on its peak demand - which may occur for only a few hours once or twice each year, but which establishes demand charges for the rest of the months. Institute a thorough load shedding program to assign levels of priority to the equipment in the building. Manually shut off some of the equipment with large motors during periods of high electrical usage to reduce peak demand, save on operating costs, and conserve energy. Practiced on a large scale, load leveling reduces the peak demand of the generating plant and lessens the chance of brownouts, blackouts, and low voltage problems.

Consult with the utility which serves the building to analyze the rate structure and to determine when the peaks occur and how they can be reduced. The loads shed most readily include electric water heaters, domestic hot water circulators, air conditioning refrigeration units, large pumps, noncritical lighting, and elevators.

ECO 40 REDUCE TRANSFORMER LOSSES

Transformers reduce transmission and distribution voltage to equipment operating voltage. Heat generation and dissipation, due to electrical resistance in the transformer, results in electrical energy losses. Even when equipment served by the transformer is inoperative, some energy is lost unless primary power to the transformer is switched off. When the transformer serves loads which are not required for relatively extensive periods of time, complete disconnection from the primary power may be feasible. Take care, however, to avoid disconnecting transformers that feed clocks, heating control circuits, fire alarms, or critical process equipment. Potential savings are 3 to 4 watts per kilovolt amp (KVA).

Example:

Switch off a 150 KVA transformer for 12 hours/week-day at night, and 48 hours over the weekend (or a total of 108 hours/week and 5616 hours/year). Assuming savings of 4 watts per KVA, energy saved in one year will be:

4 watts/KVA x 150 KVA x 5616 hours = 3,369,600 watt hours or approximately 3370 kwh.

At an electricity price of 4¢ kwh, savings will be:
3370 kwh x 4¢/kwh = $135.80

ECO 41 IMPROVE THE EFFICIENCY OF MOTORS

Because original load calculation estimates are usually conservative, and when loads have been reduced through conservation measures, most motors are oversized for the equipment load they are serving. If the ratio of the motor loading to the motor's horsepower rating is small, the power factor will be low and the motor will operate inefficiently.

Undertake a comprehensive study of all the motors in the building to determine their load factors.

Gather and record the following information:

1. The equipment served by the motor.

2. The motor nameplate information for each motor, including: brand, type, frame, horsepower, speed or speeds, voltage phases and frequency, mounting, full load rating.

3. Measure with an ammeter and record the full load running current of each phase leg.

4. Measure with voltmeter and record the voltage at the motor when measuring running current.

5. Record pully size and type, i.e. V belt, chain, etc.

Find motor loading by multiplying the current draw, as recorded in item 3, above, by the voltage, item 4, and then dividing by 1000 to convert to kilowatt input. Convert nameplate H.P. rating (item 2) to kw by multiplying H.P. by 0.746. Then take the ratio of actual input to nameplate rating to determine the load factor on the motor.

Motors that are not loaded to at least 60% of their potential are relatively inefficient and reduce the power factor of the entire electrical system. Exchange underloaded motors with others on the premises to achieve as close to full loading on each motor as possible. If a motor needs replacement, consider interchanging motors then, too. For equipment which cycles on and off at short intervals heat build-up is a less critical problem than in other cases. The motor service factor establishes the maximum overload possible without exceeding the motor's temperature rating. Utilize this information (available through the manufacturer) when in the process of exchanging and interchanging the motors currently in use, and, for cycling equipment, select smaller motors, rated slightly below maximum load requirements and allow some overloading to occur.

Small split-phase or shaded-pole motors, often used in perimeter fan coil units, have low power factors and are inefficient. Turn off these motors during mild winter nights to permit the fan coil to operate by natural convection.

When evaluating the potential for larger savings, compare the cost of replacing inefficient motors with the savings in electrical operating cost which would be achieved with new motors. Refer to ECM 2 for procedure.

GUIDELINES TO REDUCE ENERGY USED FOR POWER

REDUCE ELEVATORS AND ESCALATOR OPERATION

- Reduce the number of elevators in service during hours when majority of persons are not leaving or entering the building.
- Turn off the motor-generator set located in the elevator machine room when not in use - nights, week-ends, holidays and slack periods during the day.
- Turn off escalators to unoccupied floors of offices or retail store floors during renovations.
- Operate demand escalators only during peak periods
- Reduce speed of escalators and elevators.
- Where security arrangements permit, encourage employees to walk up and down one flight of stairs rather than to use vertical transportation systems.
- Consider turning off all down escalators during periods of light traffic.
- Consider turning off up escalators on alternate floors during periods of light traffic.

REDUCE ELECTRICAL APPLIANCE AND MACHINERY OPERATING TIME

Turn off:

- Coffee pots and food warmers when not in use.
- Refrigerated drinking fountains at the end of normal business hours.
- Refrigeration units (and consider disconnecting units at all times).
- Vending machines at the end of the week where food spoilage is not a problem. Use time clock to turn on the vending machine in time for the soft drinks to reach 45°F. by Monday morning at employee arrival time.
- Portable electric heaters, portable fans, typewriters,

calculators, and reproduction machines when not in use.
-Automatic window displays and revolving signs at the end of normal business hours (and consider further reductions in operating time).

-Electric heat tracing when there is no fluid flow in pipes and when the outdoor temperature is above freezing.
-Elevator fans where smoking is not permitted and where applicable codes allow.

Discourage Excessive Use of Equipment:

-Encourage employees to go to the cafeteria or canteen for coffee breaks rather than operating coffee percolators in offices.
-Encourage chefs to preheat ovens no earlier than neccessary and to forego preheating completely except for baked goods.

-Consider reducing the number of electrically powered business machines in use.
-Insulate cooking equipment in kitchens, when possible.
-Prohibit use of portable electric heaters and encourage employees to move to a different location on the floor if drafts or cold radiation from windows are causing them discomfort.
-Where practical, substitute manual labor for electrical power, such as using manual labor to remove snow and ice rather than electric resistance snow melting systems.

REDUCE TRANSFORMER LOSSES

-De-energize transformers supplying unused offices or other areas.
-De-energize refrigeration chiller transformers during the heating season.
-De-energize heating equipment transformers during the cooling season.
-Where there is a bank or two or more transformers, operate transformers at the most efficient loading point.
-Reduce copper losses in the wiring - which increase with ambient temperature - by ventilating transformer vaults to reduce the ambient temperature.
-Shade outdoor transformer banks from solar radiation.
-De-energize dry type transformers serving convenience outlets when there is no load - at night and during weekends and holidays.

INCREASE MOTOR EFFICIENCY

Tighten belts and pulleys at regular intervals to reduce losses due to slip.
Lubricate motor and drives regularly to reduce friction.
Replace worn bearings.
Check alignment between motor and driven equipment to reduce wear and excessive torques.
Keep motors clean to facilitate cooling.
When replacing worn or defective motors, replace with motor sized as close to load as possible, and use the highest efficiency motor available.
Where it is impracticable to replace motors which have low load and power factors, use capacitors at motor terminals to correct the power factor to 90%.

This page blank

APPENDIX A
COST INDICES

For the purpose of this study, nine cities have been considered as follows: New York, Chicago, Houston, Phoenix, Miami, Seattle, Los Angeles, Denver and Atlanta.

Labor rates, including fringes, as of January 1975 were tabulated for certain selected trades, which were likely to be used for the type of work envisaged in the study and using New York City as base 100, the relative labor index for the respective cities was calculated.

A list of basic materials was compiled and current prices for the items secured from suppliers in the respective cities, and again, using New York City as base 100, the relative material index was calculated.

On the assumption that labor/material content of construction is approximately 60/40 respectively, a relative combined cost index for the cities was calculated resulting in the following:

NEW YORK	100
CHICAGO	93
LOS ANGELES	93
MIAMI	89
PHOENIX	88
SEATTLE	82
DENVER	82
ATLANTA	80
HOUSTON	79

ESCALATION

Prices given for examples of work items in this volume are generally New York City area prevailing costs as of January 1975 (that is they have been calculated using materials and wages costs for the metropolitan area as of that date).

ESCALATION (Cont'd)

Continually increasing costs of both labor and material will rapidly outdate this information. The user is therefore advised to obtain updated local costs tailored to his specific work project by calling upon the services of a local professional consultant (Architect, Consulting Engineer, Cost Consultant) or general contractor.

Prior to taking this action however the user may wish initially to investigate the cost impact of any increase himself.

This may be done by consulting one of the professionally prepared construction cost indices such as those prepared by Boeckhs; F.W. Dodge Company; Marshall & Swift; Turner Construction Company or by consulting articles on cost escalation in trade journals such as Buildings; other building management journals and Engineering News Record.

The latter magazine, a McGraw-Hill publication (issued approximately every 8 to 10 days) would be available in the local technical library. Each issue contains a feature entitled "Construction Scoreboard" which details cost index data.

Using the Building cost index column the user can readily calculate the average percentage increase since January 1975 and multiply the adjusted example costs accordingly to establish an updated price.

APPENDIX B
ENERGY CONSERVATION EXAMPLE

A. ASSUMPTIONS:

OFFICE BUILDING CHICAGO ILLINOIS. Gross area per floor = 100' x 100' = 10,000 sq.ft., 10 stories (total gross floor area = 100,000 sq.ft.). Floor-to-floor height = 10 ft., 33% window/wall area ratio.

CONDITIONS BEFORE OPERATIONAL CHANGES:

Clear single glazing, U=1.1 Negligible interior shading assumed during heating season. Shading by curtains during cooling season, SC=.67. Wall U=0.3 (excluding glazed areas), roof U=0.2. Occupied 40 hours per week. Outdoor ventilation air = 19,800 cfm @ 30 cfm/person x 660 occupants. Average infiltration = 1/2 air change/hour. Interior heat gains: Light 4.0 w/sq.ft., Office equipment = 0.5 w/sq.ft., Fans = 1.0 w/sq.ft. Total = 5.5 w/sq.ft.

No cooling during unoccupied periods.

Indoor temperature = 75° heating season
75° cooling season = 50% RH

Domestic hot water flow rate = 2 gallons/day/person @ 140°F

B. SUMMARY OF ENERGY USED IN CHICAGO OFFICE BUILDING WITHOUT ENERGY-CONSERVING OPTIONS

1. SITE ENERGY

	BTU/sq.ft./yr.
HEATING (OIL)	106,400
COOLING (ELEC)	8,200
LIGHTING	31,400
POWER	10,600
DOM. HOT WATER	4,000
TOTAL	160,600

2. RAW SOURCE ENERGY

	BTU/sq.ft./yr.
HEATING	106,400
COOLING*	27,340
LIGHTING*	104,660
POWER*	35,330
DOM. HOT WATER	4,000
TOTAL	277,730

*The figures for raw source energy reflect the energy conversion for electricity.

C. ANALYSIS OF ENERGY SAVINGS FOR CHICAGO OFFICE BUILDING

	ANNUAL SITE ENERGY SAVINGS	BTU SAVINGS PER SQ.FT./YR	% REDUCTION IN THIS SEGMENT OF ENERGY	% REDUCTION TOTAL ENERGY
HEATING				
12 deg. night setback	27,000 gal.#2 oil	38,000	35%	24.0%
7 deg. day setback	1,000 gal.#2 oil	1,400	1%	0.9%
Increase boiler efficiency by 10%	6,600 gal.#2 oil	9,200	9%	6%
Reduce outside air during occupied periods to 8 CFM/PSN	1,800 gal.#2 oil	2,500	2%	1.5%
Caulk windows	12,000 gal.#2 oil	16,700	16%	10.0%
* Add storm windows	26,000 gal.#2 oil	36,400	34%	23.0%
* Add night barrier (U=.1)	29,150 gal.#2 oil	40,000	38%	25.0%
Selected combinations:				
Increase boiler eff. 10% Plus 12° night setback	31,000 gal.#2 oil	43,000	40%	27.0%
Increase boiler eff. 10% Plus 7° day setback	9,000 gal.#2 oil	12,600	12%	8.0%
Increase boiler eff. 10% Plus caulk windows	17,800 gal.#2 oil	25,000	23%	15.5%
COOLING				
Economizer Cycle	47,700 KWH	1,600	20%	1%
Reduce lighting to 2.0 watts/sq. ft.	31,500 KWH	1,100	13%	0.7%
LIGHTING				
Reduce lighting to 3.0 watts/sq. ft.	226,000 KWH	7,700	25%	5.0%
Reduce lighting to 2.0 watts/sq. ft.	450,000 KWH	15,300	49%	9.5%
HOT WATER				
Lower water temperature to 100°F	1,070 gal.#2 oil	1,500	37%	1.0%
Reduce flow rate to 1 gal/PSN/DAY	920 gal.#2 oil	1,250	32%	0.8%
Combined 100°F temperature Plus reduced flow rate	1,990 gal.#2 oil	2,800	70%	2.0%

* Change requires some capital investment - See ECM 2

Note: For multiple changes the savings are not directly additive.

PART 2

ENGINEERS, ARCHITECTS AND OPERATORS MANUAL

ECM-2

Introduction and Scope

ECM-2

INTRODUCTION AND SCOPE

The opportunities to conserve energy by setting the levels of indoor environmental conditions, and through operational and maintenance practices which are detailed in ECM-1 for existing buildings are equally effective in new buildings. There are also other measures to conserve energy in existing buildings which are as practical and significant as similar practices in new building design and construction, such as reclamation of waste heat, efficient lighting lamps and fixtures, insulation and special glazings, and control systems to optimize energy use, although initial costs may be greater in existing buildings.

Energy conservation options which are available in new buildings such as site selection and building orientation, configuration, height, materials and in some cases, the choice of the most efficient mechanical systems are not available in existing buildings. However, by understanding the implications of energy use which are dependent upon those factors, an energy conservation consultant can select conservation options which are available to compensate for them.

Energy conservation opportunities in existing buildings can be identified and quantified more readily, in many cases, than for proposed new structures, for in existing buildings, the magnitude of the building load, the actual number of occupants, their activities and requirements, and the present energy usage, can be determined. The full and part-load conditions, load factor and performance of the mechanical and electrical systems can also be identified. The systems can be tailored to the known requirements of the building without the need to include safety margins which are usually part of a new design. It is these safety margins, which are included as a contingency against unknown conditions, that in the final analysis contribute heavily to excessive energy use.

An energy conservation program is most effective when a specific target or goal is established. The alternative opportunities to conserve energy can be judged on a benefit/cost basis to meet the goal. As energy consultants to the G.S.A. for the energy conservation demonstration building in Manchester, N.H., Dubin-Mindell-Bloome Associates, in collaboration with the National Bureau of Standards conducted a thorough analysis showing that it was possible and feasible (cost effective) to achieve 60% savings in energy by using energy conserving mechanical and electric systems and building construction as compared to a standard practice building. As a result of this project and subsequent building designs in other climates, it is possible to set a goal of 55,000 Btu's/square foot year for new office buildings, and the G.S.A. has adopted this standard for all their subsequent new buildings. For existing government office buildings retrofitted for energy conservation GSA has established an energy budget of 75,000 Btu's/square foot year.

The condition of the existing structure and its systems, and the degree to which energy conservation options are exercised determines whether or not a similar energy budget goal is attained.

Schools, department stores, religious buildings and apartment dwellings will be somewhat lower in energy use than office buildings, and hospitals and industrial plants generally will be higher. Until there is more definitive data available to establish energy budgets in terms of Btu's/square foot per year for each building type, in each climate for each of the many variables that exist, conservation goals for existing buildings should be set in terms of a percentage reduction of current energy use. A realistic goal for energy reduction is 10 - 20% without any initial cost, another 10% with minimal first costs and 10 - 20% more with an investment which will yield simple payback in three to ten years by lower operating expense. Energy use after these reductions will approach the GSA energy budget goal of 75,000 Btu/ft.^{2} year for existing buildings.

Additional energy savings which at present are doubtful economic investments will become economically viable as fuel costs continue to escalate and will enable the yearly energy budget to be reduced even further.

Energy conservation measures can be undertaken in any existing building to reduce fuel and electricity consumption, avoid shutdown due to fuel or electricity shortages, meet new state codes which limit energy use (rapidly being promulgated) and enable undersized equipment to handle existing or new loads. Energy conservation measures should always be analyzed and included when a building is remodeled or expanded.

-To accomodate required functional use changes

-To replace inoperative or obsolete equipment

-To serve a completely new function.

-When deteriorated building components require replacement.

-For safety or code compliance.

Purpose

ECM-2 has been prepared for use as a working tool for architects and engineers who are retained to analyze energy usage in an existing building and to recommend and design appropriate measures to conserve energy. This manual is also for use by building operating personnel who are responsible for an on-going energy management program and the maintenance and operation of the building and environmental control systems. The manual was prepared especially for existing commercial buildings, but most of the conservation opportunities and options are applicable to schools, universities, hospitals, hotels, industrial buildings (not process), auditoriums and other institutional buildings, as well as for new buildings. Where there are specific measures to change the operating conditions or modify equipment which is particularly wasteful of energy, the inferences for new buildings should be clear - do not select those systems in the first place.

Scope

The energy conservation options detailed in ECM-2 include measures which may be used to:

(1) Maintain selected indoor environmental conditions (included in suggested standards ECM-2).

(2) Reduce the heating, cooling, and lighting, loads by modifying the building structure or site.

(3) Improve the seasonal efficiency of the primary energy conversion equipment and distribution systems by modification or replacement.

Site planning, interior space utilization and design of and materials used in the building envelope construction strongly influence the quantity of energy used by the various building environmental systems. Environmental system loads and energy use can be reduced by changing their architectural and structural function and for this reason their application and effects are detailed under the appropriate mechanical and electrical systems. When making these changes architectural advice should be sought to preserve or enhance the aesthetic qualities of the building.

ECM-2 provides the elements of an energy management program with detailed instructions including the following:

-Conducting an energy audit

-Determining the potential for energy conservation

-Order-of-magnitude of costs for equipment, systems and materials

-Methods of determining long term economic feasibility of any energy conservation option by using an economic model

-Methods of shedding loads to reduce peak demands in order to reduce electrical demand charges and also conserve energy

-Special heat reclamation opportunities

-Computer program sources

-Provisions for present or future installations of solar energy and total energy projects.

-Applicable codes and standard requirements

The Federal Government must address total raw source energy considerations, as well as savings of energy at the building boundary These guidelines were prepared to meet both needs by recognizing the raw source energy implications of the use of electricity. Electric resistance heating is almost 100% efficient in the building, but its use is not recommended in this manual because it requires more raw source energy than other high quality systems.

ECM-1 can be used without reference to ECM-2 to determine the low cost measures which can be implemented promptly. ECM-2 however, cannot be used without reference to ECM-1 since it includes many graphs, tables, and charts which are required to quantify the energy savings for some of the measures outlined in ECM-2, as well as the building profiles and energy audit forms for use in ECM-2.

While ECM-1 and ECM-2 are comprehensive and include many energy conservation opportunities and guidelines, they are not all-inclusive. There are alternative approaches to many of the measures which are described.

Many of the guidelines shown in these documents are not mutually compatible. In each case, trade-offs must be analyzed thoroughly so that those measures which will be most conducive to energy conservation for any particular building are selected. It is obvious that a set of guidelines can never replace the intelligent application of judgement by architects and engineers in evaluating energy savings against costs, human requirements, and aesthetic needs.

It is suggested that the user of this document read it entirely before attempting to apply any or several of the guidelines. Careless use of or simplistic abstractions from these documents could prove counter-productive.

Engineering and architectural students and recent graduates should find these documents helpful as training manuals. Hopefully, they will be stimulated by the opportunities presented to utilize their talents in this important field.

Section 11

Establish an Energy Program

ESTABLISH AN ENERGY MANAGEMENT PROGRAM

A. PURPOSE

The purpose of an Energy Management Program for existing buildings is (1) to utilize fuels in the most effective manner to minimize the amount consumed for heating, ventilating, cooling, hot water, internal transportation, power, and other processes; (2) to use the particular type of fuel or electricity which is available at the lowest costs (and to switch from one fuel to another if the relative costs change radically in the future); and (3) to carry out these objectives within the financial capability of the owner and in the interests of developing national energy and resources policy.

An energy conservation program will be influenced by one or more of the following objectives, which the owner himself must establish:

-Reduce energy consumption as much as possible without initial costs, and/or invest more to save more.

-Reduce operating costs.

-Reduce the likelihood of shutdown due to reduced voltage or curtailment of power or fuel supplies.

-Extend the life of the mechanical and electrical equipment to reduce the frequency of equipment breakdown and replacement costs.

-Improve the performance of service and maintenance personnel to reduce operating costs and equipment failure.

-Reduce air-borne pollution.

-Meet existing Federal and State guidelines and forthcoming mandatory energy conservation standards.

-Enchance the rentability or saleability of the building through lower operating costs.

-Enhance the internal environment by such measures as reducing disturbing glare from lighting and drafts from infiltration or cold wall surfaces.

B. THE TEAM

Energy management is a team effort. The role of each member is clea., but their efforts may overlap in some instances.

The Owner

-Select objectives and institute the program;

-Hire personnel and retain professional

-Collect and tabulate operating costs, fuel and energy use and other pertinent data.

-Arrange access to the building for surveying and testing, balancing, adjusting, and retrofitting the mechanical and electrical systems and the structure;

-Control the conditions of building use;

-Pay those costs for services and goods which are not borne by the tenants.

<u>The Professional Engineer and Architect:</u>

-Analyze the existing conditions, survey the premises, identify the opportunities for conservation and select candidate systems for further analysis;

-Calculate loads and flow characteristics to determine the potential energy savings for each system or building component;

-Analyze utility rates, and perform cost/benefit analyses for each system or group of systems (analyses may be simple pay-back, cash flow, present worth, discounted rate-of-return, or life-cycle costs) as dictated by the owner's fiscal needs;

-Prepare plans and specifications for operating procedures and/or construction programs;

-Prepare an operating manual for use by personnel retained by the owner;

-Train the permanent energy management co-ordinator and operating personnel to monitor, operate and maintain the systems and structure in accordance with the developed energy management program.

<u>The Utility Companies:</u>

-Provide information on existing and anticipated utility rates;

-Loan or rent meters to measure flow rates to test, balance, and adjust systems for optimal performance.

<u>Service and Maintenance Organizations:</u>

-Perform efficiency tests before and after modifications, under the direction of the Consulting Engineer, for combustion, boilers, chillers, compressors, cooling towers, air-cooled condensers, and other types of equipment (including elevators, escalators, motors, and automatic control systems).

<u>Operating Personnel</u> working for the owner, tenant, or an <u>Operation Management Consultant</u> can:

-Administer, test and actually service equipment;

-Direct the cleaning activities to maintain maximum efficiency of filters, louvers, heat transfer equipment, lighting fixtures, windows, etc.;

-Direct activities to re-establish operating parameters in accordance with changes as they occur in occupancy and/or functional requirements of the building owner or tenants;

-Set up and maintain a data collection system to include a description of all systems, their operating characteristics, maintenance performed, and monthly fuel and energy usage;

-Develop specific operating programs and schedules, working from the operation manual provided by the engineer, to effectively monitor and control the operating systems.

Tenants or Occupants:

-Help to implement specific conservation measures by shutting off lights, closing doors, turning off hot water faucets, maintaining the specified thermostat and or humidistat settings, refrain from smoking, etc.

Professional Cost Engineers:

-Estimate costs of labor and materials for preliminary analyses. (Final costs should be based on competitive bids or advice from reliable contractors.)

C. PROCEDURE

Step 1: Select the Objective

Organize a conservation journal to record all findings, actions, calculations, and procedures. Select the specific objectives which are consistent with the building owner's interests and resources. If he must limit the investment for conservation measures to a minimal amount, consult ECM 1 for conservation opportunities and options. If he is prepared to invest more consult ECM 1 and ECM 2.

Step 2: Conduct an Energy Audit

Determine the amount of energy in Btu's/year and/or KWH's/year for each system and subsystem. Where electric or flow meters are not installed and records have not been kept, determine the yearly consumption of energy from fuel and utility company bills and make a preliminary estimate of the yearly consumption by subsystem as described in ECM 1, Section 3 , "Use and Implementati n". Use the audit form, Figure 12, to record the data. Convert the annual amount of gross energy used to Btu's/sq.ft. of floor area as described in ECM 1, Section 3 .

Step 3: Establish a Data Base

a. Building Construction - Existing Condition and Use

Examine available documents,* but in any event conduct a thorough survey of the site and building, to determine the actual dimensions and configuration of the building, the construction materials used, the physical condition of the structure, its compass orientation, and the amount of shading from adjacent buildings, plantings or solar control devices. Determine which safety, ventilating, and lighting codes are applicable. Gather data on the complete lighting systems, the occupancy, type and duration of tasks in each specific area of the building, ventilation rates, probable infiltration rates, and appliances or processes which use energy and emit heat into the building (investigation has often disclosed discrepancies between actual conditions and those originally assumed for design). For data collection refer to the profiles in ECM 1, Section 3 , Figures 3 and 10. Use this data to calculate the peak heating, ventilating, and cooling loads at the extreme outdoor weather conditions (winter and summer) for the desired indoor temperature and humidity levels. Consult the ASHRAE Handbook of Fundamentals (1973) for methods of performing calculations. Based on design load calculations, estimate the annual heating and cooling loads by manual procedures. If the building is large and contains many comolex systems it may be worthwhile, considering computer calculations. Refer to Section 24 for some of the computer programs available. The computerized procedures reflect hours of occupancy and extent and duration of indoor temperature and humidity levels, available sunshine, wind direction and velocities, and annual hourly occurrence of wet and dry bulb temperatures.

Providing the building has been operated in accordance with the assumptions used for the calculations, the annual loads can be used as a basis for estimating the amount of energy which would be required for systems supplying energy to offset heat loss and heat gain - if they were 100% efficient.

Compare the annual heat loads with the fuel or energy actually used for heating to determine the approximate seasonal efficiency of the existing heating system. For instance, if the annual "building" heating load were 100,000,000 Btu's, and the annual energy used by an oil heating system were 150,000,000 Btu's (ECM 1 profile, Figure 12), the seasonal efficiency would be 100,000,000 ÷ 150,000,000 = 66.6%. If the seasonal efficiency is less than 98% for an all-electric system, 78% for an oil or gas-fired system, or 70% for a coal-fired system, there is potential for improvement. However, as the efficiency of heating equipment may be lower at any given time that it is surveyed than it was during the entire previous heating season (or higher, if adjustments had been made), it is necessary to actually measure the current performance of the equipment** rather than to depend only on the average conditions which previously existed. In the preceding example, the seasonal loss of 33.4% was due,

* Construction and as-built drawings, shop drawings, specifications, temperature control diagrams, test and balance data, operating manuals.

**Also true for cooling equipment.

in part, to the efficiency of the boiler-burner unit and, in part, to distribution system thermal losses. Each factor must be individually analyzed. The seasonal performance (C.O.P.) of the refrigeration equipment can be estimated by dividing the calculated annual "building" cooling load by the gross amount of energy used for cooling.

For instance, if the annual amount of electric power used for cooling were 200,000 KWH, - the equivalent of 200,000 x 3413 = 682,600,000 Btu's, and the annual calculated gross "building" load were 1,365,200,000 Btu's, then the average C.O.P. would be $\frac{13,652 \times 10^5}{6826 \times 10^5}$ or

If the seasonal C.O.P. is less than 3 for electrically driven central air conditioning chillers or compressors, 2.8 for direct expansion compressors, 2.5 for unitary equipment, and .60 for absorption refrigeration equipment (.42 if it includes boiler and absorption unit), there is potential for improvement. Steam turbine driven equipment cannot be categorized so simply and must be analyzed for each individual installation.

If any of the "building" loads can be reduced, adjust the distribution systems and primary energy-conversion equipment further to match them more closely to loads and maximize the amount of energy conserved.

b. Mechanical and Electrical Systems - Operation and Performance

Examine existing documents for information, and in any event, make a thorough survey within the building of the mechanical and electrical systems which consume fuel or electricity. Determine their operating characteristics, physical condition, control sequences, materials of construction, capacity, size, and performance. Use and expand the profile (Figure 2 ECM 1) for the inventory of energy-using equipment.

Make an estimate of the amount of energy used annually by each of the subsystems . For a preliminary analysis, the procedure outlined in ECM 1, Section 3 , Figure 12 is sufficient, however, for a more accurate determination of energy use, equipment must be monitored and tested as described in Step 4, below.

Assume a maximum rate of energy consumption and a regular program of starts, stops, and part-load performance for each month to produce an estimate of the monthly and annual Btu and KWH for operation of equipment. This procedure will result in a more accurate budget of monthly and annual consumption for the following:

1) Heat Energy to Provide:

   a) Domestic Hot Water
   b) Process Heat

2) Electric Energy to Serve:

   a) Illumination, Interior

b) Illumination, Exterior
c) Air Handling Motors
d) Fluid Handling Motors
e) Transportation Motors
f) Appliances and Process Devices
g) Miscellaneous

Compare the results from the computer or hand calculations with the preliminary assessment of use by subsystem (calculated in accordance with the procedures in ECM-1, Section 3 , page 75). Reassign existing energy use to subsystems as a data base for measuring the improved performance after completion of the conservation work.

Step 4: Determine the Actual Amount of Energy Used by Each System

By calculation estimate the duration of each "building" load at specific percentages of full load, and the number of hours (at the same percentages of full load) that equipment should operate to meet the part and full conditions. If systems do or can be made to operate at their maximum efficiencies at each part load condition in accordance with the load factor schedule, annual energy use will be at a minimum.

In order to estimate the seasonal performance of each piece of equipment, measure the actual performance at peak conditions first:

-Make a full load combustion test, measuring CO_2, CO, or stack temperature, and fuel rate for all burners and boilers.

-Measure the cfm, gpm, brake horse, voltage and amperage for all air systems handling more than 3000 cfm and water systems with flow rates exceeding 10 gpm. Refer to ASHRAE Systems Handbook, (1973), Chapter 40, "Testing, Adjusting, and Balancing".

-Measure the pressure drops across each item of equipment such as air filters, strainers, heat exchangers and coils to determine the system characteristic and identify where options exist to reduce system resistances and power input to fans and pumps.

-Flow and energy consumption tests will indicate whether the flow rates and energy input are proper for the load, or whether re-balancing and adjustments are appropriate. Where adjustments are not possible, analyze the costs of replacing or retrofitting equipment to supply part loads efficiently.

By calculation compare the annual energy consumed by the equipment and systems as they are now operating to that consumed when they "follow the load." The difference between the two values will indicate the potential for conservation through equipment modification or replacement. Savings can be weighed against the initial implementation costs. Before rebalancing any system, the "building" loads should first be reduced - if analyses indicate that they can be reduced, and if the costs are acceptable to the owner.

All of the foregoing calculations can be performed manually. They can, however, be time consuming for large buildings or groups of buildings and in these cases the cost of manual calculations should be wieghed against the cost of using an appropriate computer program (See Section 24).

Retain a certified "test and balancing" firm if equipment and/or expertise available are insufficient to perform the task. Mr. Guy W. Gupton, Jr. P.E., in an address to seminar conducted by HPAC magazine, lists sources for information on testing and balancing:

1) Associated Air Balance Council (AABC) - the certifying body of the 'independent agencies'.

2) National Environmental Balancing Bureau (NEEB) - sponsored jointly by the Mechanical Contractors Association of America (MCAA) and the Sheet Metal & Air-conditioning Contractors National Association (SMACNA) as the certifying body of the installing contractors subsidiaries.

3) Construction Specifications Institute (CSI) - in the CSI Specification Series the guide specification Document 15050 entitled "Testing and Balancing of Environmental Systems".

Step 5: Parameters for Selecting Energy Conservation Measures for Analyses

Refer to the guidelines which follow Sections 13 through 20 inclusive to determine when and where to consider specific options to implement ECO's.

Step 6: Make a Systems Analysis of Applicable Energy Conservation Options

Select appropriate energy conservation options from the guidelines for each ECO in ECM 1 and 2, and follow the instructions for calculating the potential savings for each item singly and/or in combination with other options where they are interrelated - i.e., analyses for lighting savings must also include savings in cooling load and energy use as well as the loss of useful heat gain during the heating season.

NOTE: Many of the options in Section 13 through 20 provide methods of reducing annual loads in KWH or Btu's. Apply the seasonal efficiency for each system, respectively, to the loads to determine the actual amount of energy in Btu's or KWH that can be saved by each measure. Convert the energy units which are in Btu's to quantities of fuel and determine local fuel and electricity costs to develop annual costs in dollars. For single ECO's, use the Economic Model Section 23, to determine economic feasibility.

A building owner will often be faced with the situation where a number of Energy Conservation Opportunities exist but due to limited capital he can only implement one. Rather than make a full economic analysis of each alternative a "simple payback" comparison will indicate the most favorable.

To make a "simple payback" comparison first determine the following for each opportunity.

Prime cost of implementation	-	PC
Annual cost savings	-	AS
Average useful life of equipment	-	AUL

(Note: The annual cost savings is equal to the annual energy cost saving ± operating cost change)

Then apply the following formula to each opportunity,

$$\frac{PC}{AS} \times \frac{AUL_L}{AUL_A} = \text{Payback Factor}$$

where AUL_L = The longest average useful life of those opportunities considered.

AUL_L = The actual useful life of the opportunity.

The most favorable opportunity will be the one which has the <u>lowest</u> payback factor.

<u>Example</u>:

Find the most favorable choice between three ECO's

ECO	P.C.	A.S.	AUL
1	$10,000	$2,000	21 years
2	$ 4,000	$1,500	3 years
3	$ 7,000	$2,250	11 years

$$\text{ECO } 1 = \frac{10,000}{2,000} \times \frac{21}{21} = 5.0 \text{ pay-back factor}$$

$$\text{ECO } 2 = \frac{4,000}{1,500} \times \frac{21}{3} = 18.7 \text{ pay-back factor}$$

$$\text{ECO } 3 = \frac{7,000}{2,250} \quad \frac{21}{11} = 5.9 \text{ pay-back factor}$$

The most favorable choice for investment is ECO 1 and a full economic analysis of this opportunity should now be made.

Step 7: <u>Prepare Report of Energy Conservation Survey and Analysis</u>

Prepare a fully documented report with recommendations, costs, calculations and benefits for the owner's consideration. See Section 23 for methods of estimating costs and preparing economic analyses from an economic analysis model.

Step 8: Alternative Energy Sources

When approved by the owner, make a separate preliminary engineering and economic feasibility study for solar energy and/or total energy. Refer to Section 8 for guidance.

Step 9: Prepare Construction Documents

Prepare working drawings and specifications for all proposed construction work and assist owner in taking competitive bids. Analyze the bids, draw up contract agreements with the supervision of the owner's attorney, supervise construction work, and certify payments to contractors.

Step 10: Post Construction Services

Develop specific operating programs and schedules to monitor and control the operating systems for maintaining the environmental conditions conserving the maximum amount of energy, reducing operating costs, and utilizing manpower most effectively.

Train operating personnel and supervise the educational program to acquaint the building occupants with new operating conditions and goals.

Supervise the preparation of log books, which record all changes, energy use on a monthly basis (by subsystems), and costs, in order to authenticate the program and assure continuity of effort. Monthly consumption, adjusted for degree days, cooling hours, and operational changes, will indicate the effectiveness of current program.

Install suitable monitoring procedures to record the performance of modified equipment or procedures.

- Institute maintenance of equipment operation logs to identify hours of operation and responsibility for start-stop.
- Where necessary, install elapsed time meters to record hours of operation of prime movers.
- Where beneficial, install electric watthour meters to record usage of electric energy for specific operations or departments.
- Where beneficial, install fuel meters to record consumption of fuel for specific operational purposes.
- Where beneficial, install Btu meters, steam flow meters or condensate meters to identify heat input or output of processes of building environmental apparatus.
- Recording thermometers may be installed to verify performance of critical automatic temperature controls.
- Recording ammeters may be used to monitor relative load changes in electric prime movers and power distribution feeders.

Develop reporting procedures and summaries to monitor the operating and maintenance personnel execution of energy conservation program elements. Daily, weekly, and monthly logs of equipment use, meter readings and performance evaluations should be summarized and reported to operating personnel. Continued reference to performance goals and actual achievement is necessary to sustain interest and reward successful effort and skill at the level at which operating decisions are actually made.

-Start-stop equipment logs and elapsed time meters should be totalized monthly and performance compared to the Energy Budget allocation.

-Submeters on electricity, fuel, steam, condensate, chilled water, hot water, etc., should be totalized monthly and performance compared to the Energy Budget allocations.

Prepare comparative statements of energy use semi-annually to show the actual purchases of electricity and fuel compared to the Energy Budget and the Adjusted Record of Energy purchased in the past. Graphic presentations may provide a continuing synopsis of the effects of energy conservation programs.

-Semi-annual calculation of the cost effectiveness of energy conservation efforts should reflect the changes in unit costs of the energy purchased. It is necessary to identify the differences in energy units resulting from the conservation efforts and then apply current unit prices to determine the monetary value.

<u>Summary</u>:

This energy management program must include the following:

1) Design loads and system characteristics and operation documents and/or calculations.

2) Current operating conditions (measured).

3) Current loads (calculated or measured).

4) Current energy consumption by component (measured and calculated).

5) Reduced loads (selected from conservation options).

6) New system design or operational requirements (selected conservation options).

7) Predicted reduction in energy consumption.

8) Measured energy use after changes.

Section 12

Energy Conservation Opportunities Mechanical Systems Background

MECHANICAL SYSTEMS BACKGROUND

This background discusses Heating, Hot Water, Cooling and Commercial Refrigeration Systems, background of other systems is discussed specifically at the beginning of each appropriate section.

Principles of energy conservation and the relative amounts of energy used by each system are discussed in ECM 1, Sections 1 & 2 respectively.

When selecting a series of ECO's as a basis for an energy conservation program a rational approach should be followed as one action will affect subsequent decisions.

The approach should be:

First, reduce the load.

Then, reduce the distribution losses from the central plant to the terminal unit.

Finally, increase the efficiency of the prime energy user in the central plant.

Reducing load first will increase the magnitude of opportunities to reduce the energy used by the distribution systems and the "parasitic" losses. These savings are cumulative and if made in sequence reduce the load on the primary energy using equipment to a minimum. This in turn maximizes the opportunity for increasing its efficiency.

If, however, opportunities are present to immediately increase the efficiency of the prime energy using equipment by simple modification or adjustment, these may be implemented out of sequence. For example if adjustments to a boiler firing rate and draft conditions will increase efficiency these should be made immediately as readjustments can easily be made later when other ECO's have further reduced the load. If, however, the combustion chamber requires rebricking this should be left until all loads are reduced as the combustion chamber shape may well need changing or reducing in size to increase the efficiency.

Transmission of heat through the building surfaces, and the introduction of outdoor air by infiltration or through the ventilation systems directly affects both heating and cooling loads. Any steps taken to reduce heat transmission and outdoor air quantities for the purpose of reducing the heating load will also reduce the cooling load and vice-versa. This interaction must be recognized and taken into account when analyzing yearly energy savings.

Heat gains from lighting, machinery and people all contribute to cooling load and any reduction in these heat gains will reduce the cooling load. The same heat gains however are beneficial in winter and reductions for cooling will result in increased heating load during the occupied periods. The trade-off between heating and cooling must be examined to determine the effectiveness of particular ECO's.

The domestic hot water and commercial refrigeration loads are related only to the demands of the particular systems and little if any interaction occurs with other systems. There are however opportunities for recovering heat from one building system for use in another which may introduce interaction.

Energy input in the form of fan and pump horsepower is required to operate distribution systems. In addition to this energy input, losses also occur due to heat transmission between the system and ambient surroundings. Load reduction allows distribution flow rates to be reduced which in turn reduces fan and pump horsepower input. Further reduction of horsepower input can be achieved by modifying the distribution system resistance to flow. Losses due to heat transmission can be reduced by insulation.

The cumulative effect of reducing loads and distribution system losses provides opportunities for changing the operating characteristics of the prime energy using equipment to improve both full load and seasonal efficiency.

The net reduction in load is obtained by assessing the cumulative effects of reducing the load, reducing distribution losses and increasing prime equipment efficiency.

The actual fuel or energy input savings will be larger than the calculated net load savings by a factor reflecting the seasonal efficiency of the heating or cooling system.

Typically, the seasonal efficiency of an oil or gas fired boiler and hot water heating system is between 60%-70% and for coal fired boilers somewhat lower.

Typical seasonal C.O.P.'s for cooling equipment are listed below.

EQUIPMENT	SEASONAL COP
Steam turbine driven water chiller	0.95
Absorption water chiller	0.60 (0.42 with boiler efficiency included)
Steam turbine driven chiller with piggyback absorption chiller	1.25
Electric driven water chiller	2.5
Electric driven direct expansion chiller	2.3

These should be applied as follows:

$$\text{Total yearly load reduction} = \frac{\text{Net load reduction}}{\text{Seasonal Efficiency or COP}}$$

To determine the economic feasibility and benefits of load reduction and efficiency improvements, utilize the economic model from Section 9 and enter the appropriate figures for installation cost, fuel cost, and yearly fuel savings. The procedures outlined in the economic model using the calculated savings give broad range order of magnitude values. If more accurate quantification is required, then a full computer analysis of load reduction and savings should be made using one of the programs currently available. (See Section 24.)

Section 13

Energy Conservation Opportunities Heating and Ventilation

HEATING AND VENTILATION

ENERGY CONSERVATION OPPORTUNITIES

ECO 42 CONTROL SPACE TEMPERATURES AT ALL TIMES

Close control of indoor temperatures during both the occupied and unoccupied periods will reduce the heating load, particularly if settings are maintained at the levels suggested in ECM 1, Figure 14.

Install a seven-day dual setting thermostat and wire to control the operation of the boiler or other prime energy using equipment. Locate the thermostat in a typical representative space and program for temperature settings and occupied/unoccupied times. Allow time for pre-heat warm-up by subjective judgment and reduce this time by increments until building does not quite reach occupied temperature settings until after occupation. It may be necessary to adjust warm-up times for different seasons.

The cost of installing a seven-day dual setting thermostat will vary with the circumstances of each individual case and quotations should be obtained from local contractors.

Install additional thermostats controlling dampers or valves to give control of atypical zones that have different load characteristics or occupied times.

Install valves or dampers to manually shut-off areas of the building that are non-critical or lightly populated. In storage areas, for example, heating can be reduced to a minimum, compatible with frost protection and local radiant or hot air heaters can be provided for work stations.

ECO 43 REDUCE TRANSMISSION HEAT LOSSES THROUGH WINDOWS

Conversion to double glazing will halve the single glazing conduction heat loss, but actually, the reduction is modified by solar radiation and wind. A computer program ("SUNSET", see Section 24) was developed for this report to study the reaction of windows to varying weather conditions and runs were made for 12 cities chosen to give a typical cross-section of climates.

Figure 32 tabulates the heat loss in Btu/year for each square foot of glazing for the 12 selected locations.

Heat loss through windows is at its maximum in areas of high degree days but for any one geographic location is modified by orientation, heat loss being greatest through windows of north exposure and least in windows of south exposure due to the beneficial heating effect of the sun. This modifying effect of the sun is more marked in cold climates, and where winter sun altitudes are low but intensity is fairly high. The greatest economic return from double glazing will be on the facades which do not receive direct sunshine. This difference between various orientations is affected by geographic location and solar intensity as indicated in Figure 32.

Graphs (Figures 33 and 34) were developed for other climates which extrapolate loads from the calculations for 12 cities and allow prediction of annual heat loss based on inputs of heating degree days (from Figure 4), solar radiation (from Figure 5), and hours of occupancy/week. These two graphs distinguish between latitudes less than 35° and greater than 35°.

The graphs were based on internal heat gains of 12 Btu/sq.ft./hour when occupied, 10% avg. outdoor air ventilation when occupied and 1/2 air change/hour continuous infiltration. They also assume a window/wall percentage between the ranges of 30% and 70%. For other conditions the "SUNSET" program or other similar programs can be used. The program will also calculate values for triple glazing.

Storm windows may be applied either to the outside or inside of the existing window. First, determine the most suitable location for the storm window, taking into account the building construction (is there enough space within the window reveal to fit a storm window) and the change in window appearance from inside or outside the building.

If storm windows are mounted outside, multi-story buildings three stories and higher with fixed glazing, will have to be scaffolded, but if they can be fitted inside, the use of special staging will be unnecessary. On the other hand lower buildings can easily be fitted with external storm windows without scaffolding and at less cost; this would not entail moving furniture and disrupting normal work.

If the existing windows and frames are of poor construction and allow a high rate of infiltration, the storm windows should be fitted on the outside where possible. This will reduce infiltration without incurring the extra cost of caulking the window.

The space between the storm window and the existing window should be vented to outdoors and provided with drainage weepholes to prevent moisture build-up.

If storm windows are added to the existing windows, consider the use of reflective or tinted glass to reduce solar gain and glare in summer.(Refer to Cooling Section.)

Where natural lighting is desirable but appearance and visibility through the window are not important, a low cost interim solution which will provide double glazing is to use clear plastic film attached to a simple sub-frame. This will not provide a durable storm window but should be considered as a temporary means to save heating energy where money is not immediately available for a permanent installation. This could be done for 10¢/sq.ft. material cost and could be installed in 30 minutes/window by unskilled labor.

FIGURE 32: YEARLY HEAT LOSS PER SQUARE FOOT OF SINGLE GLAZING AND DOUBLE GLAZING

CITY	LATITUDE	SOLAR RADIATION LANGLEY'S	DEGREE DAYS	HEAT LOSS THROUGH WINDOW BTU/FT.^{2} YEAR					
				NORTH		EAST & WEST		SOUTH	
				SINGLE	DOUBLE	SINGLE	DOUBLE	SINGLE	DOUBLE
MINNEAPOLIS	45°N	325	8,382	187,362	94,419	161,707	84,536	140,428	74,865
CONCORD, N.H.	43°N	300	7,000	158,770	83,861	136,073	73,303	122,144	67,586
DENVER	40°N	425	6,283	136,452	70,449	117,487	62,437	109,365	59,481
CHICAGO	42°N	350	6,155	147,252	75,196	126,838	65,810	110,035	58,632
ST. LOUIS	39°N	375	4,900	109,915	56,054	94,205	49,355	84,399	45,398
NEW YORK	41°N	350	4,871	109,672	54,986	93,700	48,611	82,769	44,580
SAN FRANCISCO	38°N	410	3,015	49,600	25,649	43,866	23,704	41,691	23,239
ATLANTA	34°N	390	2,983	63,509	31,992	55,155	28,801	51,837	28,092
LOS ANGELES	34°N	470	2,061	21,059	11,532	19,487	10,954	19,485	10,989
PHOENIX	33°N	520	1,765	25,951	14,381	22,381	12,885	22,488	12,810
HOUSTON	30°N	430	1,600	33,599	17,939	30,744	17,053	30,200	16,861
MIAMI	26°N	451	141	1,404	742	1,345	742	1,345	742

dubin-mindell-bloome-associates
consulting engineers

Existing single glazed windows may be converted permanently to double glazed windows by the addition of a new glazing frame to accept the additional pane of glass. The space between the glazing should be vented and drained to outside and provisions made for cleaning both sides of each sheet of glass.

Where the existing window frame is in good condition and the glazing system permits, a single sheet of glass may be replaced by a sealed double glazing unit. Where the existing window frame is in poor condition and is scheduled for imminent renovation or replacement, a double glazed unit should be considered.

In all cases whether storm windows are added or single windows are replaced by double, the frame of the selected unit should not form a heat bridge. Each frame should incorporate a thermal break between inside and outside surfaces. Such frames are commercially available. This is particularly important in frames made of metal, which readily conducts heat. A more desirable framing material is wood which is a poor heat conductor.

Before installing double glazing check the local and national codes. Some building codes specify tempered glass. Glazing fitted outside existing windows must be fastened in an acceptable manner to prevent hazard to passers-by. Wind loads must be assessed and appropriate thicknesses and strength of glass used.

Local fire codes often specify a certain percentage or number of windows per floor that must be operable. In this case, and also where operable windows are required for ventilation in toilets, etc., the additional glazing should also be operable.

The cost of converting to double glazing varies with the methods used and the type of existing glazing. To make an accurate assessment, quotations should be obtained from local contractors. For order of magnitude, however, assume costs of $7.50 per square foot for glass and $14.00 per square foot for replacing existing frames and single glazing with sealed double glazed units.

To determine the energy saved by double glazing, refer to Figures 34 and 35 and then carry out the following procedure:

a) Determine your latitude and select the appropriate graph.

b) Determine from Figure 4 the heating degree days for your location.

c) Determine from Figure 5 the mean daily solar radiation in Langleys for your location.

d) Select the orientation and determine the area of the windows to be double glazed.

e) Enter the graph at the appropriate degree days. Following the direction of the example line, intersect with the appropriate Langleys and orientation. Read out the yearly energy requirements for one square foot of single glazed window. Repeat the procedure for double glazing and subtract this value from the

FIGURE 33: YEARLY HEAT LOSS THROUGH WINDOWS—LATITUDE 25°N-35°N

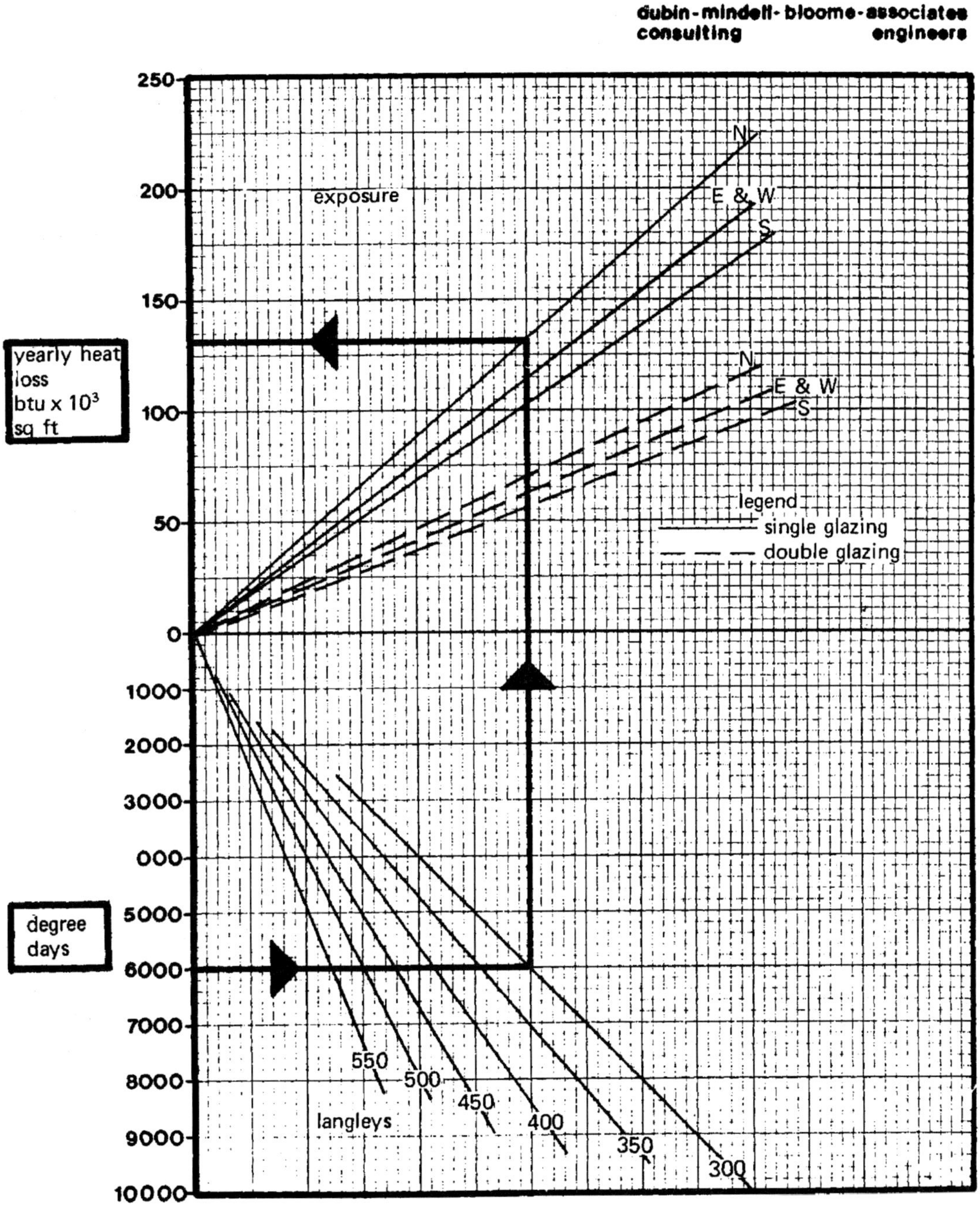

Figure #33 Engineering Data

This figure is based on the "Sunset" computer program which was used to calculate solar effect on windows for 12 selected locations. The program calculates hourly solar angles and intensities for the 21st day of each month. Radiation intensity values were modified by the average percentage of cloud cover taken from weather records on an hourly basis. Heat losses are based on a 68° indoor temperature.

Additional assumptions were: 1) Total internal heat gain of 12 Btu/sq. ft. 2) Average outdoor air ventilation rate of 10%. 3) Infiltration rate of 1/2 air change per hour. Daily totals were then summed for the number of days in each month to arrive at monthly heat losses. The length of the heating season for each location considered was determined from weather data and characteristic operating periods. Yearly heat losses were derived by summing monthly totals for the length of the heating season. These are summarized in Figure #32 for the 12 locations. The data was then plotted and extrapolated to include the entire range of degree days. Figure #33 was derived from locations with latitudes between 25°N - 35°N.

FIGURE 34: YEARLY HEAT LOSS THROUGH WINDOWS–LATITUDE 35°N-45°N

aubin-mindell-bloome-associates consulting engineers

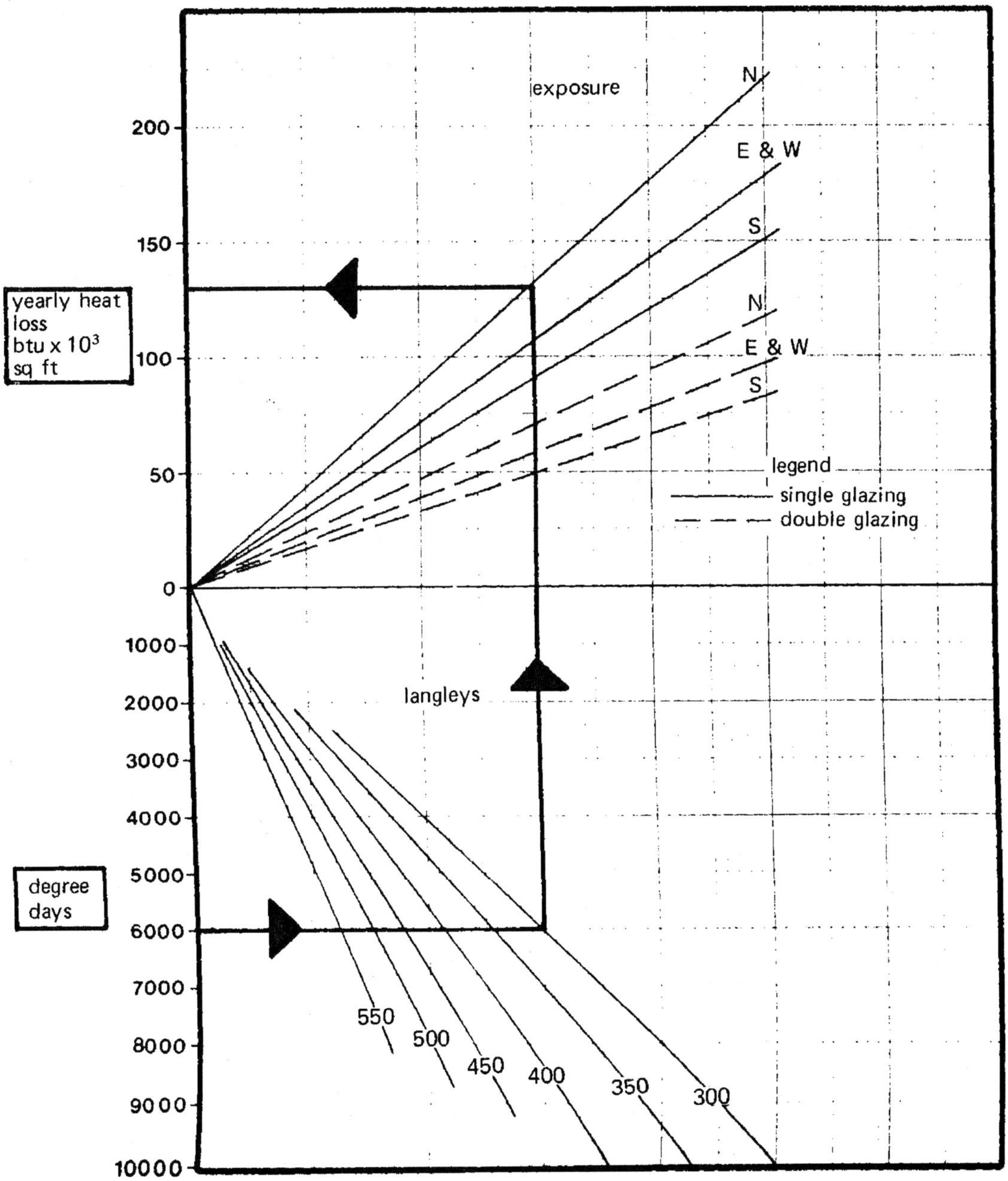

Figure #34 Engineering Data

This figure is based on the "Sunset" computer program which was used to calculate solar effect on windows for 12 selected locations. The program calculates hourly solar angles and intensities for the 21st day of each month. Radiation intensity values were modified by the average percentage of cloud cover taken from weather records on an hourly basis. Heat losses are based on a 68° indoor temperature.

Additional assumptions were: 1) Total internal heat gain of 12 Btu/sq. ft. 2) Average outdoor air ventilation rate of 10%. 3) Infiltration rate of 1/2 air change per hour. Daily totals were then summed for the number of days in each month to arrive at monthly heat losses. The length of the heating season for each location considered was determined from weather data and characteristic operating periods. Yearly heat losses were derived by summing monthly totals for the length of the heating season. These are summarized in Figure #32 for the 12 locations. The data was then plotted and extrapolated to include the entire range of degree days. Figure #34 was derived from locations with latitudes between 35°N - 45°N.

energy required for single glazing for the load reduction in Btu/yr. per sq.ft. of double-glazed window. Multiply by the area of window to determine the total yearly load reduction for the orientation considered.

f) Repeat the procedure for other orientations. Note that the heat losses derived from the graph assume that the windows are subjected to direct sunshine and are not shaded by adjacent buildings, structures, trees, etc. If shading occurs, then regardless of the actual orientation use the figures for north facing windows as these do not include a direct-sunshine component.

The total annual load reduction is the sum of all load reductions of various orientations and represents the total annual reduction in building heat load due to window treatment.

Interpolate for orientations other than the four cardinal points shown.

In addition to the energy savings made by reducing heat losses, double glazing also reduces heat gain in summer, resulting in a slight reduction of cooling loads; it is never worthwhile, however, considering double glazing for cooling benefits <u>alone</u>. (Refer to Section 15.)

Where existing single windows are very loose and allow large volumes of infiltration, the addition of double glazing will also reduce infiltration and will save both heating and cooling energy. The saving of heating energy due to lower infiltration rates can be considerable in locations experiencing more than 4,000 degree days, and in cooling energy in locations experiencing more than 8,000 wet bulb degree hours above 66° F WB, and should be evaluated in the economic study. (See ECM 1, ECO's 7 and 19 to calculate infiltration losses.)

ECO 44 INSTALL THERMAL BARRIERS TO FURTHER REDUCE HEAT LOSSES THROUGH WINDOWS DURING UNOCCUPIED HOURS

Most commercial, office, and religious buildings are unoccupied for more hours per week than they are occupied, and a major portion of the unoccupied hours occur at night when outdoor conditions are at their lowest and the potential for heat loss is highest. During unoccupied hours, daylight is either not available or not required and windows can thus be covered with thermally insulated barriers or shutters to decrease heat transmission losses. Thermally insulated barriers may be applied to the inside of the building and can be arranged to slide in a track or fold back into the window reveal or onto the face of the wall. When closed, the barriers should provide a reasonable seal over the window, but need not be air tight. Barriers may be purchased ready-made or may be custom-made to fit the windows. The major proportion of the total cost is in labor for construction and fitting rather than the cost of materials. Thermal barriers should, therefore, be selected or designed to have the best insulating value attainable for a given thickness.

Before installing the barrier, check the local fire code as some insulating materials are considered a fire or smoke hazard. Check also that barriers do not impede access required by the fire department.

The total insulating effect of the thermal barriers will vary with single glazed or double glazed windows and the type of insulation used in constructing them.

The following table shows the composite "U" values obtained by the barrier/window assembly for various thicknesses of insulation. It is assumed that the insulation has a "K" of 0.25 which is typical for glass wool, beaded polystyrene and polyurethane foam.

Insulation Thickness	COMPOSITE "U" VALUE	
	Single Glazing	Double Glazing
1/2"	0.28	0.23
1 "	0.18	0.16
1-1/2"	0.13	0.12
2 "	0.11	0.10

The cost of purchasing and installing thermal barriers will vary according to size and type of window, materials used in construction and ease of access for fitting. Quotations should be obtained from local contractors to determine the cost but for order of magnitude assume $5.00/square foot of window for a folding barrier.

To determine the energy saved by thermal barriers, refer to Figure 35 and carry out the following procedure:

a) Determine from Figure 4 the heating degree days for your location.

b) Determine the composite "U" value for the barrier/window assembly from the above table.

c) Determine the average unoccupied hours/week during the winter.

d) Enter the graph at the appropriate degree days and following the direction of the example line intersect with the appropriate composite "U" value and unoccupied hours; then read out the yearly heating load.

e) Repeat the above procedure for a window without thermal barriers and obtain the yearly heating energy required. Subtract from this the energy required when thermal barriers are fitted to obtain the yearly heating load reduction due to one square foot. Multiply this by the total area of treated windows to obtain the net yearly heat loss savings.

FIGURE 35: YEARLY HEAT LOSS FOR WINDOWS WITH THERMAL BARRIERS

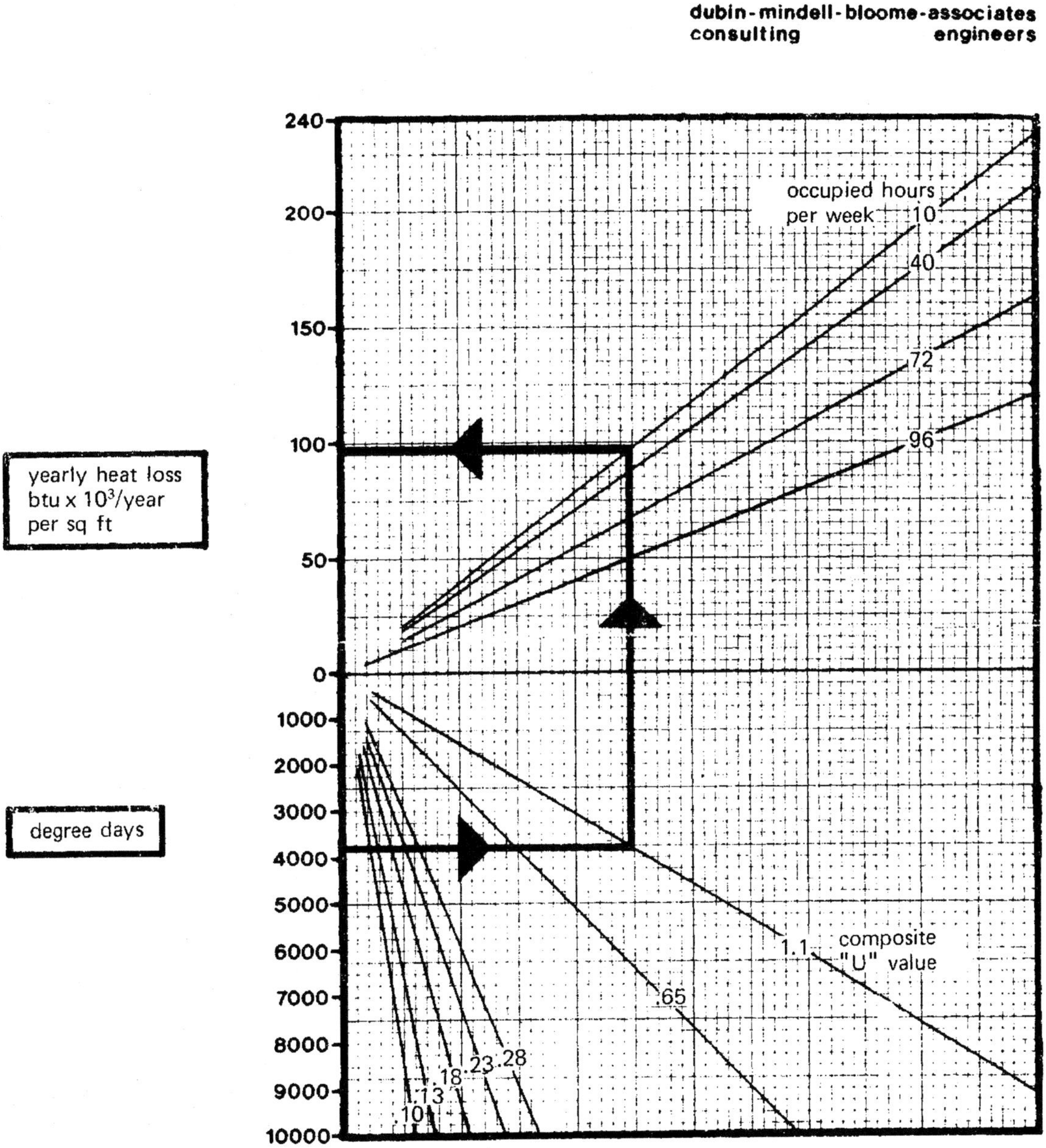

Figure # 35 Engineering Data

The development of this figure was based on the assumptions that

1) Thermal barriers are closed only when the building is unoccupied

2) The average degree day distribution is 25% during the daytime and 75% during nighttime.

The number of degree days occuring when the thermal barriers are closed (adjusted degree days - DD_A) were determined from the characteristic occupancy periods shown in the figure. This can be expressed as a fraction of the total degree days (DD_T) by the relationship:

$$DD_A = 0.25\ DD_T \left(\frac{\text{unoccupied daytime hours/week}}{\text{total daytime hours/week}}\right) + 0.75\ DD_T \left(\frac{\text{unoccupied nighttime hours/week}}{\text{total nighttime hours/week}}\right)$$

Yearly heat losses can then be determined by:

Q (heat loss/year) = DD_A x U value x 24

FIGURE 36: YEARLY HEAT LOSS PER SQUARE FOOT THROUGH WALLS

CITY	LATITUDE	SOLAR RADIATION LANGLEY'S	DEGREE DAYS	HEAT LOSS THROUGH WALLS BTU/FT.^{2} YEAR											
				NORTH				EAST & WEST				SOUTH			
				U=0.39		U=0.1		U=0.39		U=0.1		U=0.39		U=0.1	
				a=0.3	a=0.8	a=0.3	a=0.8	a=0.3	a=0.8	a=0.3	a=0.8	a=0.3	a=0.8	a=0.3	a=0.8
MINNEAPOLIS	45°N	325	8,382	74,423	70,651	20,452	19,335	70,560	62,229	19,378	16,787	66,066	51,298	18,109	13,530
CONCORD, N.H.	43°N	300	7,000	68,759	64,826	18,895	17,714	64,674	55,363	17,743	14,972	59,759	43,667	16,370	11,344
DENVER	40°N	425	6,283	57,337	53,943	15,755	14,824	53,726	44,937	14,763	12,198	48,780	34,095	13,405	8,720
CHICAGO	42°N	350	6,155	58,516	55,356	16,081	15,210	55,219	47,678	15,169	12,865	50,684	37,339	13,847	9,743
ST. LOUIS	39°N	375	4,900	45,046	42,149	12,379	11,565	41,981	35,192	11,533	9,476	38,038	26,344	10,425	6,660
NEW YORK	41°N	350	4,871	45,906	42,950	12,615	11,804	42,843	35,368	11,774	9,594	38,385	25,231	10,548	6,406
SAN FRANCISCO	38°N	410	3,015	23,258	21,120	6,392	5,803	20,916	15,631	5,748	4,118	16,948	9,812	4,645	1,743
ATLANTA	34°N	390	2,983	26,922	24,803	7,398	6,771	24,475	19,206	6,716	5,103	20,639	12,399	5,562	2,587
LOS ANGELES	34°N	470	2,061	9,900	8,549	2,720	2,349	8,392	5,758	2,306	1,316	6,139	3,040	1,520	155
PHOENIX	33°N	520	1,765	11,861	10,533	3,259	2,878	10,283	7,316	2,826	1,811	8,077	4,619	2,062	555
HOUSTON	30°N	430	1,600	14,592	12,956	4,011	3,557	12,888	9,379	3,542	2,351	10,878	6,760	2,909	1,142
MIAMI	26°N	451	141	210	106	7	0	92	0	0	0	6	0	0	0

dubin-mindell-bloome-associates
consulting engineers

ECO 45 REDUCE TRANSMISSION HEAT LOSSES THROUGH WALLS

To reduce the heat loss through walls, reduce the "U" value by increasing the insulating value of the wall.

Heat loss through a wall is a function of its resistance to heat flow modified by the effect of solar radiation (reduces heat loss) and wind (increases heat loss) on the outside surface. The effect of solar radiation is modified by the absorption coefficient of the outside surfaces; dark colors (high absorption coefficient) will reduce the heat loss more than light colors (low absorption coefficient).

The mass of the wall and its attendant thermal inertia have an overall modifying effect on heat loss by delaying the impact of outdoor temperature changes on the heated space. This time delay allows the wall to act dynamically as a thermal storage system, smoothing out peaks in heat flow and reducing yearly heat loss. Low mass walls of 10-20 lbs./square foot will have approximately 2% greater yearly heat loss than high mass walls of 80-90 lbs./square foot, assuming both walls have the same overall "U" value and absorption coefficient.

A computer program was developed for this manual to study the reaction of walls to varying climatic conditions and runs were made for 12 cities chosen to provide a typical cross section of climates.

Figure 36 tabulates the heat loss for one square foot of wall in Btu/year for two different "U" values and absorption coefficients for each of the 12 selected locations. Heat loss through walls is greatest in locations having high degree days but for any one location varies with the wall orientation; heat loss is highest through north walls and least through south walls due to the beneficial effect of the sun. The difference is more marked in walls having a high absorption coefficient. Graphs were developed which extrapolate from the calculated results and allow prediction of yearly heat loss through one square foot of wall based on inputs of degree days (from Figure 4), solar radiation (from Figure 5), "U" values, orientation and absorption coefficient. These two graphs were developed for locations having latitudes greater than 35° and locations having latitudes less than 35°. The graphs assume internal heat gains of 12 $Btu/ft.^2$ hour when occupied, 10% avg. outdoor air ventilation when occupied and 1/2 air change per hour continuous infiltration. They also assume a window/wall percentage between the ranges of 30% and 70%. For significantly different conditions, an individual computer run should be made using one of the programs listed in Section 10.

The overall "U" value of walls may be decreased by the addition of insulating material to the inside surface, to the outside surface, or to fill cavities within the wall structure.

Addition of insulation to the outside of walls can be accomplished with relative ease in one or two story buildings, but becomes increasingly difficult for the higher stories because of the requirement for access staging. Insulation added to the outside of walls must be weatherproof and vapor-sealed to prevent deterioration of its insulating properties due to the ingress of moisture. External insulation could be added in the form of prefabricated insulating panels finished to give the most advantageous absorption coefficient. A design problem likely to occur when adding insulation to the outside surface of walls is treatment of window openings and door openings; it must be resolved architecturally for each individual case. The appearance of the building will be changed and may be subject to approval to meet local fine arts standards or covenant requirements.

Addition of insulation to the inside surface of walls can be accomplishd with the same ease for single and multi-story buildings (as access is from the floor of each story treated). It may, however, entail moving furniture and fittings away from the wall and could interrup normal operation of the building. Internal insulation must be protected from moisture by a vapor barrier and from degradation by wear and tear either by having an integral finished surface or by being covered by a protective finish such as wood paneling or gypsum plasterboard. Design problems likely to occur when adding insulation to the internal surfaces of walls are treatment of window and door openings, architraves and reveals, the junction between the insulation and the floor and ceiling; and repositioning equipment such as heaters, electric receptacles, etc. that are recessed in the existing wall.

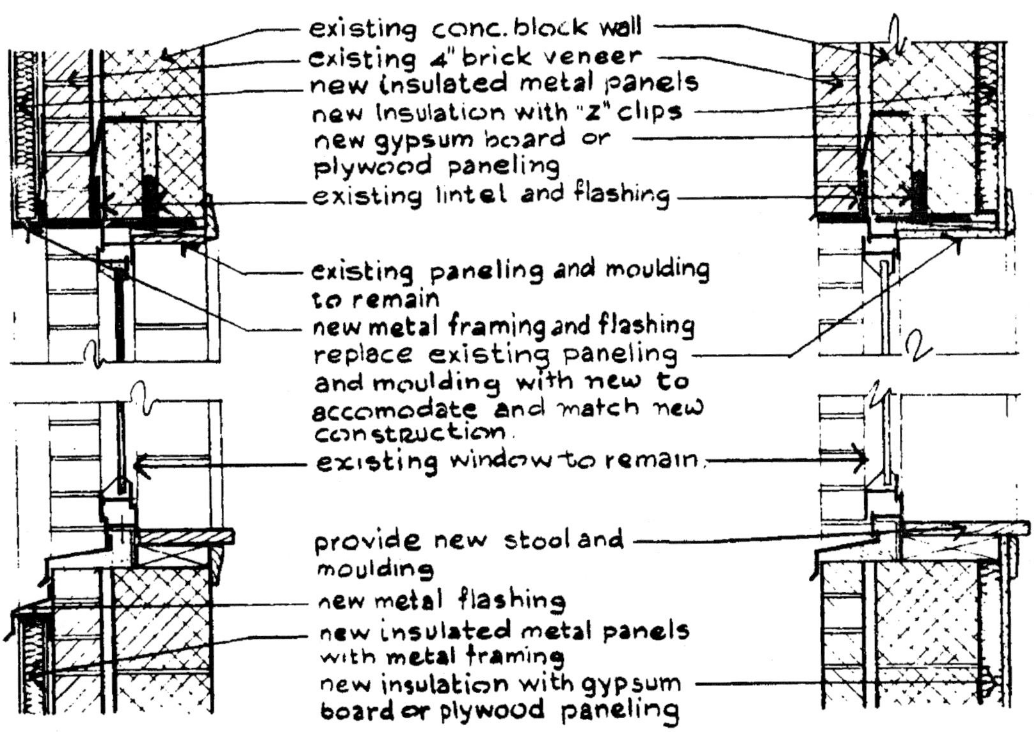

Insulation on exterior face

Insulation on interior face

Insulation can be added easily to a wall having an internal void by filling the space either with granular insulation or with plastic foam (polyurethane). Either process will not alter the appearance of the building and can be accomlished with minimum inconvenience to the occupants. Each process requires holes cut or core drilled through the wall to give access to the internal void. Granular insulation should be blown in under pressure to assure a compressed fill as it tends to pack down in time. There are many specialist firms that provide these complete insulation services. Investigation should be made to ensure that the wall materials are compatible with the type of insulation considered and that interaction will not occur that could damage either the wall structure or the insulation. Some wall cavities in older buildings were designed to prevent driving rain from penetrating to the inside skin. In such cases, use an insulating material that will not interfere with this function or coat the outside wall surface with epoxy resin or similar waterproofing.

Before installing insulation on the outside surface of walls, check the local building codes to ensure that the method of fastening is acceptable. Check the fire codes and your local fire department to ensure that the selected insulation is acceptable and is not considered a fire or smoke hazard. Fire requirements are more stringent for insulation inside the building than outside. Some processes for foaming insulation in place produce poisonous fumes; check whether the building must be evacuated while the work is being done.

The cost of insulating walls will vary with the circumstances of each particular case, and for accurate assessment, quotations should be obtained from local specialist contractors. For order of magnitude, however, the following tabulated costs may be used:

<u>Approximate Insulation Costs for Walls</u>

Rigid insulation on interior surface	\$1.06/sq.ft.
Rigid insulation on exterior surface	1.80/sq.ft.
Rigid insulation on exterior surface with stucco finish	3.20/sq.ft.
Cavity fill mineral fiber	1.70/sq.ft.
Cavity fill plastic foam	.80 sq.ft.
External epoxy coating	.42 sq.ft.

To determine whether additional wall insulation is worthwhile, it is first necessary to know the "U" value of the untreated existing wall. By examination, determine the type and thickness of the individual components of the wall. Using this information and the thermal characteristics listed in ASHRAE <u>Fundamentals</u>, compute the overall "U" value. This calculated value should be checked by the following procedure:

a) Measure the indoor and outdoor air temperature with an accurately calibrated thermometer on a sunless day or at night.

b) Measure the coincident inside surface temperature of the wall with an accurately calibrated contact thermometer.

c) Calculate the rate of heat flow through the wall from the inside air temperature, inside surface temperature and the inside film coefficient (assume R of 0.6), using the formula:

$$\text{Heat flow} = \frac{\text{Inside air temperature} - \text{Inside surface temperature}}{\text{Film coefficient}}$$

Using the calculated rate of heat flow, now calculate the overall "U" value using the formula:

$$U = \frac{\text{Heat flow}}{\text{Inside air temperature} - \text{Outside air temperature}}$$

For example, if the indoor temperature is 68° F, the outdoor temperature is 20° F and the inside wall surface temperature is 59.4° F, then:

$$\text{Rate of heat flow} = \frac{68 - 59.4}{0.6} = \underline{\underline{14.3\ \text{Btu/hr}}}$$

$$\text{Overall "U" value} = \frac{14.3}{68 - 20} = 0.298 \text{ or } \underline{\underline{\text{About } 0.3}}$$

Using this measurement method allows for normal variation of the external film coefficient, but to obtain the best results, measurements should be made on a day of average wind speed.

There will not necessarily be exact agreement between the calculated and measured "U" values and an average of the two results should be used.

Before savings can be calculated, it is necessary to know the "U" values of the walls both before and after adding insulation and having established the "U" value of the existing walls a decision must now be made as to how much insulation should be added and where it should be applied. In existing buildings, this decision is governed by the practicality of coordinating the insulation with the existing structure and requires an architectural solution based on the particular circumstances of the building.

The new "U" value can now be derived. Refer to Figure 37, and entering the graph at the present "U" value, intersect with the appropriate insulation thickness line and then read out the new "U" value. For example: a wall which had an initial "U" value of 0.3 and 2" of added insulation, would have an improved "U" value of 0.08. This graph can also be used as an aid to find the required thickness of insulation needed to improve the initial "U" value to the desired figure. For example: if it were desired to improve a wall's "U" value from 0.7 to less than 0.08, then 3" of insulation would be required.

This graph is based on insulation having a "K" of 0.25 which is typical for glass wool, beaded polystyrene, foamed polyurethane and loosely packed mineral wool.

To determine the heating energy saved by insulating walls, refer to Figures 38 and 39, then carry out the following procedure:

a) Determine your latitude and select the appropriate graph.

b) Determine from Figure 4 the heating degree days for your location.

c) Determine from Figure 5 the mean daily solar radiation in Langleys for your location.

d) Determine the orientation and area of the wall or walls to be insulated.

e) Determine the initial "U" value of the wall and the improved "U" value after insulation.

f) Determine the absorption coefficient of the outside surface of the wall.

g) Using the initial "U" value of the wall, enter the graph at the appropriate degree days and following the direction of the example line, intersect at the appropriate points for Langleys, orientation and "U" value. Read out the yearly heat loss per square foot of wall on the appropriate absorption coefficient line. Interpolate the results for absorption coefficients other than the 0.3 and 0.8 shown. Repeat this procedure using the improved "U" value and subtract the energy used before insulation to derive the energy saving/square foot of wall. Multiply this saving by the area of the wall insulated.

h) Repeat the complete procedure outlined in (g) above for each different orientation of wall to be insulated. The graph shows orientation for the four cardinal points (north, east, south, and west.) Interpolate for other orientations. Note that the heat losses derived from the graph assume that the wall is not shaded from direct sunshine. If shading occurs, then regardless of the actual orientation of the wall, treat it as a north wall, as this does not include a direct sunshine component.

Total the yearly savings of all walls insulated to obtain total saving of heat energy.

In addition to savings achieved by reducing winter heat losses, wall insulation will also reduce summer heat gains. This extra saving is small but nevertheless should be recognized when assessing the economic feasibility. (See Section 15).

FIGURE 37: EFFECT OF INSULATION ON "U" VALUE

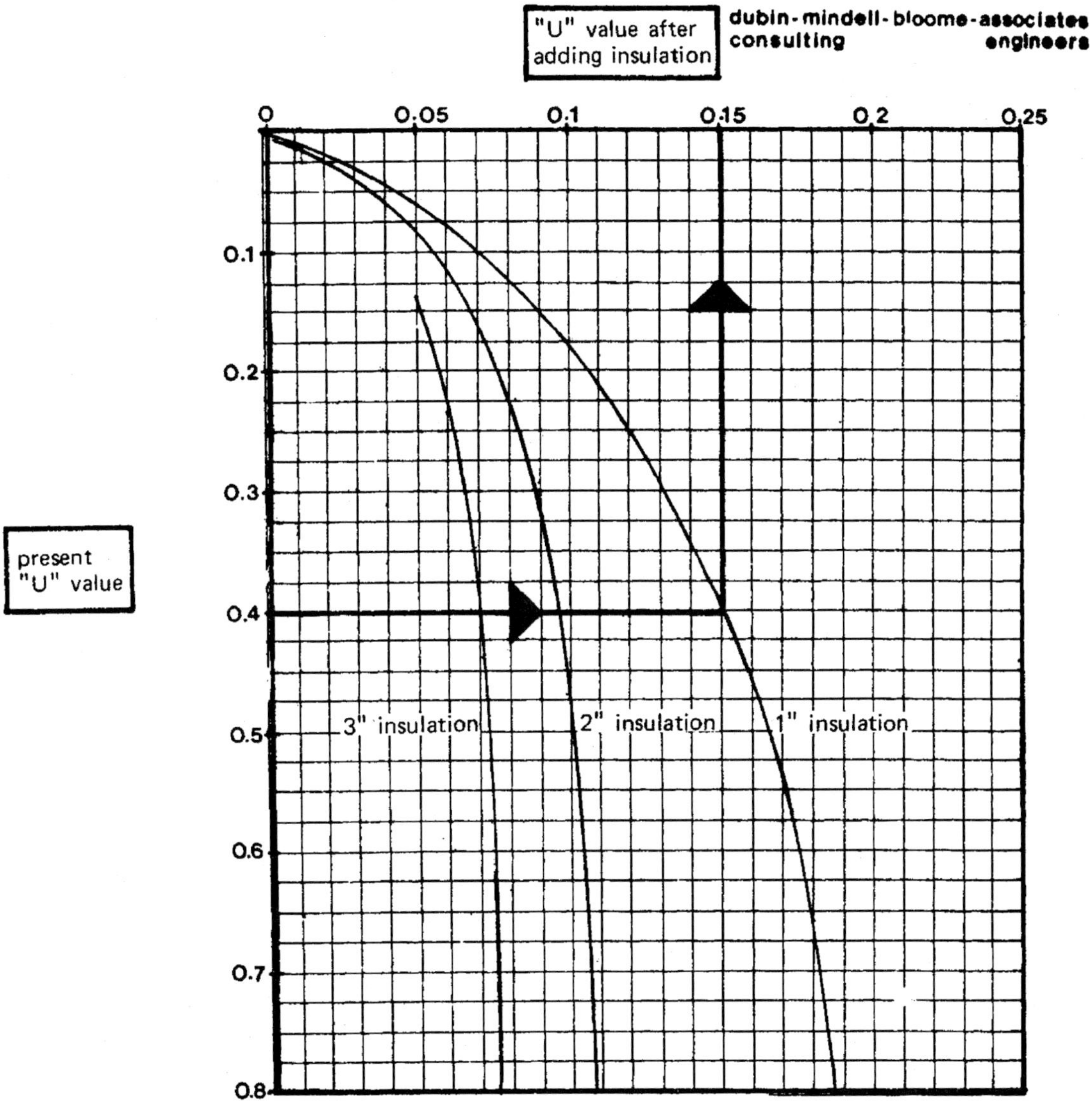

Figure #37 Engineering Data

This figure is based on the addition of insulation with an R Value of 4.0. The calculations to determine the U value after the addition of insulation assumed a combined inside/outside film coefficient of R=1. The standard formula U=1/R was used.

FIGURE 38: YEARLY HEAT LOSS THROUGH WALLS—LATITUDE 25°N-35°N

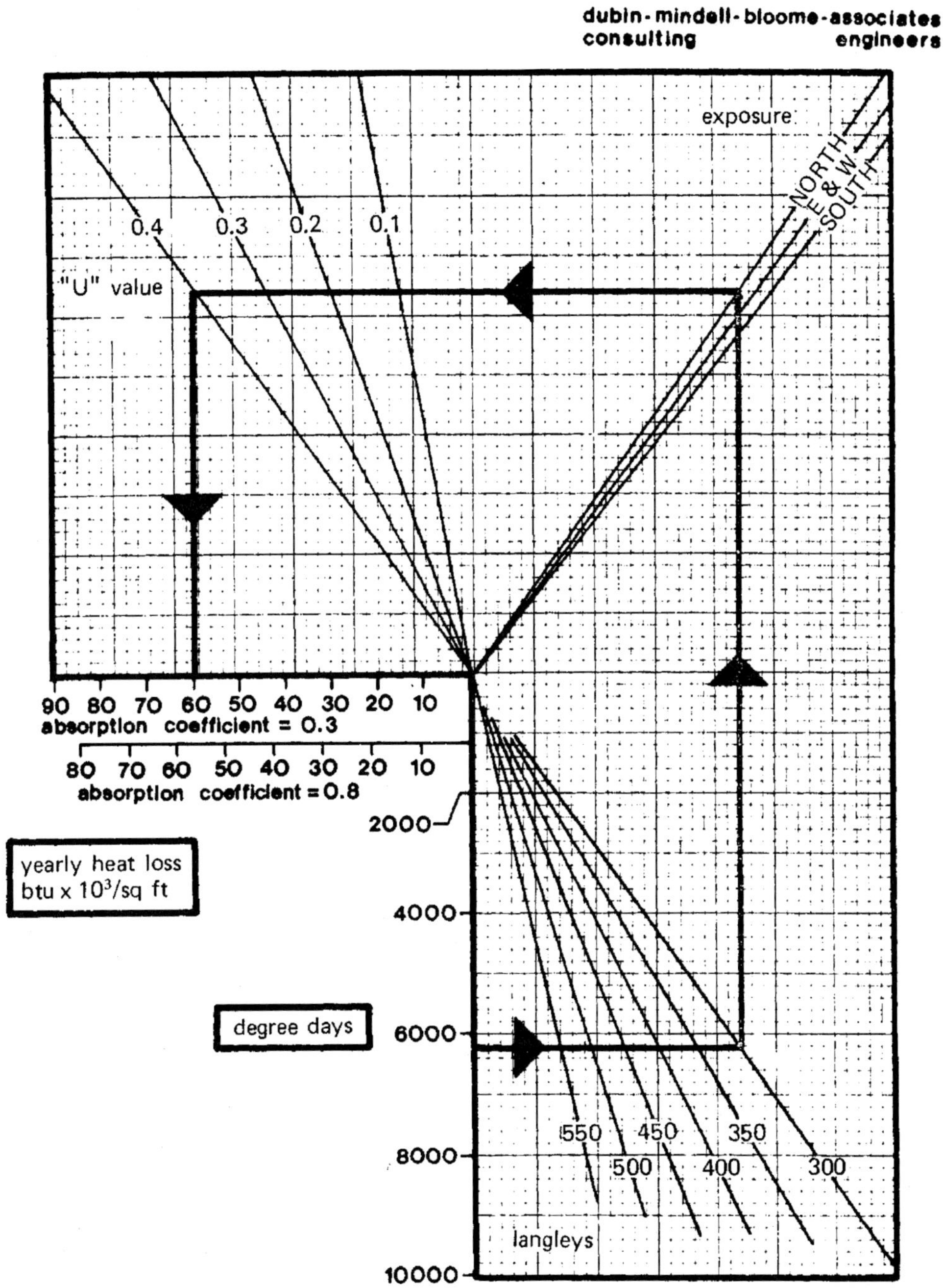

Figure # 38 Engineering Data

This figure is based on the "Sunset" Computer program which was used to calculate solar effect on walls for 12 selected locations. The program calculates hourly solar angles and intensities for the 21st day of each month with radiation intensity values modified hourly by average percentage of cloud cover taken from weather records. Heat losses are based on a 68° indoor temperature.

The solar effect on a wall was calculated by the computer using sol-air temperature and the heat entering or leaving a space was calculated with the equivalent temperature difference. Wall mass ranged from 50 - 60 lbs./sq.ft. and thermal lag averaged 4 1/2 hours. Additional assumptions were: 1) Total internal heat gain of 12 Btu/sq.ft. 2) Average outdoor air ventilation rate of 10%. 3) Infiltration rate of 1/2 air change per hour. Daily totals were then summed for the number of days in each month to arrive at monthly heat losses. The length of the heating season for each location considered was determined from weather data and characteristic operating periods. Yearly heat losses were derived by summing monthly totals for the length of the heating season.

Absorption coefficients and U values were varied and summarized for the 12 locations as shown in Figure #36. The data was then plotted and extrapolated to include the entire range of degree days. Figure #38 was derived from locations with latitudes between 25°N-35°N . The heat losses assume that the walls are subjected to direct sunshine. If shaded, losses should be read from the North exposure line.

FIGURE 39: YEARLY HEAT LOSS THROUGH WALLS—LATITUDE 35°N-45°N

dubin-mindell-bloome-associates consulting engineers

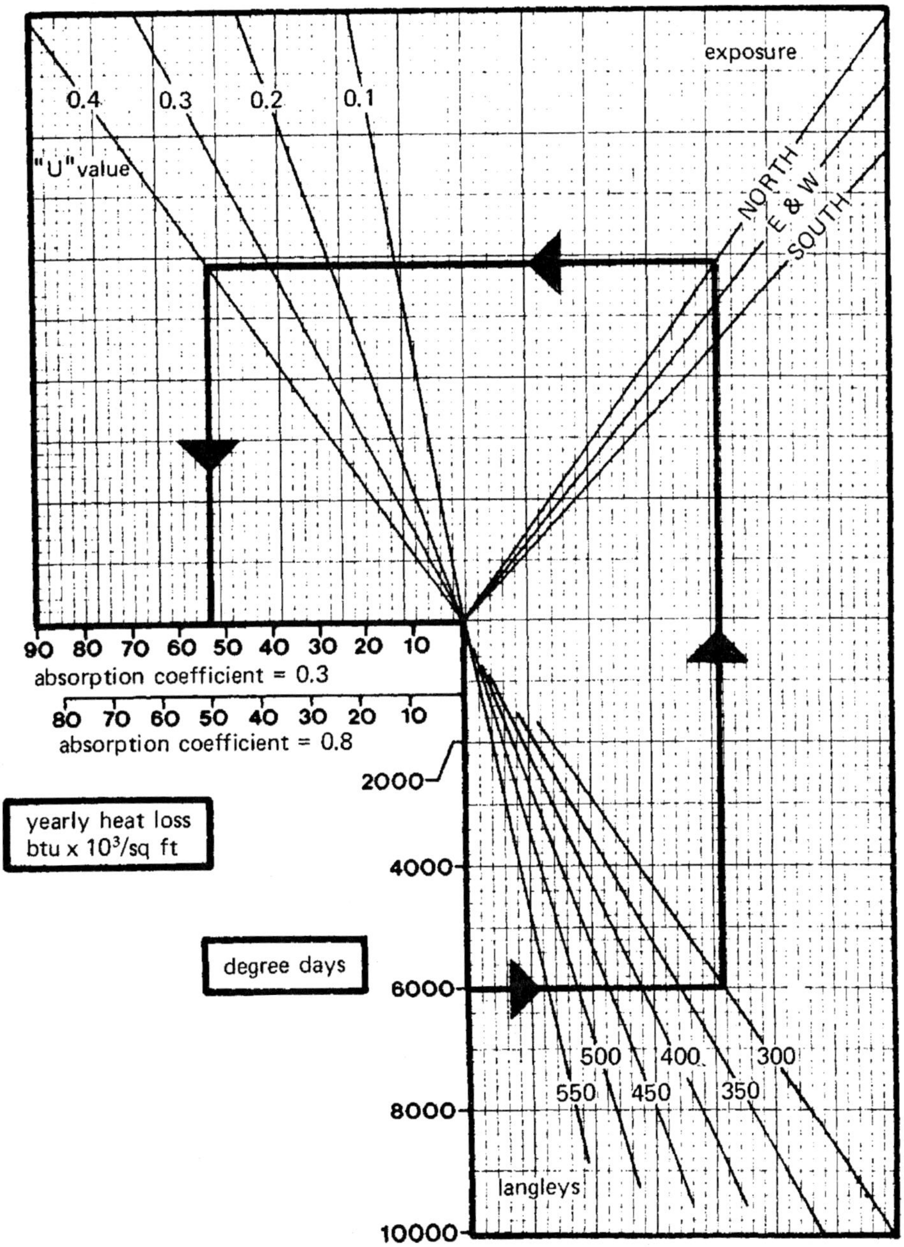

Figure # 39 Engineering Data

This figure is based on the "Sunset" Computer program which was used to calculate solar effect on walls for 12 selected locations. The program calculates hourly solar angles and intensities for the 21st day of each month with radiation intensity values modified hourly by average percentage of cloud cover taken from weather records. Heat losses are based on a 68° indoor temperature.

The solar effect on a wall was calculated by the computer using sol-air temperature and the heat entering or leaving a space was calculated with the equivalent temperature difference. Wall mass ranged from 50 - 60 lbs./sq.ft. and thermal lag averaged 4 1/2 hours. Additional assumptions were: 1) Total internal heat gain of 12 Btu/sq.ft. 2) Average outdoor air ventilation rate of 10%. 3) Infiltration rate of 1/2 air change per hour. Daily totals were then summed for the number of days in each month to arrive at monthly heat losses. The length of the heating season for each location considered was determined from weather data and characteristic operating periods. Yearly heat losses were derived by summing monthly totals for the length of the heating season.

Absorption coefficients and U values were varied and summarized for the 12 locations as shown in Figure #36. The data was then plotted and extrapolated to include the entire range of degree days. Figure #39 was derived from locations with latitudes between 35°N-45°N . The heat losses assume that the walls are subjected to direct sunshine. If shaded, losses should be read from the North exposure line.

The procedures outlined above will give order of magnitude savings which will be useful in determining the economic feasibility of adding insulation to walls. If more accurate quantification is required, then a full computer analysis should be made using one of the programs currently available. (See Section 24.)

ECO 46 REDUCE TRANSMISSION HEAT LOSSES THROUGH ROOFS

To reduce the heat loss through roofs, reduce the "U" value by adding insulation.

Heat loss through a roof is a function of its resistance to heat flow modified by the effect of solar radiation (reduces heat loss) and wind (increases heat loss) on the outside surface. The effect of solar radiation is modified by the absorption coefficient of the outside surfaces; dark colors (high absorption coefficient) will reduce the heat loss more than light colors (low absorption coefficient).

The mass of the roof and its attendant thermal inertia have an overall modifying effect on heat loss by delaying the impact of outdoor temperature changes on the heated space. This time delay allows the roof to act dynamically as a thermal storage system, smoothing out peaks in heat flow and reducing yearly heat loss. Low mass roofs of 10-20 lbs./square foot will have approximately 2% greater yearly heat loss than high mass roofs of 80-90 lbs./square foot assuming both roofs have the same overall "U" value and absorption coefficient.

A computer program was developed for this manual to study the reaction of roofs to varying weather conditions and runs were made for 12 cities chosen to give a typical cross section of climates.

Figure 40 tabulates the heat loss for one sq. ft. of roof in Btu/year for two different "U" values, two different absorption coefficients, and for the 12 selected locations.

This table shows that the beneficial effect of sunshine is more marked with roofs having a high absorption coefficient.

Graphs were developed from the computer runs which extrapolates from the calculated figures and allow prediction of yearly heat loss through one sq. ft. of roof based on inputs of heating degree days, solar radiation, "U" values, and absorption coefficient.

The graphs were based on internal heat gains of 12 Btu/sq. ft./hour when occupied, 10% avg. outdoor air ventilation when occupied, and 1/2 air change per hour continuous infiltration. For significantly different conditions, an individual computer run should be made using one of the programs listed in Section 24. The overall "U" value may be decreased by the addition of insulating material to the top of the outside surface or to the underside of the inside surface.

Where an existing roof has deteriorated and is scheduled for extensive repair or replacement, spray on polyurethane foam (or install rigid insulation). The roof surface must be cleaned of all loose debris and if necessary, cut back to provide a good bonding surface. It is not necessary, however, to provide a completely smooth surface as the foam will conform to any irregularities. The existing roof structure must be vented as necessary and protection provided against overspray. Even though the insulation is not hygroscopic, it must be protected by a weatherproof finish such as butyl. Whatever finish is selected must be applied cold as excess heat will damage the insulation.

A minimum thickness of 1" spray foam is required to provide acceptable bond strength and integrity.

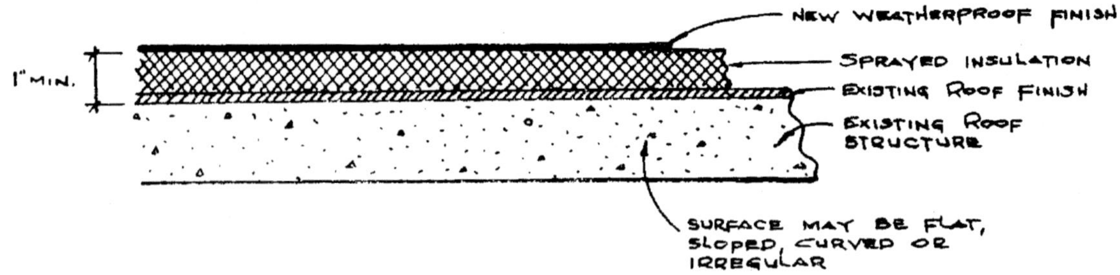

For rigid insulation, the roof surface must be cleaned of all loose debris and a smooth surface provided. Treated wood nailers should be used at all joints, roof penetrations, gravel stops, etc., and as required to fit new flashing. For sloping roofs of angles greater than 5°, treated wood nailers should be provided between insulation boards.

The rigid insulation which may be fiberglas, fiberboard, cellular glass, perlite boards, etc. should be embedded in a solid mopping of hot asphalt and then protected with a "built up" roof.

FIGURE 40: YEARLY HEAT LOSS PER SQUARE FOOT THROUGH ROOF

CITY	LATITUDE	SOLAR RADIATION LANGLEY'S	DEGREE DAYS	HEAT LOSS THROUGH ROOF BTU/FT.^{2} YEAR			
				U=0.19		U=0.12	
				a=0.3	a=0.8	a=0.3	a=0.8
MINNEAPOLIS	45°N	325	8,382	35,250	30,967	21,330	18,642
CONCORD, N.H.	43°N	300	7,000	32,462	27,678	19,649	16,625
DENVER	40°N	425	6,283	26,794	22,483	16,226	13,496
CHICAGO	42°N	350	6,155	27,489	23,590	16,633	14,190
ST. LOUIS	39°N	375	4,900	20,975	17,438	12,692	10,457
NEW YORK	41°N	350	4,871	21,325	17,325	12,911	10,416
SAN FRANCISCO	38°N	410	3,015	10,551	8,091	6,381	4,784
ATLANTA	34°N	390	2,983	12,601	9,841	7,619	5,832
LOS ANGELES	34°N	470	2,061	4,632	3,696	2,790	2,142
PHOENIX	33°N	520	1,765	5,791	4,723	3,487	2,756
HOUSTON	30°N	430	1,600	6,045	4,796	3,616	2,778
MIAMI	26°N	451	141	259	130	139	55

dubin-mindell-bloome-associates
consulting engineers

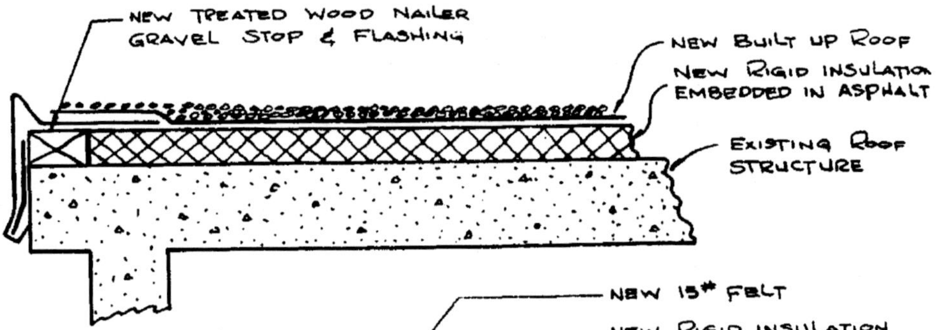

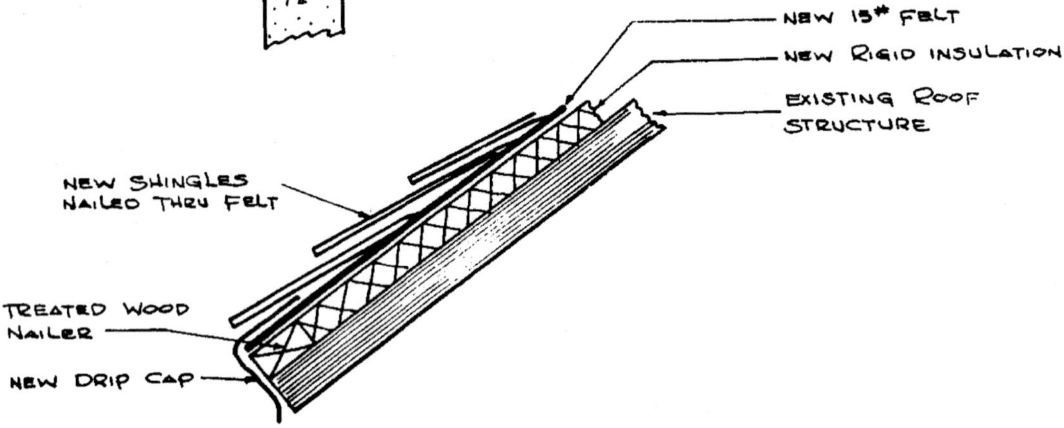

Where the existing roof is sound and the underside is directly accessible either from an attic space or ceiling void, spray on polyurethane foam, mineral fiber spray, or rigid insulation may be added to the inside surface.

For sprayed foam insulation, access must be provided over the whole area. The underside must be brushed clean of all loose debris and any ducts or pipes, etc. masked from the spray. A minimum thickness of 1" sprayed insulation is required. Difficulty may be experienced in reaching all the roof area where ducts and pipes are installed close to the roof and where structural members protrude.

Spray insulation is quick and easy to apply, but will entail either removal of furnishings or protection to prevent spoiling by overspray and droppings. The process of polyurethane spray and mineral fiber spray both entail health hazards and the immediate building area must be evacuated while the work is done. As a side benefit, mineral fiber sprays will generally improve the fire rating of the structure and provide good attenuation of noise.

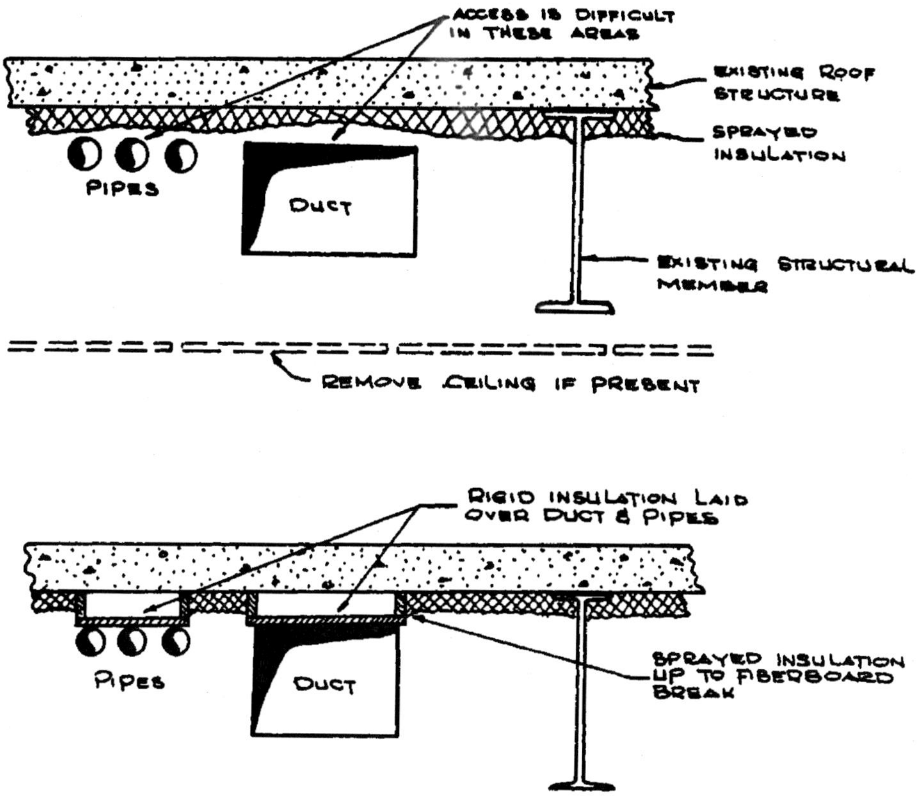

POSSIBLE SOLUTION TO DIFFICULT ACCESS AREAS

To apply rigid insluation to the underside of the roof, direct access is required over the whole area. The surface should be brushed free of all loose debris and be relatively smooth. The insulation may be glued, nailed or fixed with special clips and must be trimmed to fit around obstructions. Installation is less rapid than spray but can be restricted to small areas at any one time, causing minimum inconveneince

Rigid insulation is particularly suited for large uncluttered areas without hung ceilings and in addition to reducing heat loss, will attenuate noise.

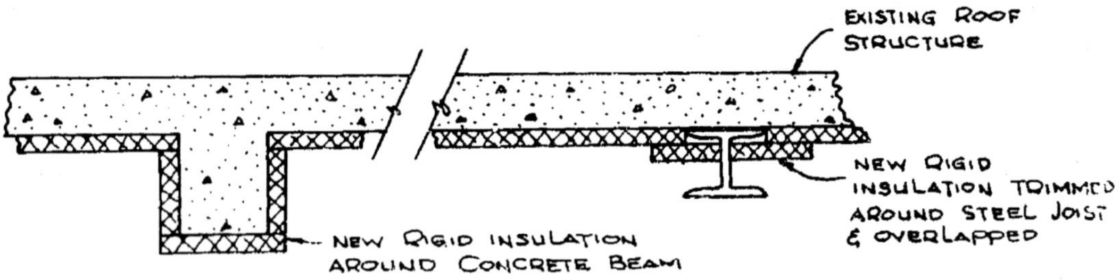

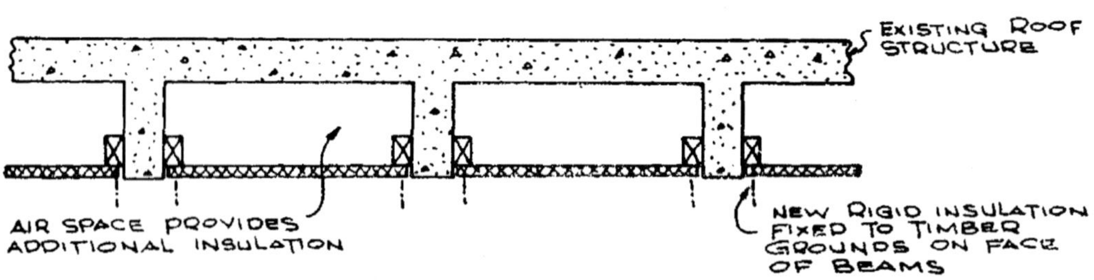

Rigid insulation, batt insulation or granular insulation may be laid on top of the ceiling in an attic space. This process requires access into the attic at one point only and is easilyand quickly accomplished. It has the effect however, of reducing the temperature in the attic; and if pipes and ducts are run through this space, their heat loss will increase and frost damage to water pipes could result in cold geographic locations.

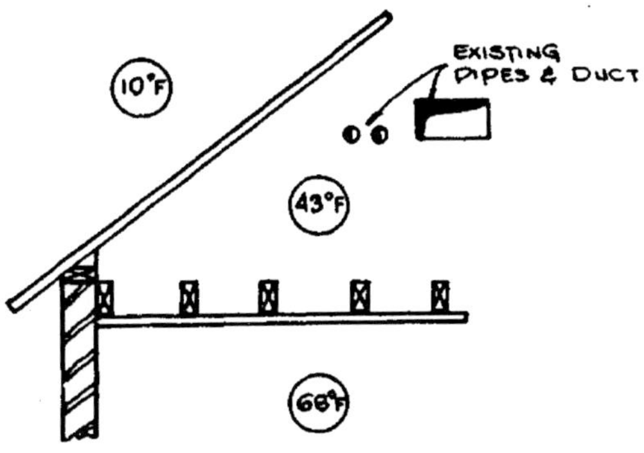

EXISTING ATTIC TEMPERATURES
OVERALL 'U' = 0.2

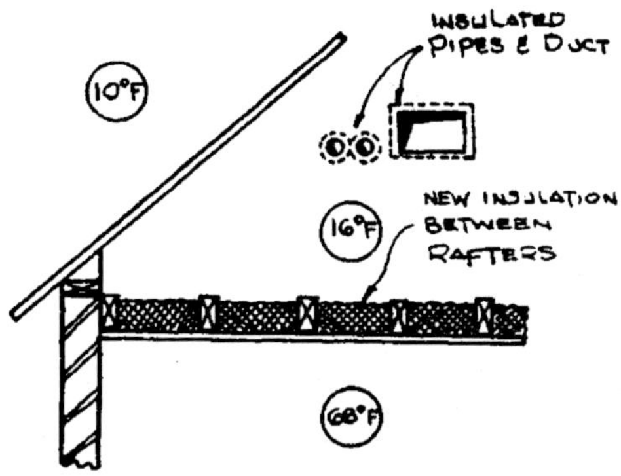

NEW ATTIC TEMPERATURES
IMPROVED 'U' = 0.06

Although most insulating materials are light weight and are unlikely to exceed the permissable loading of the existing structure, calculations should be made to check that the local building code is not infringed.

Check with the local fire department on the acceptability of insulating materials. In some areas plastic insulation is considered a fire and smoke hazard. Check with the local office of the Food and Drug Administration (Health, Education and Welfare Dept.) to determine whether the insulation considered is listed as hazardous. This is particularly important in retail stores that sell food.

Soffits of roof overhangs and roof facias are frequently left uninsulated and can result in high heat loss. If your building has these features, they should be insulated at the same time as the main roof. If the soffit is accessible, sprayed or rigid insulation may be used. If the soffit is not accessible, a hole can be cut in the structure and blown mineral fiber introduced.

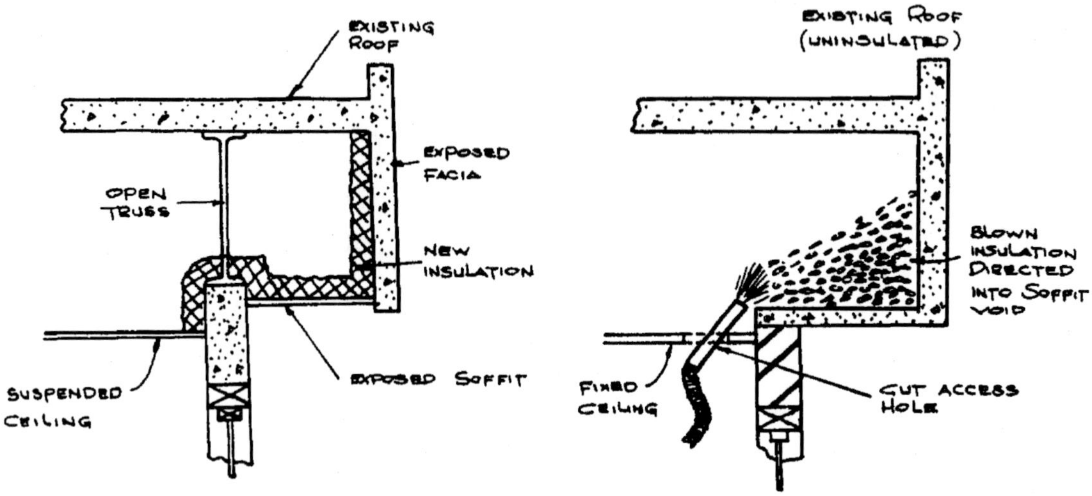

To evaluate the benefits of adding insulation to a roof, it is first necessary to know the "U" value of the existing construction. A procedure for calculating and measuring existing "U" values is outlined in this Section for walls and should be used for roofs.

Having established the "U" value of the existing roof, a decision must now be made as to whether additional insulation is required (and if so how much should be added), what type of insulation should be used and where it should be applied.

These decisions are governed by the characteristics of the particular building and should be resolved to give the optimum architectural solution.

The cost of adding insulation to roofs varies with each architectural solution and to make an accurate assessment for any particular case quotations should be obtained from local contractors. For order of magnitude, however, the following tabulated costs may be used:

Approximate Insulation Costs for Roofs

Spray insulation exterior	$2.10/sq.ft.
Spray insulation interior	3.00/sq.ft.
Rigid insulation exterior flat roof	2.50/sq.ft.
Rigid insulation exterior sloping shingle	1.80/sq.ft.
Rigid insulation interior	3.00/sq.ft.
Attic insulation	1.60/sq.ft.

To calculate savings due to roof insulation, it is necessary to know the "U" values both before and after insulation is added.

Procedures have already been described for obtaining the existing "U" values.

To determine the new "U" value, refer to Figure 37. Enter the graph at the present "U" value, intersect with the appropriate line for new insulation thickness and then read out the new "U" value.

To determine the heating energy saved by insulating roofs, refer to Figure 41, then carry out the following procedure:

a) Determine your latitude and select the appropriate graph.

b) Determine from Figure 4 the heating degree days for your location.

c) Determine from Figure 5 the mean daily solar radiation in Langleys for your location.

d) Determine the initial "U" value and the improved "U" value after insulation.

e) Determine the absorption coefficient of the roof outside surface.

f) Using the initial "U" value of the roof, enter the graph at the appropriate degree days and following the direction of the example line, intersect at the appropriate points for Langleys, and "U" value. Read out the yearly heat loss per square foot of roof on the appropriate absorption coefficient line. Interpolate to obtain answers for absorption coefficients other than the 0.3 and 0.8 shown.

   Repeat this complete procedure using the improved "U" value and subtract the answer from the energy used before insulating to derive the energy saving per square foot of insulated roof. Multiply this saving by the area of the insulated roof to determine the total yearly load reduction.

In addition to savings achieved by reducing winter heat losses, roof insulation will also reduce summer heat gains. (See Section 15.)

When adding roof insulation to the outside surface of a roof, it is possible to change the absorption coefficient at the same time. A judgment must be made whether it is more advantageous to have a dark surface to enhance the heating effect of the sun in winter or whether it is better to have a light surface and sacrifice the winter benefits in favor of reflecting unwanted solar gains in summer. (See Section 15.)

The procedures outlined above will give order of magnitude savings which will be useful in determining the economic feasibility of adding insulation to roofs.

If more accurate quantification is required, then a full computer analysis should be made using one of the programs currently available. (See Section 24.)

ECO 47 REDUCE TRANSMISSION HEAT LOSSES THROUGH FLOORS

To reduce heat loss through floors, reduce the "U" value by adding insulation.

Heat loss from slab-on-grade floors occurs mainly around the perimeter. Very little heat is lost from the center of the floor. This is due to the insulating effect of the earth which is greatest in the center and least in the perimeter. Heat loss from suspended floors over an unheated space, however, occurs evenly over the whole floor area and is proportional to the difference between inside and outside temperatures. In existing buildings, it is impracticable to add insulation to the top of existing floors and this manual will, therefore, be restricted to evaluating insulation added below suspended floors and at the edges of slab-on-grade floors.

Uninsulated slab-on-grade floors should be insulated around the perimeter. This insulation should be placed vertically on the outside edge of the floor and should extend at least 2'-0" below the floor surface.

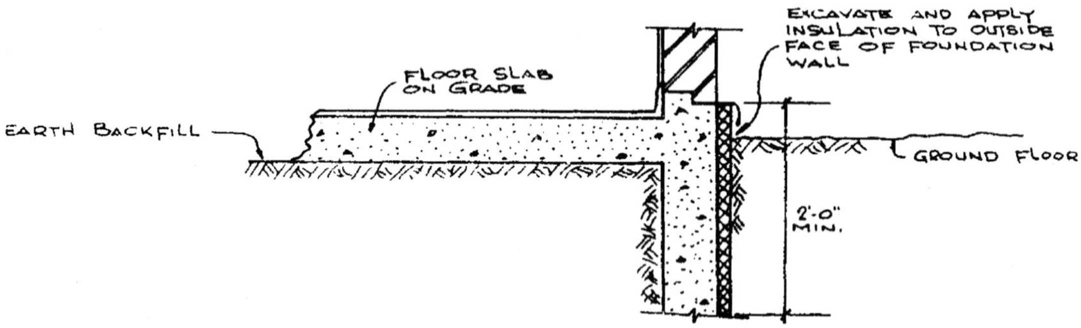

FIGURE 41: YEARLY HEAT LOSS THROUGH ROOF

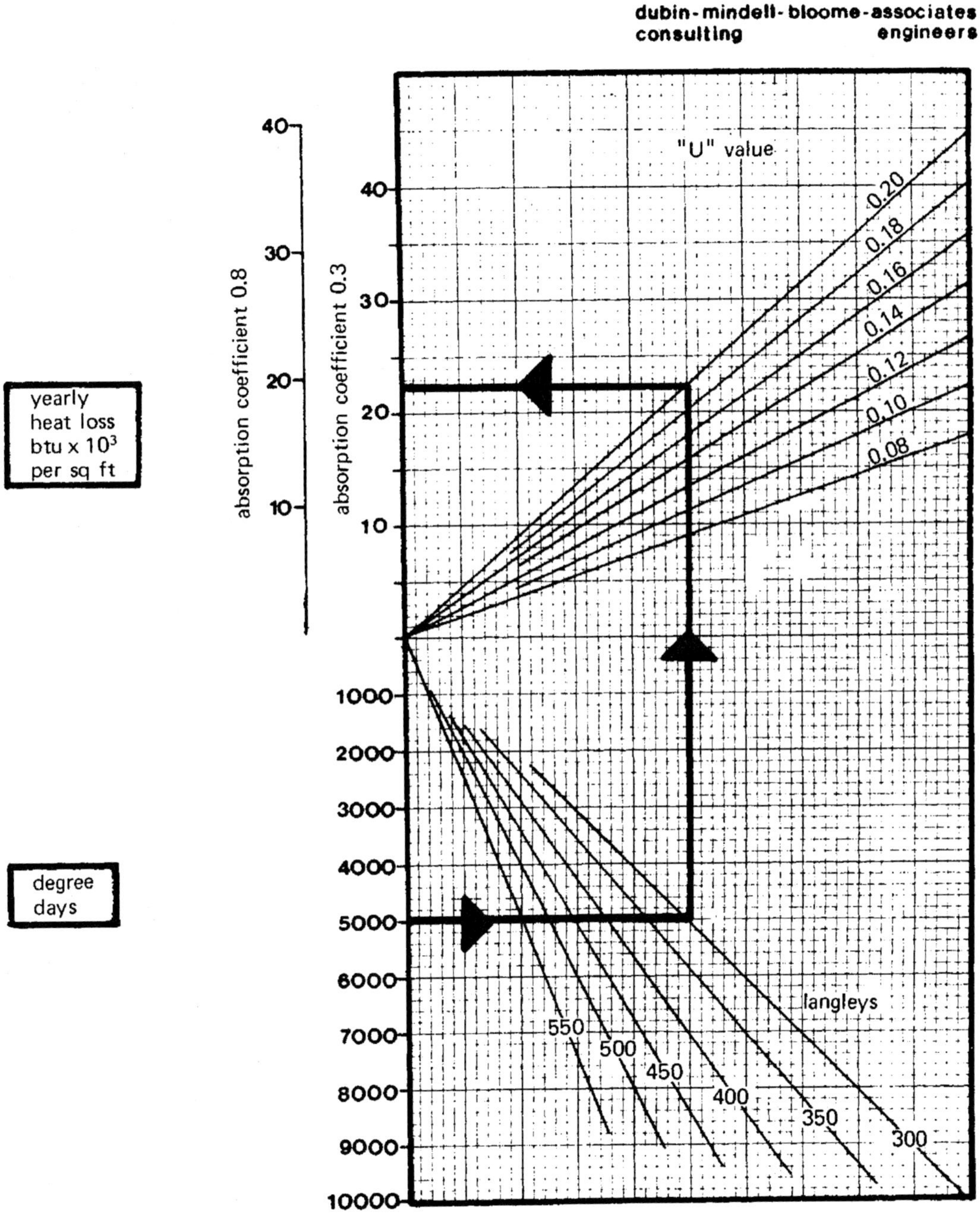

Figure #41 Engineering Data

This figure is based on the "Sunset"Computer program which was used to calculate solar effect on roofs for 12 selected locations. The program calculates hourly solar angles and intensities for the 21st day of each month with radiation intensity values modified hourly by the average percentage of cloud cover taken from weather racords. Heat losses are based on a 68°F indoor temperature.

The solar effect on a roof was calculated using sol-air temperature and the heat entering or leaving a space was calculated using the equivalent temperature difference. Roof mass ranged from 25 - 35 lbs./sq.ft. and thermal lag averaged 3 1/2 hours. Additionalassumptions were: 1) Total internal heat gain of 12 Btu/sq.ft. 2) Outdoor air ventilation rate of 10%. 3) Infiltration rate of 1.2 air change per hour. Daily totals were then summed for the number of days in each month to arrive at monthly heat losses. The length of the heating season for each location considered was determined from weather data and characteristic operating periods. Yearly heat losses were derived by summing monthly totals for the length of the cooling season.

Absorption coefficients and U values were varied and summarized for the 12 locations as shown in Figure #40. The data was then plotted and extrapolated to include the entire range of degree days.

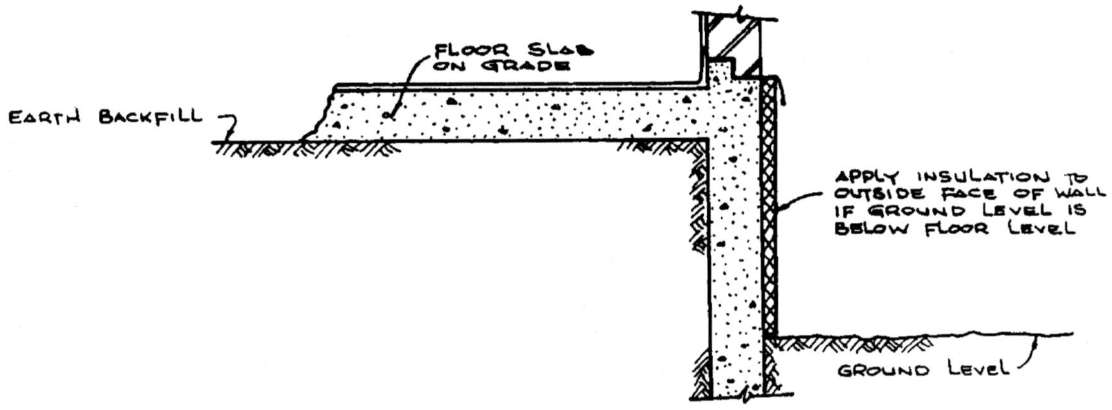

Insulation materials may be rigid board or foam as described for walls. Insulation below ground should be applied in a bedding of hot asphalt. Existing flashing at ground floor level may require extending to cover the insulation top edge.

The savings due to edge insulation cannot be predicted with any accuracy. If, however, condensation occurs on the floor perimeter in cold weather or if the floor surface temperature close to the outside wall is more than 10° F, lower than the indoor temperature, then insulation will be beneficial and should be added.

Suspended floors over an unheated space (garage, crawlway, etc.) may be insulated on the underside by applying spray foam or rigid insulation as described for roofs.

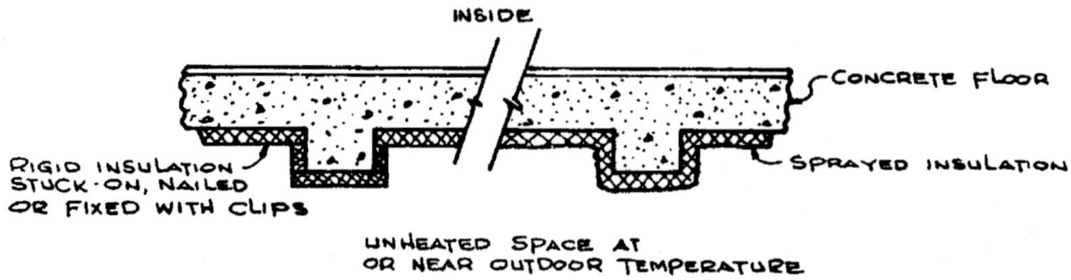

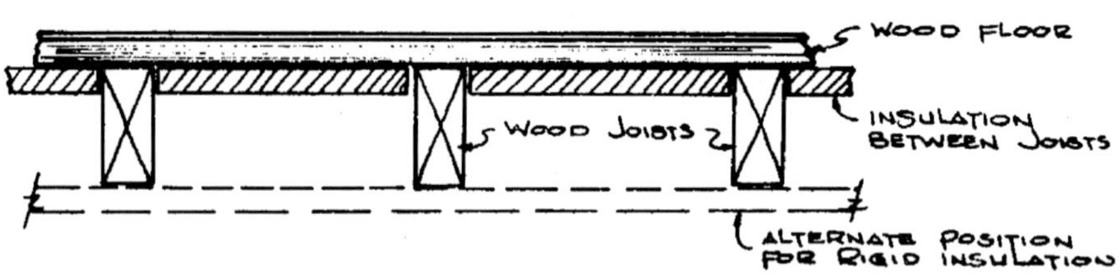

Check with the local fire department on the acceptability of insulating materials.

To evaluate the benefits of adding insulation to suspended floors, it is first necessary to know the present "U" value of the existing construction. A procedure for calculating and measuring these is outlined for walls and may be used for floors. (See ECO 45.)

Having established the "U" value for the existing floor, a decision must now be made as to whether additional insulation is required and if so, how much should be added and what type of insulation should be used.

The cost of adding insulation to floors varies with each particular case depending on circumstances and to obtain an accurate assessment, quotations should be obtained from local contractors. For preliminary evaluation, however, the following tabulated costs may be used to establish order of magnitude.

<u>Approximate Costs of Floor Insulation</u>

Edge insulation above grade	$1.70/square foot
Edge insulation below grade	2.10/square foot
Spray insulation below suspended floor	2.60/square foot
Rigid insulation below suspended floors	2.70/square foot

FIGURE 42: YEARLY HEAT LOSS THROUGH FLOOR

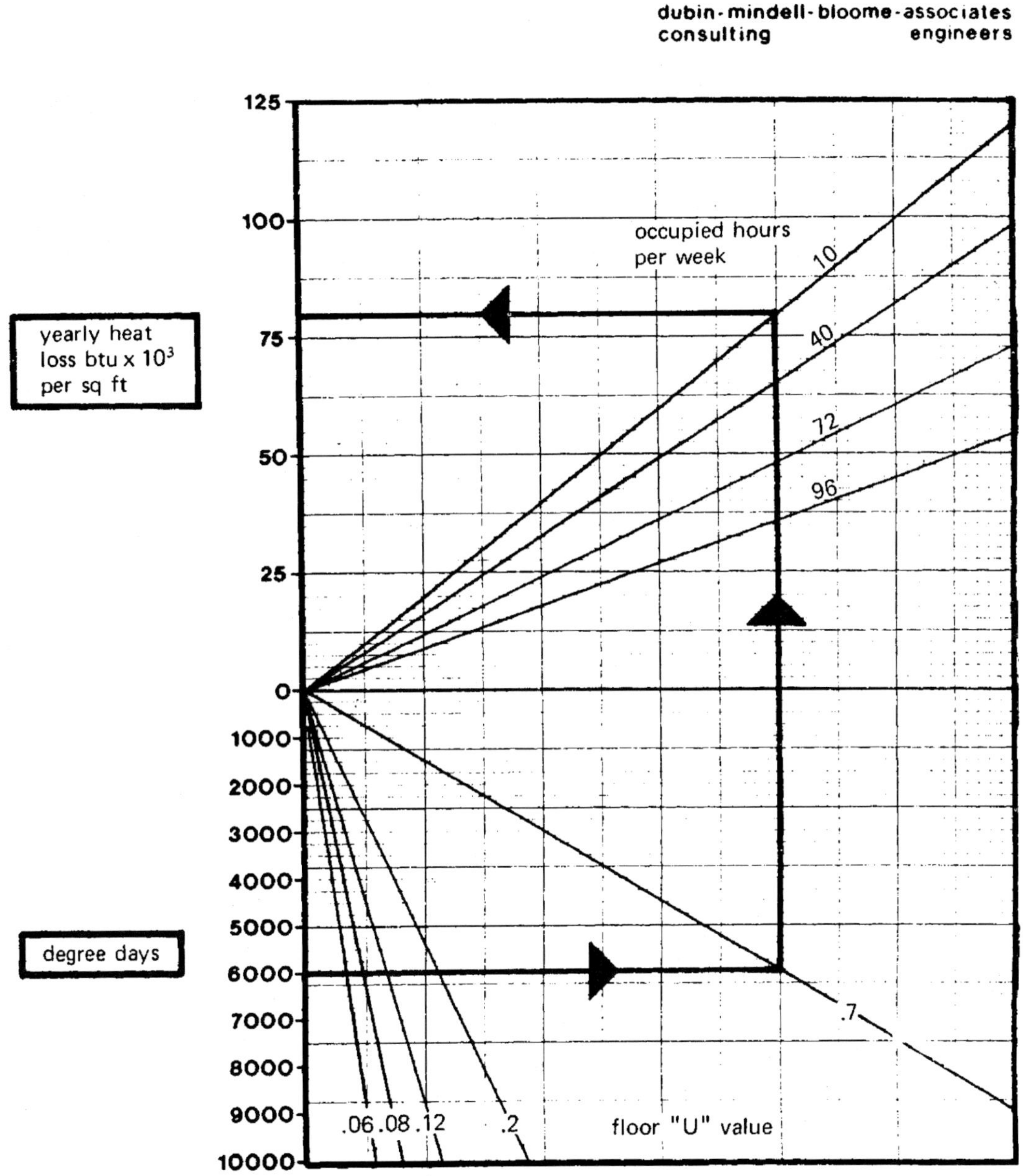

Figure #42 Engineering Data

Heat losses determined from this figure are based on the assumption that the floor is over an unheated space which is at outdoor ambient air temperature. Since the load on the heating system during occupied hours is generally a small percentage of the total annual heating load, the figure gives heat loss during unoccupied times only with 10 hours occupied time per week being the maximum (158 hours unoccupied). It was also assumed that the temperature during unoccupied time is set back to 55°F and since degree days are based on 65°F, all losses were multiplied by a factor of 55/65, thus the formula the figure is based on is: Q (heat loss) = Degree Days x 24 x 'U' Value x 55/65 x $\frac{\text{unoccupied hours}}{168}$

To calculate savings due to suspended slab floor insulation, it is necessary to know the "U" values both before and after insulation is added.

Procedures have already been described for obtaining the existing "U" value.

To determine the new "U" value, refer to Figure 37. Enter the graph at the present "U" value, intersect with the appropriate line for new insulation thickness and then read out the new "U" value.

To determine the heating energy saved by insulating suspended slab floors, refer to Figure 42 and carry out the following procedure:

a) Determine from Figure 4 the heating degree days for your location.

b) Determine the initial "U" value and the improved "U" value after insulation.

c) Using the initial "U" value of the floor, enter the graph at the appropriate degree days and following the direction of the example line, intersect at the appropriate point for "U" value. Read out the yearly heat loss per square foot of floor area.

   Repeat this complete procedure using the improved "U" value and subtract the answer from the energy used before insulating to derive the energy saving per square foot of insulated floor. Multiply this saving by the area of the insulated floor to find total yearly heat load reduction.

The procedures outlined above will give order of magnitude savings which will be useful in determining the economic feasibility of adding insulation to floors. If more accurate qualification is required, then a full computer analysis should be made using one of the programs currently available. (See Section 24.)

REDUCE SPACE LOAD DUE TO INFILTRATION

BACKGROUND

Infiltration of cold outdoor air into a building through cracks, openings, gaps around doors and windows, etc., increases the building heat load and is often responsible for as much as 25% of the yearly heating energy consumption.

The rate of infiltration is increased by the effect of wind on the surface of the building and in tall buildings by stack effect due to the difference between indoor and outdoor temperatures. This stack effect is promoted in open vertical spaces such as staircases, elevator shafts, and service shafts. In tall buildings, the potential for stack effect is always present and can only be minimized by thoroughly sealing openings through which air can enter and leave the building.

To maintain equilibrium, the quantity of air entering a building either by uncontrolled infiltration or by ventilation must equal the quantity of air leaving the building either by exfiltration or ventilation exhaust.

Buildings frequently have unbalanced ventilation systems where the rate of exhaust exceeds the rate at which air is introduced into the ventilation system and in these cases, infiltration increases (to make up the balance) beyond that level which would otherwise occur.

The fresh air ventilation and exhaust system should be examined and a balance sheet drawn up. If the exhaust exceeds the outdoor air requirements, then the rate of exhaust should be reduced, which in turn will reduce infiltration. Ideally, the quantity of outdoor air for ventilation should exceed the total exhaust from the building by 10% to promote exfiltration rather than filtration.

ECM 1 has already discussed weatherstripping doors and windows and other simple measures that can be taken to reduce infiltration.

It is assumed that these measures have been put into practice; discussion in ECM 2 will, therefore, be limited only to those items requiring capital investment.

ECO 48 FIT VESTIBULES AND/OR REVOLVING DOORS TO REDUCE INFILTRATION

In buildings that experience heavy and continuous traffic through external doors, infiltration may be reduced by building vestibules for each external door to form an "air lock".

The vestibule should be sufficiently long so that the external door closes before the internal door is opened. Depending on the particular characteristic of the building, the vestibule may be constructed inside or outside the building.

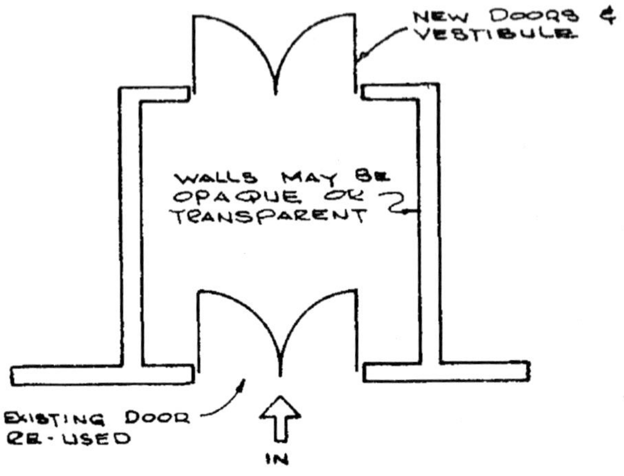

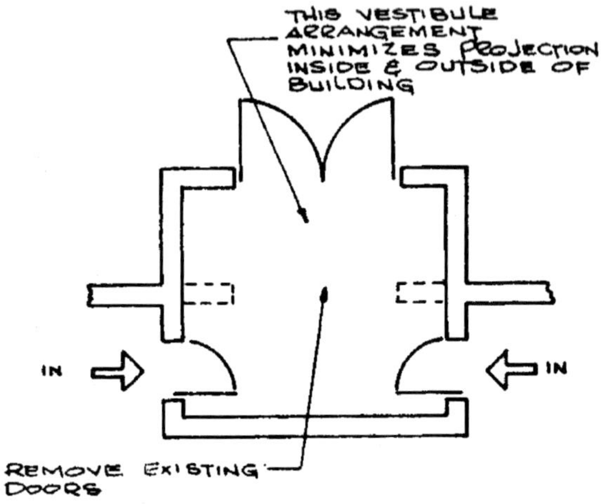

FIGURE 43: INFILTRATION DUE TO OPENING AND CLOSING EXTERNAL DOORS

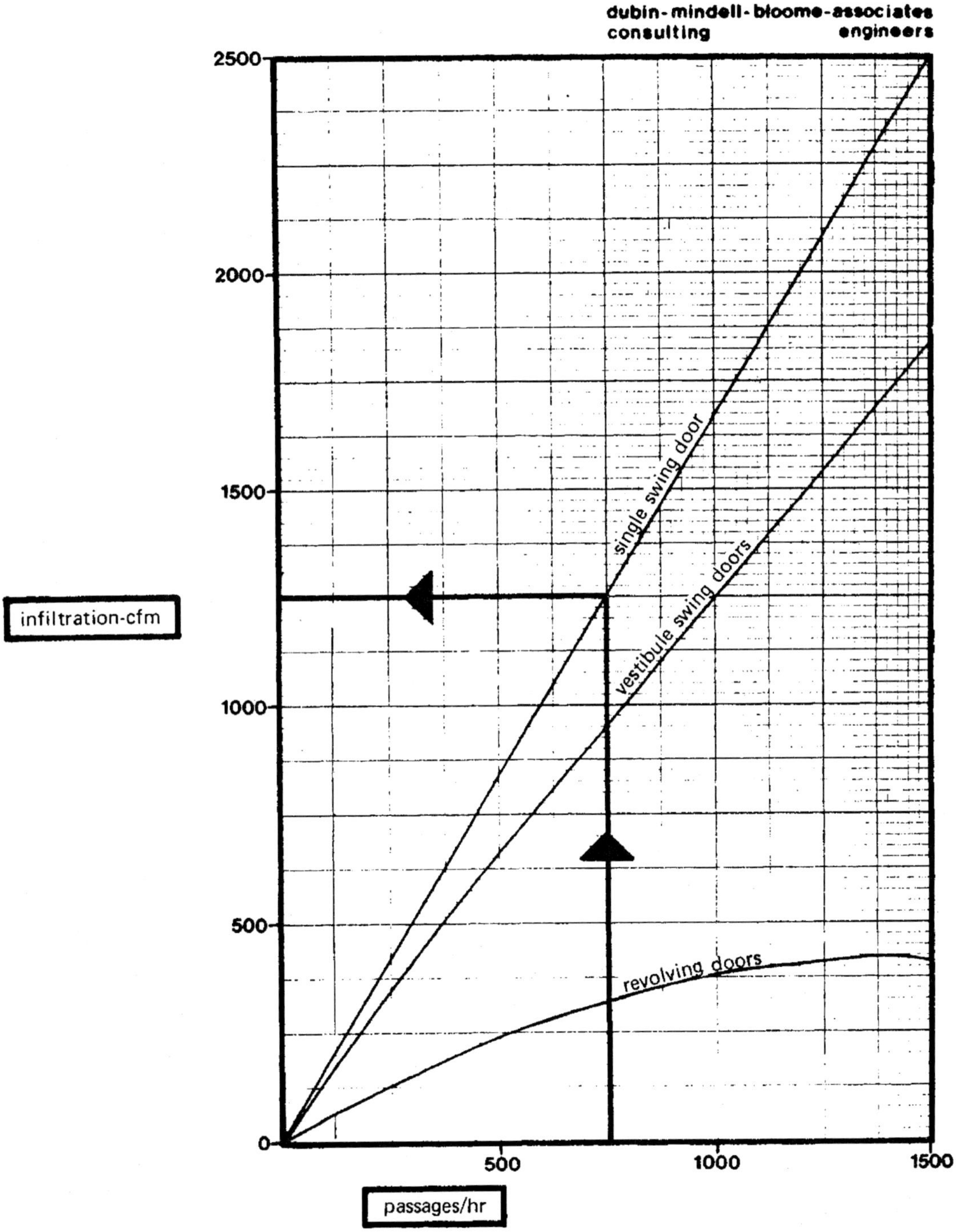

Figure #43 Engineering Data

Source: Handbook of Air Conditioning, Heating and Ventilating, 1965, Strock & Koral, 2nd Edition, Table #1, 2-105.

Vestibule doors may be either manually operated and self-closing or, if traffic is particularly dense, can be automatically opened and closed by pressure pad or photo-electric cell.

Revolving doors may be fitted in place of the existing doors and although maximum reduction in infiltration will be achieved if they are used in conjunction with vestibules, they may also be used alone. When fitting revolving doors, it is always necessary to provide a hinged door in addition for use by handicapped people and to allow bulky objects to be taken in and out of the building.

Check applicable fire codes and the local fire department to insure that installation of a vestibule or revolving door does not infringe fire exit requirements.

The cost of installing vestibules and revolving doors will vary with the circumstances for any particular installation and quotations should be obtained from the local contractors for the selected architectural solution.

For order of magnitude costs, a vestibule with double doors and automatic operators could be installed for $3,000.

Removing the existing door and replacing it with a revolving door and one single swing door would cost $10,000 without vestibule and $13,000 with the vestibule.

To determine the energy saved by fitting vestibules or revolving doors, it is first necessary to know the infiltration reduction achieved. To obtain this, refer to Figure 43 and enter the graph at the estimated passages per hour, intersect with the line for single swing doors and read out the rate of infiltration in 1000 cfm.

Intersect also with the appropriate line for a vestibule or revolving door and read out the new infiltration rate. Subtract this figure from the infiltration rate fora single door to obtain the infiltration reduction in 1000 cfm.

To determine the energy savings, refer to Figure 17 and then carry out the following procedure:

a) Determine from Figure 4 the heating degree days for your location.

b) Determine the average occupied hours/week in which the entrances are used.

c) Determine the indoor temperature during occupied periods.

d) Enter the graph at the appropriate degree days and following the direction of example line, intersect at the appropriate points with the indoor temperature and hours of occupancy.

e) Read out yearly heating energy/1000 cfm.

f) Multiply this figure by the 1000 cfm infiltration reduction to obtain the net yearly heating load reduction.

In addition to the savings achieved by reducing winter heat loss, reduction of infiltration will also reduce sensible and latent heat gains in the summer.

Depending on the building's location and the severity of the cooling season, this extra saving can be considerable and should be recognized when making an economic feasibility study. (See Section 15.)

ECO 49 <u>SEAL VERTICAL SHAFTS AND STAIRS TO REDUCE INFILTRATION</u>

To reduce the effect of potential stack action present in vertical shafts and stairs, these areas should be sealed from the rest of the building. Open stairwells that connect with circulation spaces at each floor level should be provided with walls and self-closing doors to isolate them.

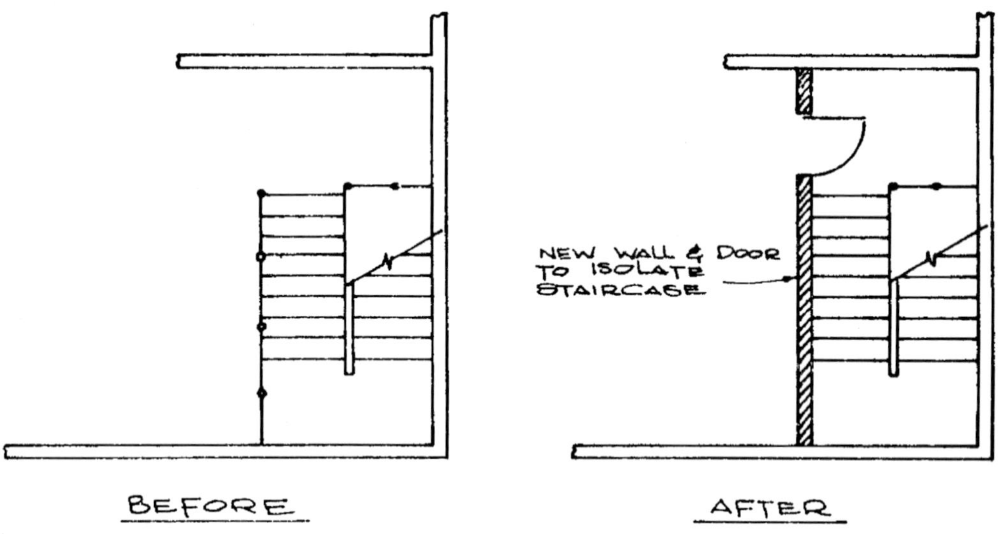

Access holes into vertical service shafts should be provided with gasketed ceiling covers. Holes at the shaft wall, to allow the passage of pipes and ducts at each floor level, should be sealed and sleeves around the pipes and ducts packed with asbestos rope or other suitable material.

Elevator shafts are usually provided with a vent at the top into the equipment room. This vent is necessary as a smoke release, but can be responsible for large air movements through the elevator doors up the shaft into the equipment room and then outside through badly fitted doors and windows in the equipment room.

Although the equipment room is designated as unoccupied space, the windows doors should be weatherstripped to reduce the effect of stack action in the elevator shafts.

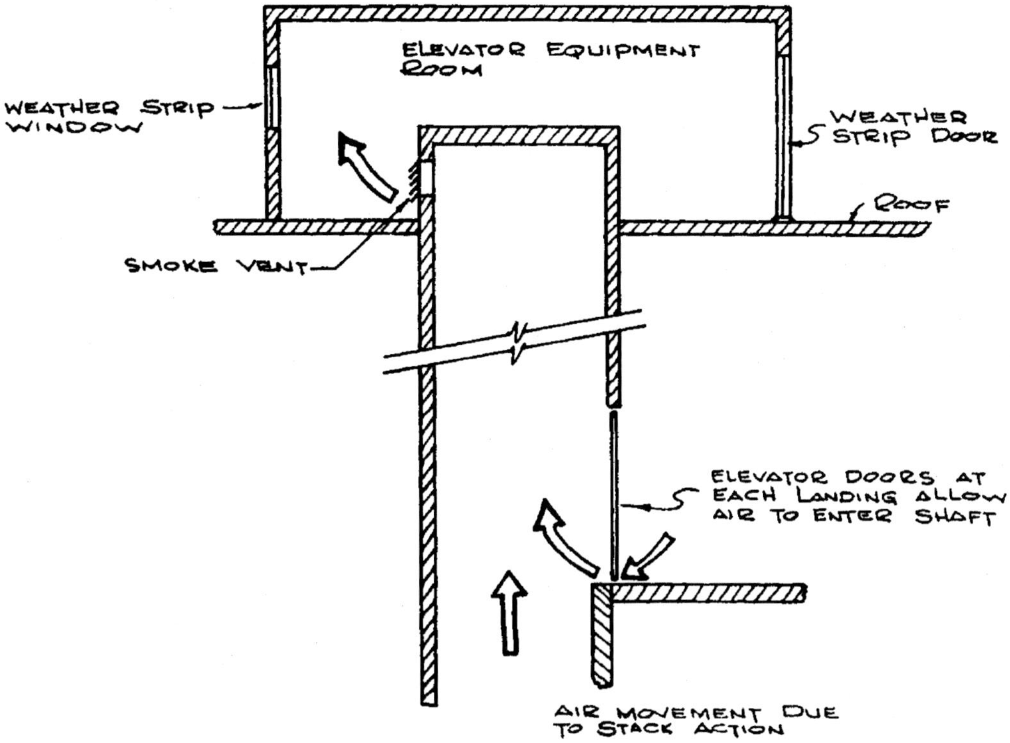

Check the local fire codes before isolating stairs from the remainder of the building to insure that fire exit requirements are met.

The walls of vertical shafts are usually required to have a high fire rating which applies equally to any access door. If access doors are replaced, make sure that these meet the fire rating requirements.

ECO 50 INSTALL SCREENS TO PROTECT EXTERNAL DOOR FROM PREVAILING WINDS

If sufficient money is not available for investment in vestibules or revolving doors, consider the installation of a wind screen to protect the external door from the direct blast of prevailing winds. Screens can be opaque, constructed cheaply from concrete block or can be transparent, constructed of metal framing with armored glass or Lexan fill. Careful positioning can reduce infiltration through external doors by about 50% of that obtained by fitting a vestibule. The cost of wind shields will vary but should be in the order of \$600 for concrete block and \$1,000 for the glazed type.

REDUCE THE SPACE LOAD DUE TO VENTILATION

BACKGROUND

The introduction of cold outdoor air through the ventilation system to meet the occupants requirements imposes a considerable load on the heating system.

ECM 1 lists optimum ventilation rates and it is assumed that the ventilation systems have now been adjusted to meet these as far as the system design permits and that further reductions can only be achieved by additions to or modification of the various ventilation systems.

ECO 51 INSTALL NEW OUTDOOR AIR DAMPERS

Many ventilation systems, particularly roof top units, are initially provided with poor quality outdoor air dampers which do not allow accurate control and, even when fully closed, leak in such quantities as to exceed the minimum outdoor air requirements.

These dampers should be replaced with a good quality opposed blade damper with seals at the blade edges and ends.

Such dampers can be installed for approximately \$20-\$25/1000 cfm. If economizer cooling is used in summer, the outdoor air dampers must be sized to pass the full volume of the supply fan. If, however, economizer cooling is not contemplated and minimum quantities of outdoor air will be used year-'round, then the outdoor air intake louvers and damper can often be drastically reduced in size and the unused portion of the outdoor air inlet blanked off.

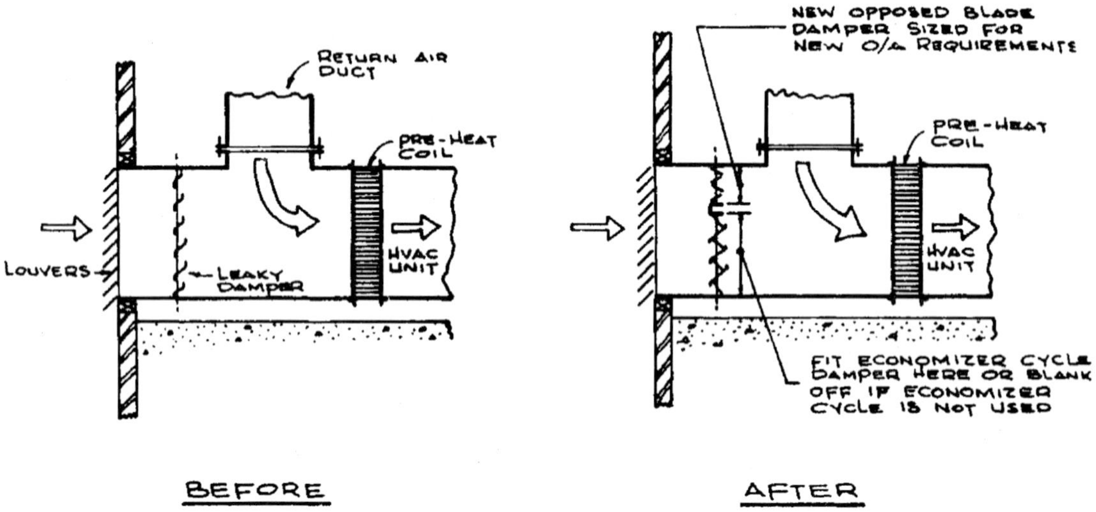

ECO 52 RECIRCULATE EXHAUST AIR USING ACTIVATED CARBON FILTERS

The quantity of air exhausted from a building often exceeds the minimum outdoor air requirements and requires excess outdoor air to be introduced to make up the difference and maintain the pressure balance of the building.

To balance the minimum outdoor air requirements and the total air exhausted from the building , the quantity of exhaust air may be reduced either by reducing the rate of exhaust or by retaining the same rate of exhaust but recirculating a proportion of the air back into the building.

Recirculating untreated exhaust air is often unacceptable because it contains unpleasant odors. This air, however, can be treated by passing it through activated carbon filters which will remove the odors. Candidates for such treatment are central exhaust systems from toilet areas, dining rooms, lounges, etc.

To recirculate the air, ducting must be installed to connect from the discharge side of the exhaust system into the return air of the HVAC system. This ductwork should contain the activated carbon filters. Recirculate that portion of the quantity of total exhaust air for the building which exceeds 90% of minimum outdoor air requirements.

Activated carbon filters remove odors from air by absorption and after a period of time they become saturated and their performance falls off. At this point, the filters must be removed and either regenerated in-house or be replaced with freshly regenerated filters supplied by the manufacturer. Heat energy is required for regeneration but can be derived from a waste heat source.

The length of time the filters can remain in service depends on the circumstances of each particular case; manufacturer's advice should be sought on this point. The cost of installing the recirculating ductwork and filters will vary for each particular case depending on the disposition of exhaust and HVAC systems and quantity of air handled. To obtain an accurate assessment, designs should be prepared and quotations solicited from local contractors.

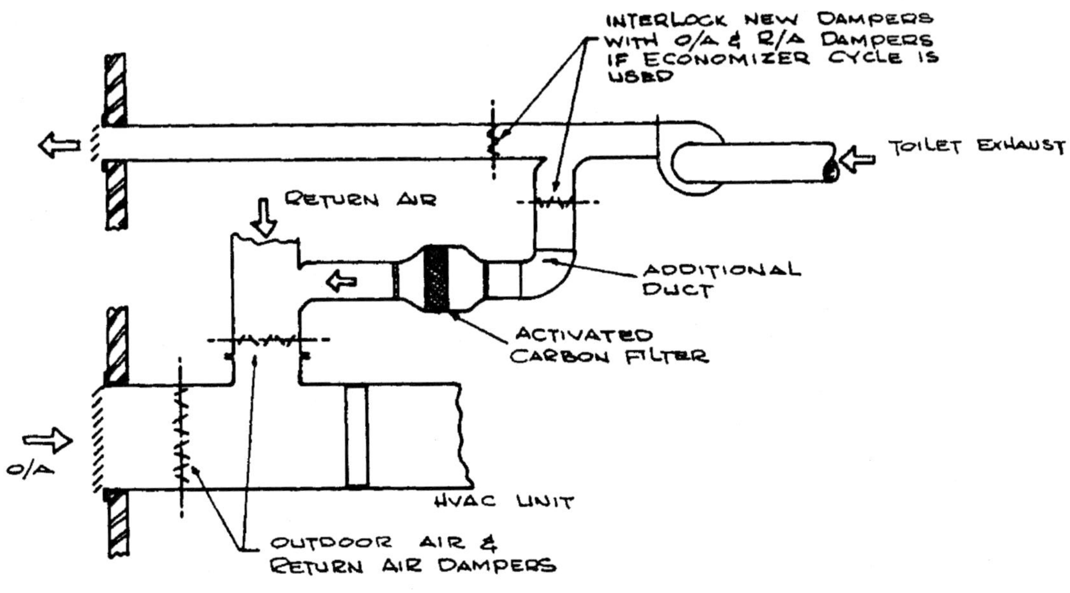

For order of magnitude, the cost of installing activated carbon filters should be approximately $190/1000 cfm.

To determine the heating energy saved by reducing the quantity of outdoor air, refer to Figure 17 and then carry out the following procedure:

a) Determine from Figure 4 heating degree days for your location.

b) Determine the average occupied hours/week during which the exhaust system is operated.

c) Determine the indoor temperature during occupied periods.

d) Enter the graph at the appropriate degree days and following the direction of the example line, intersect at the appropriate points for indoor temperature and hours of occupancy.

e) Read out yearly heating energy/1000 cfm.

f) Multiply this figure by the 1000 cfm exhaust air reduction to obtain the net yearly heating load reduction.

In addition to the savings achieved by reducing winter heat load, reducing the outdoor air will also reduce the sensible and latent heat gains in the summer. Depending on the building location and the severity and duration of the cooling season, this extra savings can be considerable (see Section 15) and should be recognized when making an economic feasibility study.

ECO 53 USE SEPARATE MAKE UP AIR SUPPLY FOR EXHAUST HOODS TO REDUCE OUTDOOR AIR VENTILATION

Kitchen equipment exhaust hoods and other process equipment hoods frequently remove large quantities of hot air from the building which must be made up by introducing outdoor air.

This make up air is often introduced through the building HVAC system which heats it in winter and cools and dehumidifies it in summer to the level required to maintain occupant comfort conditions within the building.

It is not necessary for the make up air to be treated to the same degree required by occupants; to conserve energy, separate supply of make up air for hoods should be considered. This make up supply system would consist of a fan drawing in outdoor air and passing it through a heating coil to just temper the air and maintain tolerable conditions immediately adjacent to the hood.

The temperature of the make up supply air often need only be 50°-55° providing it is introduced close to or around the perimeter of the exhaust hood. This relatively low temperature air will not cause discomfort to an occupant in the immediate area as the equipment radiates large quantities of heat which will offset any cooling effect.

The effectiveness of an exhaust hood in capturing heated air, fumes, smoke, steam, etc. is a function of the face velocity at the edge of the hood.

To maintain a satisfactory capture velocity, large open hoods require large volumes of air.

Face or capture velocities can be maintained or even increased while decreasing the total quantity of exhaust air by fitting baffles or a false hood inside an existing open hood.

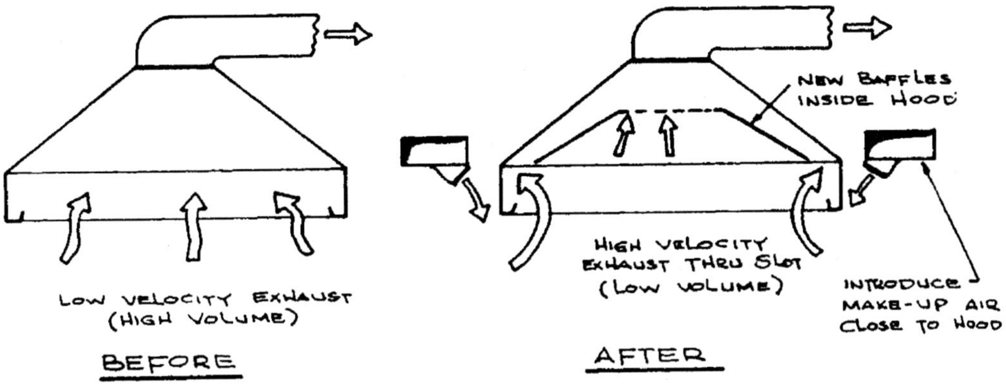

The cost of modifying exhaust hoods and installing a separate make up air system will vary with the circumstances of each application and to make an accurate assessment, quotations should be obtained from local contractors based on engineering designs to meet the particular case.

To determine the energy savings, first determine the reduction in outdoor air introduced through the HVAC systems and, using Figure 17 and the procedure described for ECO 52, calculate the energy saved.

To obtain the total yearly heating load reduction, deduct the extra heating energy that must be expended to temper the air introduced by the make up system.

In addition to this heating load reduction, savings of sensible and latent heat will also accrue during the cooling season (see Section 15) and should be recognized when making an economic feasibility study.

REDUCE PARASITIC DISTRIBUTION LOSSES

BACKGROUND

All distribution systems require energy input to transfer the heating and/or cooling medium (water-steam-air, etc.) from the point of generation (boilers-chillers-HVAC units, etc.) to the point of use (air terminals-radiators, etc.). In the case of water distribution systems, this energy input is in the form of pump horsepower, and in the case of ventilation systems, it is in the form of fan horsepower. Distribution systems also lose energy in the form of heat loss from hot pipes and ducts and heat gain to cold pipes and ducts. The energy required to operate distribution systems and the energy losses from the systems do not make any useful contribution to meeting the building load and should be considered parasitic. Any reduction of these parasitic losses is a direct saving in energy.

After the building heating load is minimized, it is possible to reduce the parasitic system loads either by reducing flow rates, or temperatures, or a combination of both.

Parasitic distribution system losses are common to both heating and cooling systems and are discussed in detail in Section 17, DISTRIBUTION & HVAC SYSTEMS.

Existing distribution systems are often oversized to meet the building load, due to engineering safety factors which were originally included to provide a contingency for unknown factors at the time of design. The systems should now be analyzed and correct flow rates calculated to meet the known building conditions. This will provide a base against which to measure all new energy conservation opportunities.

ECO 54 REDUCE AIR FLOW RATES

When all opportunities to reduce the heating load have been selected, the rate of flow of a warm air system can be reduced to meet the new load.

To determine the flow rate reduction, first, calculate the initial heating load (total heat loss/hour) at outdoor design conditions (use 97-1/2% column ASHRAE Fundamentals) giving credit to heat gains from lights, power and occupants. Using the same procedure, calculate the new heating load at the same design condition but with revised figures to account for all load reduction opportunities put into effect from both ECM 1 and ECM 2, (e.g. reduced space temperature, increased insulation, double windows, reduced ventilation and infiltration, etc.).

Determine the new supply air flow rate by using the following formula:

$$\text{New flow rate} = \text{Initial flow rate} \times \frac{\text{New heating load}}{\text{Initial heating load}}$$

Methods of achieving this flow rate reduction and calculating the energy savings are detailed in ECO 101 (Section 17). Any flow rate reduction achieved by reducing heat loads may affect a ventilating system's ability to cool in summer unless similar reductions are made in cooling load.

ECO 55 REDUCE RESISTANCE TO AIR FLOW

Reducing resistance to air flow will result in fan power savings. Methods of reducing resistance and calculating the energy savings are detailed in ECO 100 (Section 17).

ECO 56 REDUCE WATER FLOW RATES

When all opportunities to reduce the heating load have been selected, the flow rate of a hot water heating system can be reduced to meet the new load.

To determine the flow rate reduction, follow the procedure outlined in ECO 54 and apply the following formula:

$$\text{New flow rate} = \text{Initial flow rate} \times \frac{\text{New heating load}}{\text{Initial heating load}}$$

Methods of achieving flow rate reduction and calculating the energy savings are detailed in ECO 99 (Section 17).

ECO 57 REDUCE PARASITIC LOADS BY INSULATING PIPES

Insulate bare, hot pipe. to reduce non-productive heat loss. Insulation may be sectional, rigid type of fibrous material, plastic, or glass wool and should be applied in thicknesses to economically reduce the heat loss. The material used should be capable of withstanding the surface temperature of the pipe. Some plastic insulating materials deteriorate or melt at temperatures in excess of 220° F. An alternative insulating material is plastic magnesia, applied wet; it is particularly suitable where there are many changes of direction, valves, fittings, which make it difficult to apply sectional insulation. All fittings, flanges, etc. should be insulated against heat loss.

Where there is insulation on existing pipework already, but it is in a bad state of repair, it should either be stripped off and replaced or if the worn sections are restricted to a few locations, it should be replaced in those areas.

Where hot pipes are insulated but the insulation is of minimal thickness and effectiveness, it should be supplemented by additional insulation applied over it.

Do not insulate steam traps or the first six feet of the condensate discharge piping from the trap.

Check with the local fire department to determine whether the selected insulation is acceptable or is considered a fire and smoke hazard. In buildings where food is either stored, processed, or sold, avoid the use of fibrous insulating materials if there is any danger of fiber migration. Check with the local Food and Drug Administration for prohibited mate ials.

The cost of applying insulation will vary with the circumstances of each particular case and for an accurate assessment, quotations should be obtained from local contractors.

For order of magnitude, however, the following tabulated costs may be used:

Pipe Sizes	Price per L.F. Installed Insulation Thickness: 1"	1-1/2"
1/2"	$1.35	$2.05
3/4"	1.40	2.10
1"	1.45	2.15
1-1/4"	1.50	2.20
1-1/2"	1.55	2.25
2"	1.60	2.35
2-1/2"	1.65	2.45
3"	1.70	2.50
4"	2.00	3.05
5"	2.25	3.15
6"	2.55	3.30
8"	3.15	4.20
10"	3.85	5.00
12"	4.50	5.60
14"	5.20	6.45
16"	6.00	7.20
18"	6.70	7.60
20"	8.25	8.50
24"	9.00	9.70

For costing purposes, add to total lineal feet of piping, three (3) lineal feet for each fitting or pair of flanges to be insulated.

To determine the energy saved by insulating pipes, refer to Figures 44 and 45 and carry out the following procedure:

a) Select the appropriate graph for the surface temperature of the pipe to be insulated.

b) Tabulate all diameters and lengths of pipe to be insulated.

c) Enter the graph at the appropriate pipe size, intersect with the "none" line and read out the energy loss in Btu/year.

d) Repeat this procedure for the same size pipe, but intersect with the appropriate insulation thickness line and read out the annual energy loss in Btu/year.

e) The difference between these two figures is the energy saved per year for insulating ten feet of that size pipe. Multiply by the $\frac{\text{length of pipe}}{10}$ to be insulated to obtain the total yearly saving.

f) Repeat the above procedure for each different size of pipe.

FIGURE 44: HEAT LOSS FOR VARIOUS PIPE SIZES, INSULATION THICKNESSES AND WATER TEMPERATURES

dubin-mindell-bloome-associates consulting engineers

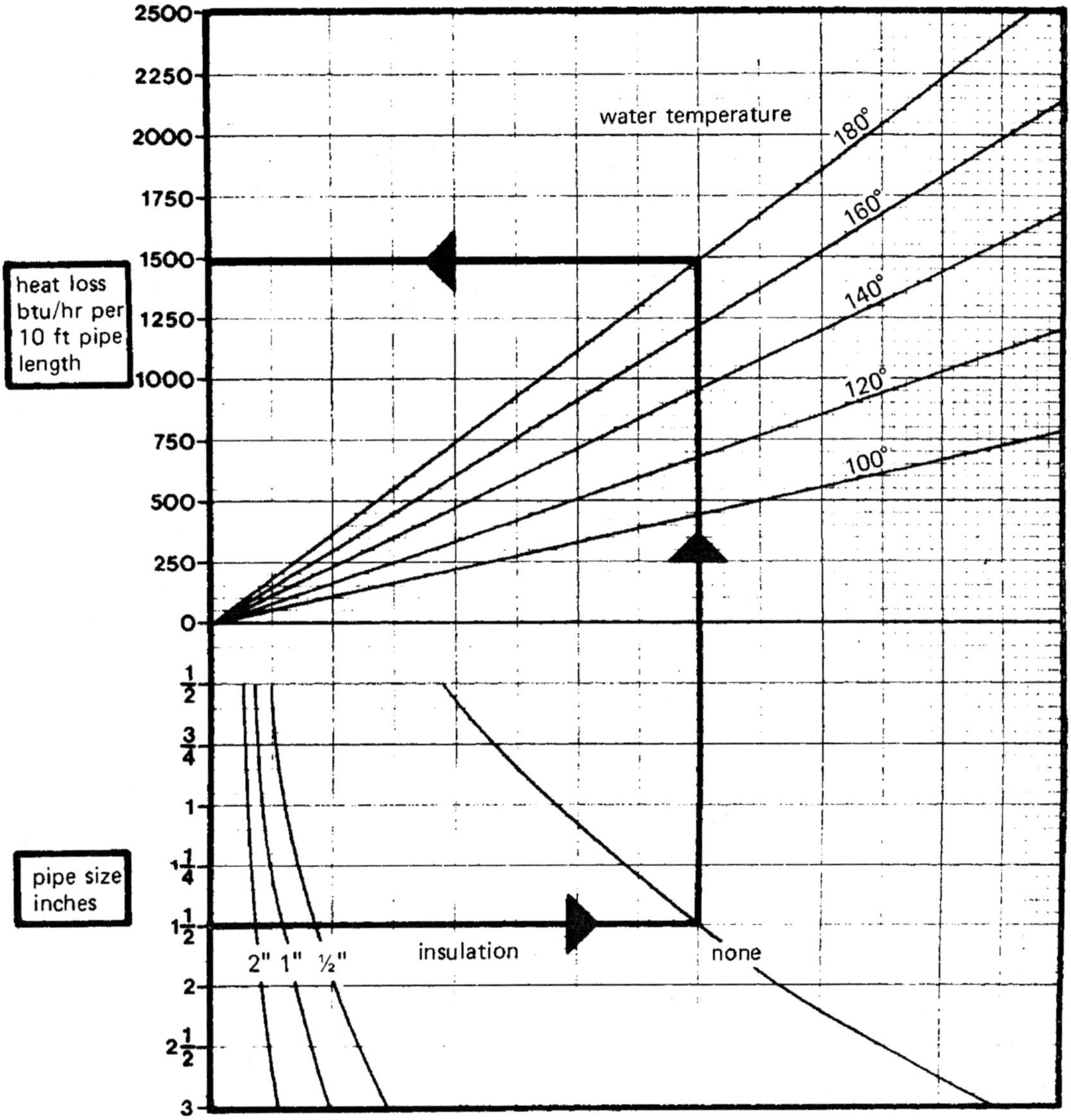

Figure # 44 Engineering Data

Figure is based on the addition of insulation with a thermal conductivity, k = 0.3 (Btu) (in.)/(sq.ft.) (hr.) (degree F temperature difference) and an ambient air temperature of 68°F.

The following formula is the basis for determining the heat emitted (heat loss) from the pipe:

$$q = \frac{T_s - Ta}{\frac{d_1}{2k}\left(\log_e \frac{d2}{d_1}\right) = \frac{d_1}{fd_2}}$$

Where:

q = heat emission from insulated pipe - Btu/(hr.) (sq.ft. of hot pipe surface)

Ts = temperature of pipe surface - degree F

Ta = ambient air temperature - degree F

d_1 = outside diameter of pipe - inches

d_2 = outside diameter of insulation - inches

k = thermal conductivity of insulating material - (Btu) (in.)/(sq.ft.) (hr.) (degree F temperature difference)

f = surface coefficient - Btu/(sq.ft.) (hr.) (degree F)

FIGURE 45: HEAT LOSS FOR VARIOUS PIPE SIZES, INSULATION THICKNESSES AND WATER TEMPERATURES

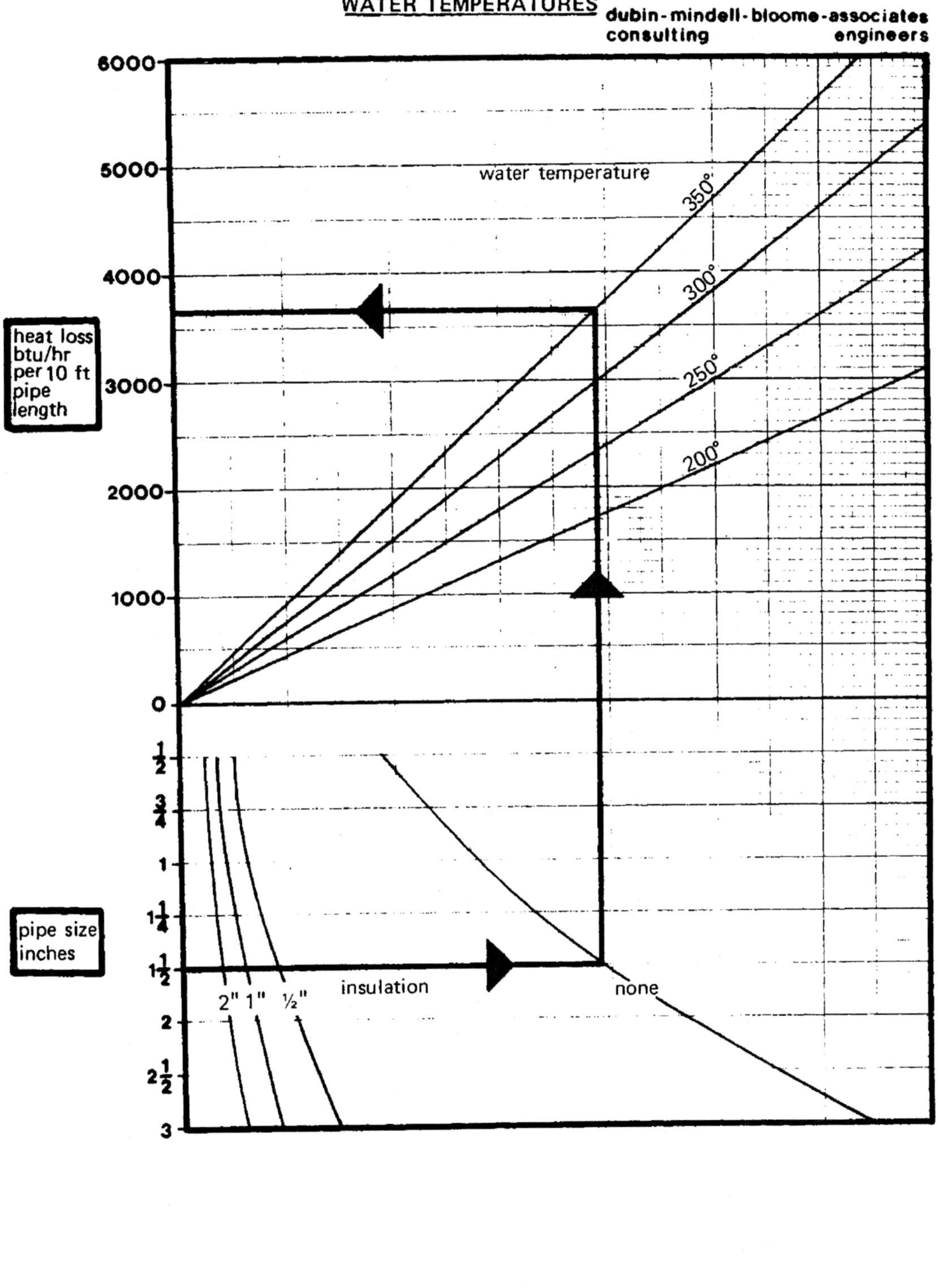

Figure # 45 Engineering Data

Figure is based on the addition of insulation with a thermal conductivity, k = 0.3 (Btu) (in.)/(sq.ft.) (hr.) (degree F temperature difference) and an ambient air temperature of 68°F.

The following formula is the basis for determining the heat emitted (heat loss) from the pipe:

$$q = \frac{T_s - Ta}{\frac{d_1}{2k}\left(\log_e \frac{d_2}{d_1}\right) + \frac{d_1}{fd_2}}$$

Where:

q = heat emission from insulated pipe - Btu/(hr.) (sq.ft. of hot pipe surface)

Ts = temperature of pipe surface - degree F

Ta = ambient air temperature - degree F

d_1 = outside diameter of pipe - inches

d_2 = outside diameter of insulation - inches

k = thermal conductivity of insulating material - (Btu) (in.)/(sq.ft.) (hr.) (degree F temperature difference)

f = surface coefficient - Btu/(sq.ft.) (hr.) (degree F)

ECO 58 REDUCE PARASITIC LOADS BY INSULATING DUCTS

Warm air ducts are commonly installed without insulation and are typically routed from the equipment room through unoccupied spaces, shafts, and ceiling voids where their heat loss is unproductive in meeting the occupied space heating load. Although the temperature difference between duct and ambient temperature is relatively small, heat loss in long duct runs can be significant.

Of equal importance is the temperature drop of supply air that accompanies heat loss. In long duct runs serving many rooms on one zone, this will result in the last room having a lower supply air temperature than the first. The tendency in this case is to heat the last room to comfort conditions resulting in overheating in each preceding room with consequent additional waste of energy over and above the duct heat loss.

Warm air ducts may be insulated with rigid fibrous material stuck on or fixed with special clips or bands. They may also be insulated with flexible mats clipped or wired on (this is particularly applicable for round or oval ducts). Ducts may also be insulated with spray-on foam or fibrous material as described for insulating undersides of roofs. It is worth considering insulating roofs and ducts as one contract.

Insulation applied to ducts supplying only warm air need not be vapor sealed. Insulation applied to ducts supplying warm air in winter and cold air in summer must be vapor sealed to prevent condensation forming within the insulation.

Check with the local fire department to determine whether the selected insulation is acceptable or whether it is considered a fire and smoke hazard.

The cost of applying insulation will vary with the circumstances of each case and for an accurate assessment, obtain quotations from local contractors. For order of magnitude, however, a cost of $\$2.00/ft.^2$ may be used.

To determine the heating energy saved by insulating ducts, refer to Figure 46 and then carry out the following procedure:

a) Calculate the exposed surface area of uninsulated ductwork.

b) Determine the temperature differential between the duct and ambient air.

c) Enter the graph at insulation thickness and following the direction of the example line intersect with the appropriate temperature differential line. Read out the heat loss in Btu/sq.ft. hour.

d) Read out the bare duct heat loss from the graph.

e) Subtract the insulated duct loss from the bare duct heat loss and multiply by the surface area of insulated duct to derive the saving in Btu/hour.

f) Determine the hours of system operation per year and multiply by the hourly saving to derive the total yearly saving in Btu's.

If the ambient temperature varied by more than 5° F over the year then determine and use the average temperature for the time considered in the calculation.

In addition to heating energy savings, duct insulation will also reduce cooling energy if the same duct is used to supply cooled air in summer. It will also enable closer control of single zone room conditions to be obtained year'round, thus reducing the need to overheat or overcool some rooms.

These additional energy savings should be recognized when making an economic analysis.

INCREASE EFFICIENCY OF PRIMARY ENERGY CONVERSION EQUIPMENT

BACKGROUND

When all applicable ECO's to reduce load and parasitic system losses have been applied, the required quantity of heat generated by the primary equipment will have been reduced. Even if the primary equipment was operating at optimum efficiency to meet the original load, it will now be operating at lower efficiency with the reduced load. Opportunities exist to conserve energy by readjustment for reduced load.

ECO 59 ADJUST FUEL/AIR RATIOS OF FIRING EQUIPMENT

Boilers sized to meet the original unmodified heat load and with fuel/air ratios adjusted accordingly will now be oversized for the new reduced load and will have incorrect fuel/air ratios.

Typically a heating system load/time curve will be as follows:

FIGURE 46: HEAT LOSS/GAIN FOR VARIOUS INSULATION THICKNESSES

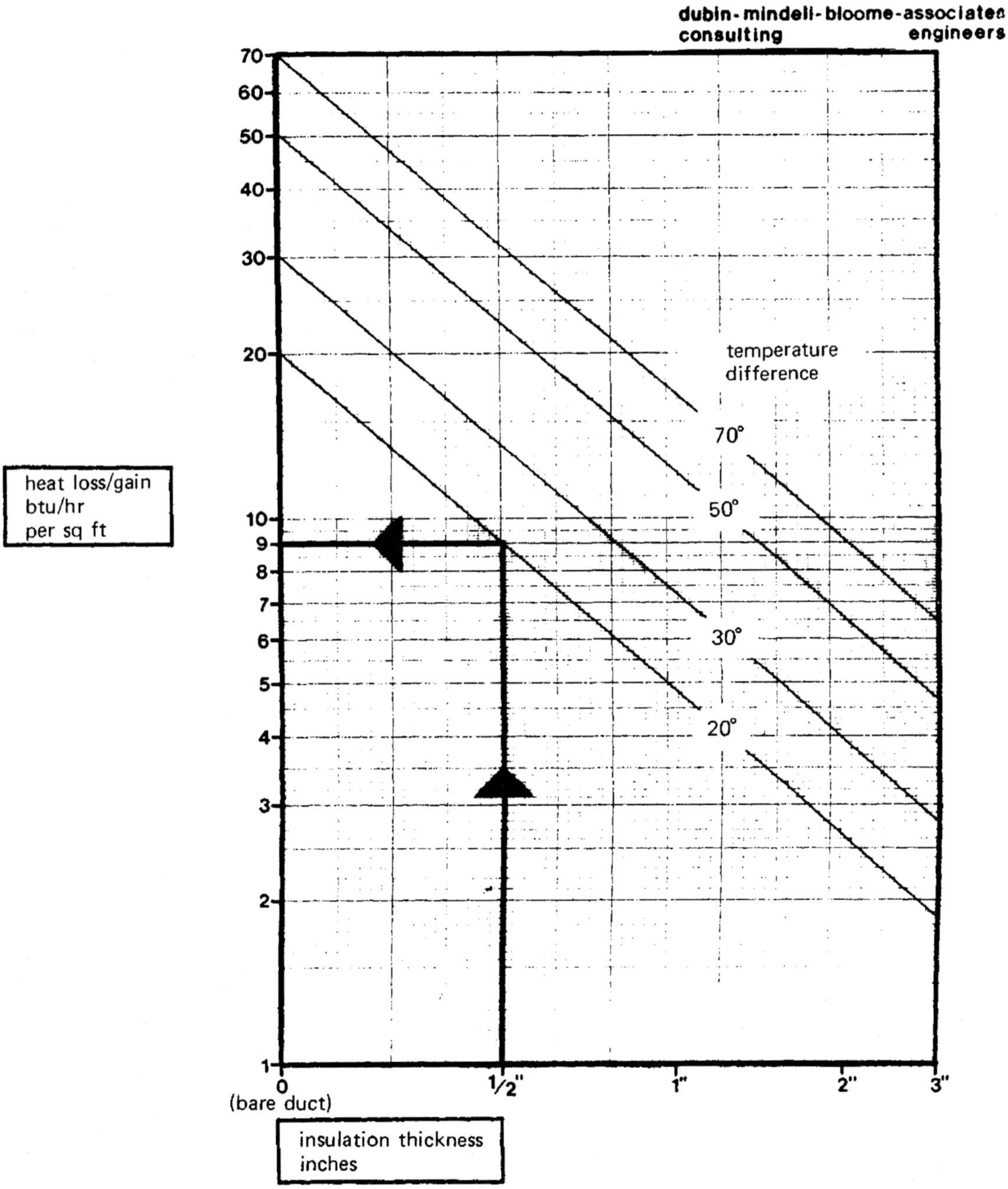

Figure # 46 Engineering Data

Source: ASHRAE Handbook of Fundamentals

Assumes Rigid Insulation with K value of 0.27 at 75°F.

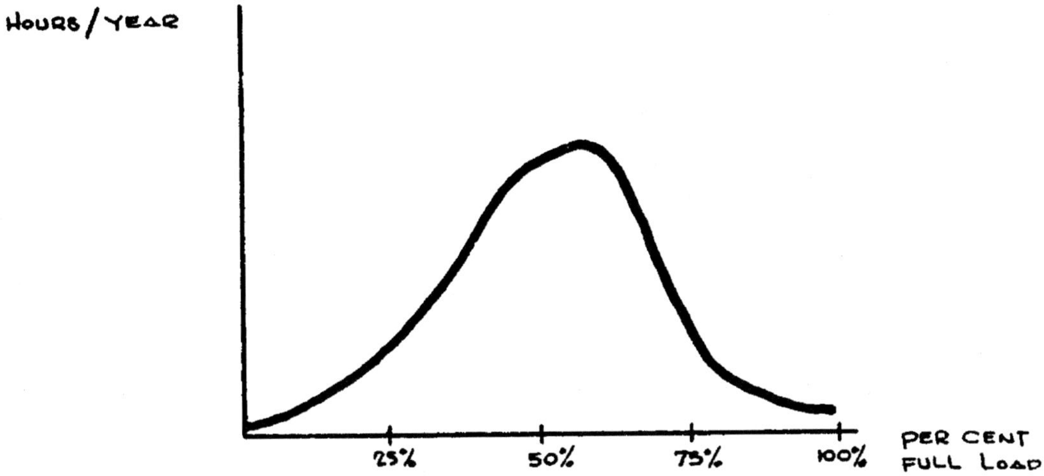

The heating system operates between 50% and 60% full load for more hours/year than at higher or lower loads and the optimum operating point of the heating boilers should be selected accordingly to give the highest efficiency spread over this range.

For example, if the installation comprises three 8×10^6 Btu/hr boilers and the maximum heating load is reduced to 18×10^6 Btu/hr then the boilers should be adjusted for maximum efficiency at 70% of the rated output. This gives three possible operating points at maximum efficiency:

i.e.: one boiler at 8×10^6 Btu/hr x 0.7 = 5.6×10^6 Btu/hr or 31% full load
two boilers at $16 \times 10^6 \times 0.7 = 11.2 \times 10^6$ Btu/hr or 62% full load
three boilers at $24 \times 10^6 \times 0.7 = 16.8 \times 10^6$ Btu/hr or 93% full load.

Indicators of maximum combustion efficiency are stack temperature, percentage CO_2 and percentage O_2.

Refer to Figure 47 to determine the optimum fuel/air ratio for any given combination of circumstances.

Primary and secondary air should only be allowed to enter the combustion chamber in regulated quantities and at the correct place. Defective gaskets, cracked brickwork, broken casings, etc. will allow uncontrolled and varying quantities of air to enter the boiler and will prevent accurate fuel/air ratio adjustment. If spurious stack temperature and/or oxygen content readings are obtained, then inspect the boiler for air leaks and repair all defects before final adjustment of the fuel air ratio.

When substantial reductions in heating load have been achieved, the firing rate of the boiler may be excessive and should be reduced. Consult the firing equipment manufacturer for specific recommendations. Reduced firing rate in gas and oil burners may require additional bricking to reduce the size and shape of the combustion chamber.

The cost of fuel/air ratio adjustments and firing equipment modifications will vary with each application and quotations should be obtained from qualified contractors.

ECO 60 INSTALL FLUE GAS ANALYZER

The efficient combustion of fuel in a boiler requires an optimum fuel/air ratio providing for a percentage of total air sufficient to insure complete combustion of the fuel without overdiluting the mixture and thereby lowering boiler burner efficiency.

Optimum combustion efficiency varies continuously with changing loads and stack draft and can be closely approached only through analysis of flue gases. Information required to allow continuous adjustment of fuel/air ratios is a) Flue gas temperature and b) Flue gas CO_2 or O_2 content.

Indicators are available which continuously measure CO_2 and stack temperature and give a direct reading of boiler efficiency. These indicators provide boiler operators with the requisite information for manual adjustment of fuel/air ratio and are suitable for the smaller installations or where money for investment in capital improvements is limited.

A more accurate measure of efficiency is obtained by analysis of oxygen content rather than measuring other gases such as carbon dioxide,and carbon monoxide.

As shown in Figure 47, the cross checking of O_2 and CO_2 concentrations is useful in judging burner performance more precisely. Due to the increasingly wide-spread need for multifuel boilers, however, O_2 analysis is the single most useful criterion for all fuels since the O_2 total air ratio varies within narrow limits.

For larger boiler plants consider the installation of an automatic continuous oxygen analyzer with "trim" output that will adjust the fuel/air ratio to meet changing stack draft and load conditions.

Most boilers can be modified to accept automatic fuel/air mixture control by flue gas analyzer, but a gas analyzer manufacturer should be consulted for each particular installation to be sure that all other boiler controls are compatible with the analyzers.

It is important to note that some environmental protection laws might place a higher priority on visible stack emissions than on efficiency optimization of fuel combustion especially where fuel oil is concerned. The effect of percent total air on smoke density might prove to be an overriding consideration and limit the approach to minimum excess air. All applicable codes and environmental statutes should be checked for compliance.

FIGURE 47: EFFECT OF FLUE GAS COMPOSITION AND TEMPERATURE ON BOILER EFFICIENCY

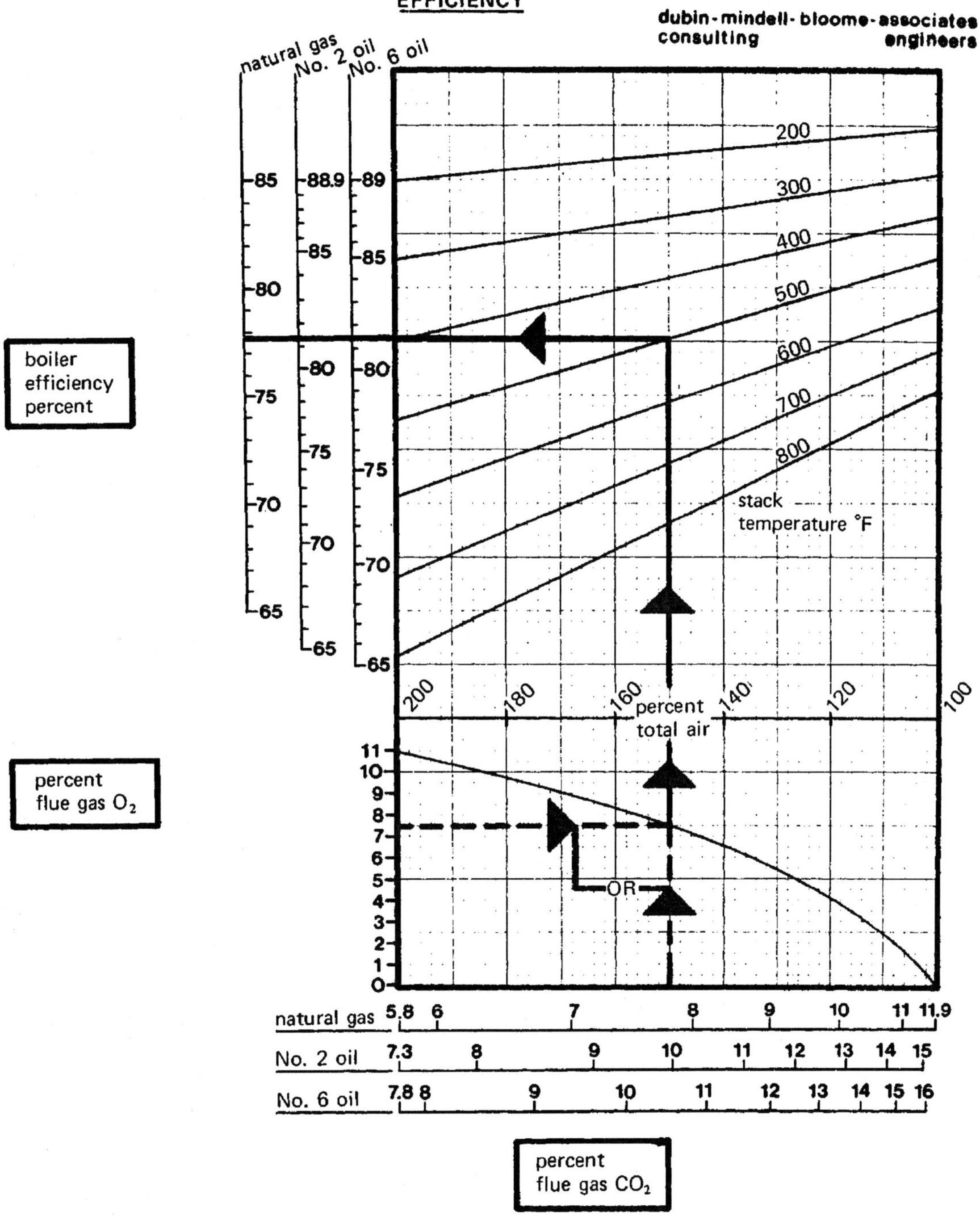

Figure # 47 Engineering Data

Sources: Data from various boiler and burner manufacturers was analyzed and integrated to make up one graph.

The cost of adding flue gas analysis instrumentation and controls will vary with each particular case but for order of magnitude the following tabulated costs may be used.

O_2 analyzer w/meter readout	$ 2,000 - 3,000
O_2 analyzer w/chart recorder	3,000 - 5,000
O_2 analyzer w/trim link control	4,000 - 6,000
Full metering w/O_2 trim link	11,000 -16,000

Refer to Figure 47 to determine efficiency improvement by optimizing boiler fuel/air ratios.

Enter the graph at either the O_2 content or CO_2 content and following the direction of the example line intersect with the appropriate stack temperature and read out the combustion efficiency against the appropriate fuel type. Percent total air may also be obtained from an intermediate line.

Savings are determined by determining boiler efficiency under present operating conditions, and then under improved conditions with optimization of the fuel/air ratio.

Example

1. A boiler burning No. 2 oil rated at 20,000,000 Btu/hr. has a yearly oil consumption of 425,000 gallons.

2. A CO_2 meter is installed and reads 11% CO_2 in the flue gas.

3. Stack temperature is 650° F.

4. Entering the graph at the 11% CO_2 point for No. 2 oil and intersecting with a stack temperature of 650° F gives an efficiency of 78%. Air percentage of 138 indicating 38% excess air and an O_2 percentage of 6 can also be read.

5. The installation of an O_2 analyzer with automatic fuel/air ratio control will reduce O_2 content to 3.5% and stack temperature to 530° F.

6. Using the graph again the new efficiency is 83%.

7. Yearly savings due to increased boiler efficiency

$$= \text{Original yearly consumption} \times \frac{\text{New Efficiency} - \text{Original Efficiency}}{\text{New Efficiency}}$$

$$= 425{,}000 \times \frac{83 - 78}{83} = 25{,}600 \text{ gallons/year}$$

at 33¢/gallon = $8,448

8. The installation cost is approximately $16,000 and is obviously economically advantageous without further economic analysis.

ECO 61 ISOLATE OFF LINE BOILERS

Light heating loads on a multiple boiler installation are often met by one boiler on line with the remaining boilers idling on stand by.

Idling boilers consume energy to meet stand by losses which can be further aggravated by a continuous induced flow of air through them into the stack and up the chimney.

Unless a boiler is scheduled for imminent use to meet an expected increase in load, it should be secured and isolated from the heating system by closing valves and from the stack and chimney by closing dampers.

Large boilers can be fitted with by-pass valves and regulating orifice to allow a minimum flow through the boiler to keep it warm and avoid thermal stress when it is brought on-line again.

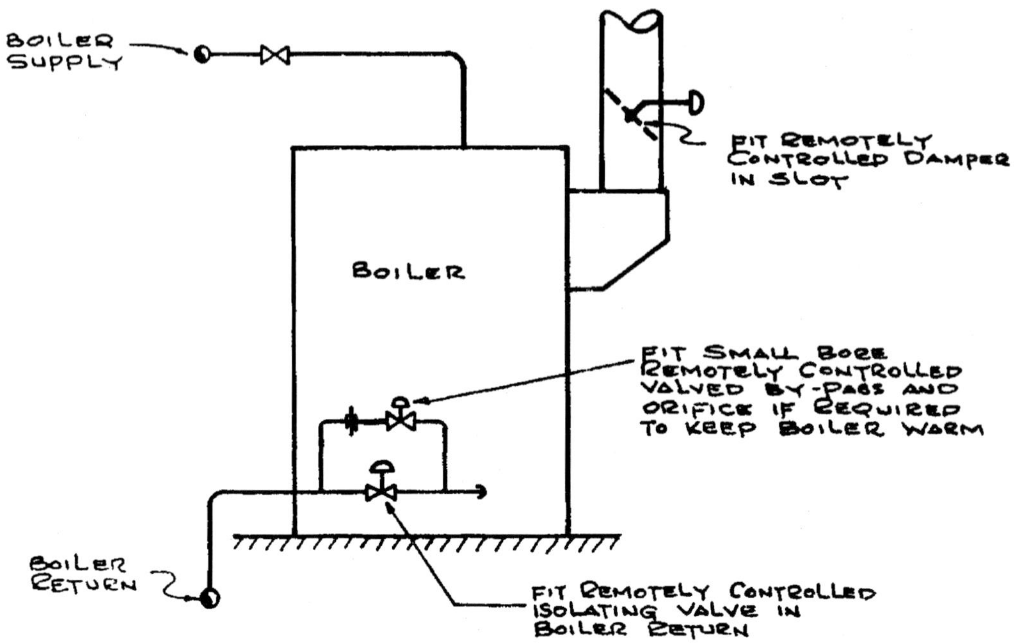

If a boiler waterside is isolated, it is important that air flow through the stack is prevented; it is possible for back flow of cold air to freeze the boiler.

The cost of isolating off-line boilers will vary according to the circumstances of each case and quotations should be obtained from local contractors.

ECO 62 REPLACE EXISTING BOILERS WITH MODULAR BOILERS

Heating boilers are usually designed and selected to operate at maximum efficiency when running at their rated output, and have lower efficiencies at reduced output.

Typical heat load distribution over a heating season shows that full boiler capacity is only required for small periods and that for 90% of the time the heat load is 60% or less than full load.

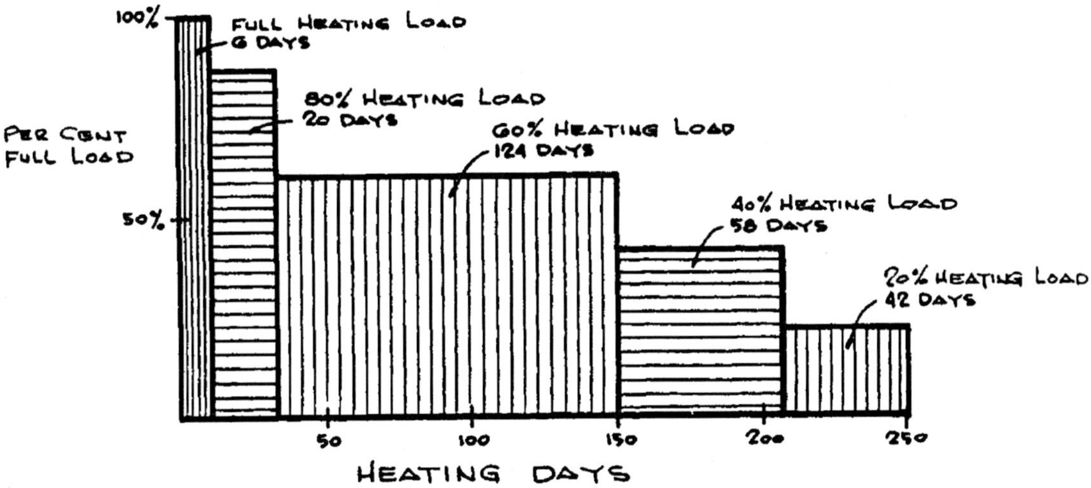

TYPICAL HEAT LOAD DISTRIBUTION
250 DAY SEASON IN 6000 DEGREE DAY ZONE

Large capacity boilers in single units must operate intermittently for the major part of the heating season and although hi-low firing capabilities may reduce cycling, the boilers can only reach their design efficiency for short periods resulting in low seasonal efficiencies.

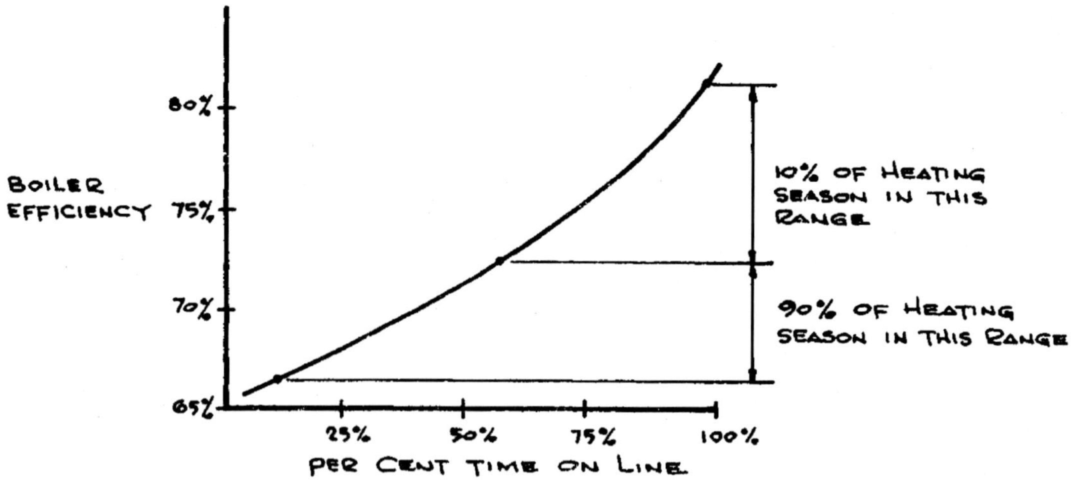

EFFECT OF CYCLING TO MEET PART LOADS

A modular boiler system comprising multiple boiler units each of small capacity will increase seasonal efficiency. Each module would be fired at 100% of its capacity only when required and fluctuations of load would be met by firing more or less boilers.

Each small capacity unit has low thermal inertia(giving rapid response and low heat-up and cool-down losses) and will either be running at maximum efficiency or will be turned off.

Boiler seasonal efficiency may be improved from 68% to 75% in a typical installation where single unit large capacity boilers are replaced by modular boilers. This represents a saving of approximately 9% of the present yearly fuel consumption.

Where the present boiler plant has deteriorated to the point where it is at or near the end of its useful life, consider replacement with modular boilers sized to meet the reduced heating load resulting from other ECO's.

Modular boilers are particularly effective in buildings that have intermittent short time occupancy such as churches. They provide rapid warm up for occupied periods and low stand by losses during extended unoccupied periods.

ECO 63 PREHEAT COMBUSTION AIR TO INCREASE BOILER EFFICIENCY

Preheating primary and secondary air will reduce the cooling effect when it enters the combustion chamber and will increase the efficiency of the boiler. It will also promote more intimate mixing of fuel and air which will also add to the boiler efficiency.

In most boiler rooms, air is heated incidentally by hot boiler and pipe surfaces and rises to collect below the ceiling. This air can be used directly as pre-heated combustion air by ducting it down to the firing level and directing it into the primary and secondary air inlets.

Waste heat reclaimed from boiler stacks, and blowdown or condensate hot wells can also be used to preheat combustion air.

Boiler efficiency will increase by approximately 2% for each 100° F increase in combustion air temperature.

Combustion air can be preheated up to 600° F for pulverized fuels and up to 350° F for stoker fired coal, oil and gas. The upper temperature limit is determined by the construction and materials of the firing equipment and manufacturers recommendations should be obtained.

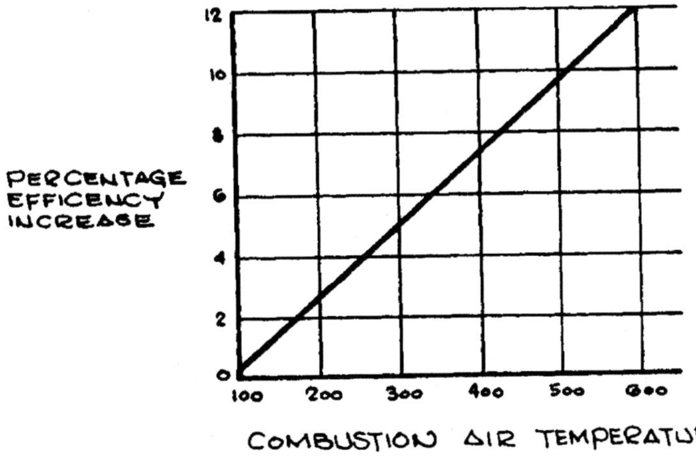

EFFICENCY INCREASE WITH PRE-HEATED AIR

Heat exchange from flue gases to combustion air may be made directly using static tubular or plate exchangers or rotary exchangers. Heat exchange may also be made indirectly through run-around coils in the stack and combustion air duct.

Waste heat from flue gases, blowdown, condensate, hot wells, etc., or from solar energy may also be used to preheat oil either in the storage tanks (low sulfur oil requires continuous heating to prevent wax deposits) or at the burner nozzle.

Oil must be preheated to at least the following temperatures to obtain complete atomization:

No. 4 oil - 135° F

No. 5 oil - 185° F

No. 6 oil - 210° F

Heating beyond these temperatures will increase efficiency but care must be taken not to overheat, or vapor locking could cause flame-outs. The increase in efficiency obtained by preheating oil could be as high as 3% but depends on the particular constituents of the oil and recommendations should be obtained from the oil supplier.

ECO 64 CHANGE STEAM ATOMIZING BURNERS TO AIR ATOMIZING BURNERS

Air atomizing burners are considerably more efficient than steam atomizing burners and should always be considered as an alternative when replacing obsolete or defective steam atomizing burners. They may in some circumstances be cost effective on their own merit without waiting for scheduled replacement times.

ECO 65 REDUCE BLOWDOWN LOSSES

The purpose of blowing down a boiler is to maintain a low concentration of dissolved and suspended solids in the boiler water and to remove sludge in the boiler in order to avoid priming and carryover. There are two principal types of blowdown, intermittent manual blowdown and continuous blowdown. Manual blowdown or sludge blowdown is necessary for the operation of the boiler regardless of whether or not continuous blowdown is installed. The frequency of manual blowdown will depend on the amount of solids in the boiler makeup water and the type of water treatment used.

Continuous blowdown is a steady drain on the energy; as the makeup water must be heated. Energy loss of blowdown can be minimized by the use of monitoring with automatic blowdown control systems and heat recovery systems.

Automatic blowdown controls monitor the conductivity and PH of the boiler water periodically and blow down the boiler only when required to maintain acceptable water quality. Further savings can be realized if the blowdown water is piped through a heat exchanger or a flash tank with a heat exchanger.

In a 100,000 lbs /hr., 600 psi boiler with a maximum boiler water concentration of 2500 ppm total dissolved solids, the blowoff will be 8% of the makeup water or 3500 lbs/hr. The total heat in the blowoff is 1660 MBH. A system using a heat exchanger only and with a 20° F terminal difference will recover 90% total heat in the blowoff or 1494 MBH. Adding a flash tank operating at 5 psig with a heat exchanger having 20° F terminal difference will recover 93% of the total heat in the blowoff or 1544 MBH. The per cent of heat recovery will change with boiler operating conditions. The best recovery range is 78% for a 15 psig boiler to 98% for a 300 psig boiler.

GUIDELINES TO REDUCE ENERGY USED FOR HEATING

1. Lower Indoor Temperature and Relative Humidity Levels in the Heating Season

Refer to ECM 1, Figure 14 "Recommended heating season indoor temperature and humidity levels". The following guidelines list the options for implementing the recommendations.

a) Install a seven-day dual stat to operate the oil, gas, stoker, or electric heating elements. The stat should be set to maintain the temperature levels during unoccupied periods, and to reduce levels when the building is unoccupied at nights, weekends and holidays.

b) Install a seven-day stat to control the operation of pumps for forced circulation hot water systems, automatic control valves for hot water or steam systems when boilers are operated by aquastat or pressure-trol.

c) Install a seven-day timer control to reset operating aquastat or pressure-trol for dual level settings.

d) Install additional thermostats for individual zones where duct or piping systems permit control of individual zones and where necessary at zone control valves. Operate controls on a seven-day cycle.

e) Provide room stat and automatic damper controls and dampers in supply air duct systems for additional zoning to permit further reduction of temperature levels in non-critical areas.

f) Where missing, install manually operated radiator valves, or duct dampers to control or shut off heat supply to non-critical areas in portions of the building.

g) Install occupied/unoccupied, on-off switch to control water supply to humidifiers to permit shutdown at night.

h) Remove and relocate central humidifier serving the entire building and install duct type or package room humidifier to serve only those zones which require humidification. Shut off humidifiers at night as per g) above.

i) Do not operate refrigeration systems or introduce outdoor air for the purpose of cooling in winter to reduce indoor temperature levels to the heating set point of 68°F. If gains exceed losses, allow space temperature to rise to 78°F.

j) Where room stats control both heating and cooling systems, exchange them for stats with a dead band level.

k) Refer to Section 6E for adjustments to HVAC equipment and systems for further guidelines to reduce temperature and humidity levels during the heating season.

l) Provide locking devices on all room stats to prevent tampering.

m) Where supply duct dampers are adjusted to reduce air flow into the space during the heating season, provide damper to reduce return air from the space.

n) Relocate room stats to the most critical area and rebalance the air or water system to reduce temperature and humidity levels in the other less critical areas.

o) Where window air conditioner units or through-the-wall units provide heating as well as cooling, provide seven-day temperature control stats to operate the heating elements.

2. Reduce the Heating Load Due to Ventilation and Makeup Air

Refer to Figure 16 "Recommended Ventilation Standards". ASHRAE Standards 62-73 sets 5 cfm per occupant as a basic ventilation rate. Where Codes require more ventilation, appeal that requirement with Code Authorities.

Energy required for humidification will be reduced when outdoor air quantities are reduced. Refer to Figure 15 for the energy required for humidification of outdoor air.

a) Provide seven-day timer to operate automatic fresh air damper control (install controller where missing), or to shut off all outdoor air for ventilation during unoccupied periods where separate outdoor air supply fans exist.

b) Install automatic damper control where missing, for operation with seven-day timer.

c) Add separate ventilation fan for zones requiring outdoor air and shutoff damper to separate unit serving the entire building.

d) Reduce operating time of exhaust systems and where interlocks with outdoor air makeup systems have been or will be installed, control the exhaust--outdoor air makeup unit or damper with seven-day stats.

e) Utilize waste heat or recovered heat from exhaust systems and other equipment to temper the outdoor air supply to reduce the heating load. (See Section 18 "Heat Recovery")

f) Install charcoal filters or other devices to control supply air quality and reduce the amount of outdoor air for ventilation. Check with Code Authorities before making the change.

g) Modify exhaust hoods to reduce the ventilation as follows:

-By inspection, determine whether the existing exhaust hood is of the simple open type and by measurement, determine the quantity of air exhausted. If the exhaust hood is of the open type, install baffles to allow reduction of exhaust air quantity without reducing the face velocity at the edge of the hood.

-If baffles are installed, reduce the rate of exhaust and by measurement determine the new quantity.

-Install a make up air system to introduce outdoor air equal in quantity to that exhausted.

-Do not heat make up air to more than 55°F and introduce as close as possible to the hood in several positions around the perimeter to promote even air flow.

-Consider using reclaimed hood exhaust heat or solar heat as the energy source for make up air.

-Do not cool or dehumidify make up air.

3. <u>Reduce Heat Loss Due to Infiltration</u>

a) Instead of caulking, add tight fitting storm windows.

b) Weatherstrip all doors with a copper interlocking type weatherstrip. In buildings where carpeting extends to the doors, use a compression type weatherstrip attached to the bottom of the door to clear the carpeting.

c) In locations where strong winds occur for long durations, use external wind baffles at external surface of the building to shield windows and doors.

d) In areas where operable windows are not required, close them permanently and caulk cracks and/or block off windows entirely with masonry or insulated wall panels. Where insulated wall panels are used, fit them to effectively block infiltration.

e) Where wall construction is particularly porous--cinder block walls or other similar porous materials, cover the exterior surface with epoxy paint.

f) Install vestibules in stores, office buildings, and/or religious buildings where doors are open frequently throughout the day.

g) Install revolving doors with, or inside of vestibules where exceptionally heavy traffic occurs, and infiltrated air circulates freely to occupied spaces. When installing revolving doors, maintain a sufficient number of operable single or double doors to meet Code requirements for safety exit. Where operable windows are not used during the winter, removable tapes can be used to seal cracks.

h) Caulk cracks around window air conditioning units or through-the-wall units which remain permanently installed year round, and cover window air conditioning units with plastic covers outside. Covers can be bought for about $5.00 each.

i) Install automatic door closers on exterior doors.

j) To reduce infiltration through leaky or jalousie windows where operable sash is not required in the winter, cover the windows with a 5-mil plastic sheet, tacked in place to the exterior of the building, making sure that the plastic shield extends to cover cracks around the window frames. Covers can be taken down in the summer time.

k) When installing new screens on windows and doors, select the units in accordance with ASTM E283 Standards and limit air leakage to 0.5 cfm per foot of crack when subjected to a wind pressure of 25 miles per hour.

l) Seal off large openings in stair towers with masonry partition or with a tight fitting door in accordance with ASTM E283 Standards.

m) Pressurize building with outdoor air, properly controlled and treated, to reduce uncontrolled infiltration.

n) In tall buildings, provide a sealed platform every 5 floors in vertical service shafts.

o) Weatherstrip doors and windows in elevator equipment rooms located at the top of the elevator shaft.

p) Reduce air volume to toilet rooms to 1 cfm per square foot, to reduce air make up requirements. In making the calculation include only those areas in the toilet room proper. Do not include janitor storage, vestibules, and lounges.

4. <u>Reduce Heat Loss Through Fenestration</u>

a) Where the heating season is 7500 degree days or greater, use triple glazing.

Options:

Remove existing single window and frame and replace with triple glazing.

Add double glazed storm window to existing single glazed window.

b) Where heating season is 4500 degree days or greater, install a storm window to reduce both conduction loss and infiltration; or where window is tight, remove glazing from window frame and install double glazing unit.

c) Where the heating season is between 1500 degree and 4500 degree days, use double glazing--where heating season is less than 1500 degree days, double glazing cannot be justified by heating energy saving alone, but may be desirable to reduce noise.

d) When considering storm windows or other options listed above, north facing windows should be given priority followed by east or west windows, whichever is on the prevailing wind side of the building and/or most heavily shaded,and then the south facing windows.

e) When replacing or adding glazed surfaces, use one layer of reflective glass or reflective coatings to reduce solar heat gains, if windows are in direct sunlight for more than three hours per day during the cooling season.

f) Where windows are subjected to high winds for a long duration of time in the winter and to sunlight in the summer, use prefabricated sunscreen on the exterior of the window to serve a dual purpose: Minimize the effect of heat loss due to wind, reduce solar heat gains in the summer.

g) Install thermal barriers whenever unoccupied hours exceed 70 per week and heating degree days exceed 4500 if existing windows are single glazed, and 6500 or greater if double glazed; consider thermal barriers also in areas with fewer degree days, particularly in windy areas.

h) Operate thermal barriers in the same manner as conventional drapes. Close them even in occupied hours whenever daylight cannot be used and when vision through the window is not rerequired.

NOTE: Do not close thermal shutters at night in the cooling season except on the west side when there is considerable sunshine after the work day ends.

i) Install storm doors in all locations where degree days exceed 3000 and provide not more than 50% glass area in the storm door. Storm doors should be wood rather than metal, unless a thermal break is provided.

j) Where exterior doors are not used during the heating season and are not required as fire exits, caulk the doors or seal them with removable strips, or insulate the areas between the storm door and the existing door. Although double windows or storm doors are not cost effective to reduce cooling loads alone, and are not recommended for that purpose alone, when installed to reduce heat loss, they will also reduce heat gain to a smaller extent in the summer.

k) Install temporary plastic shade cover over windows to reduce conduction loss as well as to reduce infiltration rates.

l) Where operable storm sash is desired, provide hopper type windows with 25% of the total sash operable, or sliding windows, casement windows, or double hung windows to most nearly match existing windows.

m) In new additions, minimize the amount of glazing area on the north facade. Where energy required for heating exceeds that for cooling, use minimal glazing.

n) In new additions, where energy for cooling exceeds the amount of energy required for heating on a seasonal basis, minimize the amount of glazing in the west first, then east, south, and north exposures in that order.

o) In new additions, where the outdoor air conditions are close to desired indoor conditions for a major portion of the year, do not caulk or install double glazing, and provide the greatest amount of operable windows to permit natural ventilation and cooling.

p) Install insulating draperies to reduce heat loss through windows.

q) Treat all skylights similarly to windows. The order of priority, for treating skylights, should follow treatment of glazing: first on the north wall in cold climates, and the west, east, and south wall in warm climates.

5. Reduce Transmission Heat Loss by Conduction Through Opaque Roofs

Due to sun angles and latitudes, the flat roofs receive the least amount of solar energy in the winter time compared to east, west and south exterior walls. On a square foot basis, insulation for roof should be considered immediately after that for the north facing walls, and, since the greatest heat gain in the summer time per square foot of building envelope opaque surface due to solar radiation is on the roof, the energy and economic benefit of insulating the roof will be most favorable on a year-'round basis.

a) The following table shows the minimum acceptable "U" value for roofs in various heating degree day zones. Existing roofs below these standards should be insulated.

HEATING DEGREE DAYS	MINIMUM ACCEPTABLE "U" VALUE
1,000	0.3
2,000	0.2
3,000	0.2
4,000	0.15
5,000	0.15
6,000	0.1
7,000	0.1
8,000 and above	0.06

b) The cost of insulation materials is small in comparison to the labor costs for installation. Once the decision is made to insulate the roof, add at least enough insulation to improve the "U" value to better than 0.06.

c) If priorities to insulate roofs are necessary, then insulate a 15'-0" wide band around the perimeter, rather than the entire roof. The roof area over the perimeter zones experiences the greatest heat loss.

d) When adding insulation to the roof outside surface, select a finish with a high absorption coefficient in areas where the yearly heating load exceeds the yearly cooling load. Select a low absorption coefficient in areas where the yearly cooling load exceeds the yearly heating load.

e) If the ceiling void is used as a return air plenum, insulation applied to the underside of the roof should be bonded or sealed to prevent migration of fibers and should not extend over the lighting fixtures.

f) If possible, include an air space between insulation boards and the underside of the roof to gain added insulating effect.

g) Insulate all pipes and hot ducts in an attic space if insulation is added on top of the ceiling below (attic, floor), as the attic temperature will tend to approach outdoor temperatures.

h) When adding insulation, provide vapor barriers on the interior surface of the roof of sufficient impermeability to prevent condensation.

i) Seal all corners to prevent infiltration.

j) Do not treat the roof with a black surface to absorb more radiation in the winter time. It is relatively ineffective to do so, especially with good insulation. Refer to the Cooling Section for exterior finish treatment.

k) Where equipment room or pipe space occurs directly under the roof, do not consider additional insulation.

l) If solar collectors are being considered for present or future insulation on the roof, the amount and type of insulation must be taken into account. Where the existing roof is in good condition, consideration should be given to insulating the underside rather than exterior and/or the void between the ceiling and the roof.

m) If reflective insulation is used, it should be selected to have an equivalent R value of not less than four and should be installed so that the shiny surface faces an air space at least 3/4 inch in thickness. Where possible, use board, batt, or fill type insulation instead of reflective insulation to reduce heat loss.

6. Reduce Transmission Heat Loss by Conduction Through Opaque Walls

a) Determine the "U" value of existing walls by inspection, measurement and calculation. The table below shows the minimum acceptable "U" value for walls in various heating degree day zones. Walls below these standards should be insulated.

HEATING DEGREE DAYS	MIMIMUM ACCEPTABLE "U" VALUE
1,000	0.40
2,000	0.30
3,000	0.30
4,000	0.20
5,000	0.20
6,000	0.15
7,000	0.15
8,000	0.10
9,000	0.10

The cost of materials to insulate a wall is small in comparison to labor costs. Once the decision is made to insulate a wall add at least enough insulation to improve the "U" value to 0.1 or lower.

b) Insulation is most effective on the north walls which be given priority followed by east or west walls, whichever face into the winter prevailing wind, then south wall.

If any exposure is shaded, its priority for insulation should be modified according to the hours of shading experienced on an average winter day.

c) Give priority to spaces that are continuously occupied for the greatest length of time. Corridors, toilets, elevator shafts, storage rooms, etc. should be assigned a lower priority regardless of wall orientation.

d) Reflective insulation is more effective on the walls than on the roof, but is not as effective as other types of insulation with higher "R" values.

e) Where possible, install insulation on the exterior surface of masonry walls. For other types of construction, provide insulation on the interior surface or in the cavity, if there is one.

f) Check all codes before installing insulation on the interior surface of the building. Some insulation materials are not acceptable due to low ignition temperatures and/or toxic or poisonous fumes when burned.

g) When blowing insulation into a masonry wall cavity or stud space, the cavity should be completely filled under slight pressure,since fill insulation often settles with time.

h) Provide vapor seal on the exterior surface of insulation when it is applied to the exterior surface of the wall. Provide vapor barrier on the room side of the insulation when the insulation is applied to the interior surface of exterior walls.

i) When constructing exterior walls for new additions, the mass should not be less than 80 lbs. per square foot, in order to take full advantage of energy conservation due to thermal mass.

j) When constructing walls for new additions, the combined "U" factor for opaque wall and glazing should not exceed the "U" values listed in the above table by more than 0.05. This can be accomplished by a combination of insulation and double or triple glazing in correct proportions.

Exceptions:

When expansion occurs in retail stores and display window is required, the space between the show window and the sales area should be blocked off by an insulated partition with a "U" value not less than 0.20.

7. <u>Reduce Transmission Heat Loss by Conduction Through Floors</u>

a) The following table shows the minimum acceptable "U" value for suspended floors with exposed undersides in various degree day zones. Existing floors below these standards should be insulated.

Heating Degree Days	Minimum Acceptable "U" Value
1,000	0.40
2,000	0.35
3,000	0.30
4,000	0.22
5,000	0.22
6,000	0.18
7,000	0.18
8,000 and above	0.12

b) If insulation is added, improve the "U" value to 0.08 or better.

c) Insulate slab-on-grade floors that are subject to condensation or that have low surface temperatures.

d) Insulate slab-on-grade floors with edge insulation and protect insulation with a waterproof seal.

e) Where suspended floors are over a closed-in, but unused, space, apply insulation to the underside of the floor. If, however, there may be some future use for the space, or if it contains pipes and ducts, then apply insulation to the inside walls of the space leaving the floor uninsulated.

8. Reduce Distribution Systems Heat Loss

Refer to ECM 1, Section 4, "Heating" for additional discussion of heat loss fromdistribution systems and equipment.

Heat loss from boilers and furnaces may comprise as much as 15% of the total heating energy used in the building. While some of the heat loss from the boiler may serve to reduce heat loss from adjacent occupied spaces, more often it is lost directly through basement walls and floors without contributing useful heat to the building. Roof top units are exposed to very low ambient temperatures with corresponding greater losses and require better insulation than the equipment located within the building. Insulation of roof top units that are also used for cooling will reduce the cooling loads as well.

a) Repair, increase, or replace the insulation on hot water and steam piping. (Heat loss from piping should not exceed 0.25 Btu's per square foot per degree difference between fluid temperature and ambient air temperature.)

b) Insulate valve bodies, fittings, and other pipe appurtenances, except steam traps and condensate leg within 5 ft. of the trap.

c) Insulate the exterior surfaces of boilers, hot air, and forced warm air furnaces, and air handling units with insulation of sufficient R value, so that exterior surface temperatures do not exceed 90°F at full load operating conditions for units located in the building.

d) For roof top units, the R value should not be less than 20 in all areas which have fewer than 7,000 degree days and not less than 30 where degree days exceed 7,000.

e) For hot air furnaces, insulation should be done after a forced air blower is installed since the loss will be greatly reduced with the addition of the blower.

9. Reduce Energy Consumption for Combustion Devices

The seasonal efficiency of primary energy conversion equipment such as boilers, furnaces, and oil and gas burners or hand-fired coal or stokers depends upon:

1. The conversion efficiency from fuel into heated gases.

2. The transfer of heat from the hot gases into hot water, steam or warm air.

In order to improve the seasonal performance of these systems and reduce energy consumption, consider the following:

a) Analyze the reduced requirements due to energy conservation measures and modify the combustion chamber to accommodate the lower firing rate made possible by load reduction.

b) If stack temperatures exceed 450°F, install baffles or turbulators to improve heat transfer efficiency. Consult the boiler manufacturer for advice.

c) If stack temperatures are over 300°, install a heat pipe or air heat recovery unit to capture waste heat from the surface of the flue pipe to heat other spaces, make up air, or combustion air. Refer to Section 6F for heat recovery devices.

d) Disconnect automatic blow down controls and schedule blow down as needed. This requires close surveillance of boiler burner operation .

e) Install automatic draft control to shut off the breaching 30 seconds after burners go off and restart automatically 30 seconds before next firing cycle to prevent heat loss up the chimney between firing cycles.

f) Modify piping as necessary to permit on-line boiler to maintain temperature in idle boilers rather than starting and stopping burner when the boiler is not needed to supply the load.

g) Fit a "shell-head" adapter to the end of the gun of oil burners firing No. 2 oil to increase combustion efficiency.

h) Add a blower and return air duct from occupied areas and convert gravity hot air furnace to force warm air units.

i) Install fan and duct, if necessary, to utilize warm air from the ceiling of the boiler or furnace room to preheat combustion air.

j) Install an interlock between combustion air unit and burner operating with proper prestart and after-firing control to reduce the amount of cold air introduced into the building for combustion.

k) Install airtroll fittings on hot water systems to vent air from the hot water piping.

l) Where electric coils are used in furnaces, install a control to reduce the period of time that the coil is energized.

m) Install additional condensate return lines to the boiler for feed water system where uncontaminated condensate is being wasted. Where condensate is being wasted, consider installation of heat exchanger to preheat domestic hot water.

n) Adjust the firing rate to suit the loads when they differ widely by season. Absorption cooling units and domestic hot water may require more or less energy than heating alone or heating and domestic hot water.

o) Analyze the loads and seasonal operating efficiency and replace inefficient heating equipment with new modular sized equipment to permit efficient operation at all part-load conditions.

p) Reset unoccupied period frost protection controls so that heating pumps operate only when the outdoor temperature is less than 35°F rather than a higher setting.

q) Eliminate gas pilots and install electric pilots for boilers and furnaces.

r) For condensate pump systems use heat exchangers to reclaim waste heat and reduce the temperature of the condensate and prevent cavitation in the condensate pump.

s) Install an economizer for preheating boiler feed water.

t) Use automatic viscosity controls on fuel oil systems for purposes of attaining best atomization for efficient combustion. This control also permits flexibility of mixing and the use of any grade fuel oil either distillate or residual.

u) Use proven additives to improve the efficiency of combustion, elimination of water in oil storage system, and reduction of soot deposits in the furnace and convection tubes in the boiler.

v) Install soot blowers on all residual-oil burner systems. For new installation, select the size of multiple boiler installations so that they can be operated at 75-100% of rated capacity for both summer and winter load conditions.

w) In very large installations, consider converting high temperature water to high temperature-low pressure fluid systems.

x) Where air conditioning is needed during the heating season for portions of the building, use condenser water heat for reheat and for temporary make up air for space heating, instead of operating the heating boiler.

y) When replacing obsolete #6 oil burners, consider air atomization instead of steam atomization to improve performance and efficiency.

z) Consult boiler manufacturers for feasibility of operating your existing boiler without a separate combustion chamber and allow the flame to impinge directly upon boiler surfaces. Caution: Must be done with approval of boiler manufacturer.

aa) Consider the use of solar energy for heating to reduce boiler operation. See Section 22.

bb) Maintain continuous monitoring and keep record of boiler efficiency at various full and part-load conditions.

cc) Install control to prevent burner short cycling.

dd) When replacing boilers, provide for dual fuel operations as a contingency against shortages of one particular fuel.

ee) Where high pressure steam is required for turbines or other equipment which only operate for a portion of the year, reduce the system steam pressure when these pieces of equipment are not in use.

ff) Install and maintain proper steam regulators to reduce steam consumption. Bypasses should only be used during emergencies and during repair of regulators. On large installations, install flow meter and record changes so that an operator can analyze and maintain maximum efficiency.

gg) Regulate electric heating elements in steps or stages rather than full-on, full-off staging controls. Limit the demand on the electric supply system and eliminate wasteful override.

hh) When district steam is used in place of on-site boilers, utilize the condensate return to heat make up air, domestic hot water and space heating.

ii) Investigate the possibility of new burners which inject steam or water vapor into the oil stream to improve efficiency.

Section 14

Energy Conservation Opportunities Domestic Hot Water

DOMESTIC HOT WATER

ENERGY CONSERVATION OPPORTUNITIES

ECO 66 INSULATE PIPING

Heat losses from uninsulated hot water system distribution piping can be substantial. The magnitude of these losses depends on the temperature differential between pipe and ambient air, pipe size, and length of system piping.

Exposed piping in basements and equipment rooms is relatively simple to insulate. Piping in ceiling spaces may also be readily accessible by removing ceiling panels. Preferably, the entire piping system should be insulated, but inaccessible portions may be left bare providing they are a small percentage of the total, as this will have little effect on the total possible savings.

Cost of installation varies in each building and should be estimated by an insulation contractor. For order of magnitude, use the installed costs for sectional insulation with jacket listed in the following table.

Pipe Size	Price Per L.F. Installed Insulation Thickness: 1" $	1-1/2" $
1/2"	1.35	2.05
3/4"	1.40	2.10
1"	1.45	2.15
1-1/4"	1.50	2.20
1-1/2"	1.55	2.25
2"	1.60	2.35
2-1/2"	1.65	2.45
3"	1.70	2.50
4"	2.00	3.05
5"	2.25	3.15
6"	2.55	3.30
8"	3.15	4.20
10"	3.85	5.00
12"	4.50	5.60
14"	5.20	6.45
16"	6.00	7.20
18"	6.70	7.60
20"	8.25	8.50
24"	9.00	9.70

For costing purposes, add to total lineal feet of piping three (3) lineal feet for each fitting or pair of flanges to be insulated.

Savings achieved by insulating hot water piping are determined from Figure #44 for domestic hot water temperatures ranging from 100° to 180°. The following example illustrates the use of Figure #44 to determine savings.

A hot water system temperature is 160°F and piping runs have been measured and are as follows:

2-1/2 inch - 125 ft.	1-1/4 inch - 350 ft.
1-1/2 inch - 30 ft.	1 inch - 150 ft.

There is presently no insulation installed. The present heat loss is calculated by entering Figure #44 at each pipe size intersecting with curves labeled "none" and 160°F and reading the heat loss in Btu per hour per 10 ft. pipe length. Multiply these quantities by the respective lengths of each size pipe giving a total system heat loss of 72.175 Btu/hr. The hot water system is operating 8760 hours per year and the total yearly heat loss is 8760 x 72,175 = 632 x 106 Btu.

1" insulation is applied to all pipes. Using Figure #44 and intersecting with the 1" insulation and 160°F line, the total yearly heat loss is 94.4 x 10^6 Btu.

Savings realized by installing insulation is the difference in heat loss of (632 x 10^6) - (94.4 x 10^6) = 537.6 x 10^6 Btu/year.

A separate oil fired hot water heater with an efficiency of 70% provides the domestic hot water. The gallons of oil saved is 5,485, and at $0.36 per gallon, this equals $1974/year.

Order of magnitude costs are:

125	x	$1.65	=	$206.25
30	x	$1.55	=	46.50
350	x	$1.50	=	525.00
150	x	$1.45	=	217.50
				$995.25

An investment of approximately $1,000 will save $1,900/year. No further economic analysis is necessary.

ECO 67 INSULATE STORAGE TANKS

The loss of heat from the domestic storage tank must continuously be offset by the addition of heat to maintain a ready supply of hot water. This heat loss occurs 24 hours/day whether the building is occupied or not.

Storage tanks should be covered with a minimum of 3 inches of insulation (K=0.3). Insulate bare tanks and apply additional insulation to tanks having less than 3 inches. Replace or repair all missing or torn insulation as required.

Applicable codes should be checked to determine acceptability of various insulation materials.

Obtain quotations for accurate cost assessment but for order of magnitude, tanks can be insulated for $3.60/ft.^{2} surface area.

Calculate energy savings by determining the heat lost from the tank before and after insulation. Assume water temperature and ambient air temperature are constant. Multiply the savings in Btu/hr. x 8760 to obtain the total yearly energy savings due to insulation.

ECO 68 REPLACE CENTRAL SYSTEM BY LOCAL HEATING UNITS

Commercial hot water systems frequently require hot water for short periods of heavy use at various locations within the building. It is often more efficient to provide water heaters close to the usage points rather than central generation and long runs of hot water piping.

Analyze the hot water use within the building to determine the patterns of demand to determine whether installation of local units is advantageous. Estimate the energy losses of the existing system and calculate the savings derived by installing local units. The energy saved is the sum of reduced distribution losses and the increase in the average generation efficiency of local units as compared to central.

Before installing local hot water generation units, review applicable codes to see if there are any restrictions on their locations or modifications to the building required such as fire walls around fuel burning equipment.

ECO 69 INSTALL TEMPERATURE BOOSTERS

When multiple temperature requirements are met by a central domestic hot water system, the minimum generation temperature is determined by the maximum usage temperature. Lower temperatures are attained by mixing with cold water at the tap. Where the majority of hot water usage is at the lower temperatures and higher temperatures are required at a few specific locations only, install booster heaters or separate heaters for high temperatures.

ECO 70 INSTALL SEPARATE BOILER FOR SUMMER GENERATION OF DOMESTIC HOT WATER

In many buildings, the heating system boilers provide primary heat for the domestic hot water system. While this is satisfactory during the heating season when the boiler is firing at high efficiency, demand for boiler heat in summer will probably be limited to hot water generation only. Operating large heating boilers at light loads to provide domestic hot water results in low boiler efficiency.

To reduce energy losses due to low boiler efficiency in summer, install a separate hot water heater or boiler, sized for the hot water demand. Shut down the heating boiler in the summer and generate hot water at improved efficiency.

Obtain quotations from a local contractor to determine the cost of this modification.

To determine the energy savings, analyze the system operating characteristics. Items investigated should include operating efficiencies of existing equipment during warm and cold months and hot water demand relationship to boiler capacity. Calculate savings by comparing the operating costs of both arrangements.

ECO 71 HOT GAS HEAT EXCHANGER

A typical refrigeration machine with a water-cooled condenser rejects approximately 15,000 Btu/hr. for each 12,000 Btu/hr. of refrigeration. An air-cooled condenser rejects up to 17,000 Btu/hr. Up to 5,000 Btu/hr. of the heat rejected from either system can be recaptured. To recover heat of compression, install a heat exchanger in the hot gas line between the compressor and condenser of the chiller. (See Section 18.) A typical arrangement in conjunction with a domestic hot water system is shown below. Hot gas temperature depends on head pressure but is usually in the order of 120 - 130°F.

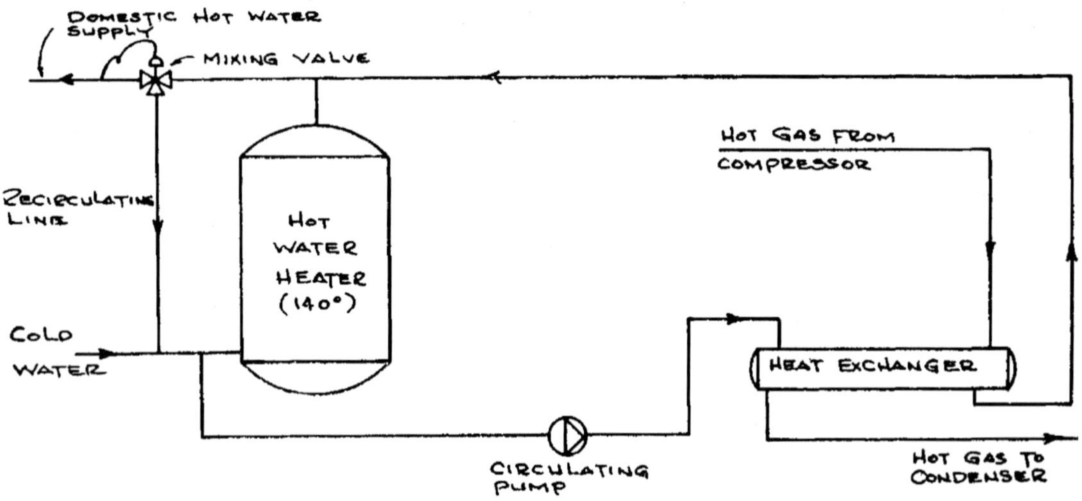

Cold water is circulated through the heat exchanger by the circulating pump. When hot water is not being used, water is pumped back through the heat exchanger through the recirculating line. When hot water is needed, it is fed from either the heat exchanger, the hot water heater, or both. A mixing valve is provided to maintain the desired temperature.

ECO 72 HOT DRAIN HEAT EXCHANGER

Buildings with kitchens, laundries, and other service facilities which utilize large quantities of hot water, in many cases discharge hot waste water to drains. By installing a heat exchanger, heat can be recaptured for use to preheat domestic hot water.

In general, it is economical to preheat water from 50°F to 105°F without excessive equipment cost. The hot water at 105°F can then be fed into the domestic hot water tank for further heating to utilization temperature, if required. The heat reclamation system saves the heat required to heat water from 50°F to 105°F which otherwise would have been provided by other heat sources.

For example, in a laundromat, the average daily wash load is 2,000 lbs./day. Water usage is 2.5 gallons of 170°F water and 1.0 gallon of 50°F water per lb. of wash load. The waste water discharged to drain is 3.5 gallons of 136°F water per lb. of wash load.

Assuming that the laundromat operates 365 days/yr. and that domestic hot water is generated by an electric hot water heater, the yearly saving is 553 x 10^6 Btu. At 90% efficiency and $0.03/KWH, this represents $5,400 cost saving/year.

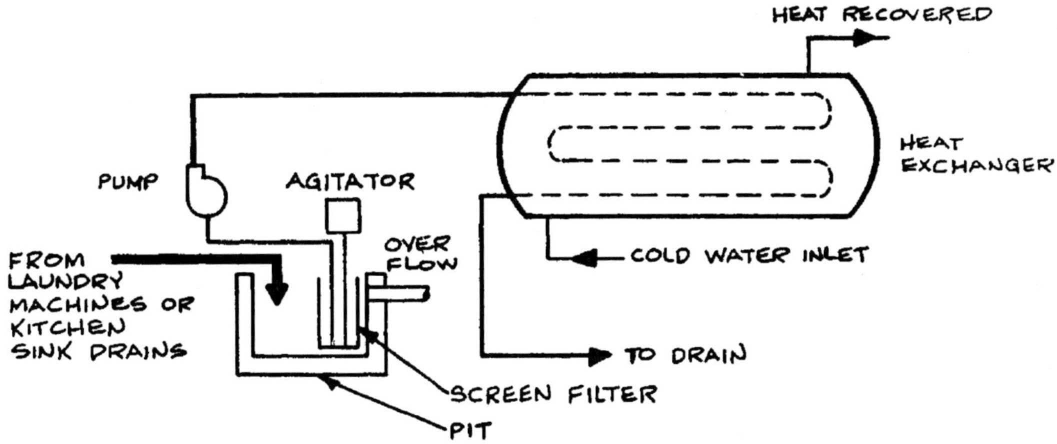

SCHEMATIC OF LAUNDRY AND KITCHEN HOT WATER HEAT RECOVERY SYSTLM

ECO 73 HOT CONDENSATE HEAT EXCHANGER

The condensate return portion of many steam systems exhaust large quantities of heat in the form of flash steam when the hot condensate is reduced to atmospheric pressure in the condensate receiver. Recover waste heat by installing a heat exchanger in the condensate return main before the receiver to reduce condensate temperature to approximately 180°F. Use the heat recovered to preheat water as required. The diagram below shows the schematic installation of a hot condensate heat exchanger.

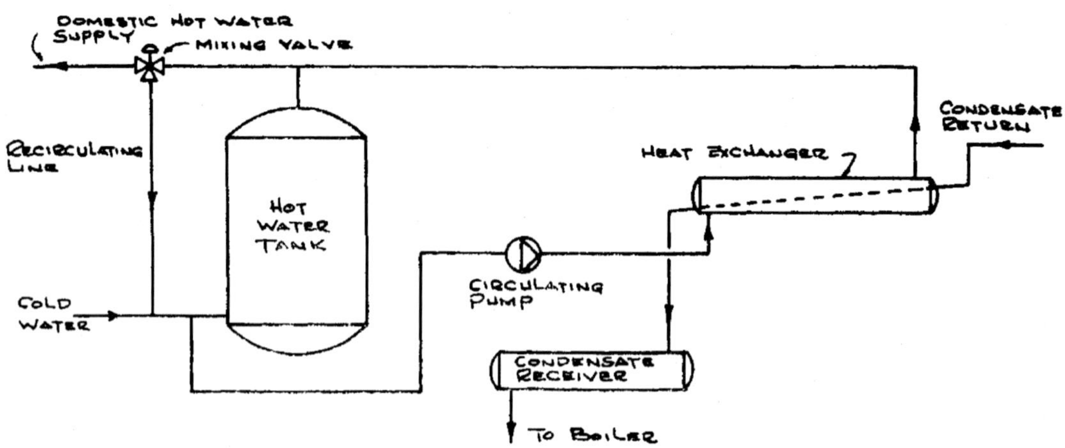

The quantity of heat recovered depends on the pressure and temperature characteristics of the boiler. For example:

Steam has a condensate return volume of 6 gpm at 260°F. A heat exchanger is installed to reduce the condensate temperature from 260°F to 180°F and the quantity of heat recovered is 240 x 10^3 Btu/hr. The average heat input required for the generation of domestic hot water is 1.1 x 10^6 Btu/day. In this instance the entire domestic hot water load can be met the major part of the year by the heat recovered in the hot condensate heat exchanger.

GUIDELINES TO REDUCE ENERGY USED FOR DOMESTIC HOT WATER SYSTEMS

Refer to ECM 1, Section 5, "Domestic Hot Water" for guidelines to reduce energy consumption by domestic hot water systems with minimal investment. Consider the following options for further reductions in energy consumption (Some of these options are addressed in ECM 1)

1. Reduce the Consumption of Domestic Hot Water

   a) Reduce the cold water pressure with an automatic control valve, but do not set lower than pressure needed to flush toilets on top floors or for pressures required for fire protection.

   b) Take measures to reduce consumption when usage rates exceed 1-1/2 gpd/person in offices and 1 gpd/person in stores or religious buildings.

   c) Install flow restrictors in the supply branch to groups of taps when existing faucets have flow rates greater than 1-1/2 gpm.

   d) Install spray type faucets that use only 1/4 gpm instead of 2 or 3 gpm.

   e) For new additions, install foot operated peddle valves.

   f) Install self-closing faucets on hot water taps.

   g) Install a flow meter on the cold water line supplying the water heater when hot water consumption is more than 3000 gallons per day.

   h) Replace obsolete kitchen equipment such as dishwashers with new ones that have minimal water requirements.

2. Reduce the Temperature of Domestic Hot Water

   a) When water temperature exceeds 100°F.

   b) When the majority of domestic hot water usage is at lower temperatures, install an electric, gas, or small oil-fired boiler to boost the temperature for kitchens, laundries, or processes requiring higher temperatures. Consult code authorities for minimum allowable temperatures for commercial kitchen use.

c) Install a seven-day timer clock to operate a water heater to further reduce temperatures during unoccupied periods.

3. Reduce the Thermal Loss from Hot Water Piping, Appurtenances, Storage Tanks, and Hot Water Generators.

a) Insulate hot water storage tanks when insulation is less than the equivalent of 3" fiberglass and/or when insulation is in need of repair.

b) De-energize hot water circulating pumps to reduce heat loss from piping within the building. Where extremely long runs require that the circulator be in operation during occupied periods, install a seven-day timer to automatically shut off the pump during unoccupied periods.

c) Insulate the exterior jacket of tankless or tank heaters which are not immersed in the hot water boiler or hot water tank.

d) When installing new storage tanks or making major modifications to the building, relocate hot water tank as close to the load as possible.

e) Insulate hot water piping whenever there is less than the equivalent of 1" fiberglass.

4. Increase the Efficiency of the Hot Water Generator and Equipment or Install More Efficient Equipment.

a) Replace gas pilots with electrical ignition.

b) Refer to Section 13 "Heating" for guidelines for boilers and apply them to hot water heaters as well.

c) Install a storage water heater for summer use when the existing space heating boiler is used for hot water generation, there is little or no demand for steam or hot water during the summer, and hot water is 20% or more of the load.

d) Replace electric hot water heaters with heat pumps to improve the coefficient of performance from 1 to approximately 3. Use hot drain water as a heat source for the heat pump, or use an air-to-water heat pump.

e) Repipe hot water storage tanks if the cold water make-up supply is connected to the upper half of the tank and/or if the hot water outlet from the tank is in the lower portion of the tank.

f) When hot water demands are increased due to expansion or change in occupancy, provide either oil- or gas-fired water heaters or heat pumps rather than electric resistance heating. If the heating boiler has sufficient capacity and is in operation year-round for air conditioning as well, install a tank or tankless heater in place of a separate hot water generator. If the additional requirements for hot water are to serve facilities remote from the boiler and usage is small, install a separate heater in the cold water line directly at the fixtures rather than serving them from a central system.

g) Install local hot water heating units when domestic hot water usage points are concentrated in areas distant from the central generation and storage point.

5. <u>Utilize Waste Heat from Other Processes to Preheat Domestic Hot Water</u>.

a) Install a hot gas heat exchanger in air conditioning refrigeration compressor systems or commercial refrigeration equipment to heat hot water.

b) Where diesel or gas engines are in use, install a heat exchanger and utilize waste heat from the engine.

c) Where incinerators are handling more than 1 ton of solid waste per day, utilize the waste heat from the incinerator to generate hot water.

d) Install a heat exchanger in hot water drains from kitchens and laundries where the flow exceeds 2,000 gallons per day.

e) Use hot condenser water from refrigeration systems to preheat domestic hot water.

f) Install a heat exchanger in condensate lines from steam equipment to preheat domestic hot water.

g) Install heat pipe or heat exchanger to extract heat from boiler breechings to preheat hot water.

6. <u>Use Solar Water Heater</u>

a) Install a solar water heater to replace or supplement the existing hot water generator. Refer to Section 22, "Solar Energy".

This page blank

Section 15

Energy Conservation Opportunities Cooling and Ventilation

COOLING & VENTILATION

ENERGY CONSERVATION OPPORTUNITIES

REDUCE TRANSMISSION GAINS THROUGH WINDOWS

Direct and diffuse solar radiation through windows is the largest single component of window heat gains. The conduction component, by comparison, is relatively small and it is rarely, if ever, worthwhile converting single glazing to double glazing for the sole purpose of reducing the heat gain by conduction. It is, however, worthwhile converting single glazing to double for the benefits that can be obtained in the heating season (see Heating Section page 305) and if this conservation opportunity has been implemented, there will an additional advantage of reduced conduction gains in the summer, and these should be included in any economic analysis on a yearly basis made for double glazing.

A computer program was developed for this manual to study the reaction of single and double glazed windows under varying weather conditions and runs were made for 12 cities chosen to give a typical cross section of climates. Figure # 48 tabulates the yearly heat gain in Btu/sq. ft. of glazing for the 12 selected locations. Heat gain through windows is maximum in areas that experience high levels of solar radiation and a large number of cooling degree hours, but for any one geographic location is modified by orientation, summer heat gain being greatest through windows on east and west facades followed by south and least in windows on the north.

Two graphs, Figures #49 & #50, were developed for other climates which extrapolate heat gains from solar sources only from the calculations for the 12 cities, and allow prediction of annual solar heat gain, only, based on cooling degree hours above 78°F D.B., solar radiation in langleys, and orientation. These two graphs are prepared for latitudes greater than 35° and latitudes less than 35°. The graphs are based on solar gains received from 9 a.m. to 3 p.m. sun time, 5 days/week when cooling systems are in operation.

Where occupancy varies significantly, adjust the yearly gain, pro-rata. For example: if a building is occupied 6 days/week, 12 hours/day, increase the solar gain obtained from the graph by 20% to allow for the extra day.

Graph Figure #51 indicates the conduction component of heat gain through windows and allows prediction of annual conduction gain based on cooling degree hours above 78° D.B., occupied hours and single, double, or triple glazed sash.

Total heat gain through a window in any given location is the sum of the solar and conduction loads.

FIGURE 48: YEARLY HEAT GAIN/SQUARE FOOT OF SINGLE GLAZING AND DOUBLE GLAZING

CITY	LATITUDE	SOLAR RADIATION LANGLEY'S	D.B. DEGREE HOURS ABOVE 78°F	HEAT GAIN THROUGH WINDOW BTU/FT.² YEAR NORTH SINGLE	NORTH DOUBLE	EAST & WEST SINGLE	EAST & WEST DOUBLE	SOUTH SINGLE	SOUTH DOUBLE
MINNEAPOLIS	45°N	325	2,500	36,579	33,089	98,158	88,200	82,597	70,729
CONCORD, N.H.	43°N	300	1,750	33,481	30,080	91,684	82,263	88,609	76,517
DENVER	40°N	425	4,055	44,764	39,762	122,038	108,918	100,594	85,571
CHICAGO	42°N	350	3,100	35,595	31,303	93,692	83,199	87,017	74,497
ST. LOUIS	39°N	375	6,400	55,242	45,648	130,018	112,368	103,606	85,221
NEW YORK	41°N	350	3,000	40,883	35,645	109,750	97,253	118,454	102,435
SAN FRANCISCO	38°N	410	3,000	29,373	28,375	88,699	81,514	73,087	64,169
ATLANTA	34°N	390	9,400	59,559	50,580	147,654	129,391	106,163	87,991
LOS ANGELES	34°N	470	2,000	47,912	43,264	126,055	112,869	112,234	97,284
PHOENIX	33°N	520	24,448	137,771	97,565	242,586	191,040	211,603	131,558
HOUSTON	30°N	430	11,500	88,334	72,474	213,739	184,459	188,718	156,842
MIAMI	26°N	451	10,771	98,496	79,392	237,763	203,356	215,382	179,376

dubin-mindell-bloome-associates
consulting engineers

ECO 74 REDUCE SOLAR GAINS THROUGH WINDOWS WITH SOLAR CONTROL DEVICES

Various solar control devices are available which can be installed outside the window, inside the window or on the window surface itself, to reduce solar gain in proportion to their shading coefficient (refer to Figure #22 for shading coefficients). External sun screens which prevent direct sunlight from falling upon the glass surface are the most effective method of controlling solar heat gain. Internal shading devices are least effective as sunlight has already entered the building and only a proportion can be reflected back through the window. External shading may be achieved fitting eyebrows over the tops of windows, or fins at the sides of windows. The size and disposition of these must be determined for each individual case based on the sun's altitude and azimuth for that particular location. Eyebrows are most effective over windows with a southern orientation and can be designed to provide total shading from high altitude sun during the middle of the day. They are not effective, however, for east and west orientations where vertical side fins are more effective.

Methods of installing external sun screens must conform to building codes for fire access and in certain areas may be subject to architectural review or approval by arts commissions or historic preservation groups. Elements protruding from the building facade may violate zoning set back requirements.

External louvered sun screens may be fitted close to the outside surface of the windows and, where possible, within the window reveal itself. Operable glazing may permit installation of external sun screens without the use of special staging. The louvered sun screens may be fixed permanently in position or can be arranged to slide in channels or made removable so that advantage can be taken of the sun's beneficial heating effect in winter.

Tinted or reflective glass may be either in addition to the existing glazing to convert it from single to double (see Heating ECO 43), or may be used to replace the existing glazing. New glazing must be of adequate strength and thickness to meet building code requirements. Wired glass may be necessary to meet fire rating requirements. Reflective polyester films may be applied direct to the inside surface of the glass. These films are self-adhesive and require careful application to avoid bubbles or trapped air and formation of Newton Rings. The film is easily abraded which reduces performance and appearance. Film on the inside of the building subjected to normal wear and tear will have a reduced useful life. If double glazing is used, apply the film to the inside surface of the outer layer of glass.

Internal shades of various types, i.e., venetian blinds, roller shades, drapes, etc., may be fitted to the inside of the windows. If drapes are used, they should be fire resistant and preferably of woven fiberglass which has a higher reflectance. Although internal shades are not as effective as other methods of solar control, they are relatively inexpensive and are easily adjustable. Shades need only be drawn when excessive solar gains are present, and at other times can be opened to allow full view through the window and maximum use of natural light.

The cost of fitting the various shading devices will vary with size, type and control devices. Obtain quotations from local specialized contractors.

ORDER OF MAGNITUDE COSTS

Solar Control Device	Average Installed Cost Per Sq. Ft.
External Louvered Screens	$5.50
Tinted or Reflective Glass	6.00
Reflective Polyster Film	1.65
Venetian Blinds	1.70
Vertical Louver Blinds	2.75
Roller Shades	.85

To determine the energy saved by shading devices, refer to Figures #49, #50 and #51.

a) Determine from Figure #7 the annual degree hours when the DB temperature is greater than 78°F.

b) Enter Figure #51 at the appropriate degree hours and following the direction of example line, intersect with the appropriate occupancy and glazing type.

c) Determine from Figure #5 the mean daily solar radiation in langleys.

d) Select the orientation of window.

e) Select from Figure #49 or #50 the appropriate graph for latitude, and entering the graph at degree hours, follow the direction of the example line to intersect with the appropriate langleys and orientation.

   Read out the yearly cooling energy requirements due to solar gain only for 1 sq. ft. of window, for occupancy of 5 days/week, 12 hours/day. Modify this yearly energy pro-rata if occupied time is different.

f) Multiply the solar gain by the shading coefficient of the selected window treatment to obtain the annual reduced solar gain and subtract this value from original solar gain value to obtain yearly cooling load reduction for each sq. ft. of treated window. Add the conduction heat gain from paragraph b.

g) Repeat this procedure for other orientations.

Interpolate for orientations other than the 4 cardinal points.

ECO 75 REDUCE HEAT GAINS THROUGH WALLS

A computer program was developed for this manual to study the effect of weather conditions on heat gain through walls. Runs were made for 12 cities, chosen to give a typical cross section of climate.

Figure #52 tabulates the heat gain in Btu/year for each sq. ft. of wall for the 12 selected locations for "U" values of 0.38 and 0.1, and outside surface absorption coefficients of 0.3 and 0.8.

Graphs were developed to show solar and conduction components for other climates to allow predications of annual heat gain, based on cooling degree days, solar radiation, orientation, "U" values, and absorption coefficients. It is rarely, if ever, economically worthwhile to insulate a wall for the sole purpose of reducing heat gains but if ECO 45 is implemented and walls are insulated to reduce heat loss during winter, the coincidental reduction of heat gain should be taken into account when making an economic analysis of cost/benefit on an annual basis.

Figure #52 shows that absorptivity of external surface has a marked effect on the heat gain through walls, particularly on south, east, and west facades.

If the building has walls of high "U" value and high absorption coefficient, it is worthwhile to consider reducing the absorption coefficient by refinishing the outside surface of the wall with light colored material or coating.

To determine savings due to insulation, refer to Figure #53, 54 and 55.

a) Determine from Figure #7 the annual cooling degree hours above 78° DB

b) Determine the initial "U" value of the wall, and the improved "U" value after insulation.

FIGURE 49: YEARLY SOLAR HEAT GAIN THROUGH WINDOWS-LATITUDE 25°-35°N

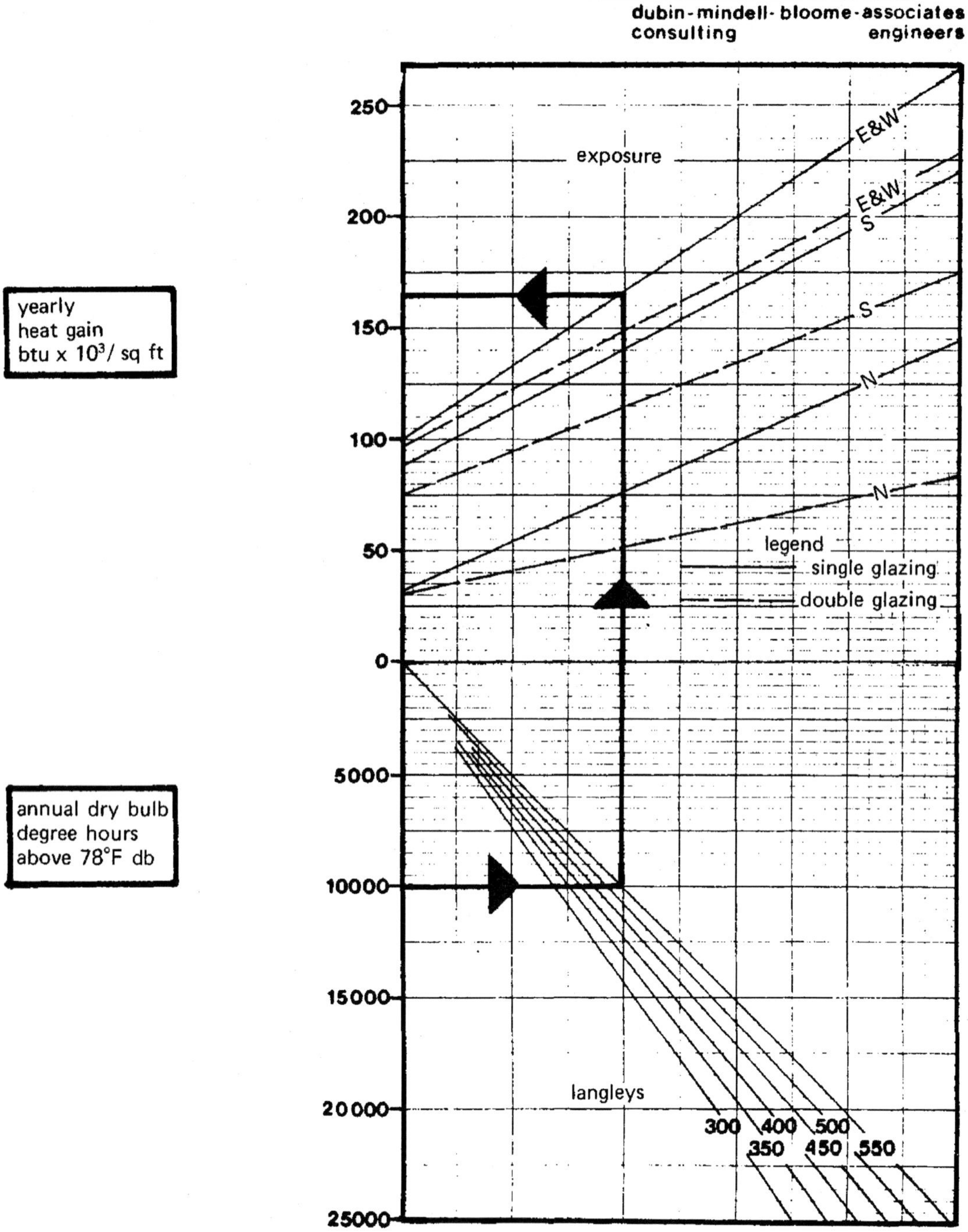

Figure #49 Engineering Data

This figure is based on the "Sunset" Computer program which was used to calculate solar effect on windows for 12 locations. The program calculates hourly solar angles and intensities for the 21st day of each month. Radiation intensity values were modified by the average percentage of cloud cover taken from weather records on an hourly basis. Heat gains are based on a 78°F indoor temperature. During the cooling season, internal gains, ventilation, infiltration and conduction through the building can create a cooling load. The additional load caused by heat gain through the windows was calculated for each day. Daily totals were then summed for the number of days in each month to arrive at monthly heat gains The length of the cooling season for each location considered was determined from weather data and characteristic operating periods. Yearly heat gains were derived by summing monthly totals for the length of the cooling season. These are summarized in Figure #48 for the 12 locations. Gains in Figure the conduction heat gain component through windows read from Figure #51 were deducted from the total heat gains to derive the solar component. The solar component was then plotted and extrapolated to include the entire range of degree hours. Figure #49 was derived from locations with latitudes between 25°N - 35°N. The heat gains assume that the windows are subjected to direct sunshine. If shaded, gains should be read from the north exposure line. The accuracy of the graph diminishes for location with less than 5,000 degree hours.

FIGURE 50: YEARLY SOLAR HEAT GAIN THROUGH WINDOWS-LATITUDE 35°-45°N

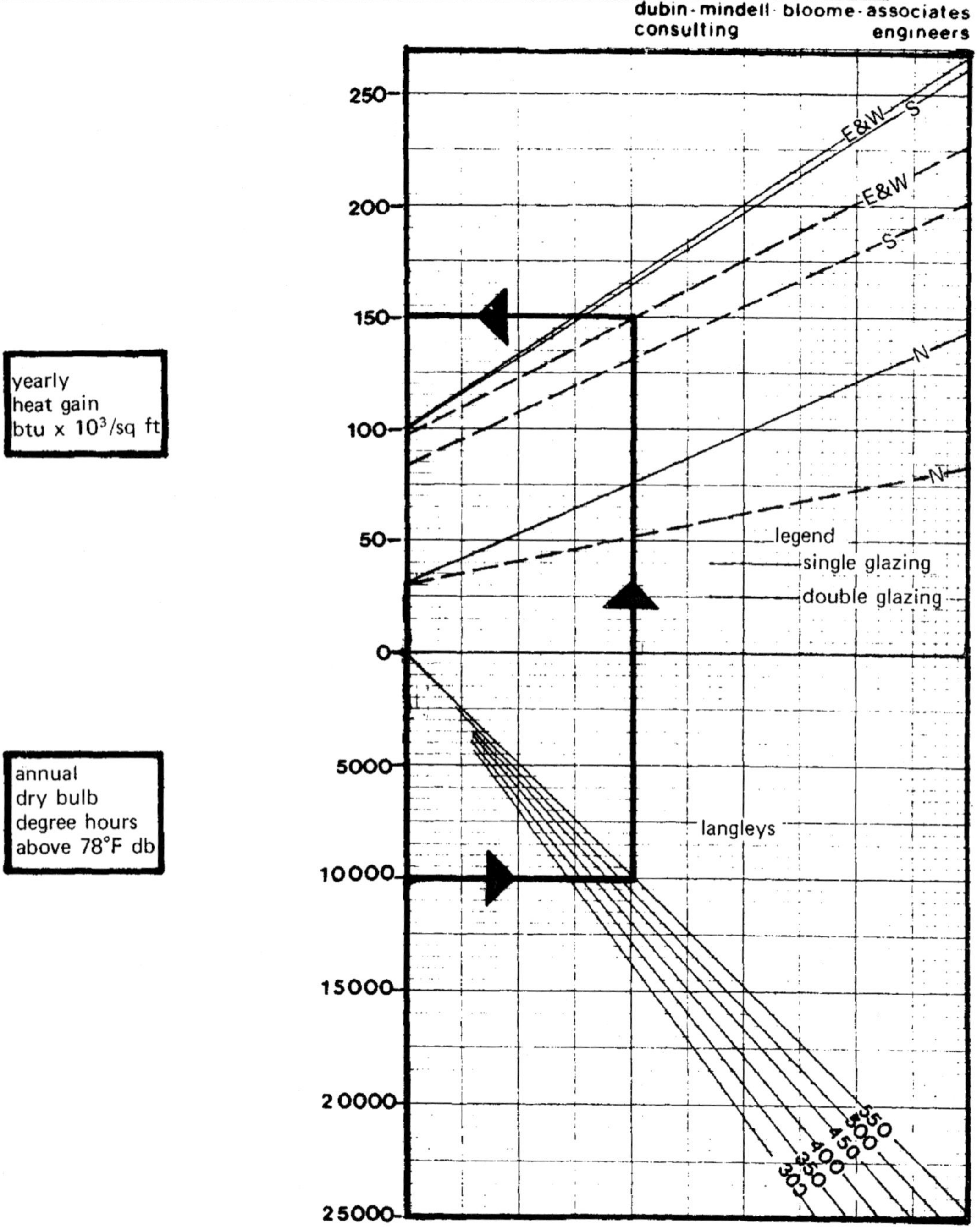

Figure #50 Engineering Data

This figure is based on the "Sunset" Computer program which was used to calculate solar effect on windows for 12 locations. The program calculates hourly solar angles and intensities for the 21st day of each month. Radiation intensity values were modified by the average percentage of cloud cover taken from weather records on an hourly basis. Heat gains are based on a 78°F indoor temperature. During the cooling season, internal gains, ventilation, infiltration and conduction through the building can create a cooling load. The additional load caused by heat gain through the windows was calculated for each day. Daily totals were then summed for the number of days in each month to arrive at monthly heat gains The length of the cooling season for each location considered was determined from weather data and characteristic operating periods. Yearly heat gains were derived by summing monthly totals for the length of the cooling season. These are summarized in Figure #48 for the 12 locations. Gains in Figure the conduction heat gain component through windows read from Figure #51 were deducted from the total heat gains to derive the solar component. The solar component was then plotted and extrapolated to include the entire range of degree hours. Figure #50 was derived from locations with latitudes between 35°N - 45°N. The heat gains assume that the windows are subjected to direct sunshine. If shaded, gains should be read from the north exposure line. The accuracy of the graph diminishes for location with less than 5,000 degree hours.

FIGURE 51: YEARLY CONDUCTION HEAT GAIN THROUGH WINDOWS

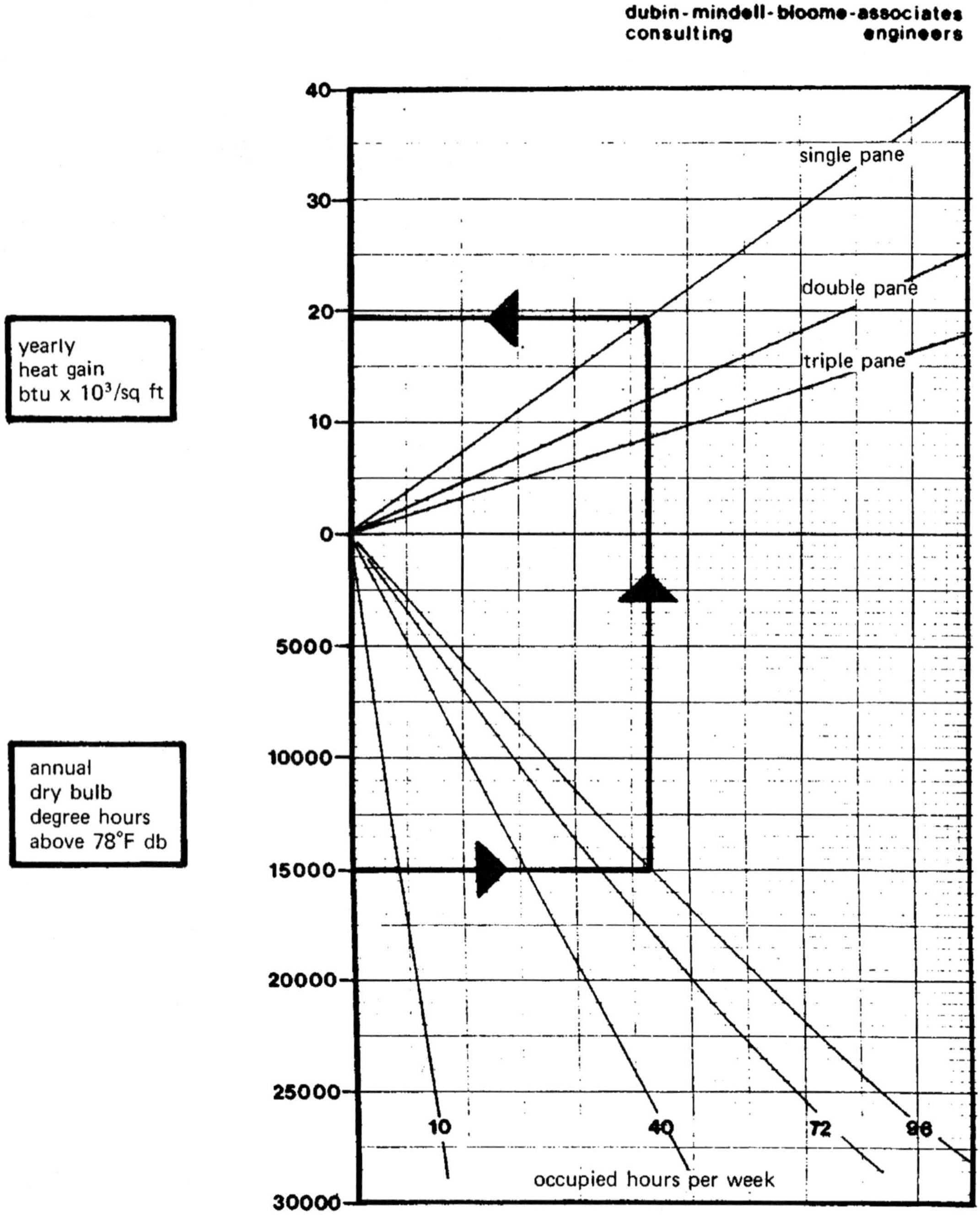

Figure #51 Engineering Data

This figure is based on degree hours read from Figure #7 which has a base of 56 hours/week. The figure is based on the formula: Q (heat gain)/yr. = degree hours/yr. x 'U' Value. 'U' Values assumed were 1.1 for single pane, 0.65 for double pane and 0.47 for triple pane. The major portion of degree hours occur between 10 a.m. and 3 p.m. Hence, for occupancies between 10 and 56 hours/week, the degree hour distribution can be assumed to be linear. However, for occupancies greater than 56 hours/week the degree hour distribution becomes non-linear, particulalry in locations with greater than 15,000 degree hours. This is reflected by the curves for 72 and 96 hour/week occupancies.

FIGURE 52: YEARLY HEAT GAIN/SQUARE FOOT THROUGH WALLS

CITY	LATITUDE	SOLAR RADIATION LANGLEY'S	D.B. DEGREE HOURS ABOVE 78°F	HEAT GAIN THROUGH WALLS BTU/FT.^{2} YEAR											
				NORTH				EAST & WEST				SOUTH			
				U=0.39		U=0.1		U=0.39		U=0.1		U=0.39		U=0.1	
				a=0.3	a=0.8	a=0.3	a=0.8	a=0.3	a=0.8	a=0.3	a=0.8	a=0.3	a=0.8	a=0.3	a=0.8
MINNEAPOLIS	45°N	325	2,500	364	2,442	19	390	1,346	7,665	164	1,747	1,601	7,439	164	1,574
CONCORD, N.H.	43°N	300	1,750	141	1,950	0	180	787	6,476	41	1,264	1,222	7,093	59	1,179
DENVER	40°N	425	4,055	321	2,476	0	291	1,361	8,450	66	1,597	1,513	8,138	78	1,301
CHICAGO	42°N	350	3,100	503	2,500	46	429	1,492	7,889	233	1,835	1,698	8,088	225	1,793
ST. LOUIS	39°N	375	6,400	2,246	5,966	419	1,386	4,165	14,116	950	3,571	3,994	12,476	779	3,074
NEW YORK	41°N	350	3,000	906	3,751	103	820	2,394	10,278	477	2,651	2,626	11,185	420	2,707
SAN FRANCISCO	38°N	410	3,000	0	0	0	0	0	3,268	0	262	43	3,459	0	297
ATLANTA	34°N	390	9,400	1,901	5,806	309	1,301	3,882	14,658	812	3,609	3,422	12,085	634	2,897
LOS ANGELES	34°N	470	2,000	0	774	0	10	180	6,575	0	889	527	7,182	0	980
PHOENIX	33°N	520	24,448	17,448	24,423	4,749	6,526	21,461	36,937	5,784	9,868	20,880	34,728	5,502	9,322
HOUSTON	30°N	430	11,500	5,002	10,687	1,178	2,643	7,895	22,431	1,981	5,521	6,985	20,893	1,605	4,713
MIAMI	26°N	451	10,771	7,507	15,717	1,912	4,052	12,358	31,745	3,164	8,416	11,778	29,906	2,814	8,057

dubin-mindell-bloome-associates
consulting engineers

c) Refer to Figure #55. Enter the graph at degree hours and, following the direction of the example line, intersect at the appropriate points for occupancy and "U" values.

d) Refer to Figure #53 and 54 depending upon latitude.

e) Estimate the absorption coefficient of the outside surface of the wall.

f) Using the initial "U" value, enter the graph at the selected degree hours and, following the direction of the example line, intersect at appropriate points for langleys, orientation, and "U" value.

Read out the yearly heat gain due to solar radiation only per sq. ft. from the appropriate absorption coefficient line. Interpolate the results for absorption coefficients other than the 0.3 and 0.8. Add the heat gain due to conduction from paragraph c to the heat gain due to solar radiation to obtain the total heat gain for the insulated wall.

g) Repeat steps a - f using the improved "U" value, and subtract the improved heat gain from the initial heat gain to obtain the annual load reduction due to insulation.

h) Repeat the complete procedure outlined in a - g above for each different orientation of wall where insulation is being considered.

Interpolate for orientations other than the four cardinal points.

The total yearly reduction is the sum of all walls treated.

Changing the external absorption coefficient will change the magnitude of the solar radiation component of heat gain through walls.

Walls of dark color having a high absorption coefficient (0.7 or greater) may be resurfaced or coated to reduce the absorption coefficient. Use light colored paint or finish; silver is best but white may be more acceptable if reflections on the surrounding properties are objectionable. Theoretically, it is possible to reduce the external absorption coefficient to 0.1, but in practice dirt accumulates on building surfaces and an absorption coefficient of 0.3 for white and 0.5 for light colors (pastel shades of yellow, green, etc.) is more realistic.

To determine the reduction of heat gain through walls due to a reduction in absorption coefficient use Figure #53 or #54 and interpolate as necessary for changes in coefficient.

FIGURE 53: YEARLY SOLAR HEAT GAIN THROUGH WALLS-LATITUDE 25°N-35°N

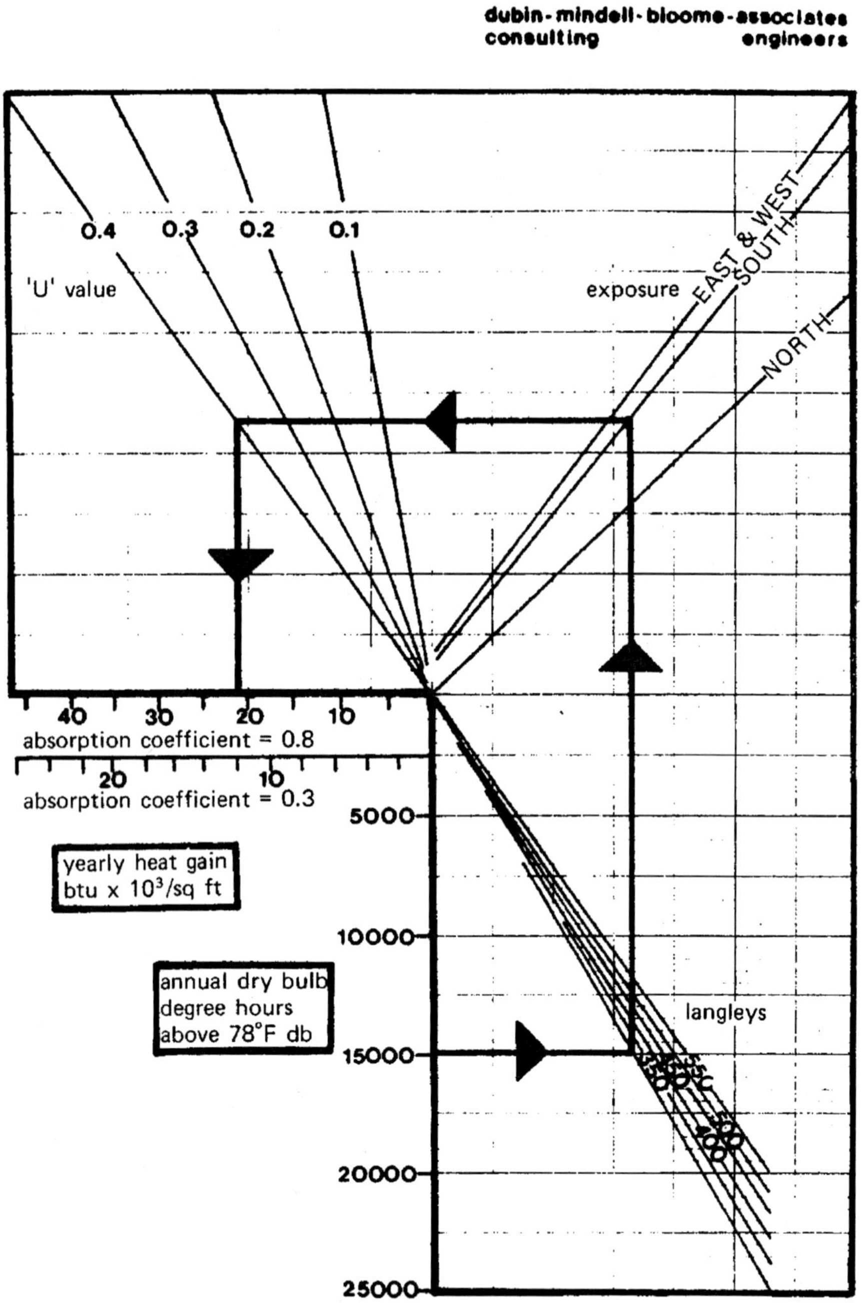

Figure #53 Engineering Data

This figure is based on the "Sunset" computer program which was used to calculate solar effect on walls for 12 selected locations. The program calculates hourly solar angles and intensities for the 21st day of each month. Radiation intensity values were modified by the average percentage of cloud cover taken from weather records on an hourly basis. Heat gains are based on a 78°F indoor temperature.

The solar effect on a wall was calculated using sol-air temperature and the heat entering or leaving a space was calculated with the equivalent temperature difference. Wall mass ranged from 50 - 60 lbs/sq.ft. and thermal lag averaged 4 1/2 hours. During the cooling season internal gains, ventilation, infiltration and conduction through the building skin create a cooling load. The additional load caused by heat going through the walls was calculated for each day. Daily totals were then summed for the number of days in each month to arrive at monthly heat gains. The length of the cooling season for each location considered was determined from weather data and characteristic operating periods. Yearly heat gains were derived by summing monthly totals for the length of the cooling season.

Absorption coefficients and 'U' values were varied and summarized for the 12 locations as shown in Figure #52. Gains in Figure #52 include both the solar and conduction components of heat gain. Values of the conduction heat gain component through walls were deducted from the total heat gain to derive the solar component. The solar component was then plotted and extrapolated to include the entire range of degree hours. Figure #53 was derived from locations with latitudes between 25°N - 35°N. The heat gains assume that the walls are subjected to direct sunshine. If shaded, gain should be read intersecting with the North exposure line.

FIGURE 54: YEARLY SOLAR HEAT GAIN THROUGH WALLS—LATITUDES 35°-45°N

dubin-mindell-bloome-associates
consulting engineers

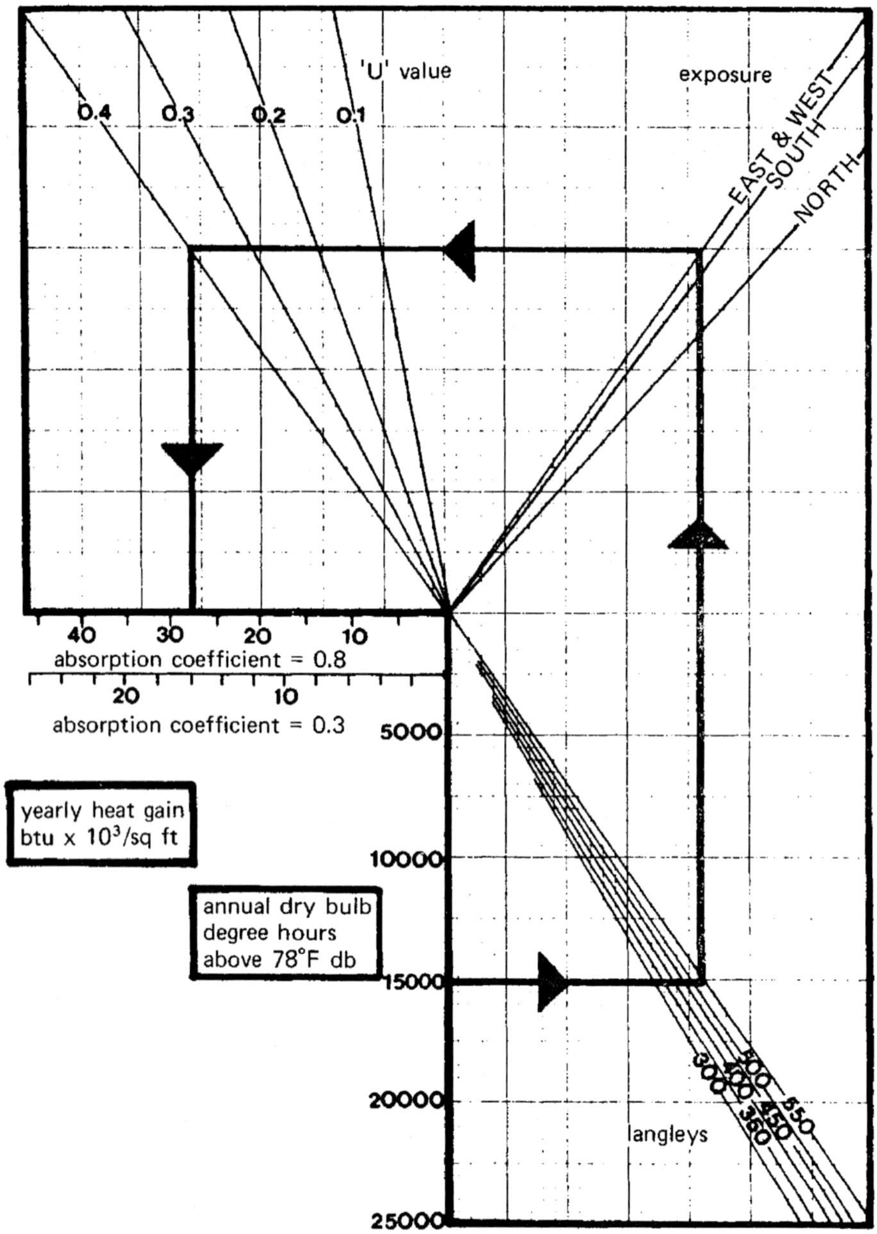

#54 Engineering Data

This figure is based on the "Sunset" computer program which was used to calculate solar effect on walls for 12 selected locations. The program calculates hourly solar angles and intensities for the 21st day of each month. Radiation intensity values were modified by the average percentage of cloud cover taken from weather records on an hourly basis. Heat gains are based on a 78° indoor temperature.

The solar effect on a wall was calculated using sol-air temperature and the heat entering or leaving a space was calculated with the equivalent temperature difference. Wall mass ranged from 50 - 60 lbs/sq.ft.and thermal lag averaged 4 1/2 hours. During the cooling season internal gains, ventilation infiltration and conduction through the building skin creates a cooling load. The additional load caused by heat going through the walls was calculated for each day. Daily totals were then summed for the number of days in each month to arrive at monthly heat gains. The length of the cooling season for each location considered was determined from weather data and characteristic operating periods. Yearly heat gains were derived by summing monthly totals for thelength of the cooling season.

Absorption coefficients and 'U' values were varied and summarized for the 12 locations as shown in Figure #52. Gains in Figure #52 include both the solar and conduction components of heat gain. Values of the conduction heat gain component tnrough walls were deducted from the total heat gain to derive the solar component. The solar component was then plotted and extrapolated to include the entire range of degree hours. Figure #54 was derived from locations with latitudes between 35°N - 45°N. The heat gains assume that the walls are subjected to direct sunshine. If shaded, gain should be read intersecting with the North exposure line.

FIGURE 55: YEARLY CONDUCTION HEAT GAIN THROUGH WALLS, ROOFS AND FLOORS

dubin-mindell-bloome-associates
consulting engineers

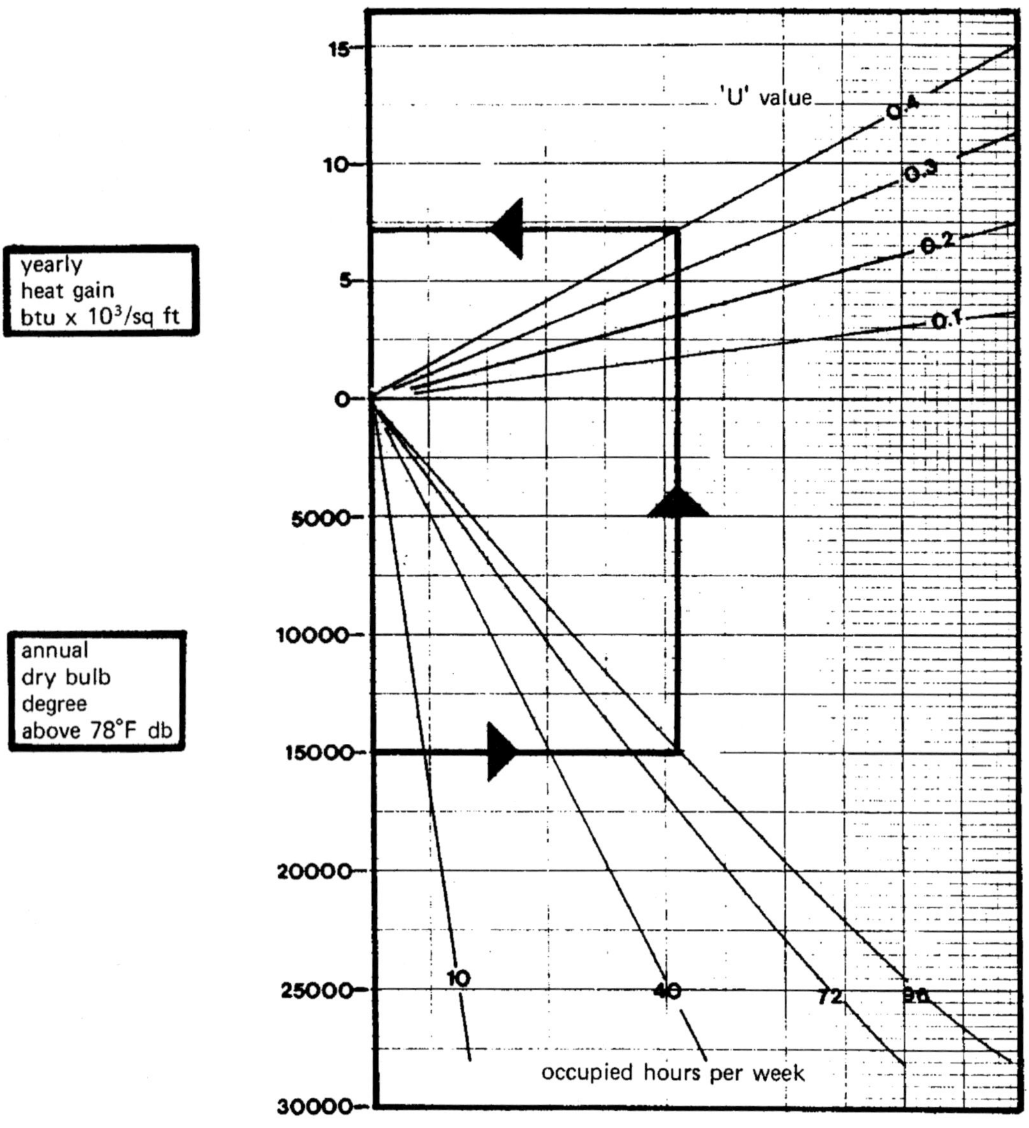

Figure #55 Engineering Data

This figure is based on degree hours read from Figure #7 which has a base of 56 hours/week. The figure is based on the formula: Q (Heat Gain)/yr. = Degree Hours/yr. x 'U' Value. The major portion of degree hours occur between 10 a.m. and 3 p.m.. Hence for occupancies between 10 and 56 hrs./week, the degree hour distribution can be assumed to be linear. However, for occupancies greater than 56 hrs./week the degree hour distribution becomes non-linear, particularly in locations with greater than 15,000 degree hours. This is reflected by the curves for 72 and 96 hour/week occupancies.

Note: It should be recognized that although reducing the absorption coefficient in summer will reduce the cooling load, it will have an adverse effect in winter by reducing the beneficial effect of sunshine in cold weather, but the net effect will normally favor the cooling load.

ECO 76 REDUCE TRANSMISSION GAINS THROUGH FLOORS

Transmission gains through floors are small in magnitude and it is not economically worthwhile insulating them to reduce cooling load. If, however, floors have been insulated to reduce heating load (see ECO 47)added benefit is obtained during the cooling season.

Methods of insulating floors and calculating initial and improved "U" values are described under ECO 47

To determine the cooling energy saved by insulating floors, refer to Figure #55 and carry out the following procedure:

a) Determine from Figure #7 the annual degree hours above 78°F DB for location.

b) Determine the initial "U" value and the improved "U" value after insulation.

c) Using the initial "U" value of floor enter the graph at the selected degree hours and following the direction of the example line, intersect at the appropriate points for occupied hours and "U" values.

   Read out the yearly heat gain per sq. ft. of floor area.

d) Repeat this complete procedure using the improved "U" value and by subtraction, obtain the reduction in heat gain/sq. ft. of insulated floor.

ECO 77 REDUCE TRANSMISSION GAINS THROUGH THE ROOF

Conduction and solar gains through roofs can form a considerable part of the cooling load. Consider insulating the roof or changing the absorption coefficient, and/or adding roof sprays to reduce annual load.

A computer program was developed for this manual to study the heat gain through roofs under varying weather conditions. Runs were made for 12 cities chosen to get a typical cross section of climates. Figure #56 tabulates the heat gain for 1 sq. ft. of roof in Btu/year for two different "U" values and for two different absorption coefficients, for each of the 12 selected locations. The table indicates the extent of the effect on heat gain that absorption coefficients have on roofs. It is very great at high "U" values, but of lesser influence on roofs with low "U" values.

It also shows that although "U" values modify solar gains to some extent, their major effect is to change the conduction component.

For buildings located in high degree hour zones and which have large roof areas in proportion to floor areas, such as supermarkets, and other single-story buildings, more consideration should be given to both insulating the roof and reducing its absorption coefficient.

Graphs were developed from the computer runs to show separately the solar and conduction components allow prediction of yearly heat gain through 1 sq. ft. of roof based on cooling degree hours, solar radiation, "U" value, occupied time and absorption coefficient. Methods of insulating roofs are described under ECO 46 (Heating Section).

Any paint or reflective finish must be compatible with the existing roof and capable of withstanding abrasion. The absorption coefficient of roofs may also be reduced by adding a surface layer of white pebbles or gravel. Check that the weight of the additional layer does not exceed the structural bearing capacity, and that gravel stops are fitted around rain water drains and the perimeter of the roof. If insulation is added to the surface of the roof for the purpose of reducing heat loss and/or heat gains, select the most desirable absorption coefficient. Change in roofing materials to effect changes of absorption coefficient are subject to building or fire code compliance; in certain cases, color and texture change may require architectural review and approval. Order of magnitude costs for adding insulation to roof are tabulated under ECO 46 (Heating Section).

The cost of altering the roof absorption coefficient will vary with the circumstances of each individual case and quotations should be obtained from local contractors. To determine the cooling energy saved by insulating roofs and/or changing absorption coefficient, refer to Figures #55, and 57 and carry out the following procedure:

a) Determine from Figure #7 the cooling degree hours above 78°DB for location.

b) Determine from Figure #5 the mean daily solar radiation in Langleys for location.

c) Determine the initial "U" value and improved "U" value after insulation.

d) Determine the initial absorption coefficient and the improved absorption coefficient.

e) Refer to Figure #55, use the initial "U" value of the roof, enter the graph at the selected degree hours and follow direction of example line, intersecting at the appropriate points for occupancy and "U" value. Read out the yearly heat gain due only to conduction/sq. ft. of roof.

f) Now, refer to Figure #57, and use the initial "U" value of roof and/or the initial absorption coefficient of the roof, enter the graph at the selected degree hours and following the direction of the example line, intersect at the appropriate points for langleys and "U" value. Read out the yearly heat gain per sq. ft. of roof on the appropriate absorption coefficient line. Interpolate to obtain answers for absorption coefficients other

FIGURE 56: YEARLY HEAT GAIN/SQUARE FOOT THROUGH ROOF

CITY	LATITUDE	SOLAR RADIATION LANGLEY'S	D.B. DEGREE HOURS ABOVE 78°F	HEAT GAIN THROUGH ROOF BTU/FT.^{2} YEAR			
				U=0.19		U=0.12	
				a=0.3	a=0.8	a=0.3	a=0.8
MINNEAPOLIS	45°N	325	2,500	2,008	8,139	1,119	4,728
CONCORD, N.H.	43°N	300	1,750	1,891	7,379	1,043	4,257
DENVER	40°N	425	4,055	2,458	9,859	1,348	5,680
CHICAGO	42°N	350	3,100	2,104	7,918	1,185	4,620
ST. LOUIS	39°N	375	6,400	4,059	12,075	2,326	7,131
NEW YORK	41°N	350	3,000	2,696	9,274	1,543	5,465
SAN FRANCISCO	38°N	410	3,000	566	5,914	265	3,354
ATLANTA	34°N	390	9,400	4,354	14,060	2,482	8,276
LOS ANGELES	34°N	470	2,000	1,733	10,025	921	5,759
PHOENIX	33°N	520	24,448	12,149	24,385	7,258	14,649
HOUSTON	30°N	430	11,500	7,255	20,931	4,176	12,369
MIAMI	26°N	451	10,771	9,009	24,594	5,315	14,716

dubin-mindell-bloome-associates
consulting engineers

than the 0.3 and 0.8 shown.

g) Repeat procedure a - g, using the improved "U" value and/or the improved absorption coefficient, and by subtraction, obtain the cooling load reduction due to insulation and/or modified absorption coefficient/sq. ft. of roof.

ECO 78 REDUCE HEAT GAIN THROUGH ROOFS BY INSTALLING ROOF SPRAYS

The use of roof sprays is advisable:

a) On buildings with large roof to floor area ratio, i.e. probably one or two story office buildings with large roof area or store where heat gains through the roof constitute a major portion of the cooling load.

b) Where there is more than 70% possible sunshine*during the cooling season,

Other climatic conditions where roof sprays may be economic:

-High dry bulb conditions accompanying the solar condition.

-Low wet bulb coincident with high dry bulb and sunshine.

ECO 79 REDUCE COOLING LOAD DUE TO OUTDOOR AIR/VENTILATION AND INFILTRATION

Outdoor air introduced for ventilation imposes a cooling load on the refrigeration equipment only, but infiltration, the leakage of unwnated air directly into conditioned spaces imposes a load on both the refrigeration cooling system and the distribution systems. Air introduced directly in the space increases the room sensible heat gain thus requiring an increase in chilled water or cool supply air with corresponding increase in power consumption. Refer to ECM1, Figure 16 for "Recommended Ventilation Standards", and the relavent ECOs in ECM1 and 2 for methods to reduce both infiltration and ventilation rates.

Refer to ECM1, Figure 21 to determine the cooling load due to ventilation and infiltration.

In Section 6A "Heating & Ventilating", costs and installations are included for charcoal filters to reduce outdoor make-up air and exhaust make-up systems. Measures taken to reduce heating loads also will reduce the cooling load as well. In areas in the country where there are 10,000 wet bulb degree hours or more above 66°F.W.B. with coincidental DB temperatures higher than 78°F the measures to reduce infiltration and ventilation should be considered for cooling regardless of the benefits during the heating season.

* From NOAA, See Reference

FIGURE 57: YEARLY SOLAR HEAT GAIN THROUGH ROOF

dubin-mindell-bloome-associates consulting engineers

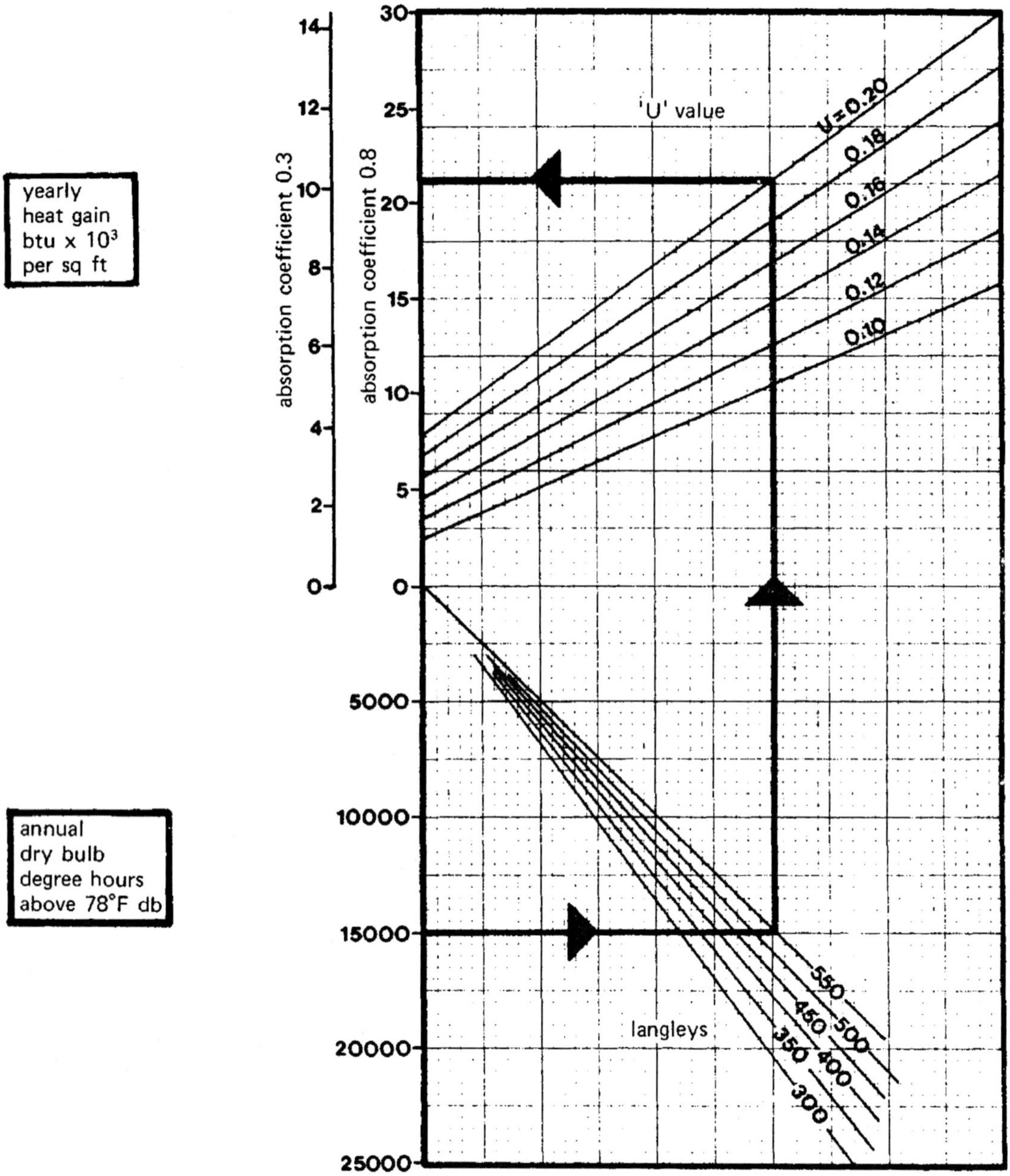

Figure #57 Engineering Data

This figure is based on the "Sunset" computer program which was used to calculate solar effect on roofs for 12 selected locations. The program calculates hourly solar angles and intensities for the 21st day of each month. Radiation intensity values were modified by the average percentage of cloud cover taken from weather records on an hourly basis. Heat gains are based on a 78°F indoor temperature.

The solar effect on a roof was calculated using sol-air temperature and the heat entering or leaving a space was calculated with the equivalent temperature difference. Roof mass ranged from 25 - 35 lbs./sq.ft. and thermal lag averaged 3 1/2 hours. During the cooling season, internal gains, ventilation, infiltration and conduction through the building skin create a cooling load. The additional load caused by heat gain through the roof was calculated for each day. Daily totals were then summed for the number of days in each month to arrive at monthly heat gains. The length of the cooling season for each location considered was determined from weather data and characteristic operating periods. Yearly heat gains were derived by summing monthly totals for the length of the cooling season.

Absorption coefficients and 'U' values were varied and summarized for the 12 locations as shown in Figure #56. Gains in Figure #56 include both the solar and conduction components of heat gain. Values of the conduction heat gain component through roofs were deducted from the total heat gains to derive the solar component. The solar component was then plotted and extrapolated to include the entire range of degree hours.

The major opportunities to reduce ventilation/infiltration loads are summarized:

a) Add vestibules and/or revolving doors.

b) Seal vertical shaft and stairs.

c) Install weather stripping.

d) Recirculate exhaust air through activated charcoal filters.

e) Install auxiliary air supply directly to exhaust loads.

f) Repair or install new outdoor air dampers.

g) Install separate outdoor air supply fans for areas of the building with ventilation requirements in excess of the average building requirements.

h) Use heat exchanger between hot and humid ventilating air and cooler and dryer exhaust air (refer to Section 18, "Heat Reclamation").

To determine cooling energy saved by reducing the quantity of air, refer to Figure 21 and

a) Determine from Figure #8 the annual WB degree hours above 66°F WB.

b) Determine the average hours of cooling system operation/week (occupied periods)

c) Determine the indoor relative humidity during occupied periods.

d) Enter the graph at the selected degree hours and following the direction of the example line, intersect at appropriate points with indoor relative humidity and hours of system operation.

e) Read out the yearly cooling load/1000 cfm.

f) Multiply this figure by the 1000's of cfm infiltration and ventilation reduction to obtain the net cooling load reduction.

ECO 80 USE AN ECONOMIZER CYCLE TO REDUCE COOLING LOAD

Economizer cooling involves introduction of outdoor air into conditioned areas to remove internal heat gain without operating the refrigeration compressor,

A OPERATION FOR COOLING DURING UNOCCUPIED PERIODS

a) Shut down compressors and chillers

b) Provide control to sense outdoor conditions and open outdoor air dampers fully when the DB temperature is at least 5° lower than indoor space temperature and is above 65°F.

c) Open exhaust dampers or exhaust air fans, if available; otherwise partially open windows or install wall type exhaust fans or return air-exhaust air fans to provide pressure relief.

B OPERATION FOR COOLING DURING OCCUPIED PERIODS

There are two suitable economizer systems:

1) System monitors and responds to dry bulb temperature only. It is suitable where wet bulb degree hours are less 8000 per year.

2) System monitors and responds to the WB and DB temperatures (enthalpy), and is most effective and economic in locations which experience more than 8000 WB degree hours.

System 1 - Economizer Cycle Cooling

Provide controls, dampers and interlocks to achieve the following control sequence:

a) When the outdoor air DB temperature is lower than the supply air DB temperature required to meet the cooling load, turn off the compressor and chilled water pumps, and position outdoor air -- return air -- exhaust air dampers to attain the required supply air temperature.

b) When the outdoor air DB temperature is higher than the supply air temperature required to meet the loads, but is lower than the return air temperature, energize the compressors and chilled water pumps and position dampers for 100% outdoor air.

c) Use minimum outdoor air whenever the outdoor dry bulb temperature exceeds the return air DB temperature.

d) Whenever the relative humidity in the space drops below desired levels and more energy is consumed to raise the RH than is saved by the economizer system, consider using refrigeration in place of economizer cooling. This condition may exist in very cold climates and must be analyzed in detail.

System 2 - Enthalpy Cycle Cooling

Provide the equipment, controls, dampers and interlocks to achieve the following control sequence:

The four conditions listed in 1 above are similar for this system with the exception that enthalpy conditions are measured rather than dry bulb conditions.

If changes to outside air intake are contemplated, take careful note of all codes bearing on ventilation requirements. Fire and safety codes must also be observed.

APPLICATIONS OF SYSTEMS 1 AND 2

- -Single duct constant volume systems.
- -Variable volume air systems.
- -Induction systems.
- -Terminal reheat systems, dual duct systems, and multi-zone systems only after making modifications outlined in Section 17, HVAC Systems and ECM1 Section 4 "Heating".

Economizer and Enthalpy systems are less effective if used in conjunction with heat recovery systems. Trade-offs must be analyzed:

It is not possible in this manual to provide blanket rules for determining energy saved or costs for economizer and enthalpy control systems since there are too many variables which affect each individual application.

Using methods described in these manuals, the amount of cooling which can be achieved with outdoor air cooling can be assessed. The costs of installation and efficiency of the systems must be determined on an individual basis.

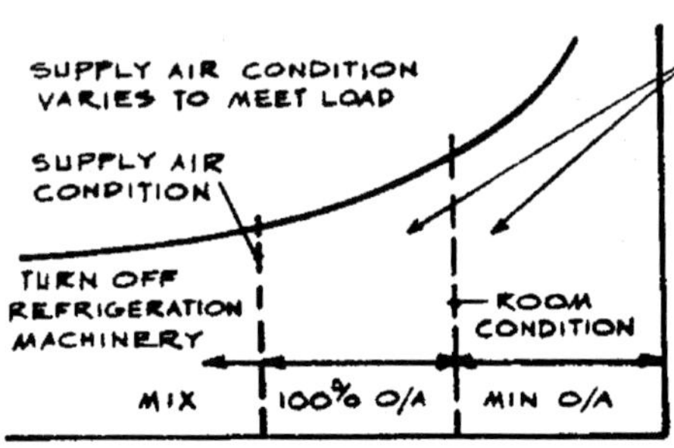

ECONOMIZER CYCLE
SYSTEM 1

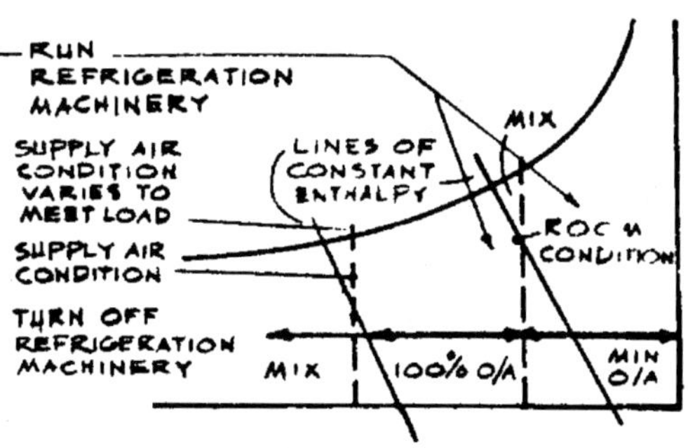

ECONOMIZER CYCLE
SYSTEM 2

ECO 81 REDUCE INTERNAL HEAT GAINS FROM LIGHTS

Heat gain from lights often form the major part of the cooling load in an office building or store. The level of illumination is often far in excess of that required for visual accuity and many steps may be taken to reduce the lighting level and increase the efficacy of the lighting systems, ranging from simply switching off unnecessary lights to changing lighting fixtures for more efficient types. Refer to Section 19 for details.

One KW of lighting input produces a heat output of 3,413 Btu/hr. Depending on the type of lighting fixture, between 60% and 80% of this heat appears as a room load. The remainder is rejected from the ballast and top surface of the lighting fixture into the ceiling void where it raises the return air temperature.

Reducing power for lighting by 1 KWH will reduce the room sensible cooling load by approximately 3400 Btu. Cooling supply air quantity is dependent upon the room sensible heat, so a reduction of supply volume results in savings of fan HP. Refer to Section 17 for methods of determining load reductions. Refer to Section 19 "Lighting" for a discussion of heat-of-light systems which reduce the cooling room loads, and in some cases, refrigeration load.

ECO 82 REDUCE HEAT GAIN FROM BUILDING EQUIPMENT

Electrically operated business machines, such as typewriters, photo copiers etc., all generate heat in direct proportion to the quantity of electricity they consume. The heat output of office equipment cannot be significantly reduced while they are being used, but hours of operation can be reduced by making sure that each piece of equipment is turned off while it is not required for immediate use.

Equipment such as ovens, ranges, fryers, and other kitchen equipment emit heat. Where hoods are provided, see ECO 53 for auxiliary air supply to reduce make-up air load. Install hoods where necessary.

Provide shields, baffles and insulation for other equipment whose surface temperature exceeds 100°F for 2 or more hours per day.

ECO 83 EMPLOY EVAPORATIVE COOLING TO REDUCE COOLING LOAD

In geographic areas which experience high DB temperatures but low coincidental WB temperatures (65°F WB or less) evaporative cooling may be used to reduce the temperature of the outdoor air. For instance, outdoor air at 95°DB, 65°WB can be adiabatically cooled by water sprays to 77°DB, 65°WB giving RH of 50%. This effectively eliminates the ventilation cooling load and under certain circumstances will allow the use of economizer cooling, even though the outdoor DB temperature exceeds the design DB temperature within the building. To determine the benefit of evaporative cooling an analysis must be made of weather data for each particular case to assess the number of hours of possible use and the quantity of energy saved.

Evaporative cooling can be achieved by installing sprays within the main HVAC units or in duct extensions together with drain pan and eliminators. If the design of the cooling coil is suitable, it is often effective to spray the water directly onto the upstream face of the coil and use the existing downstream eliminators.

Spray water can be continuously recirculated from the drain pan through the spray headers and back to the drain pan by a small pump but continuous make up and blow down should be provided to limit precipitation of solids. The quantity of blow down and make up will depend on local water analysis. If low temperature well water is available, this can first be used in the cooling coil and then used as feed to the evaporative cooler sprays.

Even in relatively humid climates, there are a significant number of hours per year that adiabatic cooling can be used and operation of refrigeration equipment can be minimized.

ECO 84 TO REDUCE COOLING LOAD USE DESICCANT DEHUMIDIFICATION

Desiccants, such as Silica Gel can be used as adsorbents to reduce the moisture content of air. Dehumidification is achieved adiabatically, and the air DB temperature increases proportionally with the quantity of moisture removed. i.e., air at 85° DB - 75°WB would change adiabatically to 122°DB - 75°WB with a corresponding reduction in moisture content of 56 grain/lb. Air can now be reduced to 60° - 50% RH by sensible heat removal only with relatively high temperature cooling medium other than chilled water in the cooling coils. Cooling tower, well water, or waste cool water can be utilized in the coils.

Desiccant dehumidification is particularly applicable in geographic locations experiencing long periods of high WB temperatures. The desiccant eventually becomes fully charged with moisture and has to be regenerated by heating to drive the moisture off. When calculating the effectiveness of desiccants, this heating energy requirement must be taken into account. If possible, recaptured waste heat or solar energy should be used for regeneration. Refer to Section 22 for solar energy applications.

Determine the economic feasibility of desiccant dehumidification. Analyze local weather data and the system operating conditions to ascertain the hours of effective use.

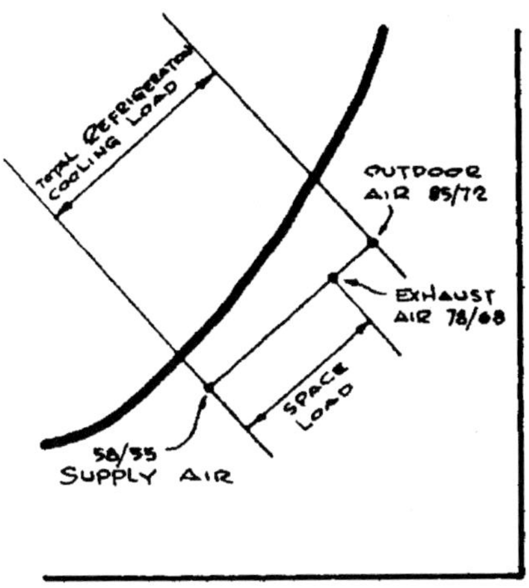

PSYCHROMETRIC CYCLE FOR 100% REFRIGERATION COOLING

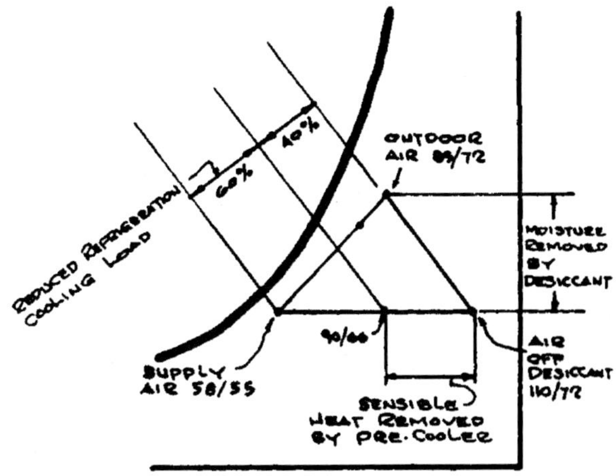

PSYCHROMETRIC CYCLE FOR DESICCANT DEHUMIDIFICATION

Desiccant dehumification can be usefully applied to buildings or areas of building subject to high internal latent heat gains or areas with requirements for a high percentage of outdoor air.

For example, refer to psychrometric diagrams. A space requires supply air at 58°F DB/55°F WB to maintain air at 85°/72°. The refrigeration cooling load to reduce 5,000 cfm of air from 85°/72° to 58°/55° is 22 tons. Using desiccant dehumidification, the outdoor air condition changes from 85°/72° to 110°/72°. High temperature dry air is now sensibly cooled to 90°/66° by a precooling coil (cooling medium can be well water, condenser water, waste water, etc.) The cooling load to reduce 5,000 cfm of air from 90°/66° to 58°/55° is 12.75 tons, resulting in a refrigeration load reduction of 9.25 tons or 40%.

Basically, desiccant dehumidification converts latent cooling load to high temperature sensible cooling load, a proportion of which can easily be handled without resorting to mechanical refrigeration.

FIGURE 58: HEAT GAIN FOR VARIOUS PIPE SIZES AND INSULATION THICKNESSES

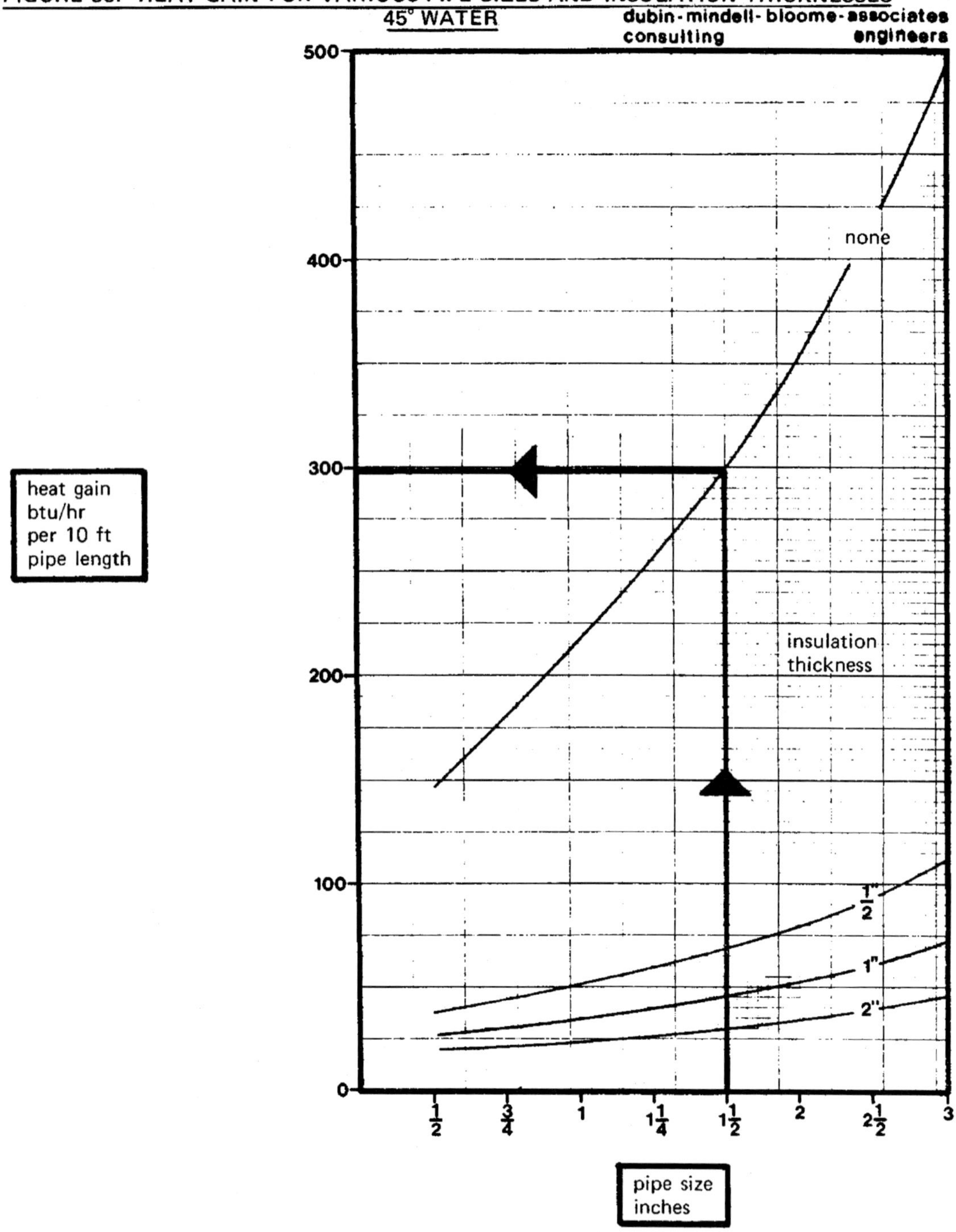

Figure #58 Engineering Data

This figure is based on the addition of insulation with a thermal conducitity, k = 0.3 (Btu) (in.)/(sq. ft.) (hr.) (degree F temperature difference) and an ambient air temperature of 68°F.

The following formula is the basis for determining the heat emitted (heat loss) from the pipe:

$$q = \frac{T_s - T_a}{\frac{d_1}{2k}\left(\log_e \frac{d_2}{d_1}\right) + \frac{d_1}{fd_2}}$$

Where:

q = heat emission from insulated pipe - Btu/(hr.) (sq. ft. of hot pipe surface)

T_s = temperature of pipe surface - degree F

T_a = ambient air temperature - degree F

d_1 = outside diameter of pipe - inches

d_2 = outside diameter of insulation - inches

k = thermal conductivity of insulating material - (Btu) (in.)/(sq. ft.) (hr.) (degree F temperature difference)

f = surface coefficient - Btu/(sq. ft.)(hr.)(degree F)

ECO 85 INSULATE PIPES AND DUCTS TO REDUCE HEAT GAIN

Refer to ECO's 57 and 58 (Heating Section 13) for criteria on insulating pipes and ducts. Note: All insulation on cold pipes and ducts should be provided with a vapor barrier to prevent condensation within the insulation.

Refer to Figure #46 for insulating ducts and Figure #58 for insulating pipes.

Calculate energy savings as described in ECO 57 and 58 .

INCREASE EFFICIENCY OF PRIMARY EQUIPMENT

BACKGROUND

When all applicable ECO's from both ECM 1 and 2 are applied to reduce the load and the distribution system losses, the quantity of cooling required from the primary equipment will have been reduced.

ECO 86 INCREASE EVAPORATOR TEMPERATURE

The COP of any refrigeration machine increases, or less power is used, as the evaporator temperature is raised. Any reduction in "building" cooling load or thermal losses from the distribution system(s) results in a reduced load on the refrigeration machine which can be met at higher chilled water or cooled air as the case may be. The efficiency (COP) of a DX refrigeration unit increases by about 1-1/2% for each degree use in evaporator temperature. Refer to Figure #59 for increase in COP per degree rise in chilled water temperature.

Provide controls to achieve the following:

- -Raise chilled water temperature to "follow the load". Install a limit switch in each modulating or diversion valve to measure whether the valve is fully open or only partially open. Arrange the control circuits so that when all coil control valves are either closed or in a partially open position (indicating light load conditions), the chilled water temperature supply set point is raised until 1 or more coil control valves returns to the fully open position.

- -Raise supply air temperature to follow the load.

- -Permit relative humidity to rise in the conditioned air space (to 60% or more).

- -Decrease the superheat on DX coil (consult manufacturer).

- -Fully load one compressor before starting the next refrigeration unit module.

- -Install a new smaller refrigeration unit when existing modules are all oversized for light loads.

- -Repipe chillers in series. See ECO 88.

Measures to permit chilled water temperatures to be raised for part loads are described in ECM 1, Section 6, "Cooling" and Section 15 herein.

FIGURE 59: EFFECT OF CHILLED WATER TEMPERATURE ON CHILLER COEFFICIENT OF PERFORMANCE

dubin-mindell-bloome-associates
consulting engineers

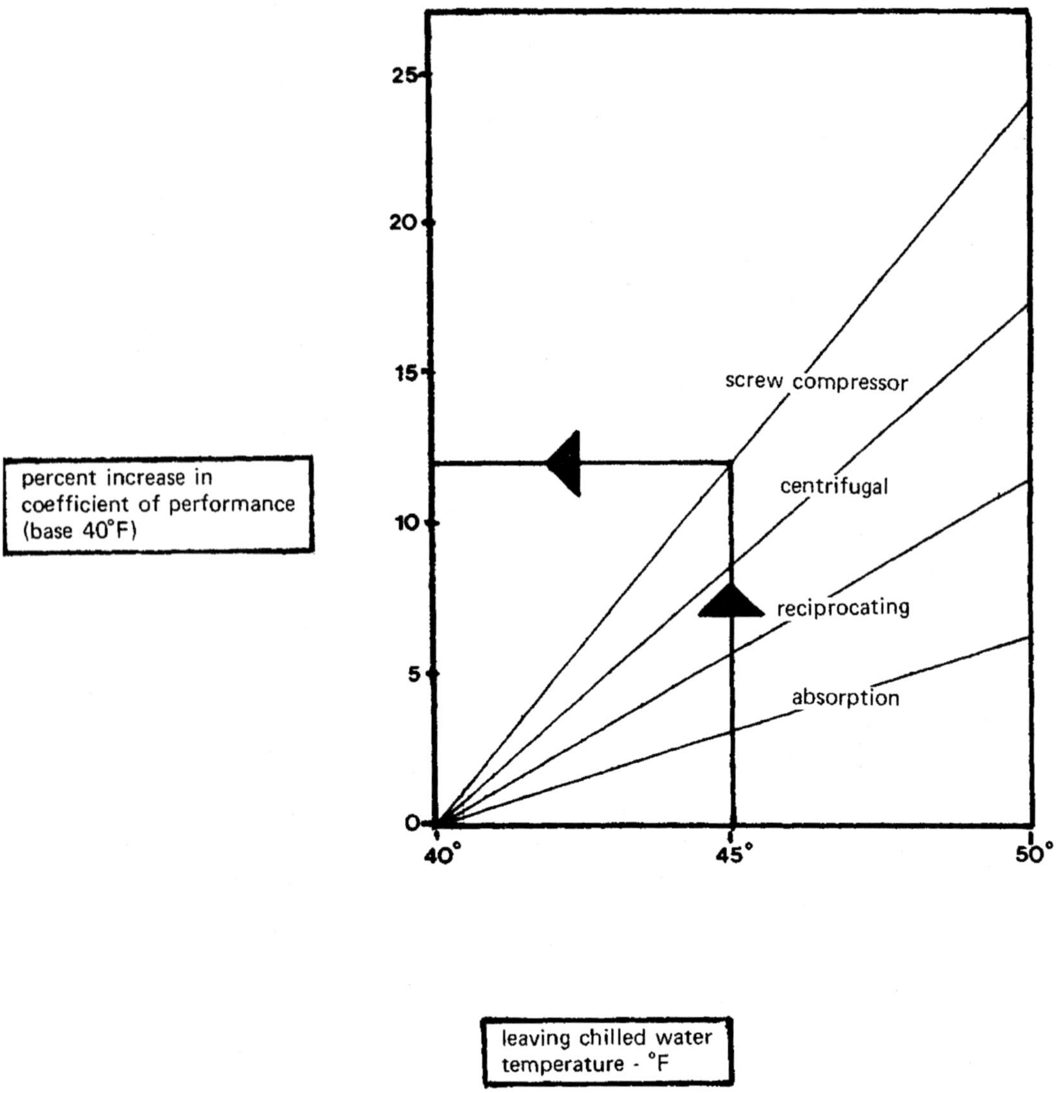

Figure # 59 Engineering Data

Source: Manufacturers data for different sizes and types of chillers.

COP at various chilled water temperatures was determined by the relationship:

$$COP = \frac{\text{output - Btu}}{\text{input - Btu}}$$

The change in COP was expressed in terms of the COP at nominal conditions and then plotted as a function of chilled water temperature.

For commercial refrigeration units, a very substantial savings can be realized if freezer temperatures are maintained at just the required temperature and not set too low. A unit operating at -30°F uses more than twice the energy than at -10°F.

ECO 87 ISOLATE OFF-LINE CHILLERS

Light cooling loads on a multiple chiller installation are often met by circulating chilled water through all chillers even though only one chiller is operating. This not only wastes pump energy by maintaining an unnecessarily high flow rate water through the system, but due to the bypass effect of off-line chillers, forces the remaining on-line chiller to produce chilled water at low temperatures to offset the mixing effect of the bypass to meet the desired supply temperature and causes the on-line chiller to operate at low evaporator temperatures (COP drops). Under light loads, isolate the off-line chillers and reduce the chilled water flow rate. off unnecessary chilled water pumps when multiple pumps are installed.

If multiple CW pumps are not installed, consider the installation of additional pumps or install a multi-speed drive on the present pump.

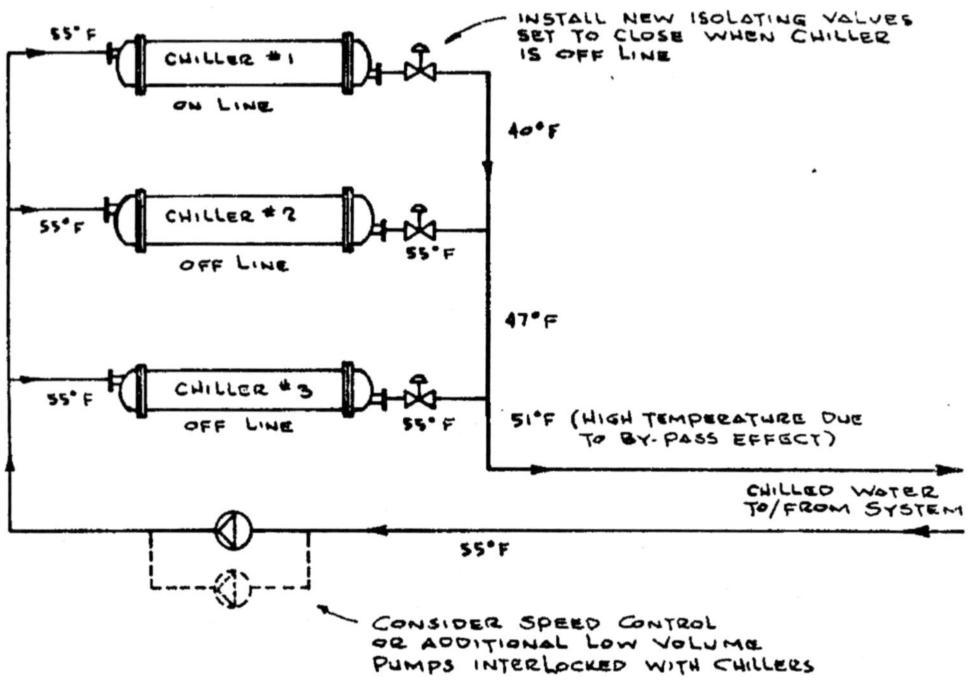

EFFECT OF BYPASS THROUGH OFF-LINE CHILLERS

ECO 88 PIPE CHILLERS IN SERIES

System efficiency is affected by the piping arrangement of the chillers, series or parallel. Generally, rather than piping chillers in parallel so that each must produce the coldest water in the system, pipe them in series as shown below. Chiller No. 2 operates at higher suction pressure and uses less KW/ton. One possible drawback to series arrangement is that chilled water is pumped through the off-line chiller, adding unnecessary resistance to the piping system. However, this can be avoided by adding a bypass around the off-line chiller. Where it is possible, chilled water volume can be reduced (coils can take higher temperature differential) to reduce pumping resistance.

The following diagrams illustrate typical series and parallel piping arrangements for a two-chiller installation.

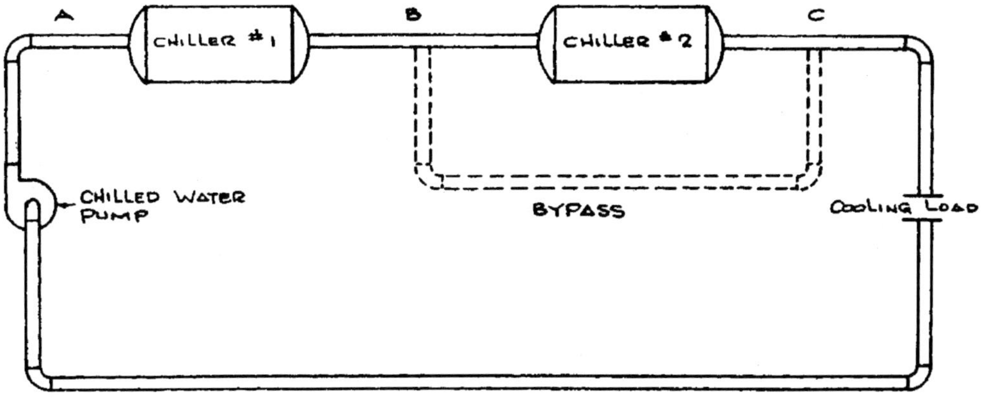

SERIES

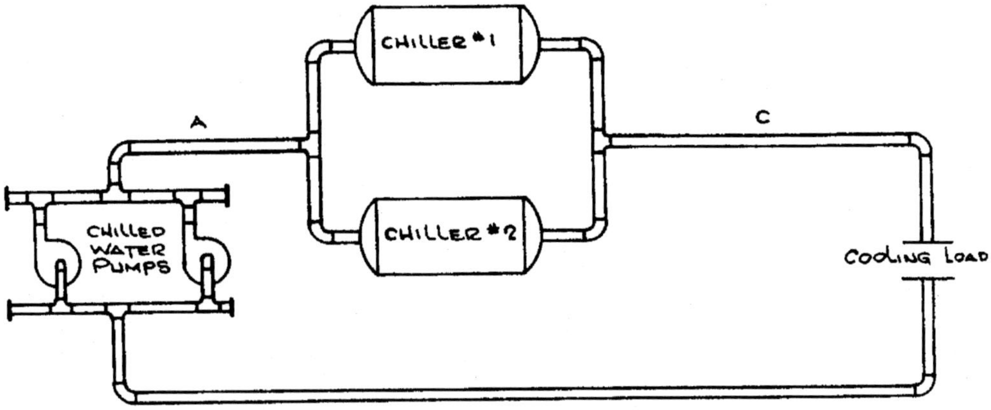

PARALLEL

An analysis of power required for each arrangement was done under various load conditions. Each chiller is rated at 250 tons output at 44°F leaving chilled water temperature and 95°F leaving condenser water temperature. The table below summarizes the operation of each system under full- and part-load conditions.

	SERIES			PARALLEL		
LOAD	FULL	1/2	1/4	FULL	1/2	1/4
Temp. A. (Sketches)	54°F	54°F	54°F	54°F	54°F	54°F
Temp. B. (Sketches)	49°F	49°F	51.5°F	-	-	-
Temp. C. (Sketches)	44°F	49°F	51.5°F	44°F	44°F	49°F
Ch. Water GPM	1200	1200	1200	1200	600	600
Chiller #1 KW Input	187	187	88.7	212	212	93.3
Chiller #2 KW Input	212	-	-	212		
Total KW	399			424		
Pump KW Input*	37.3	(33.2) 37.3	(33.2) 37.3	40.1	20.1	20.1
Total KW Input	436.3	(220.2) 224.3	(121.9) 126.0	464.1	232.1	113.4

*Can be reduced by modifying pumping arrangements.

All power inputs are KWH.

Numbers in parenthesis represent KW consumption for series arrangement with bypass installed.

It can be seen from the above table that the series arrangement saves 464.1 - 436.3 or 27.8 KW per hour of operation under full load conditions, a savings of 6%. Depending on the length of the cooling season, energy and dollar savings can be substantial over a year of operation.

ECO 89 TO INCREASE CHILLER EFFICIENCY, USE LOW TEMPERATURE CONDENSER WATER

The efficiency and the COP of chillers and compressors increase as condensing temperature is decreased. Each type of machine has practical limits of the lowest acceptable condensing temperature and this limit should not be exceeded, particularly with some of the older type absorption machines. (See manufacturers for specific data.)

If low temperature well water is available or can be made available, consider its use for condenser water rather than using cooling towers. The energy required to pump the well water is likely to be approximately the same for the condenser water pump and cooling tower fans but energy savings in compressor HP will be worthwhile.

Where existing cooling installations operate at constant condenser water temperature and cooling tower fans are cycled on and off, modify the controls to operate cooling tower fans continuously whenever there is a machine on-line, and allow the condenser water temperature to drop until it reaches a predetermined low limit, at which point the cooling tower fans can be cycled on and off.

To reduce condensing temperatures and maintain low fouling factors, install automatic tube cleaners. One such device comprises a cylindrical brush in each tube which is periodically forced from one end of the condenser tube bundle to the other by a reversal in the direction of water flow. It has been demonstrated that this type of system can be very effective in keeping condenser tubes clean and depending on the existing level of condenser tube fouling, can yield substantial energy savings.

To determine the energy savings due to reduced condensing temperature, refer to Figure #60.

Enter the graph at the present condensing temperature, intersect with the appropriate line for the particular type of machine and read out the percentage increase in COP.

Repeat this procedure using the lower condensing temperature and obtain the new percentage increase in COP. By subtraction, obtain the actual increase in seasonal COP, apply this to the present seasonal COP of the machine and calculate the energy savings using the recorded yearly energy costs.

FIGURE 60: EFFECT OF CONDENSER TEMPERATURE ON CHILLER COEFFICIENT OF PERFORMANCE

dubin-mindell-bloome-associates
consulting engineers

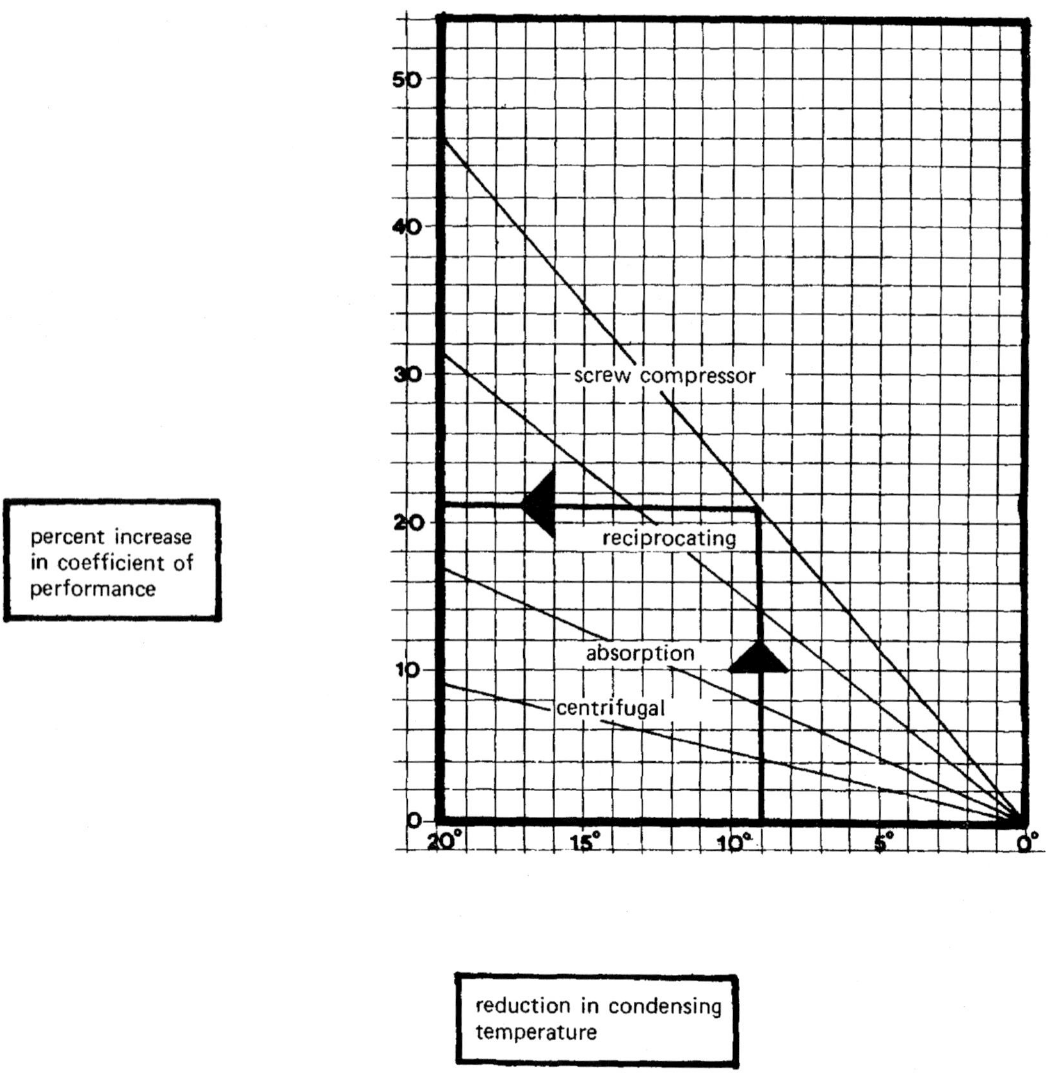

Figure #60 Engineering Data

Source: Manufacturers data for different sizes and types of chillers

COP at various condensing temperatures was determined by the relationship:

$$COP = \frac{\text{output} - \text{Btu}}{\text{input} - \text{Btu}}$$

The change in COP was expressed in terms of the COP at nominal conditions and then plotted as a function of condensing temperature.

Additional measures to lower condenser water temperature include the following:

- Increase fan volume in cooling tower by changing speed of rotation or increasing the pitch of propeller fan blades. This may entail replacing motor with larger size. Determine the trade-off between increased fan horsepower and decreased refrigeration horsepower.

- Increase water flow for maximum evaporation. Caution: Carryover may be objectionable.

Where air-cooled condensers are installed, increase air volume through condenser, add additional air-cooled condenser in parallel to increase surface area, replace condenser coil with larger heat transfer surface, remove obstructions to air flow, duct exhaust air from building to air inlet, remove high resistance strainers, fittings, and elbows in refrigerant lines to reduce head pressure (Caution: Obtain manufacturers recommendations on oil return), shade condenser from direct sunlight. For further suggestions, refer to ECM 1, Sections 8 and 6, "Commercial Refrigeration" and "Cooling".

Where dry bulb temperatures are low for a large portion of the cooling season, consider replacing air-cooled condenser with cooling tower.

Consider a water spray in air stream before inlet side of condenser coil.

ECO 90 REPLACE AIR-COOLED CONDENSERS BY COOLING TOWERS

The condensing temperature of a refrigeration machine may be significantly lowered by replacing the air-cooled condenser by a cooling tower in locations with fewer than 15,000 wet bulb degree hours.

This can be effected on a one-to-one basis if the air-cooled condenser is large or one cooling tower can replace a number of small air-cooled condensers. Whichever method is adopted, each refrigeration machine will require the addition of a water/refrigerant heat exchanger (shell and tube condenser).

Air-cooled condensers commonly operate at temperature differences of 20°F between air and refrigerant, thus if the outdoor air temperature is 95°F DB, then the condensing temperature will be 115°F.

Cooling tower performance is related to wet bulb temperature and they commonly operate at temperature differences of 10°F between the air wet bulb and condenser water. Thus, if the outdoor air is 72°F WB, the condenser water temperature would be 82°F off the tower and at 2 gpm/ton would be 9[illegible]°F giving a mean condensing temperature of 90°F.

For example, if an air-cooled condenser used with a reciprocating refrigeration machine were replaced by a cooling tower in an area that experienced 95°F DB, 72°F WB design conditions, then the condensing temperature would be reduced from 115°F to 97°F or 18°F difference.

This reduction of condensing temperature will increase the COP of the refrigeration machine. Refer to Figure #60 and read from the graph a percentage increase of 28% in COP due to the 18°F reduction in condensing temperature.

ECO 91 USE PIGGYBACK ABSORPTION SYSTEM

Where existing centrifugal chillers are driven by steam turbines and the exhaust steam is condensed before being returned as feed water to the boiler, an absorption chiller can be substituted for the condenser to produce additional chilled water from the waste heat.

For example, a low pressure turbine driving a centrifugal chiller requires 12 lbs. of dry steam at 145 psig for each ton of refrigeration. Exhaust steam from the turbine at 15 psig and 20% moisture content will yield approximately 780 Btu's/lb. when cooled and condensed to 220°F giving a total usable heat of 12 x 780 = 9360 Btu/ton. An absorption chiller used as a steam condenser requires a heat input of 18,000 Btu/ton and will yield 9360/18,000 = 0.52 tons for each ton of refrigeration produced by the centrifugal chiller. (Newer units in development will require only 13,000 Btu/ton.) In this case the addition of an absorption piggyback chiller to condenser waste steam will decrease the steam/ton of refrigeration by 30%.

EXAMPLE

Assume a centrifugal chiller with a capacity of 600 tons and operating at an equivalent of 1400 full capacity hours/year or 1400 x 600 = 840,000 ton hours/yr.

For turbine driven machine only, at 12 lbs. steam/ton = 10.08×10^6 lbs. steam/yr.

If this is reduced by 30% by using exhaust steam in an absorption machine,
Saving = $10.08 \times 10^6 \times 0.3 = 3.024 \times 10^6$ lbs. steam/year.

From records, 1 gallon of oil produces 96 lbs. of steam.

Fuel savings = $\frac{3.024 \times 10^6}{96}$ = 31,500 gallons oil/year, and at 33¢/gallon =

$10,395/year.

The capital cost of a 200-ton absorption chiller and all peripheral equipment = $54,000. This investment gives a 5-year pay back and is worthwhile. Further economic analysis will indicate cash flow benefits.

Under part load conditions, it is more efficient to use the piggyback absorption machine to meet as much of the load as possible and to modulate the turbine driven machines to meet the remainder.

Piggyback absorption is particularly effective in buildings using purchased steam when condensate is not returned to the supplier. In such cases, it is worth considering the purchase of a larger machine and derating it by operating at outlet temperatures down to 180°F. This will reclaim the maximum possible heat from exhaust steam and condensate. Consider also using lower temperature condensate from the absorption machine as a heat source for low grade systems such as domestic hot water, ventilating air pre-heaters, etc. Refer to Section 6F "Heat Reclamation".

Low temperature absorption machines are also currently being developed for operating at inlet temperatures of 190°F or lower, and yet will retain a reasonably high COP. Consider using a low temperature absorption machine in buildings that have large quantities of low grade waste heat (e.g. cooling water from engine/generator sets).

Low temperature absorption machines are also capable of operating in conjunction with solar collectors and can extend the use of a solar system from winter only to year round. See Section 22.

GUIDELINES TO REDUCE ENERGY USED FOR COOLING AND VENTILATION

Refer to ECM 1, Section 6, "Cooling and Ventilation" for guidelines to reduce energy consumption with minimal investment. Analyze and consider the following options for further reductions of energy consumption.

1. Maintain Higher Temperature and Humidity Levels Indoors during the Cooling Season.

   a) Refer to ECM 1, Figure 20 for "Recommended indoor temperature and humidity standards".

   b) Refer to Section 6A "Heating and Ventilation" in this Manual for guidelines to reduce ventilation rates; and for stats, controls and timers, and duct dampers to increase temperature levels in non-critical areas in accordance with Figure 20 standards.

   c) Install a control to prevent operation of the refrigeration system for the purpose of dehumidification until relative humidity levels exceeds 60%.

   d) Do not operate heating system during cooling season to raise the space temperatures to 78°F, or to other temperature levels listed in Figure 20 during occupied cycles.

   e) Install a seven-day timer programmed for normal daytime operation of chillers, compressors, cooling towers and chilled water pumps, with a cam set to shut off all equipment and blowers during holidays and weekends.

2. Reduce Solar Heat Gain through Windows and Glass Doors.

   a) Exterior solar control devices are the most effective. Install horizontal fixed or movable eyebrows over south facing glass. Install fin sunscreens on the exterior of east, west, and south facing windows to reduce solar heat gain and wind loads in the winter. In locations where winds are minimum in the winter and/or existing construction and aesthetics preclude locating the sunscreen on the exterior, install them in the interior surface of the window. Screens can be hinged to permit easy access to operable windows. Sunscreens to reduce the use of available day lighting for illumination. All exterior solar control devices should provide 100% shading from May 1 through October 31.

b) Install awnings over windows to block out summer sun. Awnings are particularly suitable to shade large display windows in commercial stores.

c) Replace single clear glass on the east, west, and south facades with reflective glass when replacing broken windows if the windows are in direct sunlight 200 hours or more during the cooling season. Replace all single glass on west, east, south facade if windows are in direct sunlight 500 or more hours per year during the cooling season. Reflective glass is more effective in reducing heat gain than heat absorbing glass.

d) Apply reflective coatings to the inside surface of east, west, and south facing windows which are exposed to direct sunlight.

e) If double windows are added to reduce heat loss in the winter time, provide the inner layers with reflective glass to reduce transmission at peak conditions to 50% of maximum.

f) When buildings are altered for expansion, a zig-zag wall with the glazing in the west wall and east wall facing north in the southern climates and south in the northern climates will be self-shading in the summer, while permitting the sun to penetrate through the glass during the heating season.

g) Install milium lined drapes, venetian blinds, or shutters on the inside of all south, east, and west facing windows which are subject to direct sunlight in the cooling season and/or exposed to a large expanse of sky. These devices are not as effective as external shading devices or sun screen, but are least expensive to install.

h) Treat skylights in the same manner as windows.

3. <u>Reduce Transmission Heat Gains through Opaque Walls</u>.

a) Insulate exterior east, west, and facing walls if they are in direct sunlight during the entire year if the "U" value is greater than 0.20, or the wall mass is less than 10 lbs./sq. ft., or the peak heat gain is greater than 30 Btu/hr./sq. ft. Refer to Section 13 in this Manual "Heating and Ventilation" for additional guidelines on insulation.

b) Where the materials and aesthetics are not jeopardized, paint exterior sun walls with white reflective paint to reduce the absorption coefficient to reduce the radiation heat gain.

4. Reduce Transmission Heat Gains through Floors.

   a) Do not consider insulating floors to reduce heat gain above, however, added benefits are obtained during the cooling season when floors have been insulated to reduce the heating load.

5. Reduce Transmission Gains through Roofs.

   a) Insulate roofs to "U" values listed in ASHRAE 90-P in all locations where there are more than 500 hours per year above 85°F, if the roof or combined roof-ceiling has a "U" value in excess of 0.20.

   b) Provide insulation on top of the roof, or below the roof and above the ceiling whenever the interior surface temperature of the ceiling (roof if there is no ceiling) is above 82°F on clear days with 60% or more possible sunshine.

   c) Cover the exterior surface of the roof with white pebbles, light colored tiles, or other durable materials to reduce the absorption coefficient in direct sunlight to 0.3 or less.

   d) Install a roof spray system to reduce the temperature of the exterior surface of the roof, if calculations indicate that sol-air temperatures exceed 100°F for 250 hours or more. If the roof is insulated to a "U" value of 0.2 or better, and has a mass of 50 lbs. per square foot or more, do not consider roof sprays. For all other conditions, analyze the comparative cost benefit between roof sprays and insulation.

   e) Install a roof pond in place of roof sprays if the roof construction permits it and an analysis shows that the cost benefits are greater than for insulation.

6. Reduce Cooling Loads Due to Infiltration.

Refer to ECM 1, Section 6, "Cooling and Ventilation" for measures to reduce infiltration rates with minimal cost and Section 15 of this Manual, for methods to implement them.

Analyze and consider the following:

   a) Install revolving doors, or vestibules, under the following conditions: Retail stores, where analysis shows that outdoor air passages through door openings exceeds more than 10% of the cooling load.
   Note: It is not suggested that these measures be used in religious buildings or office buildings solely to reduce cooling loads. Where they are added to reduce heat loss in the winter time, they also will benefit during the cooling season.

b) Seal vertical stacks in office buildings, 6 stories or greater, when there are 10,000 or more wet bulb degree hours during occupied periods.

c) Weatherstrip in office buildings only, when windows are very loose and there are 10,000 or more wet bulb degree hours during occupied periods.

d) Analyze the seasonal effects of air leakage to offset exhaust air systems and reduce the amount of exhaust air and the duration and operation as described in other Sections.

7. Reduce Cooling Load Due to Ventilation

Refer to ECM 1, Section 6 "Cooling and Ventilation" for measures to reduce ventilation rates with minimal cost, and Section 15 of this Manual for additional measures to implement them.

Analyze and consider the following measures to further reduce energy consumption for ventilation:

a) Rebalance air supply system to provide a maximum of 1 cfm per square foot of supply air to toilet rooms and adjust the exhaust system accordingly. Refer to ASHRAE Standard 62-7 2-73.

b) Install charcoal filters or other air treatment devices in the exhaust air systems and circulate treated air back into the space to reduce the amount of outdoor air required for make up for all systems handling 2,000 cfm or more. Supply slightly more outdoor air than exhaust air to pressurize the conditioned space. It is necessary to obtain approval of code administrating agencies having jurisdiction in the location of the building when reducing outside air quantities below code requirements.

c) Install damper and controls to permit the ventilation system to delay the introduction of outdoor air for one hour in the morning and shutting it off 1/2 hour before closing time. A seven-day timer can be used in department stores to reduce or shut off ventilation automatically for periods during the day with light occupancy.

d) Refer to Section 22 "Solar Energy" for opportunities to use solar heat to preheat outdoor air for ventilation.

e) Modify duct systems and hoods and introduce untreated outdoor air directly to the exhaust hood. Weigh this against changing hoods to new high velocity hoods which require less make up air.

f) Install heat reclaim devices between the exhaust and supply air ducts.

8. Install Controls and Operate on Economizer Cycle and/or Enthalpy Control.

a) Install dampers in the fresh air duct and the return air duct at air handling units and interlock dampers, so that one opens when the other closes. Provide control to open the outdoor air damper when the outdoor temperatures are 5° or more below indoor conditions.

b) Install enthalpy control to open outside air damper during occupied periods with the refrigeration equipment in operation when the outdoor wet bulb temperature is below 65°F and the dry bulb temperature is below 85°F.

c) Install exhaust fan, relief louvers or for large multi-floor buildings, a return air -- exhaust air fan, with dampers in the exhaust duct interlocked with the fresh air and return air dampers. (An analysis should be made to determine that the relative humidity in the space will not drop so low as to require a greater amount of energy for humidification than is saved by reducing the refrigeration load.

9. Reduce Internal Heat Gain.

a) Wherever possible, relocate all equipment and processes that emit heat, near to the windows on the north, or east exposures to facilitate transfer of heat from indoors to outdoors.

b) Insulate hot surfaces on tanks, pipes, and ductwork which are located in air conditioned spaces.

c) Install an automatic control to deenergize dry type transformers which are located in air conditioned spaces when there is no operating load.

d) Modify the lighting system to reduce heat gain:

Relamp with higher lumens per watt.

Select light fixtures with greater coefficiency of utilization to permit lower wattage lamps.

Refer to ECM 1, Section 9 "Lighting" and Section 17 "Lighting" in this Manual for additional measures to reduce internal heat gains from lighting.

e) Locate vending machines, office duplicating equipment and other equipment and appliances out of air conditioned areas. (It may be possible to relocate them back into areas requiring heat in the winter time).

f) Install hoods, baffles, and insulated panels to minimize heat gain to the space from equipment of high heat emission characteristics.

g) If exhaust is used, analyze the trade-off between exhausting air from heat releasing equipment vs. returning it to the system based on enthalpy of make up air required.

10. Reduce the temperature of incoming outdoor air by using evaporative coolers - geographical areas with high DB temperature coincident with low WB temperatures (65°F or less).

11. Use desiccant dehumidification to reduce the moisture content of incoming outdoor air in areas with long periods of high WB temperatures.

12. Reduce Distribution Systems Heat Gain.

Refer to ECM 1, Section 7 "HVAC Systems and Distribution" ECO 27 and analyze and consider the following to further reduce energy consumption due to distribution systems heat gain.

a) Insulate the following equipment:

All chilled water piping and fittings.

All air handling units, providing a greater amount of insulation for air handling units which are in non air conditioned spaces or outdoors.

The outer shells of chillers.

Supply air ducts.

b) Relocate fan motors of 5 HP or larger out of the air handling units or airstream.

13. Reduce 'Parasitic Distribution' System Losses by Reducing Flow Rates and Temperatures.

    a) Reduce air flow rates.

    b) Reduce water flow rates.

    c) Reduce resistance to flow in piping system.

    d) Reduce resistance to flow in air duct systems.

14. Reduce Energy Consumptior by Improving Performance of Compressors, Chillers, Cooling Towers, and Air-Cooled Condensers.

    Refer to ECM 1, Section 6 "Cooling and Ventilation" for guidelines.

    Analyze and consider the following measures to further reduce energy consumption.

    1) Raise evaporator temperatures and suction pressure.

        a) Isolate off-line chillers.

        b) If in parallel, investigate repiping chillers in series.

        c) When adding additional capacity for expansion and where possible in existing systems, use blow through air handling units rather than draw through.

        d) Install sensing control to raise supply air temperature when load permits. This will raise chilled water temperature and evaporator temperatures.

        e) Install controls for multiple chillers to permit one chiller to operate at full capacity before bringing on the second chiller. (Third if installed)

        f) Provide alternator to change the sequence of operation for equal wear.

        g) Maintain higher relative humidity levels in air conditioned space. Refer to Section 17 "HVAC Systems and Distribution" for further measures which will result in lower evaporator temperatures.

        h) Vary chilled water temperature with load.

2) Lower the condensing temperature.

a) Install automatic tube cleaners in condensers.

b) Modify the exhaust ductwork to discharge cooled air from the building into the air intake of air-cooled condensers or cooling towers.

c) Install sensing controls to permit cycling of cooling tower fans under light load conditions when refrigeration load is small and the ambient wet bulb temperatures are low.

d) Investigate increasing condenser water flow rates to permit lower condensing temperatures by: 1) reducing frictional pipe losses, 2) operating on-line and standby condenser water pumps together, 3) changing pump drive on motor.

e) Use well water, ground water, sea water or lake water, if the temperatures of the water from these sources is lower than the wet bulb ambient temperature in the case of cooling towers and the dry bulb temperatures in the case of condensers.

f) Adjust fan blade pitch or fan speed of cooling towers for minimum power input.

g) Install baffles in cooling towers and air-cooled condensers to reduce or eliminate bypass of hot air around coils.

h) Install controls to permit refrigerant migration on large systems to obtain refrigeration effect without power input to the chiller.

i) Modify condenser water bypass around cooling towers to reduce condenser water temperature to the lowest level compatible with the refrigeration machine.

j) Replace air-cooled condensers with cooling towers.

k) Install air-cooled condensers and cooling towers in series where ambient wet and dry bulb conditions indicate that lower condensing temperatures can be obtained. (Not worthwhile unless reduction in temperatures is at least 6°.)

15. Use piggyback absorption systems to produce additional chilled water when existing centrifugal chillers are steam turbine driven.

16. When replacing absorption refrigeration equipment, compare the performance of new high temperature absorption units with new low temperature low pressure, low energy absorption units and select the one based on best seasonal performance. (Can only consider the former where high pressure boilers already exist. Where high pressure boilers exist, analyze turbine driven equipment vs. absorption.)

17. For expansion or modification, consider the use of standby diesel engines and generators, or new diesel engine-driven refrigeration equipment to reduce the electric loads; and consider using waste heat from the engines for other useful work.

18. Fit capacity controls to existing refrigeration equipment where possible, and when replacing units, purchase those with four steps of capacity control.

19. When replacing existing absorption units, analyze energy required for new units vs. electric centrifugal compressors as well as steam turbine driven and engine driven, and select the system requiring the least annual energy usage.

20. Install seven-day timers to program shutdown of refrigeration equipment, chilled water and condenser pumps, and cooling tower fans for unoccupied periods during the week.

21. Convert straight compressors and chiller systems to heat pump operation by adding additional condensers and evaporators and/or repiping as required.

22. Refer to Section 22 "Solar Energy" for guidelines on cooling with solar heat as an energy source.

This page blank

Section 16

Energy Conservation Opportunities Commercial Refrigeration

COMMERCIAL REFRIGERATION

ENERGY CONSERVATION OPPORTUNITIES

ECO 92 REPLACE COLD CABINET INTERNAL LIGHTS BY EXTERNAL LIGHTS

Cold cabinet internal lights heat the cold air and increase the load on the refrigerating machine. In many cases, the internal lights can be removed and the contents of the cabinet lit either from repositioned existing fixtures or from new fixtures positioned to shine into the cabinet.

The reduction in load is in direct proportion to the wattage of the lamp removed, and is affected by type (fluorescent lamps give off 65% of their rating as heat, incandescent lamps give off 90% of their rating as heat). Each KW of "lighting heat" removed will reduce the refrigeration load by 0.28 tons.

Some reduction in display effectiveness will result when internal lights are removed and the trade off between this and the energy saved should be considered in stores and supermarkets.

The cost of removing internal lights and replacing them with external lights will vary with the circumstances for each case,and quotations should be obtained from local contractors. Order of magnitude costs, however, may be derived from the appropriate lighting sections of this Manual.

ECO 93 PROVIDE NIGHT COVERS FOR OPEN COLD CABINETS

Some existing cold cabinets and deep freeze chests in stores and supermarkets, because of their design, have to be open to allow visual display and public access to the produce. Heat exchange takes place at the interface between the warm room air and the cold freezer air due to mixing and cold air spill. While this heat exchange or gain must be tolerated during occupied hours, access and display are not required in unoccupied hours, and the cabinet can be modified to close by installing custom-made thermally insulated covers.

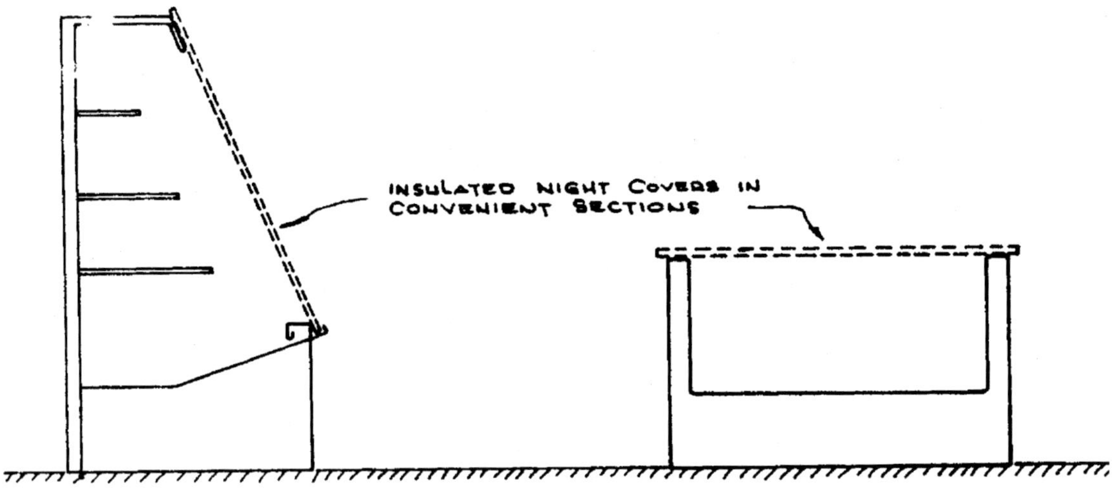

COLD CABINETS WITH NIGHT COVERS

Construct night covers in easily handled sections with sufficient thermal insulation to prevent condensation.

Institute a regular routine to cover all cold cabinets and freezers at night and during weekends both summer and winter.

The cost of night covers will vary with the configuration of cabinets. Obtain estimates from local contractors or cabinet manufacturers.

The energy savings obtained by fitting night covers will vary with each type of cabinet and will be greatest on vertical display cabinets where the cold air tends to spill over the bottom lip and spread over the floor. Subjective judgments can be made by observing near floor temperatures adjacent to the cold cabinets or by using a smoke tracer to show the magnitude of air spill.

When assessing the energy savings, bear in mind that the heat transfer into the cabinets results in a reduction of building space temperature, thus increasing the building heating load.

ECO 94 REPLACE EXISTING OPEN COLD CABINETS

Vertical display type cold cabinets and freezers are available as totally closed units with self-closing glass doors to provide visibility and access. These units have considerably lower heat gains than the open types, but still provide acceptable access and display.

Their initial cost does not warrant replacing new or almost new open types but where existing equipment is at or near the end of its useful life or where remodeling is contemplated, closed cabinets should be used in preference to open. Select new equipment on the basis of efficiency and seasonal COP.

ECO 95 MOVE CONDENSERS CLOSE TO COMPRESSORS

The resistance to refrigerant flow in long pipe runs between the compressor and condenser raises the head pressure and temperature. This results in less refrigeration output, increased power input, and longer cycles of operation to meet a given load.

To increase the efficiency of the refrigerating machines, relocate air-cooled condensers to minimize length of pipe runs. Check that oil return will not be impaired.

ECO 96 RELOCATE CONDENSERS TO RECAPTURE WASTE HEAT

In winter the hot air from air-cooled condensers can be usefully employed to meet part of the heating load.

In conjunction with ECO # 95 relocate air-cooled condensers so that hot air off the coil can be ducted into the building. Depending on the configuration, the existing condenser fan may have sufficient capacity reserve to provide adequate air flow through the new duct system or an additional fan may be required. If possible, arrange duct and dampers so that cold exhaust air from the building can be directed through the condenser coil in summer.

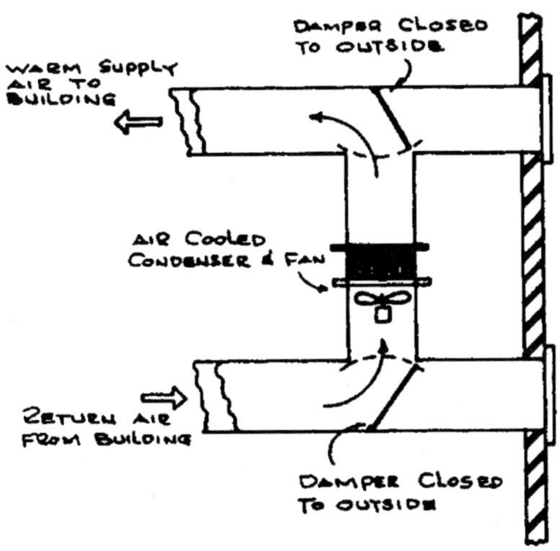

WINTER OPERATION

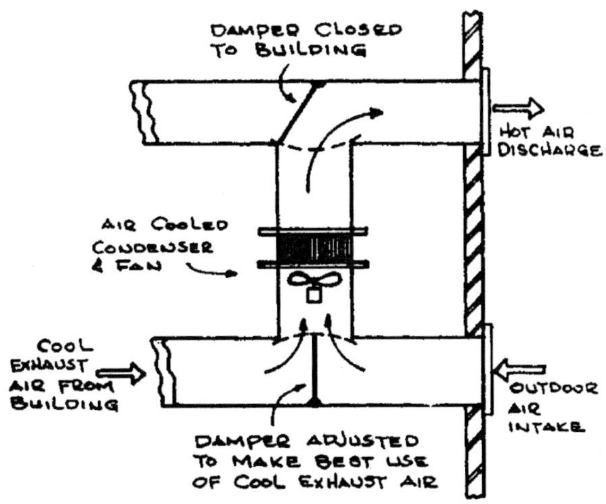

SUMMER OPERATION

ECO 97 <u>REPLACE AIR-COOLED CONDENSERS BY WATER-COOLED TYPE</u>

Refer to ECO 90 for details of installation, costs and energy savings.

GUIDELINES TO REDUCE ENERGY USED FOR COMMERCIAL REFRIGERATION SYSTEMS

Refer to ECM1, Section 8, "Commercial Refrigeration Systems" for specific measures which apply, and to all other Sections in both Manuals which address compressors and condensers.

Analyze and consider the following measures to conserve additional energy:

a) Convert air-cooled compressor and condenser installations to water-cooled systems when ambient wet bulb temperatures on a seasonal basis are favorable.

b) Install a heat reclamation system to salvage heat rejected from condensers for use in space heating.

c) When replacing frozen food cases, refrigerators and cold display cabinets, select equipment based on cooling efficiency rather than first cost.

d) When replacing refrigeration equipment, select equipment to give the highest seasonal COP.

e) Replace cold cabinet internal lights with external lights.

f) Provide night covers for open cold cabinets

g) Use closed cabinets when replacing existing open type cold cabinets.

h) Move condensers close to compressors to minimize length of pipe runs.

i) Relocate condensers to utilize hot air off the coil.

Section 17

Energy Conservation Opportunities Distribution and HVAC Systems

DISTRIBUTION & HVAC SYSTEMS

BACKGROUND

Energy transport losses are inherent in the design of any distribution system but can be minimized by reductions in the flow rate through the system and potential resistance to flow of the system.

The options for reducing distribution losses are discussed in the section. Before applying the relevant ECO's it is important to recognise that air and water system flow rates are a function of building loads. Minimizing building loads will allow flow rates to be reduced accordingly and to avoid repetitive actions every effort should first be made to reduce building cooling and heating loads.

This energy required to distribute air and water does not help to reduce the building load; it is actually counter-productive in cooling, as the heat equivalent of the power input must be removed by the refrigeration equipment.

The characteristic of any given system is pre-determined by the length and size of pipes or ducts and size and shape of fittings (bends,tees). The resistance to flow is a function of velocity (or flow rate) and the system characteristic. With any change in flow rate, the system resistance will change (according to the laws of fluid flow) and the operating point will move accordingly on the system characteristic curve.

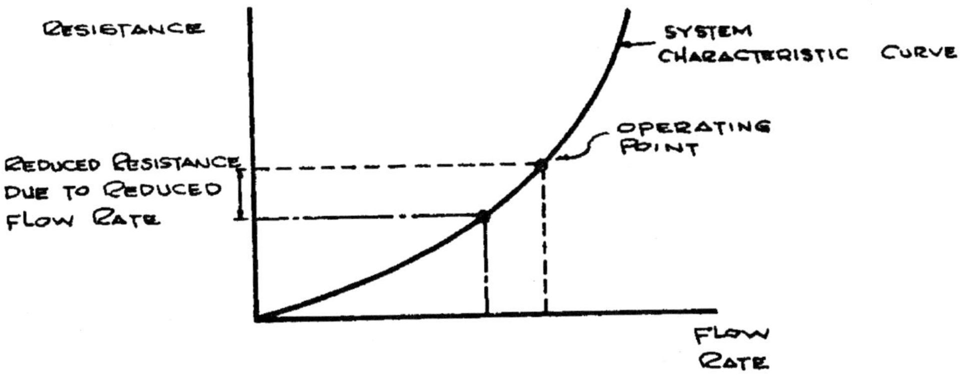

For instance, a piping system with a flow rate of 2,000 gpm at 150 ft. resistance changed to 1,800 gpm will have a new resistance of:

$$150 \times \left[\frac{1800}{2000}\right]^2 = 121.5 \text{ ft.}$$

If it is not possible to reduce the flow rate, or if it has been decreased to the maximum extent possible as a result of other ECO's, the remaining option to reduce power required for a given flow rate is to alter the system characteristic curve.

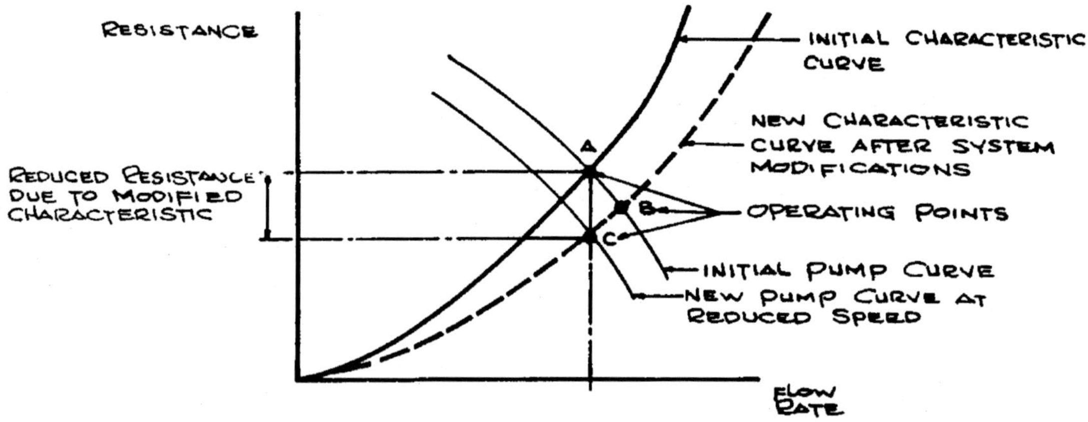

As shown in the diagram, modification of a piping system characteristic will change the operating point from A to B along the pump curve and result in an increased flow rate at a lower resistance. To reduce the flow rate back to its initial point,it is now necessary to reduce the pump speed so that the new pump curve intersects the new characteristic curve at point C giving the desired flow rate. These principles apply equally to fan/duct systems and pump/pipe systems.

When changing flow rates and system resistance, implement adjustments and modifications in the following logical sequence:

a. Measure initial flow rates and resistance and construct the system characteristic curve.

b. Reduce loads and calculate the new reduced flow rate required to meet them.

c. Modify system to reduce resistance to flow and from measurements construct the new system characteristic curve.

d. Determine from the new characteristic curve the resistance at the new reduced flow rate.

e. Reduce the fan or pump speed or reduce the pump impeller diameter so that the new fan or pump curve crosses the new system characteristic curve at the desired operating point.

ENERGY CONSERVATION OPPORTUNITIES

ECO 98 REDUCE RESISTANCE TO FLOW IN PIPING SYSTEMS

The resistance to flow of any given piping system is the sum of the resistances of all its individual parts in the index circuit. Some of the parts cannot be easily modified, but others are candidates for reducing resistance.

Heat exchangers have a high resistance to flow and are prone to fouling by scale deposits and dirt which further increases their resistance. Disassemble heat exchangers, remove scale mechanically or chemically and flush out the tubes and shell. Institute a maintenance and water treatment program for heat exchangers based on regular observations of pressure drop and temperature differentials.

Over long periods of time, formation of scale deposits which occur throughout the system may result in radically increased resistance. Determine, by inspection, whether such scaling exists and rectify it by chemical cleaning. There are many specialized contractors who have the expertise for this work.

Replace high resistance elements of filters and strainers with low resistance elements or baskets.

Pumps must develop sufficient head to overcome the resistance to flow through the longest or index circuit even though this head may exceed the requirements of the other subcircuits. If the index circuit resistance grossly exceeds resistances of the other circuits and cannot be further reduced, consider the installation of a small booster pump for the index circuit only and reduce the head of the main pump accordingly.

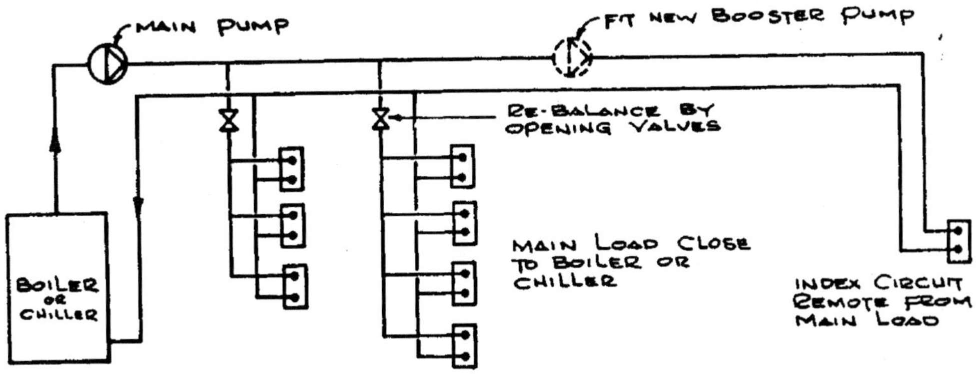

When existing piping systems were first operated, the installing contractor usually balanced flows by trial and error. He often closes balancing valves to a greater extent than is needed, imposing extra head on the pump. Designed safety margins often result in oversized pumps and the excess head is absorbed by closing down the valve on the pump discharge. To reduce resistance to flow, rebalance the system by first opening fully the balancing valve on the index circuit and the pump(s) valve or removing any orifice plates from the pump circuit. Then adjust each circuit balancing valve to achieve proportional flow rates starting with the next longest circuit and progressing to the shortest circuit. This process is trial and error - each valve adjustment will affect flow rates in circuits already adjusted but 2 or 3 successive adjustments of the whole system will give good balance. When all available options to reduce system resistance have been exercised, measure or calculate the new system characteristic and determine the new operating point taking into account the previous steps to reduce the required flow rate.

ECO 99 REDUCE WATER FLOW RATES

When the heating or cooling load is actually less than original design or is further reduced, the flow rate through the hot water or chilled water systems may be reduced in proportion, provided that the supply temperature set point is retained at its initial level.

Reduce the pump speed to decrease the water flow rate through a piping system. Derive the appropriate reduction in pump speed to achieve any given reduction in flow rate from the pump curve (obtain from manufacturer). The reduction in pump speed will be approximately in proportion to the change of flow rate.

To reduce the speed of indirect drive pumps, change the size of the motor sheave. To reduce the speed of direct drive pumps, exchange the drive motor for one of lower speed. (This is only possible within the range of commercially available motor speeds and a compromise may be necessary).

If it is not possible or economically feasible to change the speed of a pump, reduce the flow rate by changing the impeller size. Substitute one of smaller diameter, or alternatively, skim down the existing impeller. Seek manufacturers recommendations for each specific application.

Reducing the water flow rate through a piping system will also reduce the resistance in accordance with the laws of fluid flow. To determine the new system resistance, use the following formula:

$$\text{New system resistance} = \text{Initial system resistance} \times \left(\frac{\text{New flow rate}}{\text{Initial flow rate}}\right)^2$$

Any steps taken under ECO 98 to reduce system resistance should be accounted for at this point.

To determine energy consumed by pumps, refer to Figures #61 and 62. Select the appropriate graph and enter at the initial flow rate. Following the direction of the example line, intersect with the appropriate points for initial system resistance and hours of operation.

Read out the yearly energy used in Btu/year. Repeat this procedure using the new flow rate and new system resistance. Subtract the yearly energy used at the reduced flow rate from the yearly energy used at the initial flow rate. To convert Btu/year to KWH/year, divide by 3,413 to obtain yearly savings.

The cost of achieving these savings will be the sum of the costs for the individual steps taken. The cost of fitting new pump sheaves will vary with each individual case but, for order of magnitude, will be approximately \$100.00 . The cost of replacing the impeller or skimming it down in diameter should be obtained from the pump manufacturer.

FIGURE 61: YEARLY ENERGY CONSUMED-CENTRIFUGAL PUMPS UP TO 500 GALLONS PER MINUTE

dubin-mindell-bloome-associates consulting engineers

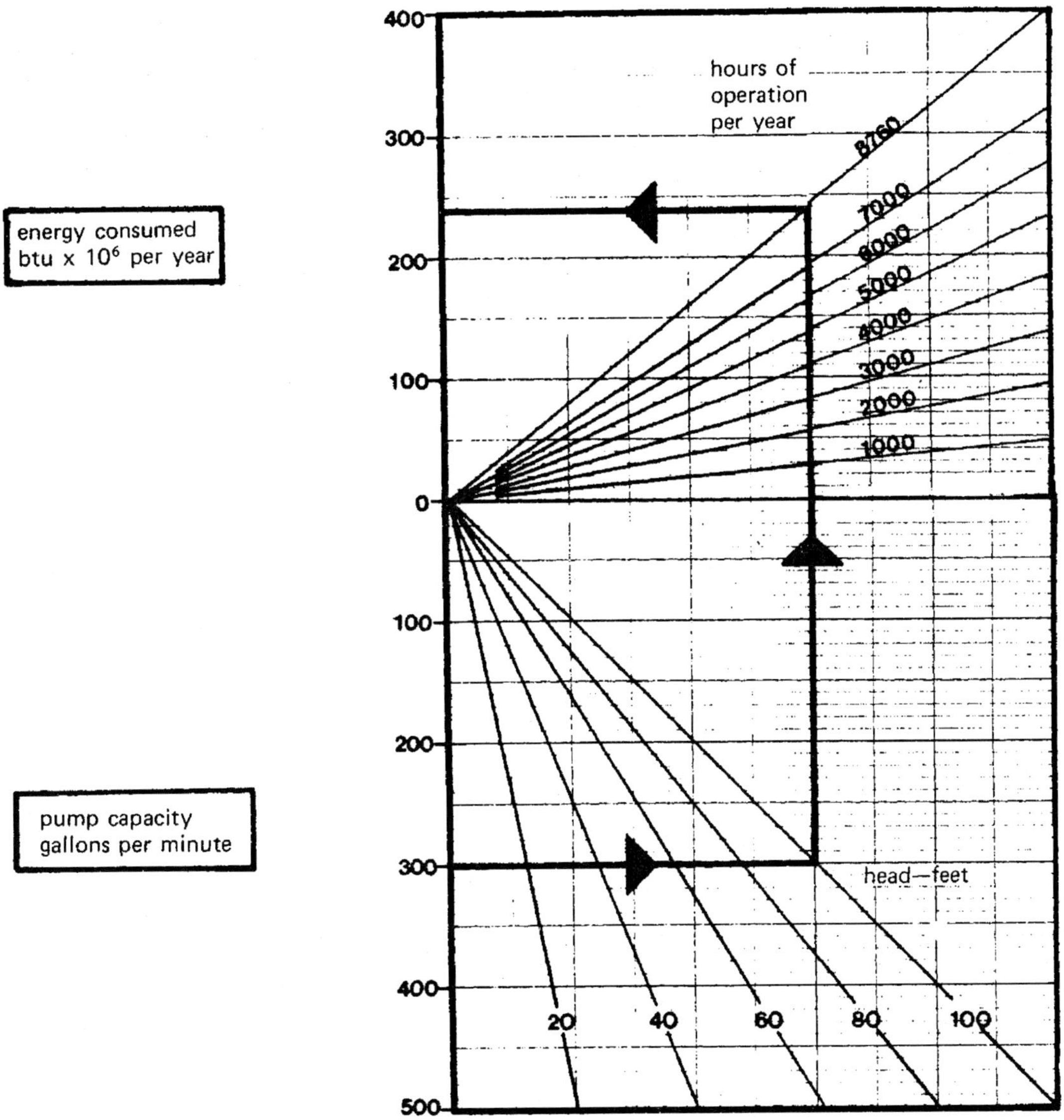

Figure #61 Engineering Data

This figure is based on the Standard Pump Formula:

$$\text{Brake Horsepower} = \frac{\text{lbs/min} \times \text{Feet head}}{33{,}000 \text{ ft/lbs} \times \text{Pump efficiency}}$$

Pump efficiency was assumed to be an average of 70%. Brake horsepower was converted to Btu's by the factor 2544.43 Btu/Horsepower Hour.

The upper half of the graph proportions the hours of pump operation per year from 100% (8760 hr.), down to 11% (1,000 hr).

FIGURE 62: YEARLY ENERGY CONSUMED-CENTRIFUGAL PUMPS UP TO 2,000 GALLONS PER MINUTE

dubin-mindell-bloome-associates consulting engineers

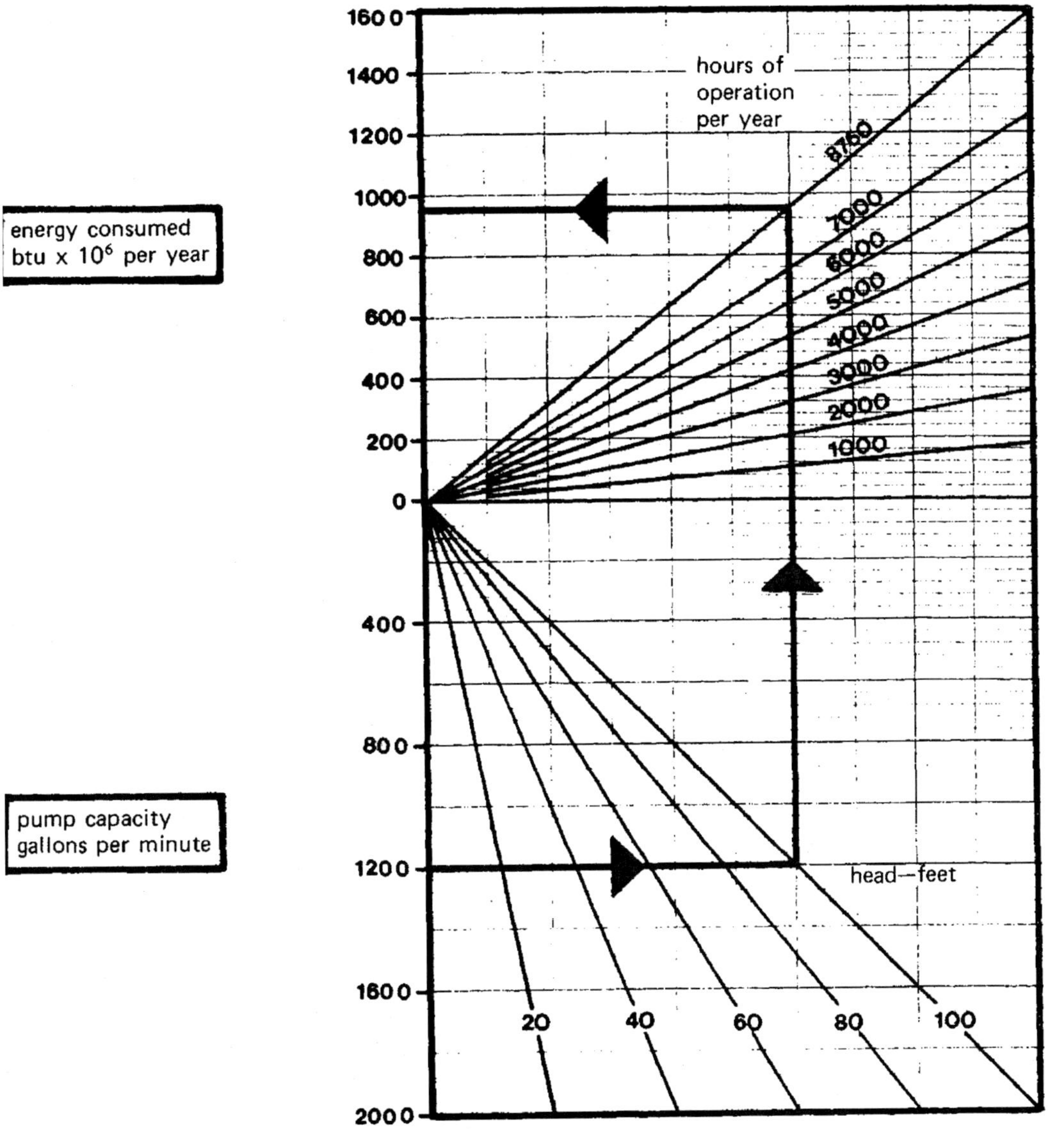

Figure #62 Engineering Data

This figure is based on the Standard Pump Formula:

$$\text{Brake Horsepower} = \frac{\text{lbs/min} \times \text{Feet head}}{\text{33,000 ft/lbs} \times \text{Pump efficiency}}$$

Pump efficiency was assumed to be an average of 70%. Brake horsepower was converted to Btu's by the factor 3544.43 Btu/Horsepower Hour.

The upper half of the graph proportions the hours of pump operation per year from 100% (8760 hr.), down to 11% (1,000 hr).

ECO 100 REDUCE RESISTANCE TO AIR FLOW IN DUCT SYSTEMS

The total resistance to air flow in a duct system is the sum of the resistances of the individual parts of the index circuit. Although straight lengths of duct generally offers little or no opportunity for resistance reductions, the potential for other items is more substantial.

The resistance to air flow through filters is a function of filter construction, type of media, area of media/unit volume (proportional to actual velocity through the media - not to be confused with face velocity) and the dirt load at any given time. In general, resistance to air flow increases with filter efficiency although there are some high efficiency filters which also have low resistance. (See Figure 63) To determine the "dirty" resistance limit, consult the manufacturers of the particular filter. Install a manometer across each filter bank to indicate when the filters should be changed. If the building contains many air systems, consider the use of an alarm system to report filter conditions at a central point. (See Section 21 Computer Control Systems.)

Many filters impose high resistance due to their inherent characteristics and/or because they exceed required filtration standards. Figure #63 tabulates 25 different filter types in common use, showing installation and operating costs. Determine your filtration requirements in terms of NBS Atmospheric Dust Spot Efficiency taking into account the ambient air conditions and space air quality requirements. Except for special applications such as computer rooms and food service, a 50% NBS dust rating is generally adequate. Then check present filter installation and determine whether an alternative type of filter will meet building needs at operating costs sufficiently reduced to provide an economic payback.

Resistance to air flow through coils depends largely upon the number of rows or depth required for adequate heat transfer. Cooling coils are usually many rows deep because of the small temperature differences between coil and air, and, consequently, offer a high resistance to air flow. Heating coils which have a higher temperature difference are usually fewer rows deep than cooling coils and impose less resistance to air flow.

Both coils are in the air stream and impose combined resistance year round, even though only one coil may be used at any given time. Where the cooling medium is chilled water and the heating medium is hot water, it may be possible to remove the heating coil and repipe the cooling coil to provide both heating and cooling (though not simultaneously). Eliminating the heating coil will reduce the system resistance and allow savings in fan horsepower to be achieved.

A side benefit of using the cooling coil for both heating and cooling is that the extra heat transfer surface of the cooling coil provides opportunities to lower the hot water temperatures and use low grade waste heat.

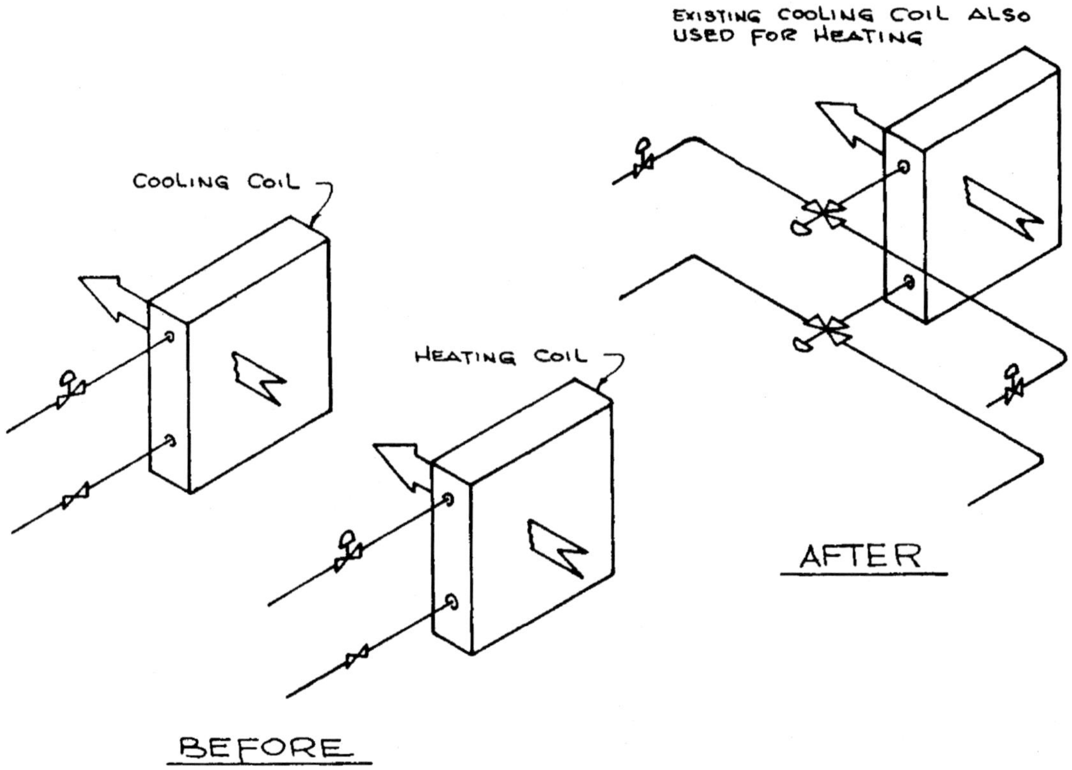

Reduce high resistance to air flow caused by duct fittings on the inlet and discharge side of fans by modifying the shape of the fitting. (See ASHRAE "Equipment Guide" for recommendations.) Where space permits modify abrupt changes in sections of duct work where velocities exceed 2000 ft./minute to long taper fittings and install turning vanes in square bends.

Supply and exhaust fans must operate at a pressure sufficient to overcome the resistance to air flow through the ducts serving the outlet farthest from the fan (index outlet). Where the index outlet is remote from other parts of the system and is served by a long run of duct, it imposes a power consumption penalty on the whole of the system. It is often worthwhile to replace this section of duct with one of larger cross section and lower resistance.

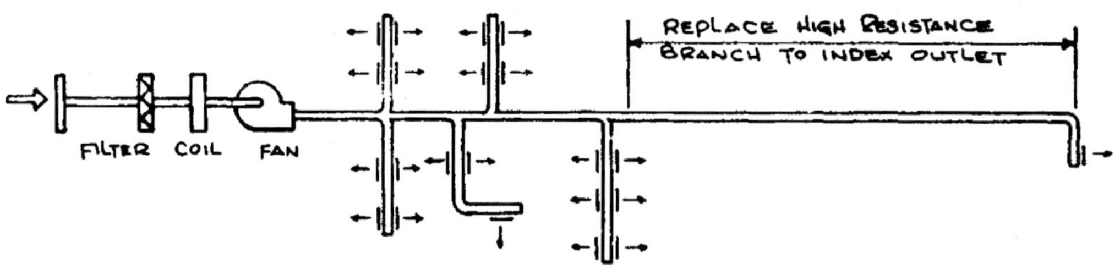

FIGURE 63: AIR FILTER TYPE COMPARISON

Filter Type	1. Cleanable 2" Thick (High Vel.)	2. Throw-away 2" Thick	3. Automatic Roll Type	4. Pleated Filter 2" Thick
Size of Filter Bank				
No. High x No. Wide	2 x 3	2 x 5	-	2 x 4
Height x Width	4' x 6'	4' x 10'	5'4" x 5'	4' x 8'
Average Efficiency				
NBS Atmospheric Dust Spot	8-10%	10-15%	20-25%	36%
Other Methods	-	-	-	-
Filter Life (Hours)				
Prefilter	600	480	3750	2000
Afterfilter	-	-	-	-
Pressure Drop (in. W.G.)				
Initial	0.08	0.10	0.40	0.14
Final	0.40	0.30	0.40	0.60
Average	0.24	0.20	0.40	0.37
Initial Installation[1]				
Equipment	$328	$107	$ 780	$124
Installation	78	91	91	91
Electric Wiring	-	-	215	-
	$406	$198	$1086	$215
Cost of One Set-Replacement Media				
Prefilter[2]	-	$ 21	$ 87	$ 49
Afterfilter[2]	-	-	-	-
Replacement Time - Man Hours				
Prefilter	4	2	2	3
Afterfilter	-	-	-	-
Annual Operating Cost				
Material	-	$381	$203	$216
Labor @ $7/hr.	$533	333	43	120
Electric Power[3,7]	126	104	210	194
Miscellaneous	-	-	-	-
Total[4]	$659	$818	$456	$530

FIGURE 63: (continued)

Filter Type	5. Bag Type Cartridge	6. Bag Type Cartridge	7. Bag Type Cartridge	8. Bag Type Cartridge w/ Prefilter
Size of Filter Bank				
No. High x No. Wide	2 x 2	2 x 2	2 x 3	2 x 3
Height & Width	4' x 4'	4' x 4'	4' x 6'	4' x 6'
Average Efficiency				
NBS Atmospheric				
Dust Spot	38-40%	45%	50-55%	50-55%
Other Methods	-	-	-	-
Filter Life (Hours)				
Prefilter	6600	2920	5500	-
Afterfilter	-	-	-	9000
Pressure Drop (in.W.G.)				
Initial	0.30	0.25	0.30	0.30
Final	0.75	1.00	0.80	0.80
Average	0.525	0.625	0.55	0.55
Initial Installation				
Equipment	$113	$182	$351	$351
Installation	78	78	78	78
Electric Wiring	-	-	-	-
	$191	$260	$429	$429
Cost of One Set-Replacement Media				
Prefilter	-	-	$306	-
Afterfilter	-	$ 46	-	$306
Replacement Time - Man Hours				
Prefilter	4	2	3	-
Afterfilter	-	-	-	3
Annual Operating Cost				
Material	$ 83	$137	$486	$298
Labor @ $7/hr.	25	55	44	26
Electric Power [3,7]	276	336	288	288
Miscellaneous	-	-	-	-
Total	$384	$528	$818	$612

FIGURE 63: (continued)

Filter Type	9. Auto-Roll Bag Type Afterfilter	10. Rigid Cartridge	11. Rigid Cartridge w/Prefilter	12. Bag Type Cartridge
Size of Filter Bank				
No. High x No. Wide	-	2 x 3	2 x 3	2 x 3
Height & Width	4'8" x 6'8"	4' x 6'	4' x 6'	4' x 6'
Average Efficiency				
NBS Atmospheric Dust Spot	50-55%	55-60%	55-60%	80-85%
Other Methods	-	-	-	-
Filter Life (Hours)				
Prefilter	3,750	3,500	-	-
Afterfilter	10,000	-	4,500	4,000
Pressure Drop (in.W.G.)				
Initial	0.70	0.20	0.20	0.37
Final	1.25	1.00	1.00	1.00
Average	0.975	0.60	0.60	0.685
Initial Installation				
Equipment	$1,175	$332	$332	$378
Installation	169	78	78	78
Electric Wiring	215	-	-	-
	$1,559	$410	$410	$456
Cost of One Set-Replacement Media				
Prefilter	$ 72	$278	-	-
Afterfilter	306	-	$278	$337
Replacement Time-Man Hours				
Prefilter	2	3	-	-
Afterfilter	3	-	3	3
Annual Operating Cost				
Material	$ 434	$ 697	$546	$ 737
Labor @ $7/hr.	68	69	53	60
Electric Power	512	316	316	360
Miscellaneous	-	-	-	-
	$1,014	$1,082	$915	$1,157

FIGURE 63: (continued)

Filter Type	13. Bag Type Cartridge w/Prefilter	14. Automatic Roll with Bag Type Afterfilter	15. Rigid Type Cartridge	16. Rigid Type Cartridge with Prefilter
Size of Filter Bank				
No. High x No. Wide	2 x 3	-	2 x 3	2 x 3
Height x Width	4' x 6'	4'8" x 6'8"	4' x 6'	4' x 6'
Average Efficiency				
NBS Atmospheric	80-85%	80-85%	85-90%	85-90%
Dust Spot	-	-	-	-
Other Methods				
Filter Life (Hours)				
Prefilter	-	3,750	-	-
Afterfilter	5,500	6,000	2,500	3,500
Pressure Drop (in. W.G.)				
Initial	0.37	0.77	0.30	0.30
Final	1.00	1.50	1.00	1.00
Average	0.685	1.135	0.65	0.65
Initial Installation				
Equipment	$378	$1,203	$367	$367
Installation	78	169	78	78
Electric Wiring	-	215	-	-
	$456	$1,587	$445	$445
Cost of One Set - Replacement Media				
Prefilter	-	$ 72	-	-
Afterfilter	$337	337	$332	$332
Replacement Time - Man Hours				
Prefilter	-	2	-	-
Afterfilter	3	3	3	3
Annual Operating Cost				
Material	$536	$ 658	$1,161	$ 829
Labor @ $7/hr.	44	83	96	69
Electric Power	360	596	342	342
Miscellaneous	-	-	-	-
Total	$940	$1,337	$1,599	$1,240

FIGURE 63: (continued)

Filter Type	17. Electrostatic Auto-Roll Afterfilter	18. Bag Type Cartridge	19. Bag Type Cartridge w/Prefilter	20. Electrostatic Bag Type Afterfilter
Size of Filter Bank				
No. High x No. Wide	-	2 x 3	2 x 3	-
Height x Width	4'8" x 8'8"	4' x 6'	4' x 6'	4'8" x 6'8"
Average Efficiency				
NBS Atmospheric Dust Spot	90%	93-97%	93-97%	93-97%
Other Methods	-	-	-	-
Filter Life (Hours)				
Prefilter	2,500	-	-	2,500[6]
Afterfilter	4,000	2,500	3,500	12,000
Pressure Drop (in. W.G.)				
Initial	0.45	0.48	0.48	0.55
Final	0.45	1.00	1.00	1.00
Average	0.45	0.74	0.74	0.775
Initial Installation				
Equipment	$2,990	$437	$437	$2,730
Installation	169	78	78	156
Electric Wiring	397	-	-	215
	$3,556	$515	$515	$3,101
Cost of One Set - Replacment Media				
Prefilter	2	-	-	2
Afterfilter	2	3	3	3
Annual Operating Cost				
Material	$157	$1,411	$1,008	$ 294
Labor @ $7/hr.	104	96	69	83
Electric Power	236	388	388	408
Miscellaneous	-	-	-	-
	$497	$1,895	$1,465	$ 785

FIGURE 63: (continued)

Filter Type	21. HEPA without Prefilter	22. HEPA with 85% Prefilter	23. Activated Charcoal
Size of Filter Bank			
No. High x No. Wide	2 x 5	2 x 5	1 x 5
Height x Width	4'0" x 10"	4'0" x 10'0"	2'0" x 10'0"
Average Efficiency			
NBS Atmospheric			
Dust Spot	100%	100%	-
Other Methods	99.97%DOP	99.97%DOP	-
Filter Life (Hours)			
Prefilter	-	-	-
Afterfilter	6,000	24,000	8,760
Pressure Drop (in.W.G.)			
Initial	1.00	1.00	0.35
Final	2.00	2.00	0.35
Average	1.50	1.50	0.35
Initial Installation			
Equipment	$1,300	$1,300	$2,600
Installation	390	390	975
Electric Wiring	-	-	-
	$1,690	$1,690	$3,575
Cost of One Set - Replacement Media			
Prefilter	-	-	-
Afterfilter	$ 975	$ 975	$ 598
Annual Operating Cost			
Material	$1,424	$ 356	$ 598
Labor @ $7/hr.	79	20	109
Electric Power	788	788	184
Miscellaneous	-	-	-
	$2,291	$1,164	$ 891

FIGURE 63: (continued)

Filter Type	24. 95% DOP Rated Filter w/Prefilter	25. 95% DOP Rated Filter w/85% Prefilter
Size of Filter Bank		
No. High x No. Wide	2 x 5	2 x 5
Height x Width	4'0" x 10'0"	4'0" x 10'0"
Average Efficiency		
NBS Atmospheric		
Dust Spot	99%	99%
Other Methods	95% DOP	95% DOP
Filter Life (Hours)		
Prefilter	-	-
Afterfilter	6,000	24,000
Pressure Drop (in.W.G.)		
Initial	0.50	0.50
Final	1.00	1.00
Average	0.75	0.75
Initial Installation		
Equipment	$1,300	$1,300
Installation	390	390
Electric Wiring	-	-
	$1,690	$1,690
Cost of One Set - Replacement Media		
Prefilter	-	-
Afterfilter	$ 975	$ 975
Replacement Time - Man Hours		
Prefilter	-	-
Afterfilter	6	6
Annual Operating Cost		
Material	$1,424	$ 356
Labor @ $7/hr.	79	20
Electric Power	394	394
Miscellaneous	-	-
	$1,897	$ 770

FOOTNOTES

1. Spare set of replacement cells was included in installed cost of cleanable filters.
2. Allowance for replacement filter price increase averaging 15% over 20 years was included.
3. Fan power was figured at $0.03 per KWH and 59% * overall efficiency.
4. Operating costs based on 8760 hours of operation per year.
5. Prefilter costs are not included in columns 8, 11, 19, 21, 22, 24, 25. Add costs from columns 1, 2, 3, 4, or 5 to arrive at the total cost.
6. Costs are for 10,000 cfm system. Cost per cfm will be slightly lower for smaller systems and higher for larger systems
7. Power absorbed by filter to overcome its resistance to air flow.

* Fan efficiency typically varies from 50 - 70%

To achieve correct volumes at each grille or register, adjust dampers in low resistance branches until all branches are of resistance equal to the index run. When systems are initially started, all dampers are often closed more than they need be, adding unnecessary resistance.

It is also common practice to reduce fan volume by closing down dampers on the fan inlet or outlet rather than by reducing fan speed.

Check whether these situations occur in the building and if necessary, rebalance the system by first opening fully any dampers which are not used to proportion air flow between branches or outlets (main dampers close to fan). Identify the index outlet and fully open any dampers between this outlet and the fan. Measure the volume at the index outlet and adjust branch dampers successively (starting with the next longest run) until proportional volumes are achieved at each outlet. This process is trial and error as each damper adjustment will affect flow rates in branches already adjusted, but 2 or 3 successive adjustments of the whole system will give good balance.

When all available options to reduce system resistance have been exercised, measure or calculate the new system characteristic and determine the new operating condition on the fan curve, taking into account the previous steps to reduce the required flow rate.

ECO 101 REDUCE AIR FLOW RATES

When the heating load and/or the sensible room cooling load are reduced, the quantity of air can also be reduced proportionately providing the set point temperature is retained at its initial level.

To reduce the volume of an air system and conserve energy, reduce the fan speed in proportion to the load change.

To reduce the speed of belt driven fans, change the size of the motor sheave. It may be necessary also to change the drive belts if adjustment cannot take up the slack.

To reduce the speed of direct driven fans, change the drive motor for a lower speed type. This may only be possible if the desired fan speed falls within the motor speed ranges available or if a compromise speed is acceptable.

Where the air system both heats in winter and cools in summer, it is unlikely that one flow rate is ideal for both seasons. The conflict between heating and cooling requirements can be resolved by changing pulley sizes for each season, fitting a two-speed motor, or by adopting a compromise air volume for both seasons and adjusting the supply air temperature to suit.

Reducing the supply air flow rate through the system will also reduce the resistance to flow. Reduction in resistance will follow the laws of fluid flow, and the new system resistance can be determined by using the following formula:

$$\text{New system resistance} = \text{Initial system resistance} \times \left(\frac{\text{New flow rate}}{\text{Initial flow rate}}\right)^2$$

ECO's adopted to reduce system resistance should also be included in "initial system resistance".

Select the appropriate graph from Figures #23 and 24 to determine energy consumed by fans.

Enter the graph at the initial system volume and following the direction of the example line intersect with the appropriate points for initial system resistance and hours of operation/week. Read out the yearly energy used in Btu/year. Repeat the procedure using the new flow rate and new system resistance.

To determine the yearly energy saving, subtract the yearly energy used by the modified system from the yearly energy used by the initial system, then convert this answer from Btu/year to KWH/year by dividing by 3,413.

The total cost of achieving savings by reducing system resistance is the sum of the costs of all the individual steps taken. The cost of reducing fan speed will vary with each individual case, but for order of magnitude, the cost of replacing a motor sheave will be approximately $ 100.00 . Refer to "Electric Power" Section 6H for the cost of replacing motors on direct drive fans.

When a fan is reduced in speed, its power requirements are also reduced,

$$\text{hp} = \left(\frac{\text{rpm } 1}{\text{rpm } 2}\right)^3$$

and it is possible that the drive motor will now far exceed the new requirements. Motors operated far below rated output are inefficient. Of equal significance, its power factor will be drastically reduced affecting the power factor of the entire electrical system. Depending on the electrical rate structure, this reduced power factor could entail extra charges. For guidelines on when and where to change the electric motors, see the "Power" Section 20.

GUIDELINES TO REDUCE ENERGY USED FOR DISTRIBUTION AND HVAC SYSTEMS

Refer to ECM-1, Section 7 "HVAC Systems and Distribution" for guidelines to reduce energy consumption with minimal investment.

Analyze and consider the following options for further reducing energy consumption:

1. Reduce the energy consumption for fans

   a) Reduce the system resistance and then reduce fan speed to meet both reduced load requirements and reduced air resistance.

      -Replace air filters with new filters which offer less resistance to air flow with filtration efficiency matched to requirements.

      -Eliminate preheat coils whenever possible without increasing freezeup possibilities.

      -Remove unnecessary dampers, high resistance elbows and fittings and sections of duct work with high resistance to lower the total system resistance.

      -Adjust variable diameter motor sheave.

      -Change motor sheave if adjustment is not possible.

      -Change motor if loading is less than 40% of the rated output.

   b) Rebalance the entire supply and return air distribution systems. Blank off unused portions of exhaust hoods in kitchens and cafeterias to reduce air volume and horsepower where direct driven fans are used and it is difficult or uneconomic to change fan speeds or motor.

   c) In large buildings where steam is available, consider turbine driven fans with variable speed control.

   d) In high bay buildings, lower the levels of supply diffusers to decrease the operating time of fans and equipment.

   e) Determine the trade off between raising supply air temperatures during the cooling season to reduce refrigeration power vs. lowering the temperature to decrease the volume of air and reduce fan horsepower. Take into account additional heat gain by lowering the supply air temperature. Make similar analysis for heating season operation of lowering the supply air temperature vs. longer operation of boiler or furnace equipment.

   f) Reduce air volume and horsepower with existing variable air volume systems or other systems converted to variable volume. Refer to "Variable Volume" guidelines below.

g) Install a control to reducė total air volume during the heating season when the same system is used for both heating and cooling. During the cooling cycle, more air is required because of lower temperature differentials.

h) Install gauges to measure the resistance across filters and across coils to indicate when cleaning is necessary.

2. Reduce energy consumption for chilled water, hot water, and condenser water pumps

a) For small systems, under 10 h.p., throttle the pump discharge to supply only the amount of water required to serve the load.

b) Reduce energy by pump and/or motor modifications:

-Open all throttling valves and remove high resistance fittings or pipe sections to reduce frictional resistance; then adjust the speed of variable speed motors to a lower speed.

-Change the speed of rotation of pumps by substituting lower horsepower and lower speed motors where the drive permits.

-Remove impeller and replace with one of smaller diameter.

-Add new smaller pump in parallel with the existing pump and provide controls to sequence the operation of both pumps, so that the smaller pump can handle the required volume of water at part loads.

-In new installations, install modular pumps sized for 1/3 and 2/3 of the load where the system permits reduction in water volume. The larger pump could be of variable volume to give complete modulation over the entire range of loads.

3. Reduce the energy consumption of steam distribution systems

a) Check all steam traps and replace those that are passing steam together with condensate.

b) Reduce line losses by installing pressure reducing valves in steam lines serving loads which do not require the same pressure as other loads.

c) Install condensate return lines from equipment where condensate is now discharged to waste.

d) In new steam turbine systems, operate at the highest possible pressure with condensing turbines, but analyze the heat balance and extract steam to operate equipment at lower pressures when feasible.

e) Use condensate to preheat domestic hot water or for other thermal use especially when steam is supplied from district systems.

DISCUSSION OF HVAC SYSTEMS

Each HVAC system is of course unique and its particular characteristics can only be identified by inspection and measurement. It will, however, have many characteristics which are common to all systems of its generic group and it is the common characteristics which are discussed in this section.

The minimum information required to understand the present operation of a system and to provide a basis for deciding which modifications are likely to prove beneficial is tabulated below. Measurement should be as nearly simultaneous as possible.

Air Flow Rates	- total outdoor air total return air total supply air trunk ducts terminal units air cooled condenser
Water Flow Rates	- through boilers through chillers cooling towers heat exchangers coils and terminal units
Temperatures	- outdoor air DB & WB return air DB & WB mixed air entering coils, DB & WB supply air leaving coils, DB & WB hot deck cold deck air at terminals conditioned areas DB & WB (typical for each functional use) boiler supply & return chiller supply and return condenser supply and return heat exchanger supply and return coil supply and return
Refrigerant Temperatures	- hot gas line suction line

Energy conservation must be approached in a systematic manner rather than considering individual items out of context. Systems do not operate in isolation but depend on and react with other systems. It is important to recognize this interaction of systems as modification to one will cause a reaction in another which may be either beneficial or counter-productive.

Buildings frequently use a combination of system types to meet the differing requirements in interior and perimeter zones. It is common practice, for instance to use a VAV system for interior spaces and multi-zone system and/or radiation for perimeter spaces. There is an interface point between the interior and perimeter zones where mixing and interaction between the two different systems takes place. In cool weather, cold supply air from the interior zone can spill into the perimeter zone resulting in overcooling and shift to a heating mode of operation and in hunting with both systems cooling and overheating. To avoid this situation, increase the supply air temperature for interior zones in cold weather.

SINGLE DUCT, SINGLE ZONE SYSTEMS

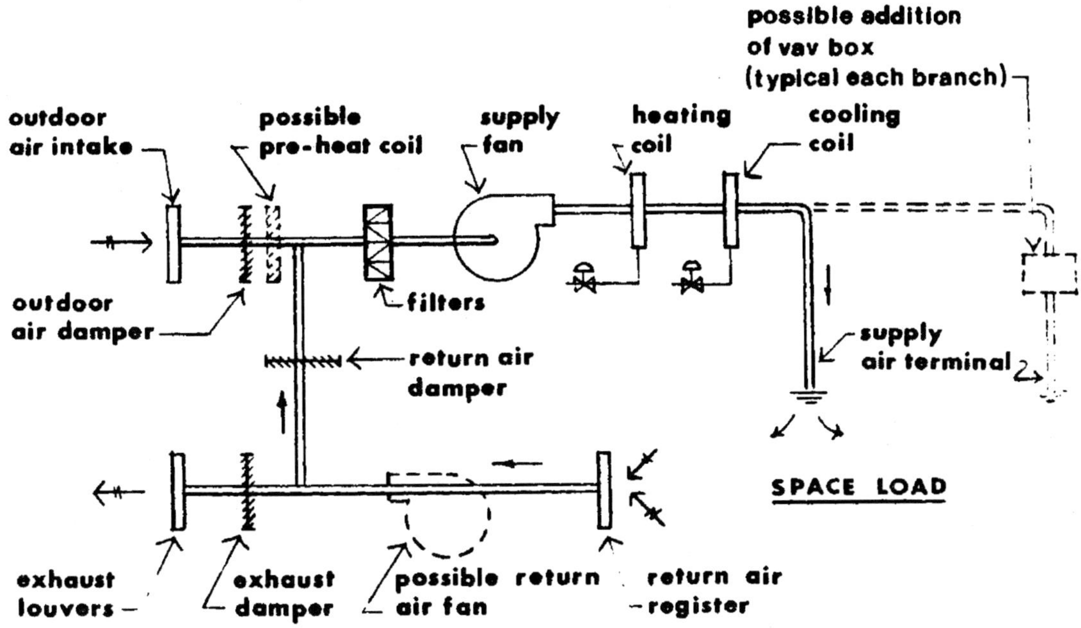

representative single duct system

These systems are the simplest and probably the most commonly used. They can comprise just a single supply system with air intake filters, supply fan and heating coil or can become more complex with the addition of a return air duct, return air fan, cooling coil, and various controls to optimize their performance. Basically, the system supplies air at a pre-determined temperature to one zone or the entire building, the quantity of heating or cooling being controlled either by modulating the supply air temperature or by turning the system on and off.

The energy output of a single duct system to meet a space load is determined by the volume/temperature differential relationship, i.e., to maintain a space temperature of 65°F, the heating load could be met by a system supplying 10,000 cfm at 105°F, or 6,000 cfm at 125°F.

To conserve energy in single zone, single duct systems, determine from measurements the system output and the building load. In the heating mode, if the system maximum output exceeds design building load, first reduce the volume of system air as this will show the greatest energy saving, i.e., 90% of the original fan volume requires only 65% of the original fan HP. Next reduce the supply temperature; this will not conserve as much energy as reduced volume flow.

In the cooling mode, if the system output exceeds the design building load, the saving by reduced air flow must be equated with the saving by increased air temperature as increased COP of the refrigeration equipment may yield savings which exceed the fan power losses.

Install controls on the heating and cooling coils to modulate the supply air temperature. Allow a 8 - 10° "Dead Zone" between heating and cooling to prevent simultaneous heating and cooling and rapid cycling. If possible, use an economizer cycle wherever the total heat of the outdoor air is favorable during occupied periods; the sensible temperature is favorable in unoccupied periods; and the size of the system makes the installation economic. Refer to ECO 80 Section 15.

The economizer cycle can be a simple arrangement of dampers controlled by outdoor and return air DB temperature or can be a more complex system based on the total heat or enthalpy of the outdoor air, return air, and the desired enthalpy of the supply air.

Single zone single duct systems can be readily converted to VAV by adding control boxes on each branch and substituting outlets with VAV type. Fan volume should preferably be controlled according to demand either by installing inlet guide vanes or by installing a variable speed motor.

TERMINAL REHEAT SYSTEMS

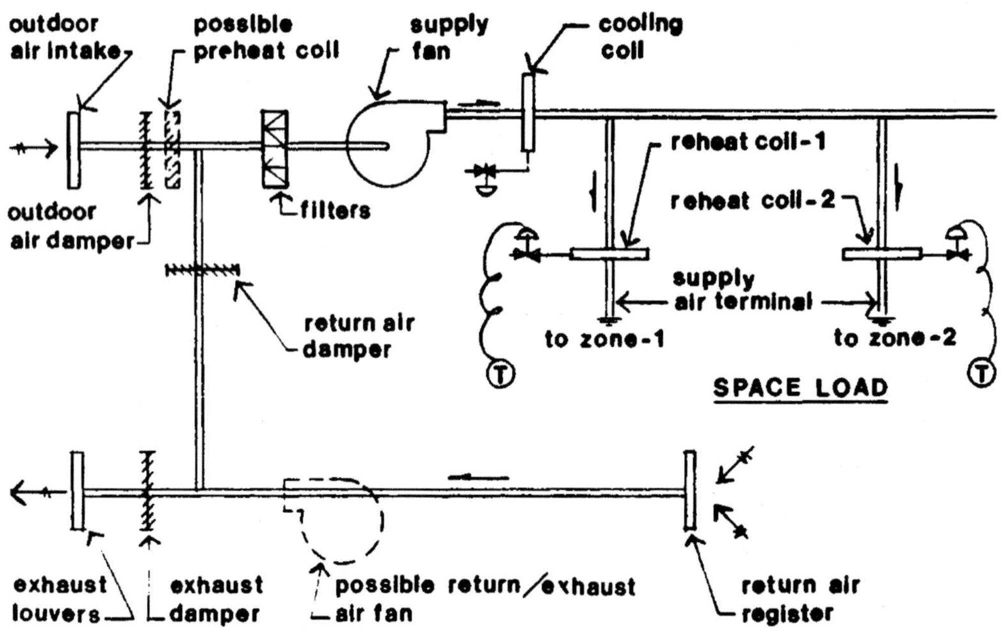

representative terminal reheat system

Terminal reheat systems were developed to overcome the zoning deficiencies of single duct systems and basically comprise a single duct, single zone system with individual heating coils in each branch duct to zones of similar loads. Reheat systems were also developed to give closer control of relative humidity in selected spaces.

Terminal reheat allows each different zone to be individually controlled but wastes energy in the cooling season as all of the supply air must be cooled to a low enough temperature to meet most critical load zone but must be reheated for zones of lesser loads to avoid overcooling.

As with single duct systems, energy can be conserved by reducing the supply air volume. Many reheat systems are controlled at a fixed supply temperature of around 55°DB/55°WB. To conserve energy, fit controls to reschedule the supply air temperature upward according to demands of the zone with the greatest cooling load. If one zone has cooling loads grossly in excess of all others, the controlling thermostat should be located in that space.

The greatest quantity of energy can be saved by adding variable air volume boxes to each of the major branch ducts. Each VAV box should be controlled by a space thermostat located in its particular zone and its associated reheat coil should be provided with controls to prevent reheat until the VAV box has reduced the zone supply air volume to 50%.

In buildings where a reheat system supplies zones of different occupancy, add dampers and control valves to enable water to be shut off during unoccupied times.

- Reduce the water flow and temperature of hot water to reheat coils.

- Use waste heat from condensate, incinerators, diesel or gas engines, or solar energy for reheat; but in all cases, raise the cold duct temperature to reduce refrigeration load.

- Modify controls to operate terminal reheat systems on a temperature demand cycle only. Caution: Control of humidity will be eliminated and zone control will be modified.

- Add or adjust controls to schedule supply air temperature according to the number of reheat coils in operation. If 90% of the coils are in operation, reset supply air temperatures to a higher level until the number of reheat coils falls to 10%.

- Install an interlock between the two valves to prevent simultaneous heating and cooling.

-If air conditioning is needed during unoccupied hours or in very lightly occupied areas, install controls to de-energize the reheat coils, raise the cold duct temperature and operate on demand cycle.

-Under an alteration or expansion program, install variable volume rather than terminal reheat or other systems.

-De-energize or shut off terminal reheat coils, raise the chilled water and supply air temperature of the central system and add re-cooling coils in ducts in areas where lower temperatures are needed.

MULTI-ZONE SYSTEMS

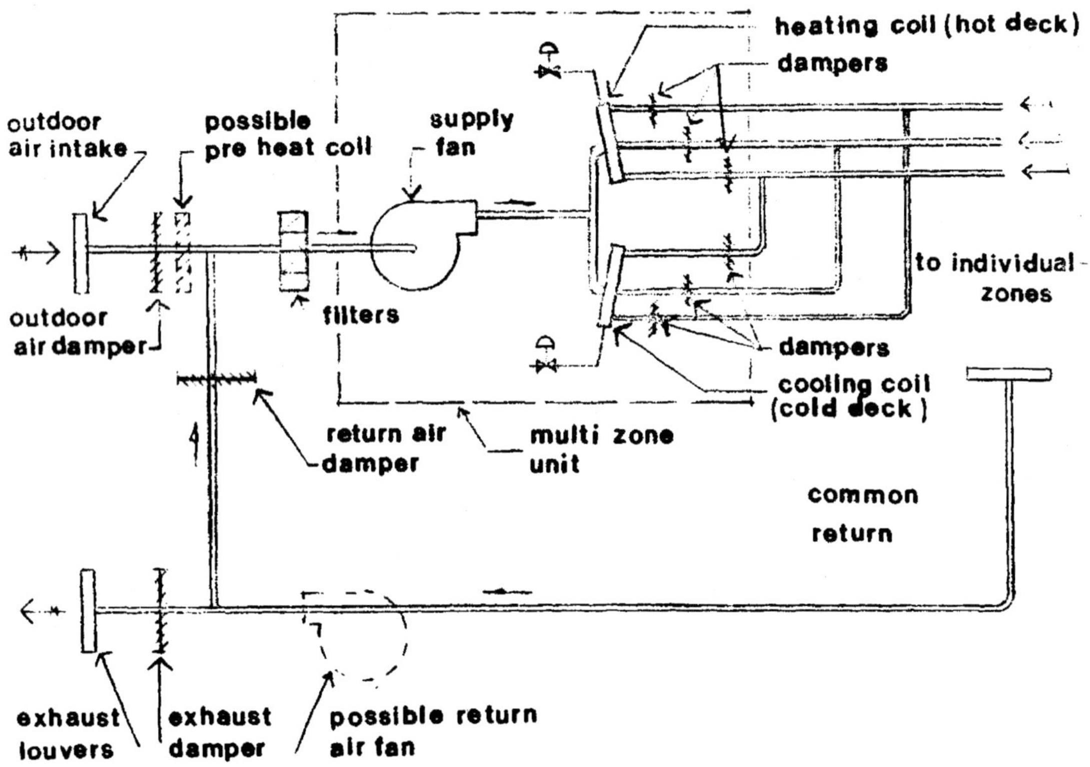

representative multi zone system

Most multi-zone units currently installed have a single heating coil serving the hot deck and single cooling coil serving cold deck. Each zone supply temperature is adjusted by mixing the required quantities of hot and cold air from these coils. With these types of units, the hot deck temperature must be sufficiently high to meet the heating demands of the coldest zone, and cold deck air must be sufficiently low to meet the demands of the hottest zone. All intermediate zones are supplied with a mixture of hot and cold air wasting energy in a similar manner to reheat systems.

New model multi-zone units are now available on the market which have individual heating and cooling coils for each zone supply duct and the supply air is heated or cooled only to that degree required to meet the zone load. These new types of units use far less energy than units with common coils and where renovations are contemplated or the existing multi-zone unit is at or near the end of its useful life, replacement should be considered, using a multi-zone unit with individual zone coils.

Analyze multi-zone systems carefully and treat each zone as a single zone system and adjust air volumes and temperature accordingly.

Hot deck and cold deck dampers are often of poor quality and allow considerable leakage even where fully closed. To conserve energy, check these dampers and modify to avoid any leakage.

Install controls or adjust existing controls to give the minimum hot deck temperature and maximum cold deck temperature consistent with the loads of critical zones.

Where a multi-zone unit serves interior zones that require cooling all of the summer and most of the winter, energy can be conserved by converting to VAV reheat. To achieve this, blank off the hot deck and add low pressure VAV dump boxes with new reheat coils in the branch duct after the VAV box. This system works well in conjunction with an economizer cycle and reheat energy is minimized. Careful analysis, however, is required of the existing system and the zone requirements to make the correct selection of equipment.

Arrange the controls so that when all hot duct dampers are partially closed, the hot deck temperatures will progressively reduce until one or more zone dampers is fully opened and, when all of the cold duct dampers are partially closed, the cold duct temperature will progressively increase until one or more of the zone dampers is fully opened.

- Install controls to shut off the fan and all heating control valves during unoccupied periods in the cooling season and shut off the cooling valve during unoccupied periods in the heating season.

- Convert system entirely or in part to variable volume by adding terminal units and pressure bypass and/or add fan coil units in specific areas requiring constant air volume.

DUAL DUCT LOW VELOCITY SYSTEM

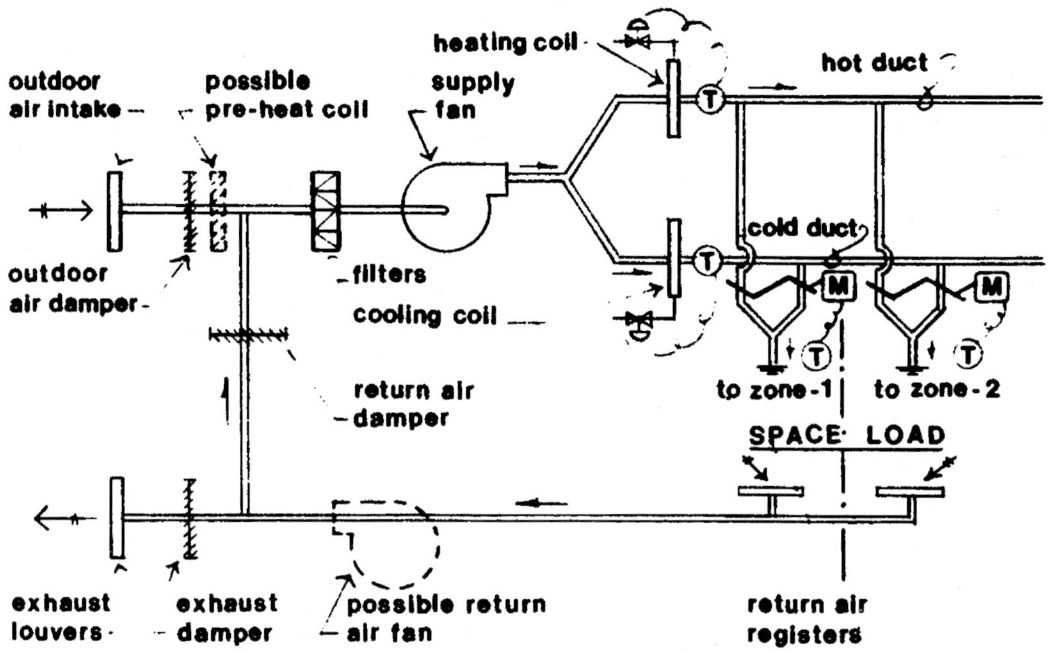

representative dual duct low velocity system

Dual duct low velocity systems supply hot and cold air in individual ducts to the various zones of the building which are supplied with a mixture of hot and cold air to maintain the desired zone conditions. Control of mixing is normally achieved by automatic dampers in the branch ducts; the damper being positioned according to the dictates of the space thermostat. Each duct should be analyzed as a single zone system to determine whether the air volume or its temperature set point can be advantageously changed to conserve energy. Reduce the temperature of the hot duct and increase the temperature of the cold duct to that point where the heating and cooling loads of the most critical zone can just be met.

Return air is a mixture of all zones and reflects the average building temperature. In some designs of central station equipment, it is possible to stratify the return air and the outdoor air by installing splitters so that the hottest air favors the hot deck and coldest air favors the cold deck. This will reduce both heating and cooling loads.

DUAL DUCT HIGH VELOCITY SYSTEMS

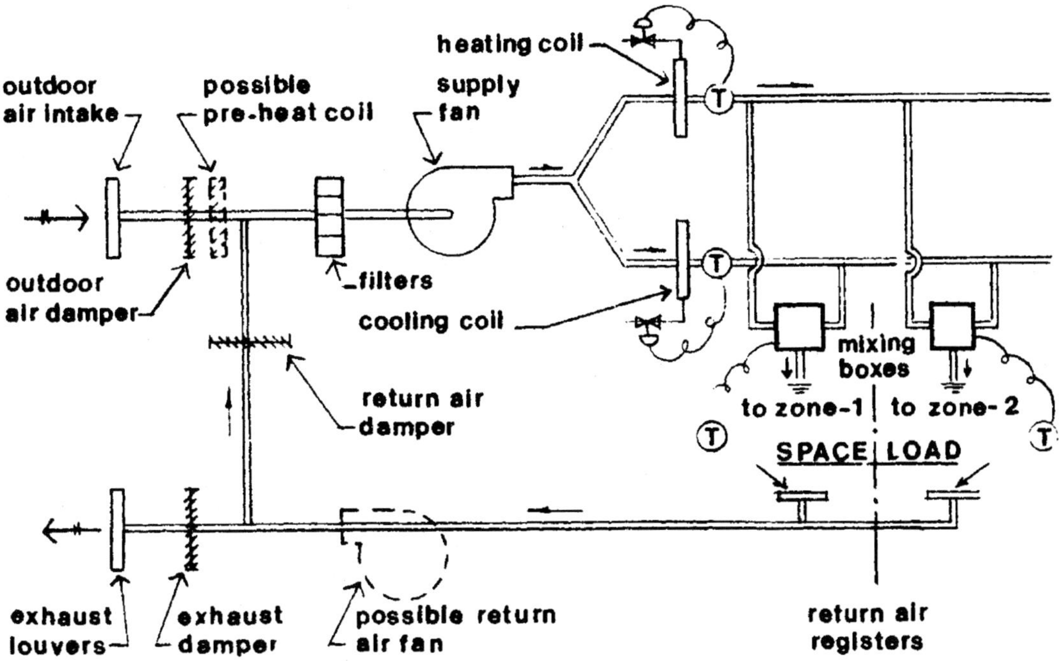

representative dual duct high velocity system

Dual duct high velocity systems operate in the same manner as low velocity systems with the exception that the supply fan runs at a high pressure and that each zone requires a mixing box with sound attenuation.

Considerable quantities of energy are required to operate the fan at high pressure and close analysis of pressure drops within the system should be made and fan pressure reduced to the minimum to operate the mixing boxes.

To reduce system pressure, replace existing high pressure mixing boxes with lower pressure types.

-In conditions when there is no cooling load, install controls to close off cold air duct; de-energize chillers and cold water pumps and operate as a single duct system; rescheduling the warmer air duct temperature according to heating loads only.

-Under conditions where there are no heating loads, install controls to close off the warm air ducts; shut off hot water,steam or electricity to the warm duct and operate the system with the cold duct air only; rescheduling supply air temperature according to cooling loads.

-Replace obsolete or defective mixing boxes to eliminate leakage of hot or cold air when the respective damper is closed.

-Provide volume control for the supply air fan and reduce capacity preferably by speed reduction when both the hot deck and cold deck air quantities can be reduced to meet peak loads. Reducing the heat loss and heat gain provides an opportunity to reduce the amount of air circulated.

-When there is more than one air handling unit in a dual air system, modify duct work if possible so that each unit supplies a separate zone to provide an opportunity to reduce hot and cold duct temperatures according to shifting loads.

-Change dual duct systems to variable volume systems when energy analysis to do so is favorable and the payback in energy saved is sufficiently attractive by adding VAV boxes and fan control.

VARIABLE VOLUME AIR SYSTEMS

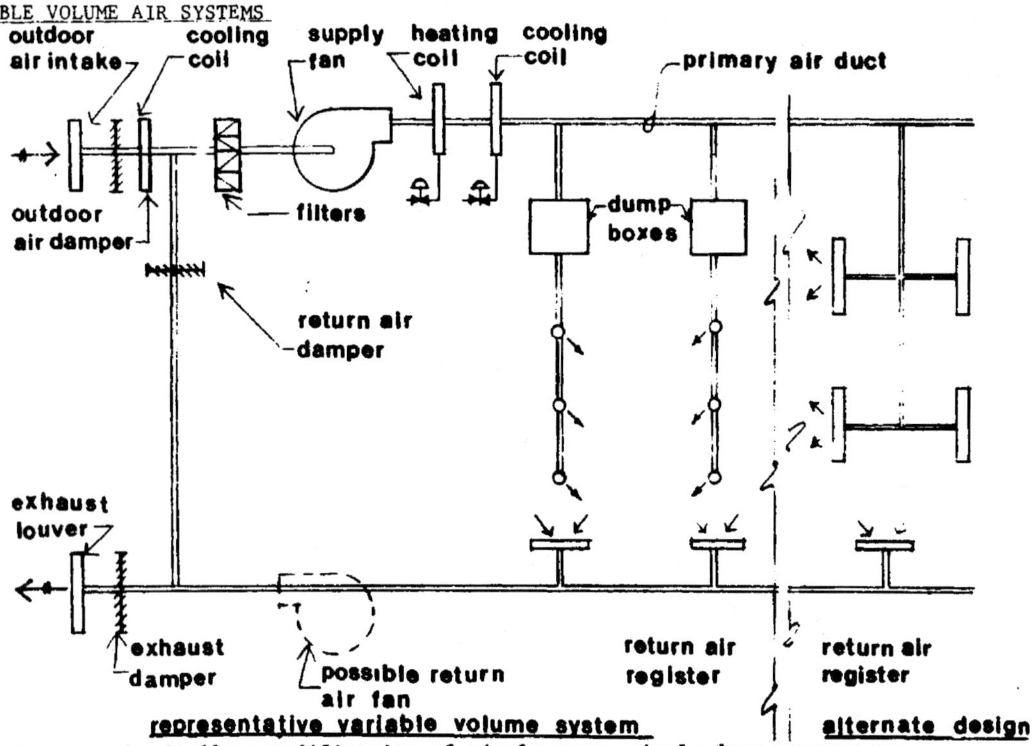

VAV Systems are basically a modification of single zone, single duct systems to allow different conditions in multiple zones to be met by varying quantities of constant temperature air. The air quantity supplied to the zone can be modulated in two different ways:

Dump boxes installed in the zone supply duct modulate the amount of air through the outlets by bleeding off unused quantities of air and dumping into the ceiling void and back into the return air system. This type of VAV system maintains constant fan volume and constant flow in all the trunk ducts and can waste energy by mixing the treated dump air with the return air. This system does, however, have practical advantages when used with direct expansion cooling coils since it maintains constant flow through the coils regardless of load, which is necessary to obtain correct operation of refrigerating equipment.

The second type of VAV system is where the air volume is controlled at the outlet by closing down dampers as the zone load decreases. This type of system does, in fact, vary the total supply volume handled by the fan by increasing the resistance to air flow under light load conditions.

To conserve energy, regulate the fan volume according to the demands of the system by a variable speed motor with SCR controller or by inlet guide vanes.

Provide a control to modify the fan speed up or down to maintain constant pressure in the supply duct. Reset the supply air temperature in accordance with zone loads.

- If not installed, provide either inlet vortex dampers or variable speed drive such as a multispeed motor or SCR control so that fan volume will be reduced when dampers begin to close until one or more VAV dampers is fully opened.

- Provide control to reschedule supply air temperature to a point where the damper of the variable air volume box serving the zone with the most extreme load is fully opened.

- Install control to reduce the hot water temperature and raise the chilled water temperature in accordance with shifting thermal demands.

- As with single duct variable volume systems with reheat, set control to delay reheat until cfm is reduced to minimum.

- Where possible, eliminate reheat coil and set the box to full close off.

- Variable volume with reheat systems should be operated and modified similarly to conditions described for terminal reheat.

INDUCTION SYSTEMS

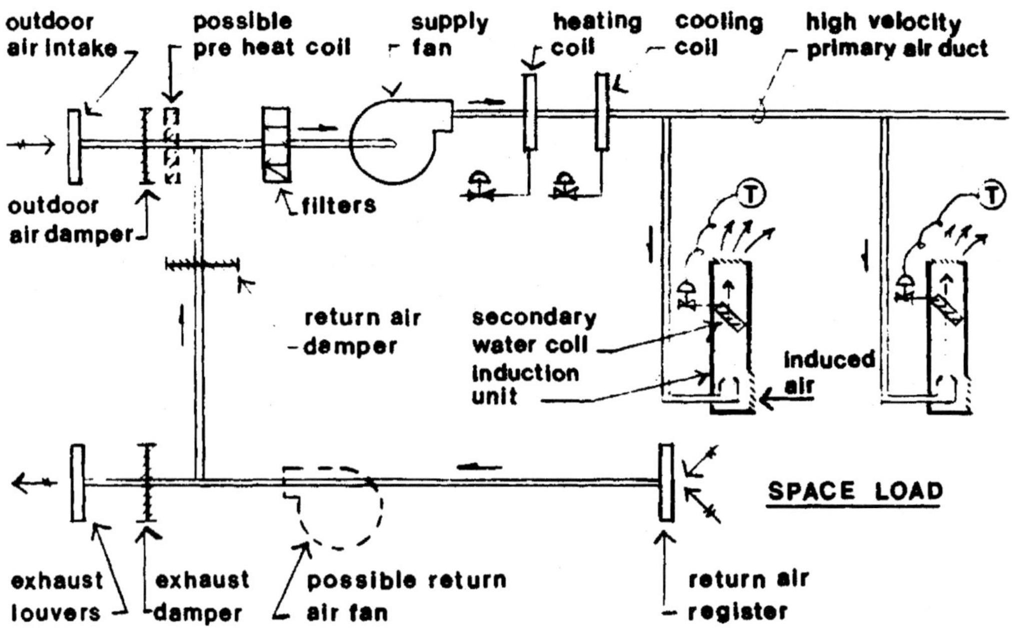

representative induction unit system

Induction systems are commonly used for heating and cooling perimeter zones where large fluctuations of heating and cooling loads occur.

Primary air is either heated or cooled and supplied at high pressure to the induction units located within the conditioned space. Primary air is discharged from nozzles arranged to induce room air into the induction unit approximately 4 times the volume of the primary air. The induced air is cooled or heated by a secondary water coil. The water coil may be supplied by a 2-pipe system, whereby either chilled water or heated water is available, but not simultaneously, or by a 3-pipe system where separate supplies of hot water or chilled water are continuously available and after passing through the unit are mixed into a common return, or by a 4-pipe system, where a supply and return of hot water and chilled water are both continuously available.

The primary supply air fan operates at high pressures requiring high horsepower input. By careful analysis, reduce the primary air volume and pressure to the minimum required to operate the induction terminal units.

Induction unit nozzles may be worn through many years of cleaning and operation resulting an increased primary air quantity at lower air velocities with lower induced air volumes. Check each induction unit and either repair the nozzles or replace them.

Provide controls to do the following:

Set the primary air reheat schedule as low as possible without causing complaints in the occupied spaces.

Lower the set point of secondary heating water temperatures in winter and schedule according to outdoor conditions.

During maximum cooling or heating periods, reduce the secondary water flow rate to the minimum necessary to obtain satisfactory room conditions.

- Raise temperature of the primary air for cooling to reduce refrigeration load, but weigh this benefit against increased cooling demand by the secondary coils.

- Reduce water temperature to the heating coils during the heating season to reduce piping loss and improve boiler efficiency.

- For night operation during the heating season, shut down primary air fans, raise the hot water temperature and operate the induction units as gravity convectors.

FAN COILS - UNIT VENTILATORS

Fan coil and unit ventilator performance can be improved by careful maintenance which in turn will allow energy savings to be made in the associated heating and cooling water distribution systems. Heating and cooling coils should be cleaned and air and water flow reduced to the minimum required to meet space conditions. Consider changing 4-pipe heating and cooling supply systems to a 2-pipe system. This will avoid changeover loss and prevent simultaneous heating and cooling within a given zone.

- For fan coils systems which have separate coils for heating and air conditioning, install a control to prevent simultaneous heating and cooling.

- Install a seven-day timer to shut down fans, close off valves, and shut off chilled water pumps, compressors and cooling towers or air cooled condensers during unoccupied periods.

- Install controls to shut off fans during unoccupied periods during the heating season and operate the fan coil unit as a gravity convector. (The same is applicable to unit ventilators as well.)

- Block off outdoor air inlets where no dampers are installed if infiltration meets ventilation requirements.

- If fan coil units are not located in conditioned areas, insulate the casings to reduce heat loss/gain.

-With three-pipe systems, set controls for minimum mixing of hot and cold water.

-Change four-pipe or three-pipe systems to two-pipe systems where possible to avoid change-over losses.

WATER-TO-AIR HEAT PUMPS

Air coils should be kept clean to maintain the highest possible seasonal COP. Heat pump units are normally controlled by integral thermostats which turn them on and off according to the demands of space load and these units are frequently left in operation 24 hours/day regardless of whether building is occupied or not.

Install a centralized system of on/off controls on a floor-by-floor basis using a remote control contactor on the power circuit feeding the heat pump. Rewire the power supply to heat pumps in such a manner that 1/3, 2/3, or all of the units in any zone can be turned on and off depending on whether the building is occupied or not and what the expected outdoor conditions are likely to be.

-When retrofitting, install one larger unit rather than multiple smaller units for greater efficiency.

-All guidelines for condensers, compressors, evaporators, fans, pumps, piping and duct work listed in other sections are applicable to unitary closed loop water source heat pumps.

AIR-TO-AIR HEAT PUMPS AND INCREMENTAL AIR CONDITIONING UNITS

-The guidelines for evaporators and condensers in Section 6C "Cooling and Ventilation", are applicable to these units.

-Where possible, direct the warm exhaust air from the building to the inlet of air/air heat pumps to raise their coefficient of performance. Waste heat from other processes can also be used as a heat source.

-Replace air-to-air heat pumps with water-to-air heat pumps where there is a source of heat such as ground water with temperatures above average ambient winter air temperatures.

-For incremental air conditioning units follow guidelines for fan coil units during the heating cycle.

-Install a seven-day timer to program operation of compressors in accordance with occupied - unoccupied periods.

-When replacing incremental air conditioning units which are equipped with electric heating coils, install a heat pump model instead.

-Replacement compressors should have an EER rating of 9 or better during cooling cycle. (Applicable also to window air conditioning units).

EXHAUST SYSTEMS

All the ECO's and guidelines in this Section detailing methods of reducing fan power requirements by reducing flow rate and resistance apply equally to exhaust systems as well as supply systems.

Balance exhaust systems so that exhaust air flow rate does not exceed supply air flow rates of associated systems. Ideally, exhaust should be 10% less than supply to maintain a positive building pressure.

Modulate exhaust fan volume in step with associated VAV supply fan by installing inlet guide vanes or variable speed control.

-Recirculate toilet room exhaust air through charcoal filters to reduce makeup air requirements.

-Install controls to operate toilet exhaust fans intermittently for 10 to 20 minutes out of every hour and to automatically shut them off during unoccupied periods.

-Install a motorized damper at inlet grilles and wiring from damper to light switches to reduce the air quantity when toilet rooms are unoccupied. Modulate fan volume according to load by monitoring pressure in main exhaust stack.

-For new installations or when existing fans and/or motors are replaced, provide variable speed control or inlet vane control to operate in accordance with static pressure.

-Shut off supply fans which serve only as makeup air for toilet rooms and install new door louvers or cut off the bottom of the door to permit air from conditioned areas to migrate into the toilet rooms as makeup for the exhaust system. Set maximum capacity to provide one cfm per square foot of toilet area.

THERMAL STORAGE

-For large buildings with interior zones that require cooling all year round, analyze the potential for conserving energy by capturing the heat from the hot condenser line and storing it in water tanks when it is not required for immediate use in the perimeter zones, but can be used during unoccupied times when the cooling machines are not running.

-Analyze the cost benefits of providing chilled water storage to reduce peak demands and permit operation of the refrigeration systems at lower condensing temperatures which occur at night or early morning. Note: Both of the above options should only be considered when making sizeable additions or alterations to existing buildings. A thorough engineering and cost analysis is required to consider these systems.

This page blank

Section 18

Energy Conservation Opportunities
Heat Reclamation

HEAT RECLAMATION

A. BACKGROUND

Heat reclamation is the recovery and utilization of heat energy that is otherwise rejected to waste. Waste heat is a substitute for a portion of the heat energy that would normally be required for heating, cooling and domestic hot water systems. Heat recovery conserves energy, reduces operating costs, and reduces peak loads.

The performance of any heat recovery system depends upon the following factors:

a) The temperature difference between the heat source and heat sink.

b) The latent heat difference (where applicable) between the heat source and sink.

c) The mass flow multiplied by specific heat of each source and sink.

d) The efficacy of the heat transfer device.

e) The extra energy input required to operate the heat recovery device.

f) The fan or pump energy absorbed as heat by the heat transfer device and which can enhance or detract from the performance.

B. AVAILABLE HEAT RECLAMATION EQUIPMENT

Refer to Fig. 63A below and relevant Sections 13-17 for specific applications. Heat recovery devices reduce the peak heating and cooling loads when used with outdoor air systems and may reduce or completely eliminate the requirements for the heating and/or cooling equipment for major building expansions.

1. THERMAL WHEELS

A thermal or heat wheel is a rotating heat exchanger driven by an electric motor with a high thermal inertia core. They are capable of transferring energy from one air-stream to another and, in very large boiler plants, from flue gas to air.

The hot and cold air-streams must be immediately adjacent and parallel to permit installation of the heat wheel. Duct modifications may be necessary. Two types of thermal wheels are available: One which transfers sensible heat only, and the other which transfers both sensible and latent heat.

Exhaust air (from the building or space) and make-up air (from outside) flowing in the opposite direction are introduced to each half of the thermal wheel through separate adjacent ducts. A thin cylinder containing heat transfer media slowly rotates between the two air-streams within the thermal wheel. Energy absorbed by the media is transferred from one air-stream to the other.

During a heating cycle, incoming low temperature fresh air is heated and humidified by heat transferred from the warm, moist exhaust air. During summer months, cooling and dehumidifying of higher temperature moist, outside air is accomplished in a similar manner by giving up heat to the cooler exhaust air-stream.

Thermal wheels are commercially available in sizes ranging from 3,000 cfm to 60,000 cfm. They can be provided with purge sections and can also be arranged for vertical and horizontal mounting. Thermal wheels of larger size can be provided with variable speed, SCR controls.

Order of magnitude costs of thermal wheels, excluding ductwork modifications, are between $700 and $1000 per 1000 cfm for sensible heat transfer and 10% more for enthalpy wheels.

The total system efficiency is equal to the heat transfer efficiency plus the heat of the driving motors in the heating mode. During the cooling cycle, the energy required to rotate the wheel, plus the additional energy required to overcome the air resistance, must be deducted from the gross load reduction due to heat transfer.

The maximum potential heat that can be exchanged between a warm exhaust air-stream and a cold outdoor air-stream can be derived from the following formula:

$Q = (Q_e - Q_o) \times M_e$ or M_o (whichever is least), where:

Q_e = heat content of exhaust air in Btu's/lb.

Q_o = heat content of outdoor air in Btu's/lb.

M_e = exhaust air flow rate in lbs/hr.

M_o = outdoor air flow rate in lbs/hr.

Thermal wheels typically have an efficiency of 60% to 80% for sensible heat transfer and 20% to 60% for latent heat transfer. These efficiencies should be applied to the theoretical maximum heat transfer Q to obtain the actual heat transfer rate in Btu's per hour. Check with the manufacturer for specific efficiencies.

FIGURE 63A: APPLICATIONS OF HEAT RECLAMATION DEVICES

ENERGY SOURCE	TEMPER VENTILATION AIR	PREHEAT DOMESTIC HOT WATER	SPACE HEATING	TERMINAL REHEAT	TEMPER MAKE-UP AIR	PREHEAT COMBUSTION AIR	HEAVY OIL HEATING	INTERNAL TO EXTERNAL ZONE HEAT TRANSFER
Exhaust Air	1a, 1b, 1c, 2,3,4,5				1a,1b,1c, 2,3,4,5	Direct		5, Direct
Flue Gas	1b, 3, 4		3, 4		1b, 3, 4	1b,2,3,4		
Hot Condensate	3, 6	3, 6	3, 5, 6	3, 6	3, 6	3, 6	6	
Refrigerant Hot Gas	6	6		6	6			
Hot Condenser Water	3, 6	3, 6	3, 5, 6	6	3, 5, 6	3, 5, 6	6	5
Hot Water Drains		6						
Solid Waste	7	7	7	7	7	7	7	
Engine Exhaust & Cooling Systems	6, 8	6	6, 8	6	6, 8	6, 8	6	
Lights	5, 9		5, 9	5, 9	5, 9			5, 9

Device
1. Thermal Wheel
   a) Latent
   b) Sensible
   c) Combination
2. Run-around coil
3. Heat pipe
4. Air-to-Air Heat Exchanger
5. Heat Pump
6. Shell/Tube Heat Exchanger
7. Incinerator
8. Waste Heat Boiler
9. Heat-of-Light

The preceding formula provides the load reduction for one hour at a specific set of conditions. To determine yearly load reduction, analyze the specific conditions for each hour of operation and integrate these for the total number of hours of operation.

The above procedure may be used either for sensible heat recovery or for enthalpy recovery providing the units are in Btu's/lb. for each respective condition. It may also be used in the same manner for the cooling cycle where the direction of heat flow is reversed.

When applying thermal wheels to ventilation systems, install a roughing filter and/or insect screen to protect the core. Icing, which can possibly occur under some outdoor conditions with enthalpy wheels, can be eliminated by preheating 10°F. When used to transfer heat from flue gases, the construction of the thermal wheel must be suitable for the temperatures encountered.

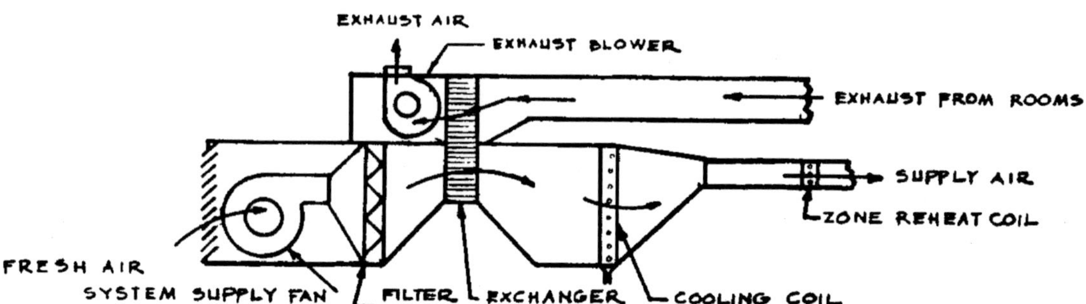

ROTARY HEAT REGENERATOR-(THERMAL WHEEL)

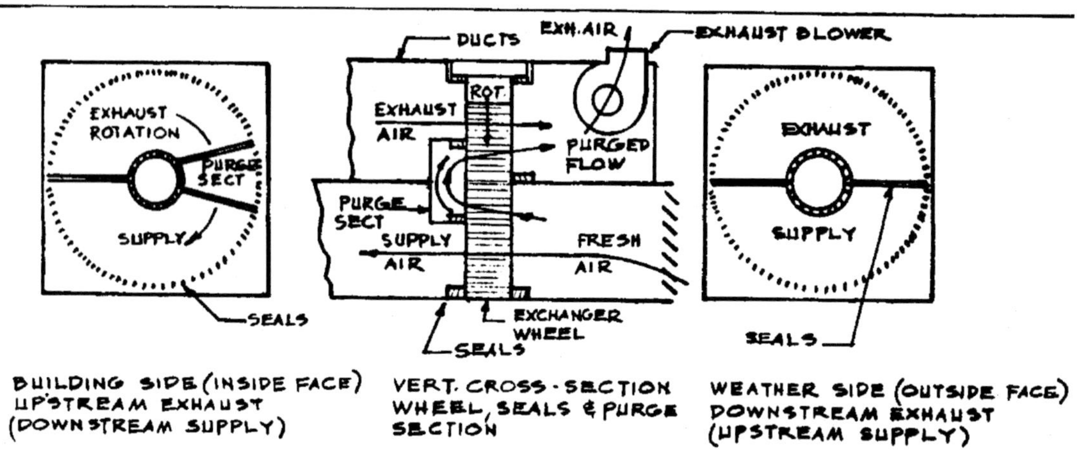

THERMAL WHEEL SECTIONS

2. RUN-AROUND-COIL SYSTEM

This system comprises two or more extended surface fin coils installed in air ducts and interconnected by piping. The heat exchanger fluid, consisting of ethylene glycol and water, is circulated through the system by a pump, removing heat from the hot air-stream and rejecting it into the cold air-stream. A run-around-coil system may be used in winter to recover heat from warm exhaust air for use in pre-heating cold outdoor air, and in summer to cool hot outdoor air by rejecting heat into cooler exhaust air. The method of determining system efficiency described for thermal wheels should also be applied to run-around-coil systems.

When calculating reductions in load due to run-around coils, the effect of coil depth on the rate of heat transfer and the resistance to air flow must be recognized. A trade-off analysis between the extra heat transfer obtained by deep coils against the extra resistance to air flow should be made to determine the optimum coil size.

Where there is no possibility of freezing, water may be used as the heat transfer medium. Where frost protection is necessary and ethylene glycol is used as the heat transfer fluid, the different specific heats and rates of heat transfer must be accounted for in the calculations.

The coil type exchanger has the advantage of permitting heat transfer between supply and exhaust systems in widely separated portions of the building, since coils are installed in each of the intake and exhaust ducts, and are connected by piping only. The liquid heat transfer medium is pumped from coil to coil. It is relatively easy to run piping long distances through the structure, interconnecting a number of separate exhaust and fresh air systems to form one complete system.

The initial cost of run-around-coils is less than the rotary type heat exchangers by about 40%. They are most economical for small systems or for heating systems only. Order of magnitude costs are $2/cfm for systems up to 10,000 cfm and $1.50/cfm for larger systems.

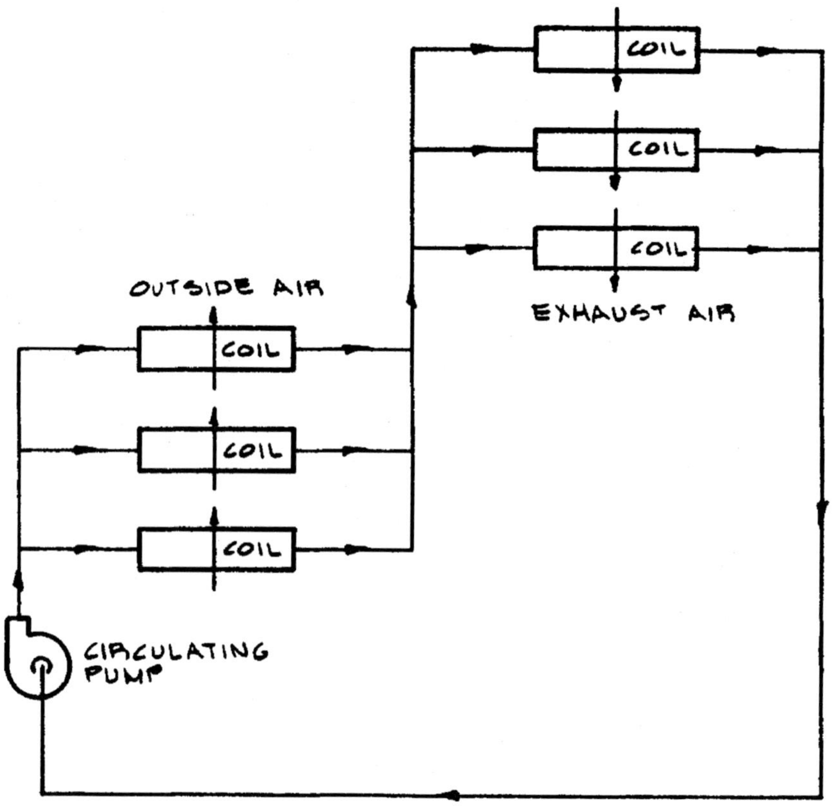

RUN AROUND HEAT RECOVERY SYSTEM WITH MULTIPLE COILS

3. HEAT PIPE SYSTEMS

Heat pipes comprise extended surface finned tubes extending between adjacent air ducts. The tubes are continuous from one duct to the other and are on the same horizontal plane. Each tube contains liquid refrigerant which evaporates at the warm end absorbing heat from the warm air-stream, and migrates as a gas to the cold end where it condenses and releases heat into the cold air-stream. The condensed liquid then runs back to the hot end of the tube to complete the cycle. To operate effectively, the heat tubes must slope down from the cold end to the hot end. Accurate control of the slope must be provided.

Heat pipes between hot and cold air ducts can only be used when these are immediately adjacent to each other, and some modification of existing ductwork may be necessary for installation.

To calculate the effectiveness and yearly load reduction of a heat pipe system, use the same procedure described for thermal wheels.

Heat pipe coils are available in sizes ranging from 2,500 cfm to 20,000 cfm. Order of magnitude cost is $2,500/1000 cfm.

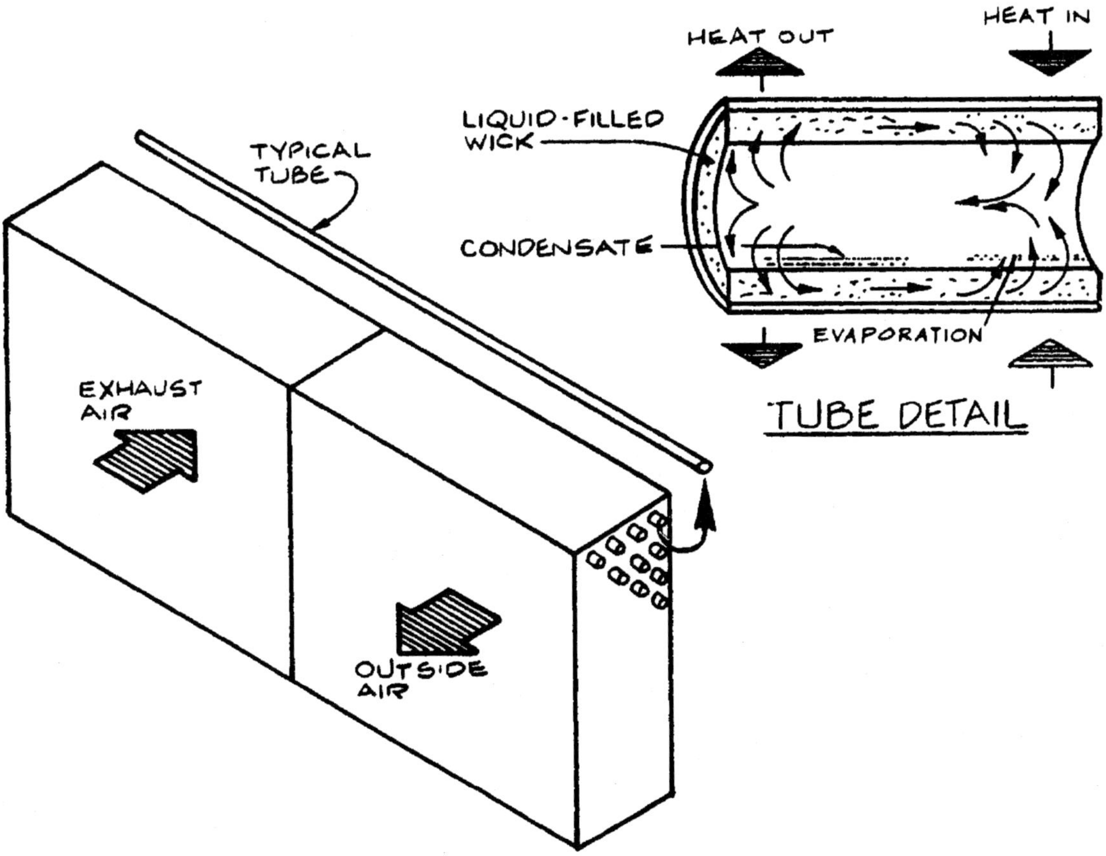

HEAT PIPE HEAT EXCHANGER

4. AIR-TO-AIR HEAT EXCHANGERS

Air-to-air heat exchangers transfer heat directly from one air-stream to another through direct contact on either side of a metal heat transfer surface. This surface may be either convoluted plate (more common for low temperature use in HVAC system) or tube (more common for boiler flue gas heat transfer). The heat exchangers may be purchased as packaged units or can be custom made. They transfer sensible heat only and are not designed for cooling applications. Size is limited only by the physical dimensions of the space available.

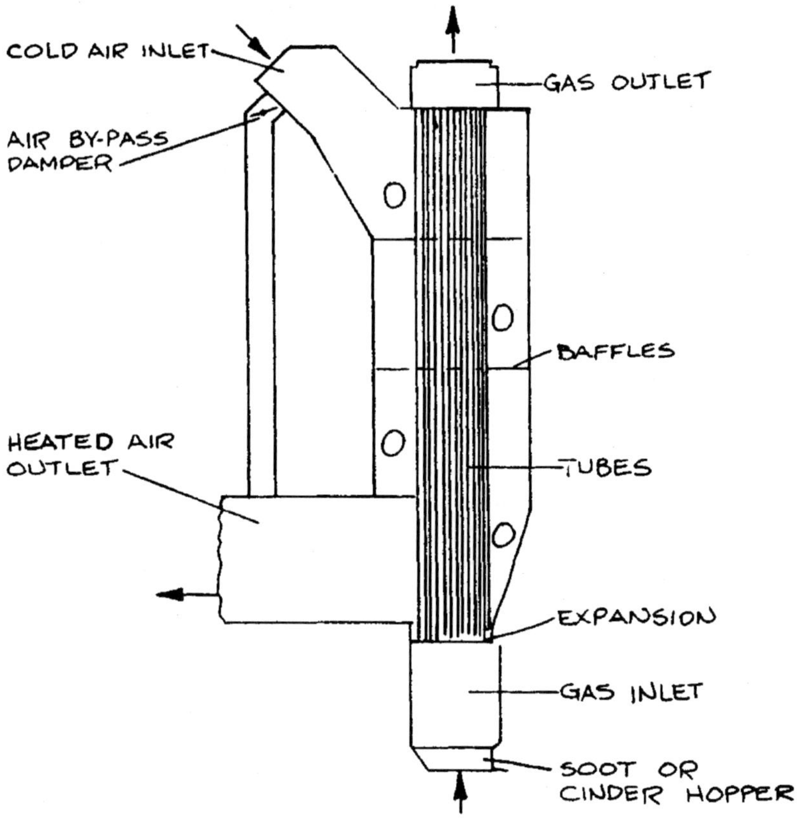

FLUE GAS HEAT EXCHANGER

The efficiency of each installation and cost must be calculated based on the particular circumstances that apply. Generally, efficiencies are lower than 50%, however, they are relatively inexpensive, have low resistance to air flow, require no motive power input, and are trouble-free and durable.

5. HEAT PUMP AS HEAT EXCHANGERS

Heat pumps are inherently heat transfer devices and, unlike those previously described, upgrade the temperature by as much as a factor of 3:1. This feature makes them particularly attractive for use with low temperature heat sources. They also have the capacity to transfer latent heat as well as sensible heat.

Heat pumps are available in the following configurations each having particular application:

a) Air to air (very limited applications)

b) Water to air

c) Water to water

d) Air to water

Sizes range from 1/2 to 25 tons as commercially available package units. Larger sizes can be obtained by simple modifications to refrigeration machines.

Heat pumps are commonly used to heat and cool spaces (Section 17). As heat exchange equipment in the heating mode they have a high COP and can be used to extract more heat from any given source than any other heat exchange device. In the heating mode the power input is approximately 1/3 to 1/5 of the heat output. Heat pumps are more favorable than electric resistance heating and compare well with oil-or gas-fired heating systems. When used in the cooling mode, performance decreases and they should not be considered for this purpose alone unless simultaneous heating and cooling will occur.

Heat pumps make heat available from low temperature sources. For example: heat can be extracted from cold drain water at approximately 50° F with rejection at temperatures high enough for use by domestic hot water systems, air pre-heating, etc.

When selecting heat pumps, choose machines that have the highest COP over the range of expected operating conditions.

Order of magnitude costs are approximately $300 to $500/ton.

6. SHELL AND TUBE HEAT EXCHANGERS

Shell and tube heat exchangers can be used to exchange heat in the following configurations:

- Liquid to Liquid
- Steam to Liquid
- Gas to Liquid

All three configurations are commercially available in a whole range of sizes and outputs and with reliable heat exchange data. Cost rises rapidly as the approach temperature is decreased and should be considered when making an economic evaluation. Do not strive to achieve the last degree of captured heat.

Particularly favorable applications are to capture energy from hot condensate, hot refrigerant gas, condenser water, and hot drain lines from kitchens and laundries.

Heat exchangers should be insulated to prevent unnecessary heat loss and should be constructed of materials to suit the application.

7. INCINERATORS

Incinerators burning solid waste produce 6,000 Btu/lb of common trash. Special heat recovery incinerators are now available in sizes down to approximately 100 lbs/hr burning rate generating about 450,00 Btu/hr. These incinerators are oil-jacketed and raise the oil temperature to approximately 450°, which can be used in a heat exchanger as a source of high and low temperature heat. The heat should be utilized at the highest practicable temperature. Typical applications are absorption refrigeration, heating, domestic hot water, etc.

The order of magnitude costs for incinerators are approximately $70/lb waste/hr. All local, State and Federal air pollution standards must be strictly complied with and environmental impact statements are required.

8. WASTE HEAT BOILERS

Waste heat boilers are used in conjunction with high temperature heat recovery such as engine exhaust and exhaust from gas turbines used to drive refrigeration equipment or electric generators. (See total energy systems).

9. HEAT-OF-LIGHT SYSTEMS

Wet and dry heat-of-light systems are commonly used to reduce sensible heat gain in overlighted spaces and can transfer heat from interior areas in stores and offices to perimeter zones. Refer to Section 19 - Lighting.

C. ENERGY CONSERVATION OPPORTUNITIES

ECO 102 RECLAIM WASTE HEAT TO REPLACE OR SUPPLEMENT PURCHASED ENERGY

Analyze and consider the following options:

1. TRANSFER ENERGY BETWEEN EXHAUST AND OUTDOOR AIR DUCTS WHEN THERE IS MORE THAN 4,000/CFM BEING EXHAUSTED

   a) When there are more than 3,500 heating degree days and more than 8,000 cooling degree hours above 78° F,DB consider thermal wheels, heat pipes and other devices listed in the matrix. Where supply and exhaust ducts are remote from each other and cannot be brought together, consider systems other than heat pipes and thermal wheels.

b) Install a thermal wheel or heat pump to recover both sensible and latent heat in locations which experience more than 12,000 wet bulb degree hours above 66° F,WB.

c) When justified for heating mode only, install an air-to-water-to-air heat pump to transfer energy from the exhaust air-stream to the fresh air-stream.

d) Utilize exhaust air heat energy to temper make-up air to pre-heat combustion air or use for space heating via heat pumps.

2. RECOVER WASTE HEAT FROM THE BOILER FLUE GASES WHENEVER THE STACK TEMPERATURE IS GREATER THAN 350°F.

a) Install a heat pipe or an air-to-air heat exchanger to transfer energy from the hot flue gas to temper ventilation air, pre-heat domestic hot water, for space heating or pre-heating combustion air.

b) Take into account the corrosive effect of flue gas when selecting materials.

c) Allow for change of draft conditions caused by heat exchanger.

d) Provide alternative source of combustion air when heat exchanger dampers are closed for cleaning.

3. RECOVER HEAT FROM LAUNDRY AND/OR KITCHEN WASTE WATER

a) When more than 30,000gal/week of water at temperatures above 120°F is discharged to waste.

b) Use as heat source for heat pump for other system requirements.

c) Consideration must be given to the characteristics of the waste water, particularly the soap/detergent content of laundry waste water and the grease content of kitchen waste water. Piping and/or material modifications may be necessary to enable the heat exchanger to handle water with high concentrations of these impurities. In addition, in order to maintain a steady flow rate through the heat exchanger when water is being sporadically discharged, a holding tank may be required.

d) Waste heat recovered may be used by any system requiring hot water, such as domestic hot water and heating systems. Installed costs range from $500/100 gal/hr for a large unit, to $2,000/100 gal/hr for a small unit.

4. RECOVER HEAT FROM ENGINE OR COMBUSTION TURBINE EXHAUST AND COOLING SYSTEMS

a) On engines larger than 50 HP.

b) Exhaust gas heat recovery is limited by practical limitations of the heat exchanger plus the prevention of flue gas condensation. The recommended minimum exhaust temperature is approximately 250°F. Depending on the initial exhaust temperature, about 50% - 60% of the available exhaust heat can be removed.

5. RECOVER HEAT FROM INCINERATORS

a) If the quantity of solid waste exceeds 1000 lbs/day.

6. RECOVER HEAT FROM CONDENSATE RETURN SYSTEMS

a) When district heat steam condensate is discharged to waste.

b) When steam condensate from equipment supplied by on-site boilers is at a temperature of 180°F or greater.

7. RECOVER HEAT FROM REFRIGERATION SYSTEM HOT GAS

a) Where there is a steady and concurrent demand for refrigeration and waste heat, and where the refrigeration systems operate 1,000 hours or more per year.

b) Do not reduce superheat to the point where liquid slugging occurs.

c) The heat exchanger must be located after the hot gas bypass or other unloading devices. If located outdoors, drains must be provided to prevent freezing.

8. RECOVER HEAT FROM CONDENSER WATER SYSTEMS

a) Install a heat exchanger or heat pipe in the hot condenser water line to temper outdoor air, pre-heat domestic hot water, or modify the piping in air handling units to utilize hot condenser water to heat air.

b) Install a coil to extract heat from the hot condenser water line to heat intake air in an air cycle heat pump which can then reject its condenser heat to the space requiring it.

c) Generally, it is not economical to replace existing condensers by double bundle condensers; however, in the event that replacement is being contemplated due to age or the installation of new refrigeration equipment, give consideration to a double bundle condenser.

9. RECOVER HEAT FROM AIR-COOLED CONDENSERS

a) In supermarkets, grocery stores, and any building with 15 HP or more of refrigerated display cases or storage boxes, and heating degree days exceeding 3,000.

b) When air-cooled condensers are located away from spaces requiring heat, condenser heat can be recovered by adding insulated ducts.

10. UTILIZE HEAT FROM INTERNAL SPACES FOR HEATING PERIMETER AREAS

a) When replacing an existing chiller, install a double bundle condenser, or cascade condensers, to recover heat that is otherwise dissipated by the cooling tower.

b) Install tanks to store hot condenser water from daytime cooling for nighttime heating. This water can be used directly for space heating or as a heat source for a heat pump system.

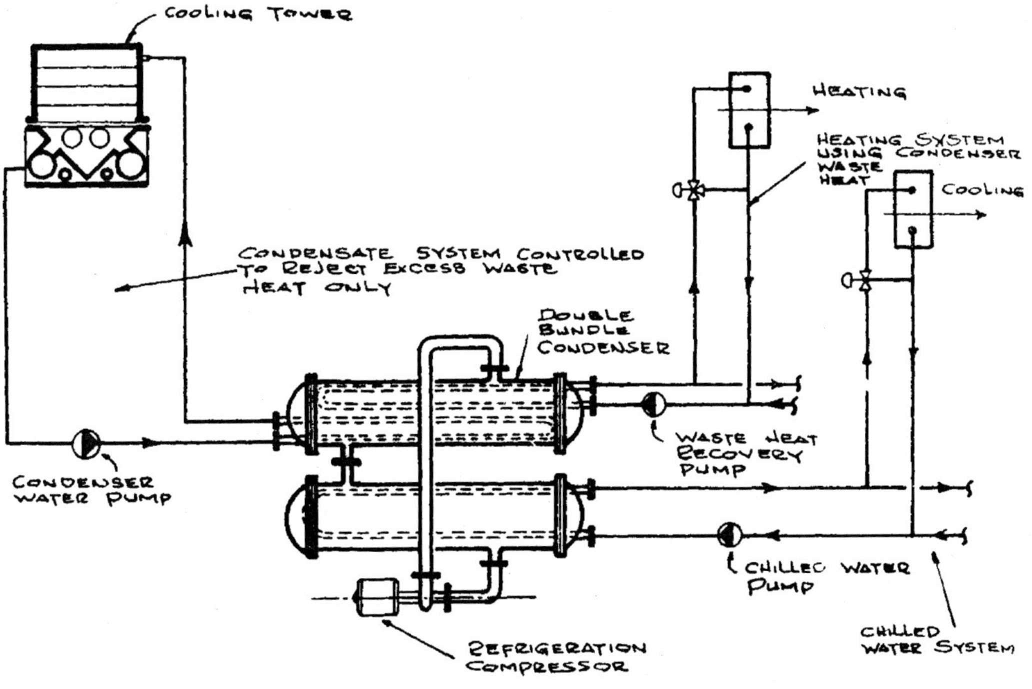

WASTE HEAT RECOVERY FROM CENTRAL CHILLER USING DOUBLE BUNDLE CONDENSER

This page blank

Section 19

Energy Conservation Opportunities
Lighting

LIGHTING

BACKGROUND

The relative amounts of energy generally used in buildings for lighting compared to the amounts used for each of the other purposes is discussed in ECM-1, Section 2 but the relative amounts for any specific building vary, and hence the magnitude of potential savings also varies widely. Nationally, about 22% of all electrical energy generated is used for lighting. Lighting in office buildings accounts for approximately 10% and retail stores consume an additional 10% of the lighting energy. However, lighting actually accounts for a greater percentage of total energy use since the heat generated from lighting in buildings contributes to the cooling load as well, and can account for as much as 60% of this load in existing buildings.

In each building, the existing conditions and interrelationship between systems must be carefully examined and understood in order to determine the full potential for energy conservation due to modification and operation of the electric lighting system. For instance, for each KWH reduction in lighting energy, there could be a reduction of an additional 1/2 KWH for cooling systems using electric refrigeration, plus some savings in the air and water distribution systems, for a total reduction of 1-1/2 KWH or more when the lighting and air conditioning loads occur simultaneously. Although the reduction in lighting also reduces the amount of useful heat available in the heating season, heat can be supplied more efficiently by the heating system than by the lighting system. Refer to Figure 64 for the effect of lighting reduction on cooling and heating.

Conservation of electrical energy in building lighting and power systems are especially important to our national energy goals; every unit of electrical energy saved in the building saves about three units of raw source energy.

Recently completed building lighting system designs have shown that office buildings and stores can be adequately lighted with an average of about 2 watts/sq. ft., and religious buildings and parking lots with 0.5 watt/sq. ft. There are many opportunities in existing buildings to reduce energy consumption by lighting to the levels suggested in ECM-1. For instance, existing windows can be used advantageously for day-lighting to reduce lighting electrical requirements by using existing or new switches to turn off lights at the perimeter when daylight is sufficient; task areas are well-defined, therefore non-uniform lighting can be provided to suit them; equipment with greater lighting efficacy is now available; and building usage has been established, permitting programming of systems to light spaces to the levels they require on a selective, as-needed basis. Caution: Some areas have minimum allowable light levels - refer to O.S.H.A. requirements as outlined in ANSI All 1-1965 (R-1973).

An energy conservation program for lighting involves two major distinct, but related areas: 1) Reducing annual energy consumption, 2) Reducing peak loads. All of the energy conservation measures outlined in the text will result in lower costs for electric consumption throughout the entire year, however, reducing the peak lighting demands which are likely to occur simultaneously with other peak electrical demands produces additional cost savings through lower established demand where such charges are included in the utility rate structure.

The most effective opportunities for reducing energy consumption with minimal cost and modification of the existing lighting system are outlined in ECM 1, Section 9, "Lighting". It is important to implement these first; in many cases, they can completely eliminate, or reduce the extent of the more costly measures outlined in ECM 2, Section 19. Many of the opportunities and options for further reducing energy consumption outlined in this volume, Section 19, can also be done at relatively low costs, i.e., switches and pilot lights, group relamping, replacing defective ballasts with more efficient ones. If no other building modifications are involved, some highly effective energy conservation options may require larger investment costs. However, if implemented at the time of building alterations or expansions for other purposes, the costs may be significantly less.

FIGURE 64: BUILDING ENERGY USAGE REDUCTION DUE TO 30% LIGHTING WATTAGE REDUCTION

dubin-mindell-bloome-associates consulting engineers

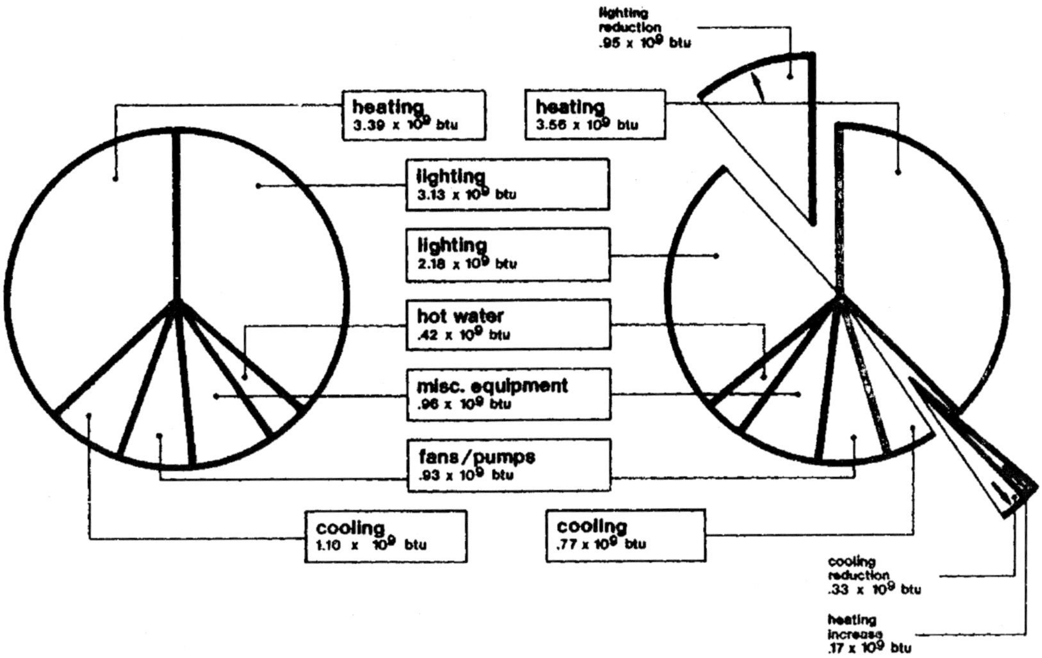

energy usage of original building	energy usage after 30% lighting reduction
9.93×10^9 btu	8.82×10^9 btu
lighting at	lighting at
3.5 watts/ft^2 for 75%	2.45 watts/ft^2 for 75%
1.0 watt /ft^2 for 25%	0.70 watts/ft^2 for 25%

note: reduction in lighting energy achieved without reducing illumination on tasks

	lighting	cooling	heating	hot water	fans/pumps	misc. equipment
	3.13×10^9 btu	1.10×10^9 btu	$3.39 \quad 10^9$ btu	$.42 \times 10^9$ btu	$.93 \times 10^9$ btu	$.96 \times 10^9$ btu
	2.18×10^9 btu	0.75×10^9 btu	3.56×10^9 btu	$.42 \times 10^9$ btu	$.93 \times 10^9$ btu	$.96 \times 10^9$ btu
difference	$-.95 \times 10^9$ btu	$-.33 \times 10^9$ btu	$+.17 \times 10^9$ btu	0	0	0

all btu's corrected for system efficiencies

Figure #64 - Engineering Data

Computer Run at Average Lighting of 2.875 W/ft^2

a. Building 135' x 115' located with long axis N-S
b. 8 floors 13'-0" floor to floor, 9'-0" floor to ceiling
c. 5 zones/floor (four perimeter - one central)
d. 15'-0" wide solid panels at
North Face - NE corner and NW corner
East Face - SE corner
West Face - SW corner
e. Lighting - 3.5 W/ft^2 for 75% floor area
1.0 W/ft^2 for 25% floor area
f. Power - 0.5 W/ft^2 for 100% floor area
g. Occupancy - 600 people 8 a.m. - 6 p.m. weekdays (except holidays)
h. Ventilation - 15,000 cfm outdoor air when occupied
i. Infiltration - 1/4 air change/hour constant
j. Indoor Conditions - 70°F, 30% RH October thru May when occupied
60°F, 42% RH October thru May when unoccupied
75°F, 50 RH June thru September
k. Equipment Loads - Ventilating Fans - 37 KW when occupied
Domestic Hot Water - 100,000 Btu/W when occupied
Pumps - June thru September, 27 KW when occupied
October thru May, 27 KW 24/hours/day
Other Equipment - 60 HP when occupied
l. Seasonal Efficiencies - Heating = 60%
Cooling - COP 2.5
m. Building Parameters
1. Walls - U = 0.06 Absorptivity - 0.9
2. Floor - U = 0.06
3. Roof - U = 0.06 50% of area covered by equipment rooms
4. Windows - double pane, clear glazing U = 0.65
5. Window/wall ratio 10% of wall exposed to room
6. Solar shading - No direct sun on windows March thru September.

No cooling allowed when building is unoccupied.
No heating allowed June thru September.

Computer Run at Average Lighting of 2.0125 W/ft^2

All parameters remain as above except:

e. Lighting - 2.45 W/ft^2 for 75% floor area
0.7 W/ft^2 for 25% floor area

Note: Reduction in lighting energy achieved without reducing illumination on tasks.

Computer Program - NBSLD

ENERGY CONSERVATION OPPORTUNITIES

ECO 103 IMPROVE LIGHTING MAINTENANCE

By maintaining a higher light output from existing fixtures, may be possible to reduce the wattage of lamps in each fixture and, in some cases, the number of fixtures in service without a reduction in illumination. Two variables affect the system maintenance:

- Lamp lumen depreciation (L.L.D.), is the drop in lamp lumen output over lamp life due to the lamp characteristics

- Luminaire (lighting fixture) dirt depreciation (L.D.D.), is the drop in lumen output due to accumulation of dirt on the interior reflective surfaces of the fixture and lens. See Figure 65.

The combination of L.L.D. and L.D.D. results in an average reduction in light output of about 25% from fluorescent fixtures (this will vary with room atmosphere and lamp types) over extended time with the maintenance procedures which are in general practice today.

Improve maintenance as follows:

- Replace existing lamps with new ones which have a lower lamp lumen depreciation over rated life (initial light output is often increased also). Fluorescent lamps with better L.L.D. factors may cost somewhat more, but the investment can generally be recovered by lower expenditures for energy and maintenance. Refer to Figure 67 for lamp lumen depreciation.

- Clean fixtures and lamps more frequently (refer to ECM 1) and institute a group relamping program (for higher maintained light output from each fixture - better L.L.F.) by replacing lamps when light output drops to 70 or 75% of initial output. Refer to Figure 66 "Fluorescent Group Relamping". (H.I.D. and incandescent lamps are similar).

A trade-off between the added investment costs (annualized) to replace lamps before the end of their useful life, and the improved performance and lower operating costs which result from shorter relamping periods must be analyzed. For example, from Figure 66, the optimum practical L.L.D. factor of a 40 watt, cool white fluorescent lamp with an 18,000 hour rated life* would be 75% of its rated life, or 18,000 hrs. x 0.75 = 13,500 hours. Lamps which were burning for 12 hours per day for 280 days per year would be energized 3,360 hours/year. Dividing 13,500 hrs. by 3,120 hrs. reveals a 4.02 year period for relamping. Relamping at longer intervals results in increased L.L.D. and more frequent burn-outs; relamping at shorter intervals results in slightly decreased L.L.D. but at the expense of added labor costs and lamp replacement costs.

*Rated life varies depending on hours operated per start. Refer to manufacturers published lamp information.

ECO 104 USE EFFICIENT LENSES

Remove lenses where they are not required for glare control (corridors, toilet rooms, non-critical task areas) hereby improving the fixture's coefficient of utilization (CU) and permitting a reduction in the wattage required (with lower wattage lamps) without an attendant reduction in the number of footcandles on the task or background.

Replace inefficient lenses and consider both efficiency and quality when choosing them for new fixtures. A more efficient lens can increase effectibe light output of each fixture thus allowing a reduction in the lamp wattage and/or decrease the number of fixtures required. To evaluate the efficiency of lenses in existing fixtures, use the fixture manufacturer's CU's for a standard fixture. Compare the CU with the present lens to the CU's resulting with contemplated replacements. The improvement in light levels will be directly proportional to improvement in CU. The relative efficiency and general characteristics of the most common lenses compared to prismatic plastic are as follows:

LENSES FOR FLUORESCENT FIXTURES

Prismatic Plastic and Glass - Generally the most efficient of its type for the degree of glare control. Best suited for all finished office space applications.

Plastic 1/2" x 1/2" or 3/8" x 3/8" Louver - Very inefficient, should be used only where extreme dirt build-up is a problem. About 33% less efficient than prismatic plastic.

Parabolic Wedge Louver - Very inefficient in transmitting light but provided excellent visual comfort and glare control. This lens is practical for use where the task requires a high level of visual acuity with low glare but may increase veiling reflections where the task is specular. About 45% less efficient than prismatic plastic.

LENSES FOR INCANDESCENT AND H.I.D.

Fresnel Type Lens - The most efficient of its type, best suited for all recessed fixture applications.

Opal or White Diffusers - Much less efficient in delivering light to a work surface. Generally 30% less efficient than Fresnel type lens. Good light diffusion.

CHOOSING LENSES

-Evaluate lens efficiencies for a particular fixture by comparing CU's. Use the same room reflectances and room cavity ratio with the CU's (from manufacturer's tables) for the various lenses available.

-Evaluate the needs of the task and the selected light source to determine the requirements for light control.

-Where visual acuity is critical, it may be better to use a less efficient lens with better glare control. The quality of lighting will be improved which may permit a reduction in footcandles (and wattage) for the same visual acuity.

FIGURE 65: LUMINAIRE DIRT DEPRECIATION OVER TIME

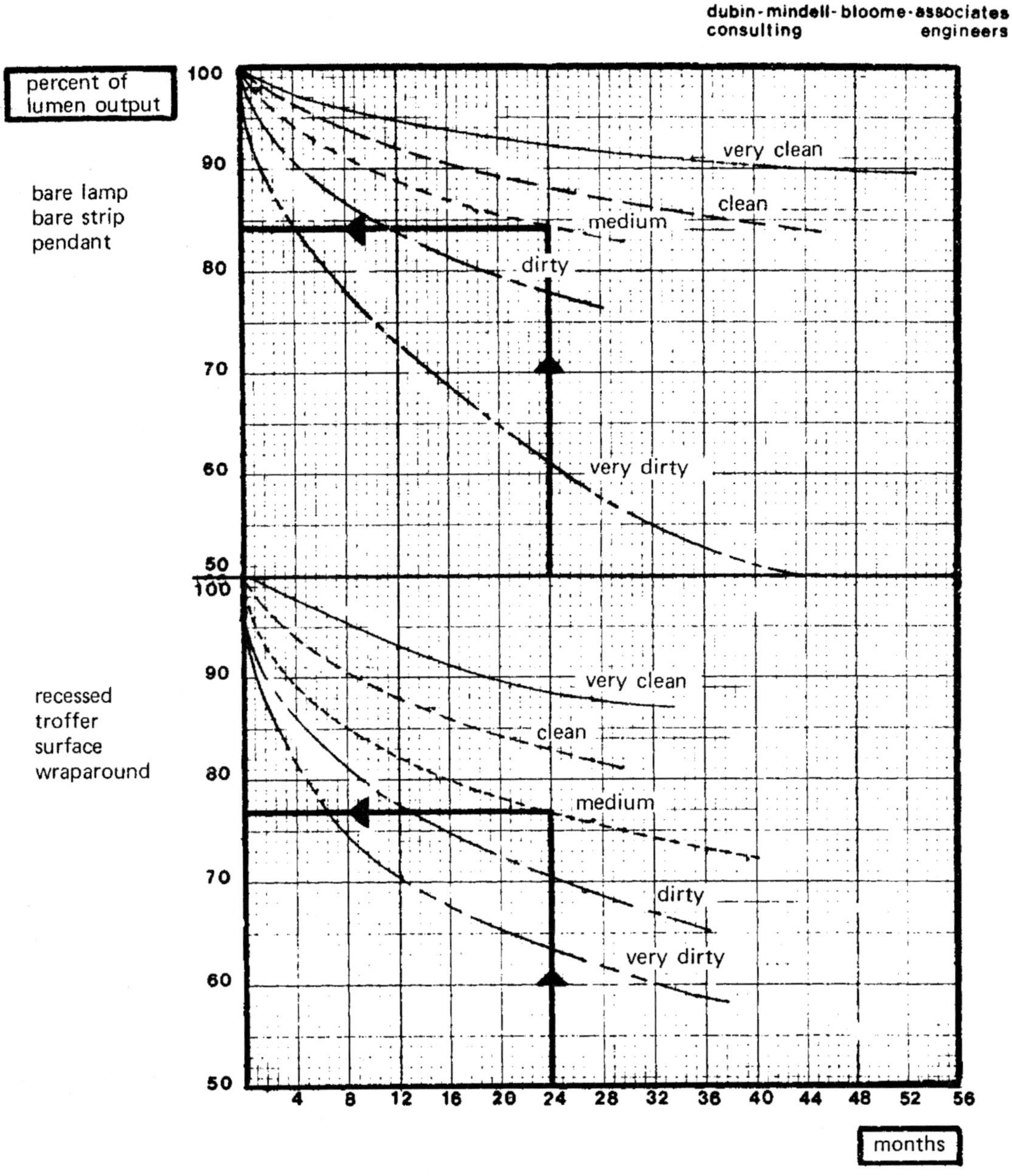

Figure # 65 Engineering data

Reference: I.E.S. Lighting Handbook, 4th edition 1968
page 9-17, figure 9-7 Illuminating Engineering
Society 345 East 47th Street New York, N.Y.

Based on actual measurements of lumen output of bare lamps in various environments at various lengths of time.

FIGURE 66: FLUORESCENT GROUP RELAMPING

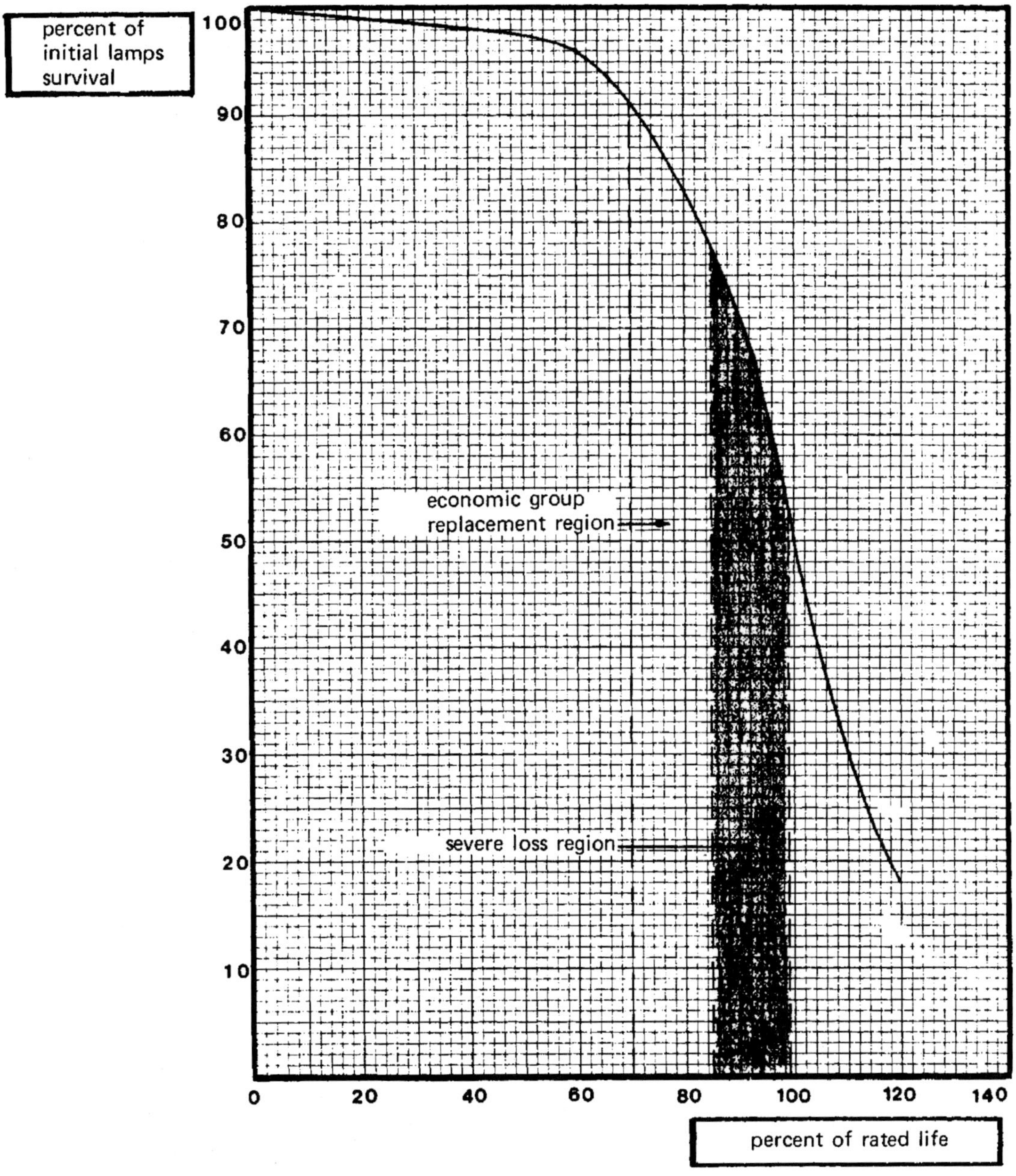

Figure # 66 Engineering data

Graph based on manufacture's information of statistics of flourescent lamp life.

FIGURE 67: LAMP LUMEN DEPRECIATION OF STANDARD LAMPS

dubin-mindell-bloome-associates
consulting engineers

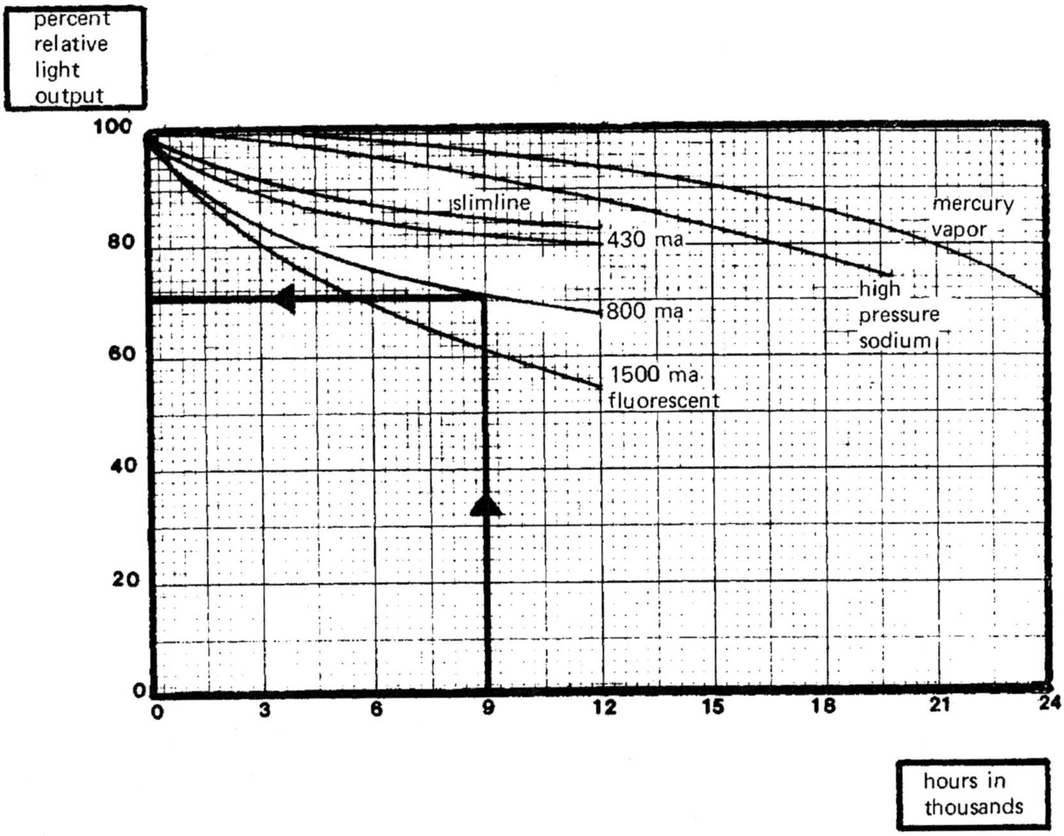

Figure # 67 Engineering data

Graph based on manufacturer's information of lamp performance, typical for the types listed.

-When higher efficiency light sources such as high pressure sodium are used, less efficient lenses providing more effective control may be required.

APPROXIMATE COSTS

Labor for lens replacement in existing fixtures is about $10.00/unit. The cost of various replacement lenses is approximately as follows:

Prismatic plastic (acrylic 2' x 4')	
- Extruded	$ 8.00
- Injection molded	12.00
Prismatic glass (2' x 4')	20.00
Louvered (2' x 4')	
- Plastic	12.00
- Aluminum	10.00
Parabolic wedge louver (2' x 4')	40.00
Dropped opal (acrylic 2' x 4')	12.00
Fresnel (8' x 8')	8.00
White diffuser (8" x 8")	5.00

ECO 105 PROVIDE SWITCHES TO TURN OFF UNNECESSARY LIGHTS

Turning off lights which are not needed will conserve energy and reduce expenditures for electricity. If switches are available, refer to ECM 1, Section 9, "Lighting", for guidelines and to Figure 27, "Effect of Switching on Energy Savings". Where local switches do not exist, consider adding them.

When surveying the building, list all room areas and the number of occupants (and their activities) in each space and the length of time that each space is occupied. A careful analysis will reveal many opportunities for reducing the intensity of illumination or shutting off lights completely. Install switches where opportunities exist for their effective use. When new building additions or extensive remodeling are undertaken, provide a sufficient number of switches so that all lights which are not needed for periods of time during the day can be turned off.

Individual light switches should be considered particularly for the following areas:

-All spaces with multiple fixtures which are not all required for periods of time.

-Areas intermittently occupied.

-Areas where daylight is available for illumination (i.e., perimeter zones).

-Halls of worship and other large areas where occupancy may take up only some of the space at any one time.

For adding switching in existing buildings, line voltage or low voltage surface mounted tape conductors are available which adhere to walls or ceilings. Ceiling switches (pull chain type) can be added where wall installations are not feasible.

For large areas, add remote control lighting contactors or lighting relays (to operate multiple circuits) above the ceiling or at the lighting distribution panel. Remote switches can be either line or low voltage.

Control site lighting and parking lot lighting with time switches and/or with photo-electric cells; the latter are also effective in controlling light fixtures at the interior perimeters of the building where daylight is sufficient for illumination.

A frequency controlled relay, now available, can be added at individual fixtures or at the circuit-breakers which control a number of fixtures. The relay is controlled by an activator which superimposes a special command frequency over the existing wiring system. At present, two command frequencies are available. Each activator will handle 300 forty watt rapid start ballasts located within 500 feet of the activator which is mounted at the panel. The activator can be controlled by remote control or local switch, time clock or photo-electric switch. One activator can control lights which are on any number of separate circuits.

Add pilot lights outside all rooms that are infrequently used where there is no other external indication that lights have been left on. Pilot lights should also be added to indicate when loads are energized at remote locations such as site lighting, sign lighting, snow melting, ovens, blueprint and reproduction equipment, mechanical spaces, penthouses, etc.

Pilot lights can be surface mounted or recessed and can usually be added in parallel with the circuit or the load to be monitored. To conserve energy, use neon type pilot lights rather than incandescent lights. In buildings where conduit runs from the switch boxes to the lighting fixtures, it may be possible to pull a neutral wire down to an external switch (if not already available there) and install a combination switch and pilot light in the existing switch box.

The cost of adding switching depends upon the type of switching system selected and the existing fixture circuiting. Time switches cost from $30.00 to $150.00. Photo-electric cells cost about $75.00. Frequency controlled relays cost about $20.00 per relay, and for the associated controller approximately $300.00 per channel, a wall switch costs about $25.00 each, low voltage relays cost as little as $6.00, and low voltage transformers cost approximately $25.00 each. A surface mounted neon pilot light can be installed over a door with wiring in surface raceway to a ceiling space and b.x. cable (12') to the nearest fixture in a room for as little as $25.00.

FIGURE 68: EFFICACY OF DIFFERENT LAMP TYPE AND WATTAGE

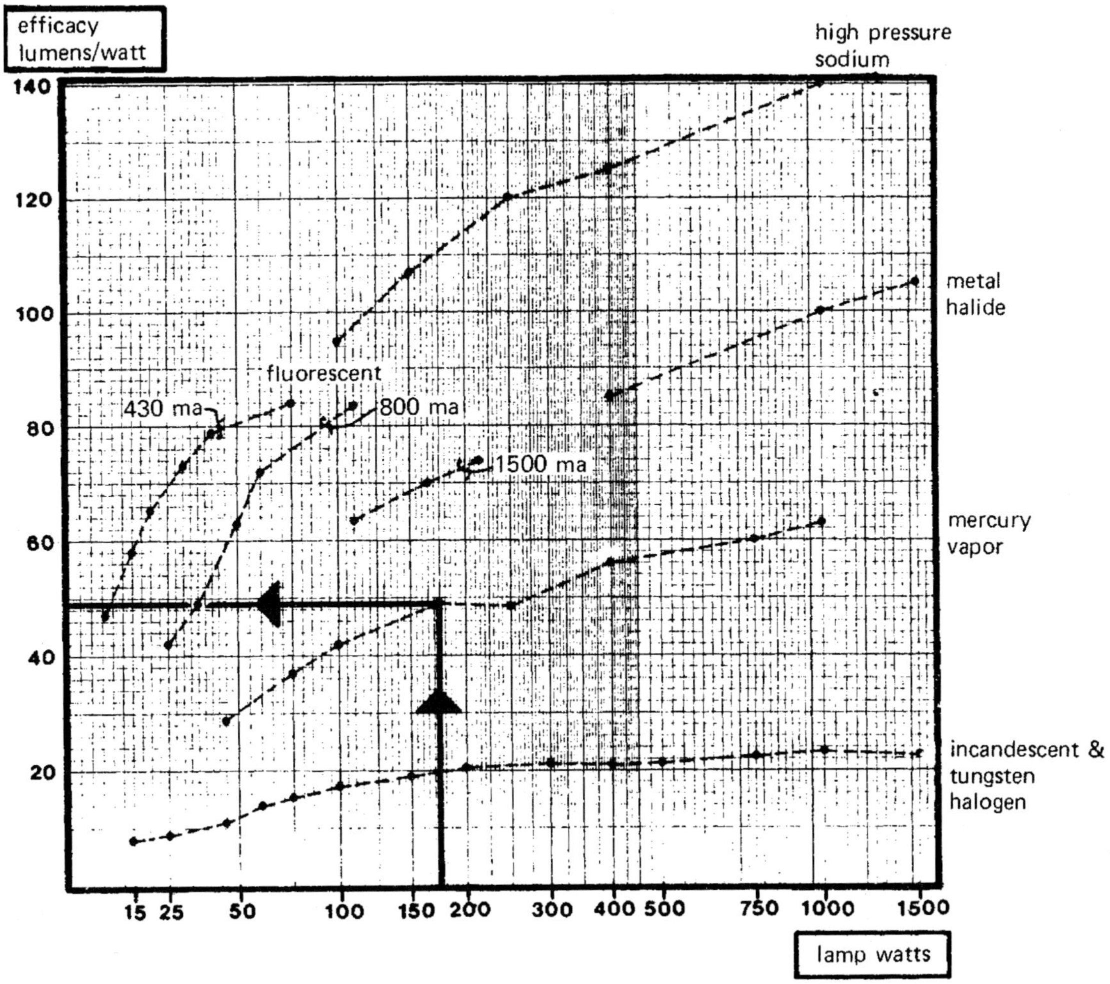

Figure #68 Engineering data

This graph is developed from manufacturer's information and is typical for lamp generic types.

Based on the following lamps

High pressure sodium - 100, 150, 250, 400 & 1000 watt size.
Metal halide - 400, 1000 & 1500 watt size, standard line lamps.
Fluorescent
430 ma - 4-90 watt size, cool white color
800 ma - 20-110 watt size, cool white
1500 ma - 110-220 watt size, cool white
Mercury vapor - 40-1500 watt sizes, delux white color
Tungsten halogen - 43-1250 watt size.
Incandescent - 3-1500 watt size, inside frosted.

ECO 106 MODIFY EXISTING FIXTURES TO ACCOMODATE MORE EFFICIENT LAMPS

ECM 1, Section 9, "Lighting", includes opportunities and guidelines for replacing lamps in existing fixtures with more efficient lamps or lower wattage lamps.

Energy savings can often be amplified by modifying an existing fixture to accept a different type and higher efficiency lamp. When this is not possible, consider replacing the fixture with one that will accept a lamp of greater efficiency. Refer to ECO 107 for fixture replacement. Figure 68 indicates the lumens per watt for various lamp types. More efficient lamps save energy by permitting use of lower wattage lamps or a reduction in the number of fixtures used.

Convert incandescent fixtures or lower efficiency H.I.D. fixtures to accommodate more efficient lamps where feasible. Conversion from incandescent to mercury vapor or other high intensity discharge source requires the addition of a ballast (when self ballasted lamps are not used). Ballast can be added at each fixture or remote mounted, (provided the wire from the fixture to ballast is sized properly). Fixtures which use mercury vapor and metal halide lamps can often be converted to high pressure sodium where color rendition is not critical. Conversion from any other H.I.D. source to high pressure sodium will require installation of a new ballast.

The tables that follow compare the efficiencies of lamps which can be used for the same general purposes and give some lamp types to consider when fixture conversion is contemplated.

Modify fixtures for site lighting, parking lots, canopies, advertising signs, display counters and other special applications as well as fixtures for interior lighting.

LAMP CHARACTERISTICS

Characteristics	Incandescent (including tungsten halogen)	Fluorescent	High-Intensity Discharge (HID): Mercury-Vapor	High-Intensity Discharge (HID): Metal-Halide	High-Intensity Discharge (HID): High-Pressure Sodium
Wattages normally available	6 to 1500	4 to 219	40 to 3000	400, 250 1000, 1500	100, 150, 250, 400, 1000
Efficacy (lumens per watt, lamp only)	15 to 25	55 to 92	20 to 63	80 to 100	100 to 130
Life (hours)	750 to 12,000	9000 to 30,000	16,000 to 24,000	1500 to 15,000	10,000 to 20,000
Light Control	Very good to Excellent	Fair	Very good	Very good	Very good
Relight Time	Immediate	Immediate	3 to 5 minutes	10 to 20 minutes	Less than 1 minute
Color Rendition	Very good to Excellent	Good to Excellent	Poor to very good	Good to very good	Fair
Comparative Fixture Cost	Low because of simple fixtures	Moderate	Higher than incandescent, and fluorescent	Generally higher than mercury-vapor	Highest
Comparative Operating Cost	High because of relatively short life and low efficacy	Lower than incandescent; replacement costs higher than HID because of greater number of lamps needed: energy costs generally lower than mercury-vapor	Lower than incandescent; replacement costs relatively low because of relatively few fixtures and long lamp life	Generally lower than mercury-vapor; fewer fixtures required, but lamp life is shorter and lumen maintenance not quite as good	Generally lowest; fewest fixtures required

INCANDESCENT AND H.I.D. - HIGHER EFFICACY LAMP REPLACEMENT GUIDE

PRESENT SOURCE	REPLACEMENT SOURCE (Lumen Outputs Will Not Be Reduced Below 85%)
Typical Incandescent	
60 - 75 Watt, A type	50 Watt, E. type, mercury vapor
100 Watt, A type	75 Watt, E type, mercury vapor
150 Watt, A type	75 Watt, E type, mercury vapor
150 Watt, Par or R type	100 Watt, Par. mercury vapor
200 Watt, A or PS type	175 Watt, E type, mercury vapor
200 Watt, Par or R type	100 Watt, R type, mercury vapor
300 Watt, PS type	175 Watt, E type, mercury vapor
300 Watt, R type	175 Watt, R type, mercury vapor
500 Watt, PS type	250 Watt, E type, mercury vapor
500 Watt, R type	175 Watt, R type, mercury vapor
750 Watt, PS type	400 Watt, E type, mercury vapor
750 Watt, R type	400 Watt, R type, mercury vapor
1000 Watt, PS type	400 Watt, E type, mercury vapor
1500 Watt, PS type	700 Watt, BT type, mercury vapor
Mercury Vapor	
175 Watt, E type	100 Watt, high pressure sodium
250 Watt, E type	100 Watt, high pressure sodium
400 Watt, E type	250 Watt, high pressure sodium
700 Watt, BT type	400 Watt, high pressure sodium
1000 Watt, BT type	400 Watt, high pressure sodium
Metal Halide	
400 Watt, E type	250 Watt, high pressure sodium
1500 Watt, BT type	1000 Watt, high pressure sodium

All of the above will require ballast addition or replacement. Before replacing lamps carefully consider the effects of color rendition where this is important.

Fluorescent Lamp Power Requirements

Lamp Designation	Nominal Length (inches)	Operating Current (milliamp)	Approx. Watts Consumed					Lamp Lumens**	
				Single-Lamp Circuit		Two-Lamp Circuit			
			Lamp	Ballast	Total	Ballast	Total	Each	2 Lamps
Preheat Types									
20T12	24	380	20	5	25	10	50	1155	2310
Instant Start									
96T12 (lead-lag)	96	425	75	-	-	40	190	5800	11600
96T12 (series)	96	425	75	-	-	22	172	5800	11600
Rapid Start 430 m.a.									
40T12	48	430	41	13	54*		92*	2805	5610
Rapid Start 800 m.a.									
48T12	48	800	60	25	85*		140*	3740	7480
96T12	96	800	112	18	130*		250*	8005	16010
Rapid Start 1500 m.a.									
48T12,48PG17	48	1500	116	24	140*		240*	5600	11200
96T12,96PG17	96	1500	218	42	260*		450*	12800	25600

*Total watts consumed includes power required to heat lamp cathodes.

** Based on cool white color, mean lumens output.

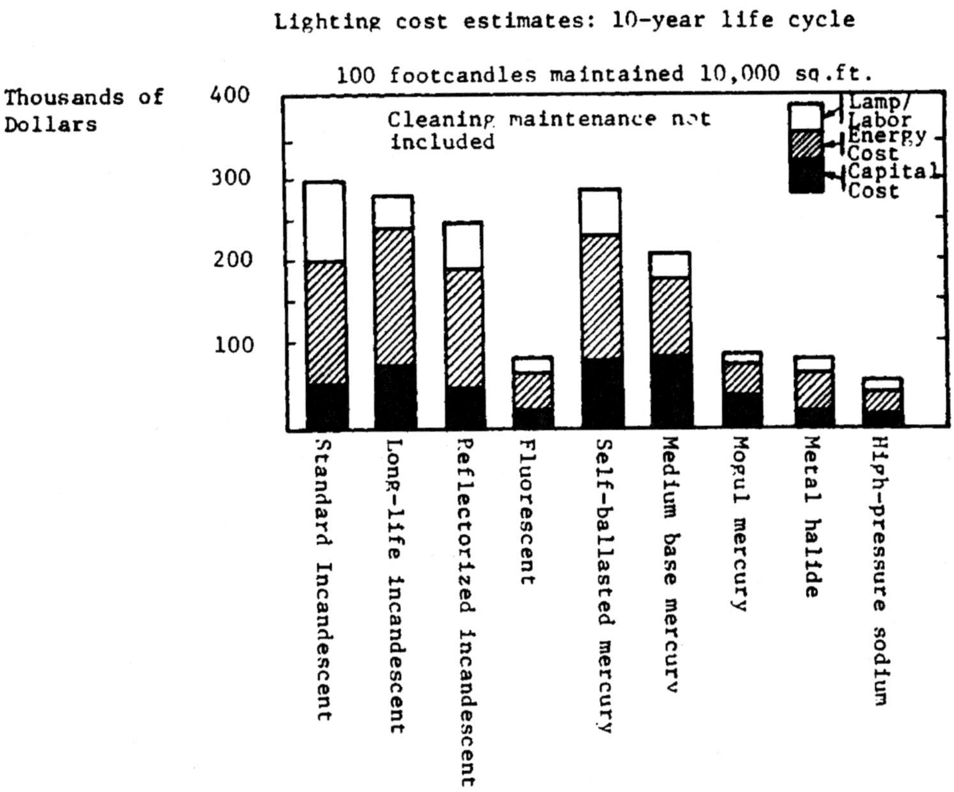

THE TOTAL 10-YEAR COSTS OF PRODUCING ONE MILLION USEABLE LUMENS

* Although there are many types of fluorescent lamps, all types generally follow the costs shown.

ECO 107 INSTALL NEW EFFICIENT FIXTURES

The lamp and fixture efficiency together determine the quantity of light transmitted into a space for each watt of power consumed. When a more efficient lamp source would suit the application but conversion of the fixture to handle this source is not possible, consider replacing the fixture itself. Select the most efficient light source for the application then choose a fixture with high efficiency performance. Fixture performance is indicated by its coefficient of utilization under specific room conditions, and includes the effect of lenses or reflectors. The total fixture performance in light output per watts input depends upon both lamp lumens per watt and the fixture CU. Consult manufacturers for CU or particular fixtures under consideration.

The price of fixtures is not directly related to efficiency, but incandescent fixtures are generally less expensive than flourescent fixtures which are less expensive than the higher output H.I.D. fixtures.

Fixture price is not an indication of efficiency, for example one manufacturer lists three recessed fluorescent troffers, the least expensive of which has the lowest CU. Choose the efficient reflector when fixtures are available with optional reflector types.

EXAMPLES OF CONSERVATION OF ENERGY BY REPLACING FIXTURES AND LAMPS

	EXISTING	REPLACEMENT
STORE	66-500 watt incandescent down lights	20-400 watt metal halide surface modules
	Total KWH/yr. = 122,100	Total KWH/yr. = 34,040
	Total savings = 88,060 KWH/yr. ...	72% savings
	17-1000 watt incandescent pendant fixtures	14-400 watt metal halide pendant fixtures
	Total KWH/yr. = 62,900	Total KWH/yr. = 24,050
	Total savings = 38,850 KWH/yr. ...	62% savings
PARKING AREA	12-1500 watt tungsten halogen lamp floodlights	4-1000 watt metal halide cluster
	Total KWH/yr. = 27,000	Total KWH/yr. = 6,480
	Total savings = 20,520 KWH/yr. ...	76% savings
SIGN	2-1500 watt tungsten halogen floodlights	2-400 watt metal halide floodlights
	Total KWH/yr. = 12,000	Total KWH/yr. = 3,680
	Total savings = 8,320 KWH/yr. ...	69% savings

	Existing	Replacement
SECURITY	6-500 watt tungsten halogen floodlights	6-150 watt high pressure sodium lights
	Total KWH/yr. = 12,000	Total KWH/yr. = 3,920
	Total savings = 8,080 KWH/yr. ... 67% savings	
SUPER MARKET	250-4 lamp 40 watt strip fluorescent fixtures	70-400 watt high pressure sodium down lights (where change in color rendition is acceptable)
	Total KWH/yr. = 146,250	Total KWH/yr. = 84,000
	Total savings = 62,250 KWH/yr. ... 43% savings	

OFFICE

Building Area	1,000,000 sq. ft.
Office Area (fluorescent)	700,000 sq. ft.
Support Areas (incandescent)	
Mechanical Areas	150,000 sq. ft.
Circulation and Storage	150,000 sq. ft.

	EXISTING	REPLACEMENT
Mechanical Area Lighting Energy Consumption	Primarily incandescent lamps used	90% fluorescent lamps used
	117,500 KWH/yr.	70,500 KWH/yr.
Circulation and Storage Area Lighting Energy Consumption	85% incandescent 15% fluorescent	15% incandescent 85% fluorescent
	518,400 KWH/yr.	324,00 KWH/yr.
Yearly Energy Consumption in these Portions of the Building	635,900 KWH/yr.	394,500 KWH/yr.

Total Energy Saved

635,900 KWH/yr.
394,500 KWH/yr.
241,400 KWH/yr. ... 38% savings

ECO 108 INCREASE ROOM INTERREFLECTANCES

The larger a room and the lighter the color of room finishes and furnishings, the lower the light absorption by these objects, and hence less watts per square foot will be required to produce the same footcandles.

ECM 1, "Lighting" includes guidelines and costs for cleaning and/or painting and redecoration room surfaces to increase interreflectances which will usually permit a reduction in the wattage for lamps in existing fixtures or may even permit turning off some of the fixtures completely without a reduction in footcandles.

Electrical comsumption can be reduced by about 15% with an improvement from 50-30-20 to 80-70-50 wall, ceiling and floor reflectance values and by almost 35% with improvement to 80-70-50 from an original level of 50-10-10 (see example below).

When remodeling, use open landscape planning to eliminate as many interior partitions as possible; even though light colored walls reflect a large percentage of lighting, they also absorb lighting energy.

An interior environment with lanscape planning can be made to be aesthetically pleasing without affecting building functions by proper use of plantings and ascoustical treatment.

Studies made by the authors for the GSA Energy Conservation Demonstration Building in Manchester, N.H. reveal that a 50' x 80' office space designed for open landscape planning would use 25% less energy for illumination than an equal amount of area divided into 30 smaller sized offices.

The following example reveals (1) the reduction in wattage possible by increasing the size of office modules with the same surface reflectances, and (2) the decrease in wattage made possible by increasing interior surface reflectances for an equal sized office.

ASSUMPTIONS:

Reflectances	- 80-50-30
Ceiling height	- 10'
Fixtures, fluorescent	- 200 fixtures
Use	- 2400 hours/year
Fixture CU	- As shown for specific conditions

(1) Effect of Room Size on Watts/Sq. Ft. to Produce 60 Footcandles, Maintained

$$\text{no. of fixtures} = \frac{\text{area x footcandles}}{\text{no. of lamps x lumens x CU x MF}} = \frac{\text{area x 60}}{4(3150) \text{ x CU x } 0.75} = \frac{\text{area x } 0.0064}{\text{CU}}$$

12' x 12' room, no. of fixtures $= \frac{144 \text{ x } 0.0064}{0.43 \text{ CU}} = 2.19$ fixtures = 438 watts

$$\frac{\text{watts}}{\text{sq.ft.}} = \frac{438}{144} = 3.04 \text{ watts/sq.ft.}$$

16' x 24' room, no. of fixtures $= \frac{384 \text{ x } 0.0064}{0.56} = 4.39$ fixtures = 878 watts

$$\frac{\text{watts}}{\text{sq.ft.}} = \frac{878}{384} = 2.28 \text{ watts/sq. ft.}$$

40' x 60' room, no. of fixtures $= \frac{2400 \text{ x } 0.0064}{0.70} = 21.9$ fixtures = 4388 watts

$$\frac{\text{watts}}{\text{sq.ft.}} = 4388 = 1.83 \text{ watts/sq. ft.}$$

The electrical energy load, in watts/sq. ft., for lighting in rooms with 1/2 or 3/4 height partitions (with or without glass from top of partition to the ceiling) falls between that for individual office layout and that for open space.

(2) Effect of Room Finish on Power Wattage Requirements

Basic formula: Assume a 16' x 24' room with a 7' cavity height

$$\text{Room cavity ratio (rcr)} = \frac{5(\text{cavity height})(\text{length \& width})}{\text{length x width}} =$$

$$\text{no. of fixtures} = \frac{\text{area x footcandles}}{\text{lamps x lumens x CU x MF}}$$

$$= \frac{384 \text{ x } 60}{4(3150) \text{ CU x } 0.75} = 2.44 = \text{no. of fixtures}$$

80-70-50 interreflectances, and CU = 0.63 therefore 2.44 = 3.85 fixtures

= 770 watts

$$\frac{\text{watts}}{\text{sq.ft.}} = \frac{800}{384} = 2.08 \text{ watts/sq. ft.}$$

50-30-20 interreflectances and CU = 0.53, therefore, $\frac{2.44}{0.53}$ = 4.58 fixtures

= 916 watts

$\frac{\text{watts}}{\text{sq.ft.}} = \frac{916}{384}$ = 2.38 watts/sq. ft.

50-10-10 interreflectances and CU = 0.47, therefore, $\frac{2.44}{0.47}$ = 517 fixtures

= 1034 watts

$\frac{\text{watts}}{\text{sq.ft.}} = \frac{1034}{384}$ = 2.69 watts/sq. ft.

ECO 109 LOWER THE FIXTURE MOUNTING HEIGHTS

Lowering the mounting height of ceiling mounted lighting fixtures can result in fewer watts required to illuminate given tasks - since the lumens/watt that reach the task increase as the distance between light source and task decreases.

Mounting heights can be decreased by (1) lowering the mounting heights of the fixture alone (either existing or replacement fixtures) and (2) lowering the entire ceiling. Consider all of the following factors when lowering mounting heights:

-Capital Cost

-Increased light output due to increased direct and diffused light and/or reduced wattage through lamp changes.

-Glare and/or quality of illumination where visual comfort is a factor.

-Fixture patterns which suit the task and fixture light distribution.

Because of high initial costs, do not consider lowering fixture mounting heights solely for energy conservation alone under the following circumstances:

-When floor-to-ceiling heights are already lower than 10 feet (except for stem or chain mounted incandescent fixtures in selected task areas).

-When fluorescent fixtures are recessed in ceilings which are lower than 13 feet.

-When building function is impaired to the extent that extensive additional construction would become necessary (moving ducts, pipes, structural elements).

It is generally less costly to rehang fixtures than to lower the entire ceiling. The choice of options depends upon the comparative costs and benefits, which should be analyzed for the alternative.

The energy saved by dropping the mounting height form 14' to 9' in a 30' x 40' space while maintaining the same illumination level would be 10% of the annual energy consumed for lighting the area.

Also consider lowering stem or chain mounted fixtures in existing stores, supermarkets, warehouses, and high ceilinged lobbies. In supermarkets, fixtures are often mounted on the ceiling, 14 to 18 feet above the floor, with the spaces lighted to 100 to 150 footcandles - above that necessary to read labels of packages on the lower shelves. Lowering fixture mounting heights and reducing the wattage of lamps may reduce lighting power and costs by 50% or more.

ECO 110 USE NON-UNIFORM LIGHTING

Uniform lighting maintained at a level necessary for the most critical tasks in a given area wastes energy when other less critical tasks within the same area do not require the same amount of illumination.

Uniform lighting is a major cause of excessive energy use, especially in offices and stores with different visual task requirements. Convert uniform lighting systems to non-uniform (or selective lighting), so that each distinct functional area within the building and the discrete tasks which occur within the same room are lighted only to the lighting levels and quality required for each task, and only for the time span when the tasks occur. Refer to ECM 1, Figure 26, "Recommended Lighting Levels".

Non-uniform lighting can be accomplished by the following measures either individually or in combination with others:

- Install switches to turn off unnecessary lights. Refer to ECO 105.
- Use lamps with light output required for specific tasks.
- Relocate or install new fixtures to suit specific tasks.
- Control lamp intensity with dimmers. Refer to ECO 113.
- Use multi-level ballasts to obtain light level required for specific tasks. Refer to ECO 112.
- Use furniture mounted lighting fixtures. Refer to ECO 111.

In order to select the most appropriate measures, make a careful analysis of task requirements, expected duration, the possible future relocation, the quality of illumination required for specific tasks, and the frequency with which they may change.

Grouping work tasks with similar requirements can reduce the total amount of illumination necessary in a space where tasks of lesser visual requirements and circulation areas are also located in the same space.

When relocating ceiling fixtures to provide task oriented lighting, there will be an opportunity to improve the quality of illumination which will result in reduced power requirements. The quality of light is most significant in any accounting, drafting, typing, reading, and clerical areas. but is also important in many retail store applications, i.e. disability glare from lighting fixtures may cause customer discomfort and loss of interest in shopping.

A lighting system characterized by high quality illumination can reduce energy consumption for illumination by as much as 25% to 33% while maintaining the same visual environment as that provided by a system of lower quality.
The efficacy of a lighting system is reduced by two factors: Discomfort glare caused by extreme uneven luminance in the field of view and veiling reflections.
Visual comfort probability (V.C.P.) is a measure of discomfort glare.
For satisfactory V.C.P. brightness ratios in the field of view should be restricted to 3:1 for immediate background and 10:1 for more remote background.
To reduce veiling reflections, lighting fixtures should be kept out of the zone from which reflections originate - the area in front of the task and in the viewing angle of incidence.

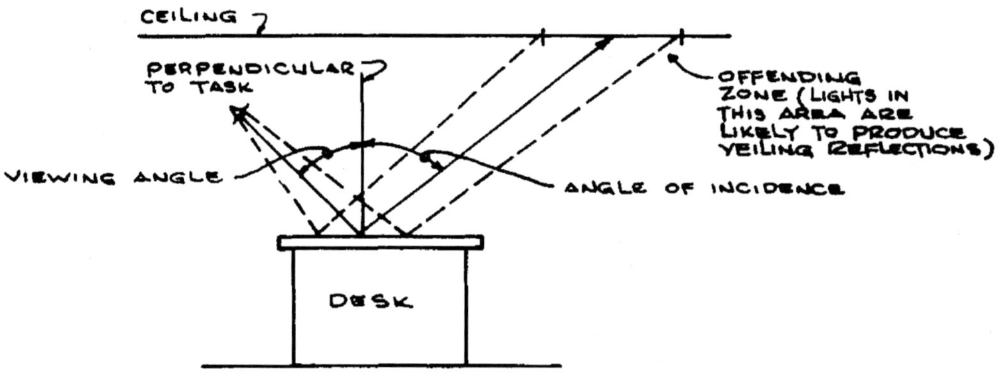

VEILING REFLECTION DIAGRAM

Refer to the I.E.S. Lighting Handbook for calculations for discomfort glare and principles for limiting vecting reflections. It may be possible to reduce energy consumption by as much as 25/35% with a better visual environment by providing higher quality illumination.

ECO 111 USE FURNITURE MOUNTED TASK LIGHTING AND REDUCE GENERAL ILLUMINATION

Furniture mounted lighting provides illumination of specific tasks to the extent necessary; circulation areas between tasks and the background can be maintained at lower levels.

At least one furniture manufacturer is currently designing modular work stations with built-in lighting as an option.

Power for the furniture mounted fixtures can be supplied from under floor electrical duct from existing lighting circuits in the ceiling and with power poles or by surface mounted raceways.

An added advantage of furniture mounted task lighting is that the light source will move with the furniture, when it is relocated.

Furniture mounted fixture applications should be considered:

- When major renovations are made to the interiors to accommodate an open landscape office plan, or

- When new desks and tables are being purchased.

The existing ceiling fixture can be used to provide additional low level general illumination when it is needed to supplement furniture mounted task lighting.

The operating costs for electricity for lighting and air conditioning when using furniture mounted fixtures may be reduced in many installations from 30 to 60%. Since the lighting fixtures are a part of the furniture, their cost can be written off for tax purposes, in a shorter period of time than ceiling mounted fixtures. When comparing the economic feasibility, include the cost of furniture with integral lighting vs. separate furniture and conventional ceiling mounted lighting system. Refer to examples on the following page.

ECO 112 USE MULTI-LEVEL BALLASTS

In office buildings or stores, portions of a single large space may be used for different purposes each week or month. Multi-level ballasts with 2 or 3 levels of lamp control are available for 430 milliamp fluorescent lighting fixtures; they allow reduction in illumination levels without sacrificing the symmetry of the lighting fixture pattern. A percentage reduction in power actually exceeds the percentage reduction in lumen output at the lower level.

Two-level ballasts can replace single-level ballasts in existing fluorescent installations without a change in the fixture and with only slight modifications, if any, to mounting and wiring. A three-level ballast can also be used with existing fixtures and requires mounting and wiring changes. At present, level switching is available only at the ballast, and wall switches with relays must be used if remote switching is desired. Other equipment for remote control of multi-level ballasts without relays is currently in development.

The lumen output of 400 watt mercury vapor lamps can be reduced by 50% through modifications to the ballast system which permit switching from a 33 microfarad capacitor to a 22 microfarad capacitor. The lumens are reduced by 50% and the wattage by about 40%. Reduced power factor militates against a one-to-one reduction. If the P.F. of the entire building is high, the lower P.F. of the lamps will not be a serious problem.

Multi-level fluorescent ballasts cost about twice as much as standard ballasts, but if the low level is used most of the time, the costs are repaid within a few years. The cost of installing a second capacitor in mercury vapor ballasts for multi-level switching varies widely, when it can be done at all.

Where new electrical fixtures are contemplated, consider using a 3 level lighting fixture in place of a 3 level ballast. These are offered as 3 lamp 3' x 4' fixtures with standard 40 watt lamps. The center lamps are wired in tandem. A separate switching circuit is required for each light level. One manufacturer offers the following data:

3 lamps: 124 fc at 3.7 watts per sq. ft.
2 lamps: 83 fc at 2.5 watts per sq.ft.
1 lamp: 37 fc at 1.2 watts per sq.ft.
Reflectances: Ceiling 0.80 - Walls 0.40 - Floor 0.6

ECO 113 INSTALL DIMMERS TO REDUCE LIGHT LEVELS

When frequent light level changes are required in areas of transient occupancy, dimmers are a better alternative to multi-level switching, multi-level ballasts, or lamp replacement. The least expensive application is dimmers for incandescent lights - materials cost about $30.00 for a 600-watt dimmer and about $125.00 for a 2000-watt dimmer. Fluorescent lamp dimming is more expensive as ballasts must be replaced and a wall control unit installed. Ballasts cost about $15.00 each, and $70.00 per dimmer control module which will handle up to twenty 40-watt lamps. Dimmers for 400- and 1000-watt mercury lamps, which also require ballast replacement and special control modules, are more costly than fluorescent dimmers. (Costs are highly variable, but may be approximately $75.00 for an additional ballast and from $400.00 to $2,500.00 for the control unit, depending on the number of lamps dimmed.

ECO 114 MAKE BETTER USE OF SKYLIGHTS AND WINDOWS FOR DAYLIGHTING

SKYLIGHTS

Reactivate skylights; modify them to provide daylight to supplement electric lighting without glare or excessive heat.

The amount of illumination available from skylights varies with the following factors:

- The duration and intensity of available daylight which is affected by amount of sunlight, shading, and ground reflectances.
- Area and location of the skylights.
- The distance from floor to skylight.
- The transmission properties of the skylight.
- The construction of the skylight in regards to orientation of transparent or translucent surfaces.
- The reflectance of the interior surface of the skylight well. Refer to Reference 40 for methods of calculating daylighting, using skylights.

If electric lighting is turned off when daylight from skylights is available, as much as 40 watts of electric power can be saved per sq. ft. of skylight area. The amount of energy saved for lighting using skylights must be analyzed in comparison to the increased amount of energy required for heating and the increase or reduction in heat gain and cooling load in order to determine the net yearly increase or reduction in energy due to skylights.

In order to reduce heat loss, heat gain, and glare, consider the following options:

- Paint skylights with white or light colored paint.
- Add prismatic lighting lens below the skylights to reduce glare.
- Provide an external louver above the skylight to reduce solar heat gain.
- Add louvers above the skylights and lens below to reduce glare, heat loss, and heat gain.
- Install a movable shutter above the skylight to reduce heat gain and glare in summer, but allow solar radiation to be used to offset heat loss in the winter.
- Paint the interior of the light well with a light color to increase illumination.
(Caution: Where skylights are used as an automatic smoke hatch, modifications must not change this function.)

Modifying skylights with louvers and/or lenses costs about $6.00 per sq. ft. for lenses and about $10.00 per sq. ft. for louvers. Shades cost about $25.00/16 sq. ft. of skylight and about $50.00 each to be motorized.

Consider installing skylights in existing buildings when reroofing, if the energy conservation balance is favorable. The cost of installation is lowest at that time.

WINDOWS

Refer to ECM 1 for opportunities to maximize the use of daylight from windows. When major building renovations or building additions are to be instituted, consider adding windows (with multiple glazing, see Section 13, and thermal barriers in cold climates).

The number, size and type of windows to be used and the shielding or screening that should be provided can be determined by evaluating and instituting the recommendations for window treatment and selection of types in both the mechanical and electrical sections of ECM 1 and 2.

The net amount of energy saved or lost due to additional windows in place of opaque walls must be analyzed as described above for skylights and detailed in ECM 1, Section 9.

As compared to opaque surfaces, windows or skylights reduce the energy required for lighting, reduce the energy required for cooling by reducing heat gain from lights, reduce heating load in the winter by permitting more solar radiation to enter (benefit of this item outweighed by heat loss in all climates), but may increase cooling load in summer due to solar radiation and conduction losses (well shaded windows reduces this liability).

ECO 115 UTILIZE REFLECTORS TO MAXIMIZE DAYLIGHT FOR ILLUMINATION

Install reflectors at windows to increase the amount of daylighting. Horizontal or vertical reflectors located on the exterior of the building can often be added without significantly changing the appearance of the building. The effectiveness of windows for daylighting can be increased by 25% or more with proper reflector devices. The light contributed by the reflecting surface should be directed onto the ceiling. Diffuse daylighting provides more lumens per watt than the artificial sources it replaces and the heat gain and load on the air conditioning system will be less than for an equal quantity and quality of electric lighting.

Where possible in light wells and interior courts, paint building walls lighter colors to increase reflectances onto the windows. Reflectors can also be added at windows to direct more light into the interior space. At windows close to grade, a border of white stone or concrete serves well as a reflector.

When using daylighting, light levels must be controllable; if they are significantly higher than required, the excess heat gain (which increases the cooling load) could offset the energy saved in electric lighting. Provide shades, drapes, or blinds at the windows to allow control of the intensity of light.

ECO 116 INSTALL HIGH FREQUENCY LIGHTING

When remodeling or expanding all or a portion of an office building or store with fluorescent lighting (or when changing from incandescent lighting to fluorescent), consider high frequency lighting. Existing fixtures can be modified to be used with high frequency systems, and new fixtures are available with high frequency ballasts.

The advantages as compared to 60 Hertz systems, include the following:

- Lamps produce about 10% more lumens/watt.

- Ballasts can be located out of the conditioned area, reducing the load on the air conditioning system.

- In a new addition to the building, fewer lighting fixtures are required.

- Ballast life is increased.

The higher frequencies most commonly considered are 400, 800, and 3,000 Hz. Future installations may include 10,000 Hz. The higher the frequencies, the greater are the advantages.

Two types of converters commonly used are (1) rotary phase converters, which are about 70-85% efficient and (2) solid state converters, which are approximately 90% efficient.

The increase in lamp efficacy, and the reduction in power for air conditioning exceed the converter losses. (Refer to Fig. 72 "Fluorescent Lamp Efficiency vs. Frequency"). Mercury vapor and incandescent lamps show no significant improvement in light output at higher frequencies.

The converter and series capacitors can be mounted in each circuit or at a central location. With a central system, the circuits serving the fluorescent lighting fixtures must be isolated from circuits supplying other loads. As a result, major wiring revisions are often required in existing installations and must be accounted for in cost and economic feasibility analyses.

If a total energy system is installed at the same time, high frequency lighting can be generated directly and the costs for converters eliminated. Refer to Section 8B.

Individual converters which can be mounted directly on a fluorescent light fixture (eliminating the need for separate wiring), are in development and may be available soon.

The capital costs to convert a system from 60 Hertz to higher frequency are quite high; undertake a preliminary analysis to determine whether the payback period for such an installation is favorable before spending time on a detailed analysis as outlined in Section 23.

ECO 117 USE THE WASTE HEAT FROM LIGHTS

The major advantage of a "heat-of-light" system lies in its reduction of heating, cooling and HVAC system and distribution loads, rather than in savings in electrical energy for lighting. However, slightly higher lamp efficiencies will result as the cooling effect on the lamps increases their output. Refer to Figure 69 for the effect of heat transfer on light output.

Although "heat-of-light" systems can conserve energy and reduce operating costs, the installation for such systems will be quite costly unless renovations to the building, requiring completely new lighting and duct work systems, are contemplated.

The two types of heat-of-light systems, "Dry" and "Wet", provide the following advantages in office buildings and stores:

- Excess heat from interior areas of the building can be collected and distributed to the perimeter areas.
- The sensible room heat component of the cooling load is decreased permitting a reduction in the quantity of air required for cooling (and saving fan horsepower).
- In the case of wet heat-of-light systems, the cooling load is reduced and less power is required for the refrigeration units.

Refer to Figures 69 and 70 for thermal performance of wet heat-of-light systems.

Both systems require special fixtures, and in addition, the wet systems also require a completely new water circulating system and connections to a cooling tower and is thus not particularly suitable for retrofitting.

"DRY" HEAT-OF-LIGHT SYSTEMS

For dry heat-of-light, room air is extracted through the lighting fixture, passes over the lamp and ballast, and is either ducted to a fan or drawn into the ceiling plenum space. The heated air from the lamp and fixture can be supplied to cooler perimeter zones during the heating season or recirculated back to an existing air handling unit or discharged outdoors during the cooling season. If a ceiling plenum system is used, a separate fan is necessary to draw the warm air from the ceiling plenum and deliver it to cooler zones or to discharge it outdoors.

When ceiling plenums are used as collection chanbers, each zone must be isolated from the others by a vertical barrier and ceilings over conditioned areas should be insulated to a "U" value of .1 or better to limit reradiation to the occupied spaces. Consult all codes to comply with fire protection requirements.

"WET HEAT"-OF-LIGHT SYSTEMS

The special lighting fixtures include built in water passages and air extract inlets. Air is extracted through the fixture, but in addition, water is circulated through the passages to the cooling tower where the heat is removed and the cool water piped back to the fixtures. The heat removed from the fixtures can be used for reheat purposes in the cooling system instead of being rejected to the outdoors. The circulating water system can also be used to pick up heat

FIGURE 69: EFFECT OF HEAT TRANSFER ON LIGHT OUTPUT AND ELECTRICAL USAGE

dubin-mindell-bloome-associates consulting engineers

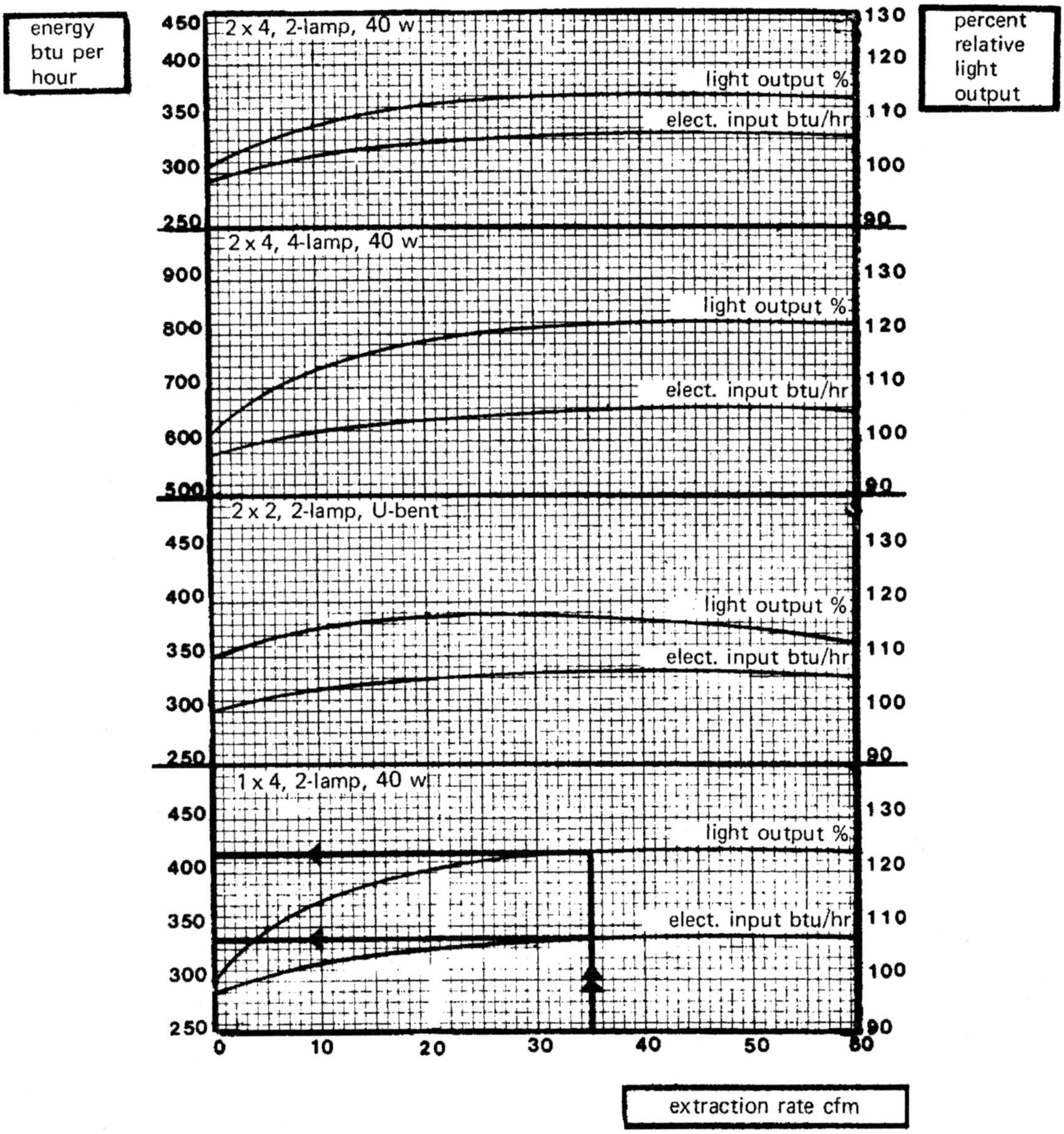

Figure #69 Engineering data

This graph is derived from manufacturers information and is typical for "dry heat of light" fixtures.

FIGURE 70: THERMAL PERFORMANCE OF "WET" HEAT OF LIGHT FIXTURE

dubin-mindell-bloome-associates
consulting engineers

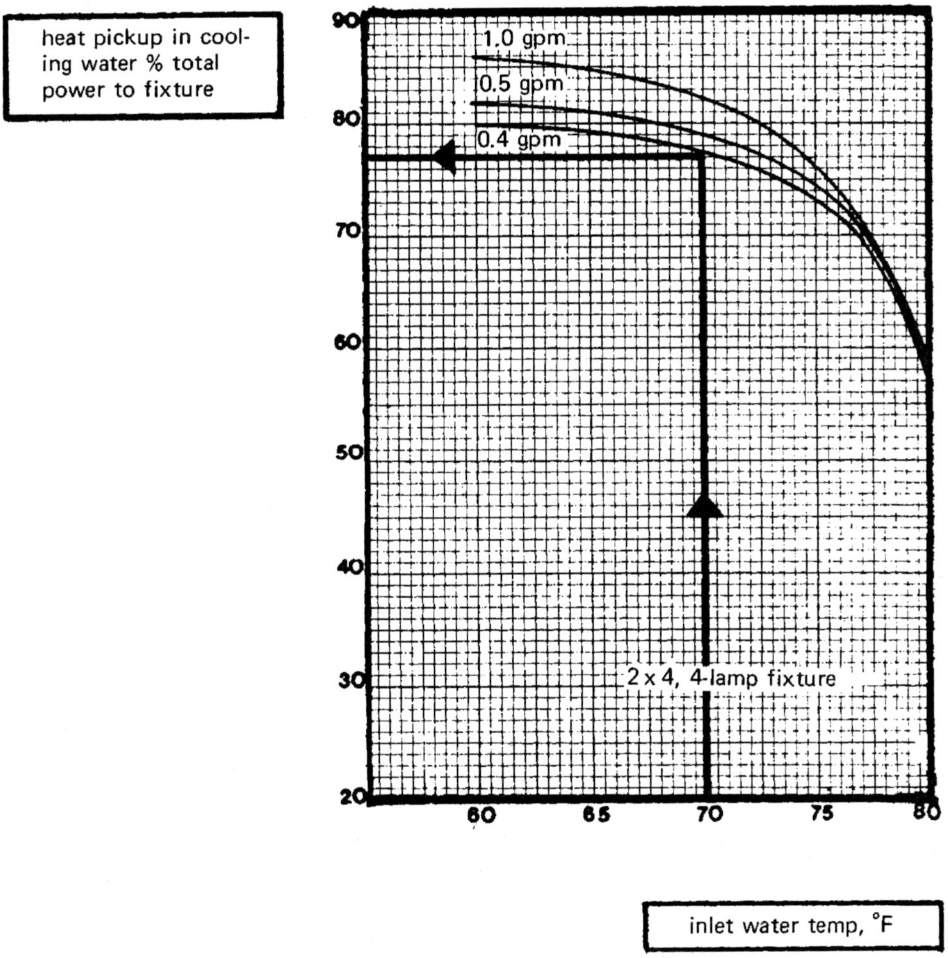

Figure #70 Engineering data

This graph is derived from manufacturers information and is typical for "wet heat of light" fixtures.

FIGURE 71: REDUCTION OF POWER USING DIMMING BALLAST

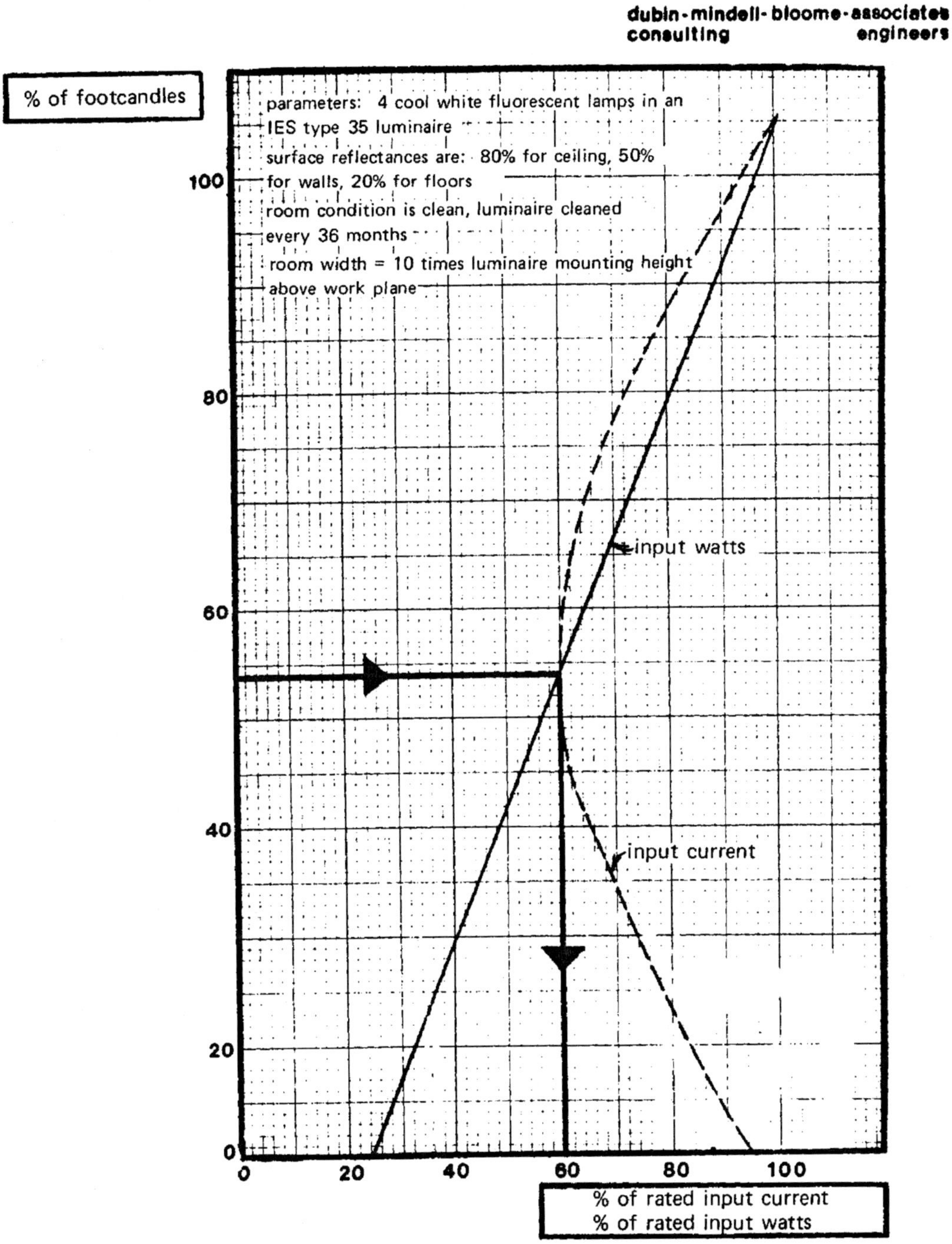

Figure #71 Engineering data

This graph is derived from manufacturers information and is typical for fluorescent dimming ballasts.

FIGURE 72: IMPROVED EFFICACY WITH HIGHER FREQUENCIES

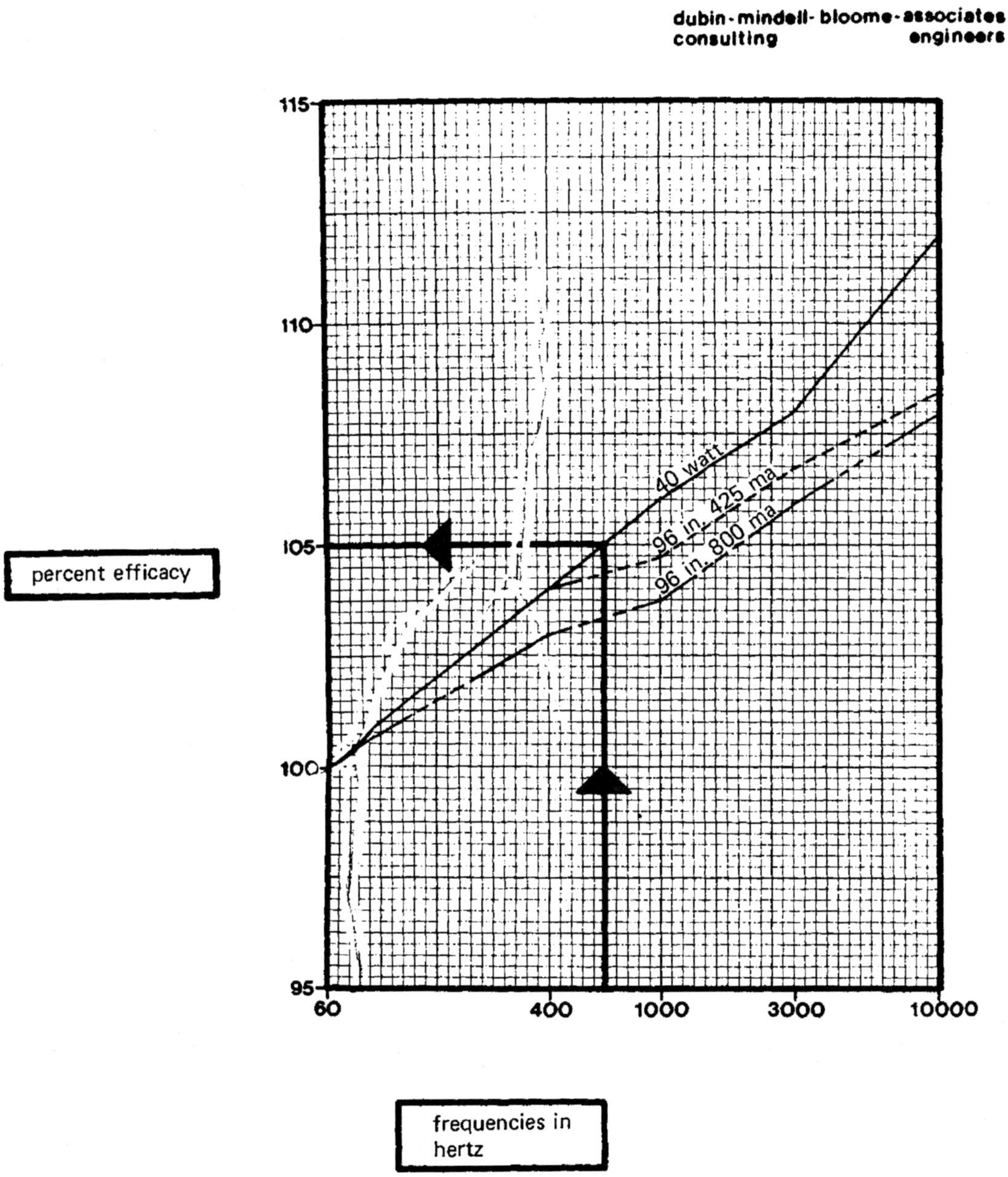

Figure #7 2 Engineering data

This graph is based on data published by the I.E.S. obtained from measurements of lumen output against watts input over a range of frequencies.

from special water cooled louvered venetian blinds to reduce solar heat gain at the windows, as well as make the heat available for other uses.

If the lower lighting levels recommended in ECM 1, Figure 21, are adopted, total heat-of-light systems will not be economically feasible for most existing buildings - even on a life cycle cost basis; accordingly, the priority of such systems should be low compared to other energy conservation measures. An exception would be where high lighting levels are required over large areas to meet specific task requirements.

GUIDELINES TO REDUCE ENERGY USED FOR LIGHTING

Refer to ECM-1, Section 9, "Lighting" for guidelines to reduce energy consumption for lighting with minimal investment. Analyze and consider the following options for further reducing energy consumption:

1. Reduce Lighting Load

   a) Refer to ECM-1, Figure 26, for suggested non-uniform illumination levels.

   b) Adopt a better lighting maintenance program to improve the L.L.F; adopt group relamping.

   c) Replace louvers and lenses and install lenses to give better quality, more efficient lighting, on the tasks. Lenses should be selected for low brightness and high efficiency.

   d) Remove all louvers and lenses in areas where CU will be improvised and the additional glare can be tolerated. Where lenses are necessary to control glare, use efficient types.

   e) Provide manual switches or photo-cells to shut off lights when available daylight is adequate to replace the required lighting.

   f) Install additional switches to permit shutting off lights in unoccupied areas of the building.

   g) Install time switches which automatically turn off lights after a preset time, in areas where occupancy is light or spasmodic, and requires manual switching to energize the lights again after that time.

   h) Provide time switches to turn off internal fixture lighting such as in food cases or over merchandise displays in supermarkets when the premises are unoccupied.

   i) Install photo-cells or electric timers to shut off outdoor parking lot lighting at night, and during weekends, and at other times when the lot is not in use. Schedule the operation of photocells and electric timers to the working shifts of employees.

   j) Install NEON pilot lights to alert personnel that lights are on in adjacent totally enclosed areas.

   k) Consider installing a master switching system, using low voltage switching, permiting an operator at one or more stations to turn off all lighting at the end of occupied periods. Seven day timers can also be used to automatically program the lights.

l) Modify existinr fixtures to accommodate higher efficiency lamps; for instance, convert incandescent lamp fixtures to high intensity discharge (HID lamps) or convert existing HID lamps to newer or more efficient ones.

m) Replace incandescent fixtures with more efficient fluorescent or mercury fixtures, depending upon application.

n) When replacing fixtures, select that type with higher coefficient of utilization.

o) Replace existing fixtures or provide new fixtures for additions with 3 lamps with individual control.

p) Replace all incandescent parking lighting with mercury vapor or HID lamps.

q) Remove the display lighting fixtures in retail stores, bulletin boards, and other areas where inefficient incandescent lamps are used, and substitute fluorescent fixtures only where required.

r) Paint or apply light colored reflective surface on interior room surfaces to increase inter-reflectances and improve the performance of both natural and artificial illumination, walls whould be first in priority then ceiling and floors.

s) When undertaking alterations for other purposes, use open landscape planning to the greatest extent possible. This will reduce the amount of lighting energy which would be absorbed by walls.

t) Where partitions cannot be removed entirely, install low partitions,with or without transparent material,to the ceiling, to reduce light absorption while permitting the use of "borrowed" light from adjacent spaces.

u) When refurnishing, select lighter colored furnishings that do not have a glossy surface or give specular reflectances.

v) When retroffiting a building, lower the fixture mounting height.

w) Relocate existing lighting fixtures to immunize veiling reflectances. Where possible light tasks from the side rather than from the front.

x) Where appropriate, install lighting fixtures which are integral with furniture, so that the light on the tasks is always uniform regardless of changes in furniture location; then reduce background lighting levels.

y) Install multi-level ballasts and integral switch to reduce the levels of lighting.

z) Install dimmers to reduce light levels in areas where lighting level requirements are constantly changing. The savings in energy will not be in proportion to the percentage of light reduction but will provide net savings of energy

aa) Modify existing skylights to use available daylight. Modifications should include removing paint or coverings which may have been installed to reduce glare or heat, and substituting proper solar control devices to keep out direct sun but allow diffuse light to enter.

bb) When remodeling or extending the building, install skylights and/or windows for the maximum practical use of natural lighting to the extent an analysis indicates that the reduction in electrical energy for lighting does not result in an increase of the heating or cooling loads by the same or larger amount of energy.

cc) Install reflectors on the exterior of the building and/or treat the horizontal surface below windows with light reflective materials to increase the intensity of daylight illumination at the window surface.

dd) When undertaking major alterations or additions to a space or when replacing groups of lighting fixtures, consider high frequency lighting to obtain more lumens/ watts.

ee) When undertaking major alterations or expansions in an air conditioned building, use air extract fixtures in a "dry heat of light" system to both reduce the room cooling load and increase the lamp efficacy.

ff) Using air extract fixtures, redestribute the captured heat from interior zones to perimeter zones to supplement heating requirements.

gg) When undertaking major alterations or expansions where 480/277 volt service is used, install 277 volt fluorescent lighting in preference to 110/120 volts and provide dry type transformers to serve convenience outlets.

Section 20

Energy Conservation Opportunities
Power

POWER

BACKGROUND

Reducing power requirements for motors for mechanical systems (as detailed in Sections 6A, C, D, & E) and lighting, (Section 6G), provides the major opportunities to conserve electrical energy. However, there are additional measures which result in significant energy savings in power distribution and utilization. ECO's 118 and 119 compare additional opportunities to reduce the power load, ECO's 120 through 126 to reduce power losses.

There are many loads , such as elevators and escalators, transformers, water coolers, etc. that are often energized 24 hrs/day, 7 days/week--even though they may be required only for 50 hours or less per week. Due to their storage capacity, other loads, such as electric water heaters, can be de-energized for periods of time, even when their function is required. Automatic devices which de-energize loads on a programmed basis are available.

Load leveling, or load shedding, to reduce peak power demands provides an opportunity to conserve energy and reduce consumption costs, as well as to reduce drastically electrical energy costs where the utility rates include demand charges. For instance, large electrical air conditioning refrigeration equipment can be unloaded during periods when all elevators are in use. Demand limiters and computer programs for load leveling are commercially available. (Refer to Section 10 for computer programs).

In current carrying conductors and equipment, resistive and reactive losses are incurred. It may not be worthwhile to increase conductor size to reduce line losses in an existing building except for major remodeling projects; there are, however, opportunities to improve the power factor of motors and lighting ballasts without major building alterations, thus reducing losses and increasing the capacity of the existing system.

Low power factor results in losses, due to high inductive loads and a reduction in the useful capacity of the electric distribution system. Resistive losses in electrical conductors increase with temperature. Refer to ECM 1, Section 4, "Power" for applicable guidelines. Motors which are excessively oversized for the loads they serve reduce the power factor for the entire electrical system, the motor efficiency is also reduced.

Opportunities to reduce energy for electric power include replacing obsolete equipment with new, more efficient equipment such as SCR controllers for elevators, and transformers with low loss factors. Refer to Section 23 "Economic Analyses Models" for substantiation procedures for measures requiring capital investment.

ENERGY CONSERVATION OPPORTUNITIES

ECO 118 DE-ENERGIZE EQUIPMENT TO REDUCE ENERGY CONSUMPTION

Provide automatic timers with remote control switching to de-energize equipment which is not required at night, and during weekends.

It is natural to think first of large motors and loads only, when embarking upon an energy conservation program, but the power consumption of many small inefficient motors in aggregate, can exceed the same amount of connected horsepower of one larger motor. Where circuiting permits, control the operation of small motordriven equipment by a single timer. Investigate recircuiting where multiple circuits supply equipment.

In addition to motors, other loads, such as resistance heating equipment, electric cooking equipment, meat preparation rooms in supermarkets and restaurants, electric signs, business machines, and mechanical and electrical equipment noted in Section 13-16, can be de-energized for a substantial portion of the week. Transformers with no load can also be denergized to save core losses.

The costs to de-energize equipment are for the control system only, unless other equipment on the same circuit must be recircuited. Automatic time switches can be installed at a distribution panel for about \$100. Remote control switches vary in cost depending upon the length of inter-connected wiring between switch and contactor. Relays or contactors vary greatly in cost depending on electrical rating and number of poles. A 100 amp. contactor and time clock combination costs about \$450 installed at the load panel.

ECO 119 REDUCE PEAK LOADS

The major purpose of load shedding is to reduce the peak electrical loads to decrease electrical demand charges.

Since many of the largest loads are deactivated for considerable periods of time, load shedding, or leveling, conserves energy in addition to reducing electrical equipment power and line losses.

Where manual load shedding, outlined in ECM 1, cannot be instituted because of cost and demand upon personnel, consider automatic load shedding.

First, tabulate all electrical loads; schedule the periods when they must be in operation, and the duration of time that they can be disconnected without impairing the safety of the occupants or condition of the structure and equipment. Then organize the load shedding program to shut down those loads by a timing device (when the hours that operations can be suspended are predictable) so that large loads are not allowed to occur simultaneously. Peak loads which are not predictable in advance can be monitored, and equipment operation can be automatically programmed to respond to load conditions by de-energizing selected loads by priority, and reactivating them when reduced demand permits. (New loads are also precluded on a priority sequence.)

Available automatic load limiting devices range in complexity from a simple thermal sensor (which works much like a circuit breaker thermal element and switches off a low-priority load when the building load reaches a preset point) to a unit that reads current and provides for shedding and restoring loads. Automatic load shedding is also commonly accomplished with a watt-hour meter, combined with time switches and load controllers.

The simplest way to control a 25-amp. circuit is to install a time clock with adjustable start/stop times. This costs about $50 installed. The larger, more complex load limiters vary greatly in cost depending on the sophisitcation of the system and the size of the loads controlled. The cost of eqiupment for a four priority load controller with automatic shed and restore is about $6,000.

ECO 120 <u>CORRECT POWER FACTOR</u>

Low power factor (P.F.) increases losses in electrical distribution and utilization equipment such as wiring, motors,and transformers, and reduces the load handling capability and voltage regulation of the building electrical system. At a unity power factor, P.F. losses are at a minimum. When P.F. is below a designated level, utilities often have a provision for a separate additional penalty charge, although the methods of computing the charge vary, the magnitude is about the same. As an example, one utility company charges a penalty of $0.25 per kilowatt-ampere reactive (KVAR) over 50% of total KW demand. Poor power factor also reduces the capacity of the electrical service. Even with a P.F. of 80% electrical service spare capacity is decreased by an additional 25% of the load, and transmission losses are increased by 56% when compared to unity P.F.

Power factor is the ratio of actual power (KW) to apparent (KVA); it also may be expressed as the cosine of the phase angle between the impressed voltage and the current. The following diagram shows the relationship between P.F., KW, KVAR and KVA and indicates the effect of corrective capacitance.

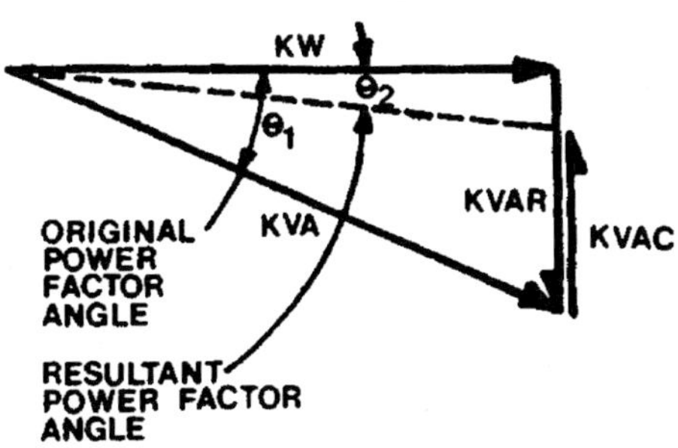

$$\text{POWER FACTOR} = \frac{\text{KW}}{\text{KVA}}$$

POWER FACTOR = COSINE OF THE ANGLE θ

KW = KILOWATT METER READING

$$\text{3 PHASE AVERAGE SYSTEM} = \frac{\text{AMMETER READING} \times \text{VOLTAGE*}}{1000}$$

KVAR = SINE OF THE ANGLE** θ x KVA

= KW x TANGENT OF THE P.F. ANGLE θ

KVAR OF CORRECTIVE CAPACITANCE = KVAR x % CORRECTION (DIVIDED BY 3 TO OBTAIN CAPACITANCE PER PHASE LEG FOR A BALANCED 3 PHASE LOAD)

* For 3 phase voltage use 1.732 x system voltage.

** A power factor reading of 75% or 0.75 for instance, is to the cosine of a 41.5° angle. The sine of that same angle equals 0.66, and tangent equals 0.88.

Partial Listing of Suggested Maximum Capacitor Rating When Motor And Capacitor Are Switched As a Unit
For use with 3-phase, 60-hertz NEMA Classification B motors to raise full load power factor to approximately 95%

MOTOR HP RATING	NOMINAL MOTOR SPEED IN RPM 3600		1800		1200		900	
	KVAR	%AR	KVAR	%AR	KVAR	%AR	KVAR	%AR
3	1.5	14	1.5	15	1.5	20	2	27
5	2	12	2	13	2	17	3	25
7.5	2.5	11	2.5	12	3	15	4	22
10	3	10	3	11	3.5	14	5	21
15	4	9	4	10	5	13	6.5	18
20	5	9	5	10	6.5	12	7.5	16
25	6	9	6	10	7.5	11	9	15
30	7	8	7	9	9	11	10	14
40	9	8	9	9	11	10	12	13
50	12	8	11	9	13	10	15	12
60	14	8	14	8	15	10	18	11
75	17	8	16	8	18	10	21	10
100	22	8	21	8	25	9	27	10
125	27	8	26	8	30	9	32.5	10
150	32.5	8	30	8	35	9	37.5	10
200	40	8	37.5	8	42.5	9	47.5	10
250	50	8	45	7	52.5	8	57.5	9
300	57.5	8	52.5	7	60	8	65	9
350	65	8	60	7	67.5	8	75	9
400	70	8	65	6	75	8	85	9
450	75	8	67.5	6	80	8	92.5	9
500	77.5	8	72.5	6	82.5	8	97.5	9

%AR-the percent reduction in line current due to capacitors. If a capacitor of lower kvar is to be used, the new %AR will be approximately proportional to the actual capacitor rating divided by the kvar value in the tables. Selection of the motor overload relay should be based on the nameplate full load current of the motor reduced by the %AR value.

There are a number of devices available for the purposes of power factor correction; three of the more common types are capacitors, synchronous motors and synchronous condensers. Synchronous condensers are the most expensive and are not a practical solution for building services applications.

Capacitor correction is relatively inexpensive both in material and installation costs. Capacitors can be installed at any point in the electrical system, and will improve the power factor between the point of application and the power source. However, the power factor between the utilization equipment and the capacitor will remain unchanged. Capacitors are usually added at each piece of offending equipment, ahead of groups of small motors (ahead of motor control centers or distribution panels) or at main services, refer to N.E.C. 1975 article 460 for installation code requirements. The advantages and disadvantages of each type of installation are listed below:

INDIVIDUAL EQUIPMENT

ADVANTAGES	DISADVANTAGES
-increased load capabilities of distribution system.	-small capacitors cost more per KVAR than larger units (economic break point for individual correction is generally at 10 HP).
-can be switched with equipment thus no additional switching is required.	
-better voltage regulation because capacitor use follows load.	
-capacitor sizing is simplified	
-capacitors are coupled with equipment thus would move with equipment if rearrangements are instituted	

GROUPED EQUIPMENT

ADVANTAGES	DISADVANTAGES
-increased load capabilities of the service.	-switching means may be required to control amount of capacitance used.
-reduced material costs relative to individual correction.	
-reduced installation costs relative to individual correction.	

MAIN SERVICE

ADVANTAGES	DISADVANTAGES
-low material installation costs	-switching means will usually be required to control the amount of capacitance used.
	-does not improve the load capabilities of the distribution system.

Where loads contributing to low power factor are relatively constant, and system load capabilities are not a factor, correcting at the main service could provide a cost advantage. When the low power factor is derived from a few selected pieces of equipment, individual equipment correction would be cost effective. Most capacitors used for P.F. correction have build in fusing; if not, fusing must be provided.

Synchronous motors can also be used for power factor correction but are relatively impractical under most conditions in commercial buildings.

The cost of capacitors for power factor correction ranges from about $200 for 1 KVAC to $550 for 60 KVAC. If more than 60 KVAC is required, use banks of capacitors. The cost if installation depends on the location and type of mounting.

In order to determine the capacitance required to correct P.F., follow one of two procedures:

1) When billings include a P.F. penalty, a KVAR meter demand reading can be obtained from the utility. Determine the required corrective capacitance (KVAC) by multiplying KVAR by % correction desired.

2) As an alternative, use a P.F. meter to determine the P.F. for individual equipment, then use an ammeter for an ampere reading (at normal loading). Use the P.F. and ampere readings to calculate KVAR and required KVAC.

3) When a P.F. meter is not available, the power factor can be determined by dividing the wattage found via wattmeter, if available, by the product of the average of the ammeter readings of the phases and the system voltage.

$$\text{P.F.} = \frac{\text{W(Wattmeter Reading)}}{\dfrac{\text{Amps (Phase A + Phase B + Phase C ammeter readings)}}{3} \times \text{System Voltage}}$$

ECO 121 <u>CHANGE OVERSIZED MOTORS</u>

The motors selected for service have often been oversized for the load they handle. Furthermore, when energy conservation measures are now implemented for the HVAC systems the load factor may be even lower. Refer to ECM 1, Section 3 "Use" for instructions on calculating "load factor" and Section 10 for general guidelines on motor replacement. (Refer to Figure 76 "Motor Power Factor & Efficiency vs. Loading")

Measure the amount that each motor is actually drawing as compared to its nameplate rating to determine load factor. If it is below 50% consider replacing it.

Factors to consider when installing new motors:

-Match the phase and frequency with the electrical distribution system characteristics.

-When motors cannot exactly match system voltage, select the following:

--For motors loaded only to 75% of their capacity, motors rated at slightly more than system voltage.

--For motors loaded to more than 75% of their capacity, motors which are rated slightly under the system voltage."(Problems may arise if nameplate rating of motor selected is in excess of system voltage by more than 5% - a likely situation in these days of brownouts and voltage cutbacks which actually increase the difference.)

ECO 122 UTILIZE EFFICIENT TRANSFORMERS

Most dry-type transformers range from 93% to 98% efficient; the losses occur from the core (magnetizing), and coils (resistance and impedance).

Select the most efficient dry-type transformer, when replacement due to load changes or breakdown is required.

Dry-type transformers are available in many temperature-rise classifications. The lower the temperature-rise rating, the lower will be the coil losses due to the larger conductors used in the winding. 80°C, 115°C and 150°C are the usual temperature-rise ratings; select the 80°C rise transformer for greater efficiency. Transformers with lower heat rise ratings also have longer life expectancy which may offset the higher cost (about 10%) of selecting them.

If larger liquid-filled transformers are required for new additions, select those with the highest efficiency.

ECO 123 REDUCE ENERGY CONSUMPTION FOR ELEVATORS

Elevators and escalators account for about 1% - 4% of the electrical energy of office buildings and large department stores. Refer to ECM 1, Section 10, "Power" for energy savings through operational changes. Select the automatic control program for the lowest speed and heaviest loading (reducing the number of elevator units traveled per year) to conserve energy. Where motor generator (MG) sets, which draw power whether or not the elevator is in operation -- are installed in existing buildings, de-energize them with 7-day timers for periods when one or more elevators are not required.

Consider changing from MG sets to more efficient SCR controllers.

When choosing new elevators, for replacements or additions, select the most efficient type for the application. The three major types of elevators are hydraulic, geared and gearless. The limitations of each must be considered in evaluating its efficiency. Hydraulic elevators are used for speeds of about 200 ft/min, geared elevators for about 300 ft/min and gearless elevators for higher speeds and high-rise buildings. Electric elevators with counter-wieghts use less energy than hydraulic elevators.

ECO 124 INSULATE ELECTRIC HEATING AND COOKING EQUIPMENT

Improving the insulation on equipment such as fryers, ovens, food warmers, kilns, refrigerators and freezers provides opportunities to conserve energy. For example, a U.S. Navy study indicated that a 9 KW insulated fryer had the same capacity as a 12 KW non-insulated unit. In addition to savings in power for operation, insulating equipment reduces the heat gain to spaces to a minimum and results in low installation and operating costs for the air conditioning system.

In existing facilities where newer types of equipment have not been installed, shielding with aluminum or asbestos barriers and insulation of hot surfaces are effective and should be utilized to reduce unwanted heat gains and eliminate sources of radiant heat.

Where heating and cooling equipment are in close proximity (inches) to one another, use rigid insulation between them.

When choosing new items, select equipment which is well-insulated, which shortens preparation time (more efficient ovens, pressure cookers), and which has a surface temperature no greater than 90°F. Additional costs will be quickly recovered and the kitchen working conditions improved.

Example: In a fast foods restaurant, a deep fat fryer was used 7 hours per day, 200 hrs./mo. If an insulated 9KW fryer were used rather than a non-insulated 12 KW unit, 600 kilowatt hours would be saved each month. The yearly savings would be $216, which could be amortized in a short period of time.

GUIDELINES TO REDUCE ENERGY USED FOR ELECTRIC POWER

Refer to ECM 1, Section 10, "Power" for guidelines to reduce energy consumption with minimal investment. Analyze and consider the following options for further reducing energy consumption:

1. Reduce The Load

   a) Install manual or automatic controls to disconnect loads when they are not required (Refer to ECM 1 for a list of equipment which can be turned off readily).

   b) Install manual or automatic load shedding devices to reduce the peak loads.

2. Reduce Losses

   a) Correct low power factor with capacitors or synchronous motors to reduce losses and increase electrical distribution system capacity.

   b) Change oversized motors to improve the system power factor and motor efficiency.

   c) Utilize efficient dry-type transformers.

   d) De-energize MG elevator sets or use SCR controllers. Select more efficient elevators and control devices when replacing elevators or expanding elevator service.

   e) Insulate electric heating and cooling equipment. Purchase new equipment with better insulation.

   f) When undertaking new additions or major remodeling involving the installation of new equipment, use 480/277 volt systems in preference to 120/208 where the electric service is approximately 500 KVA or more. Whenever possible use the highest practicable voltage available (i.e. 480V instead of 120V).

This page blank

Section 21

Central Control Systems

CENTRAL CONTROL SYSTEMS

APPLICATION

The purpose of a central control system is to provide the Physical Plant Chief Engineer and the Energy Management Team with a management tool to give constant surveillance of the building and to make the most efficient and effective use of physical plant systems and personnel.

Central control systems are applicable only to the larger more complex buildings or groups of buildings. They should be considered where net floor area exceeds 100,000 sq.ft. and where energy use is high due to extended occupancy or use of special equipment.

Central control systems vary from the relatively simple type designed to perform a few functions to the progressively more complex type performing more and more complicated functions.

When selecting a central control system, it should be tailored to the present requirements of the building and the systems to be controlled. If money for investment is limited, attention should first be given to those functions which will show the quickest and greatest return in energy saved. An overview of the central control system capacity should, however, be made at the time of the initial installation and provisions be included for expanding the system hardware and its software capacity as and when investment money permits. For instance, an initial system may well comprise a control console and central processor unit with one single printer output and may be programmed only to perform start/stop and simple reset functions and to report alarms. Optimization of system operation and load shedding in this case would be achieved manually at the central console based on decisions made by the operating staff and achieved by overriding the programmed start/stop times and reset points for the various systems.

Future hardware additions to this initial system could well be an extra printer to act as a back up and to handle alarm signals separately from system status reports; a cathode ray tube output to show in English language the status of selected systems and current modifications made to the normal operating program; a bulk storage memory unit to increase the initial core memory included in the central processing unit so that more complex programs can be used and a data file built up of operating conditions; a high speed tape input/output device to enable new programs to be fed into the computer and to extract from the computer banks historical data of systems operation; complex custom written software programs to optimize the operation of all systems, to achieve automatic load shedding on a selected rolling priority basis.

A Central Control System should be selected to:

a. Monitor all systems for off-normal conditions.

b. Monitor all fire alarm and security devices

c. Monitor operating conditions of all systems and reschedule set points to optimize energy use.

d. Monitor on a continuous basis selected portions of any systems and store this information in bulk memory for later retrieval and use in updating software. This information would also be used to determine the effectiveness of the energy management program to indicate changes or modifications in approach that should be made.

e. Limit peak electrical demand values by predicting trends of loads and shedding non-essential services according to programmed priorities.

f. Optimize the operation of all systems to obtain the maximum effect for the minimum expenditure of energy.

g. Optimize maintenance tasks to effect maximum equipment life for minimum manpower labor and costs.

h. Provide inventory control of spare parts, materials, and tools used for maintenance.

By judicious use of these functions, Engineering staff will be able to operate all systems from the central console and will have minute-by-minute control of their operation. Any physical plant critical alarm will be reported automatically at the console. The operator will then be able to scan the system in alarm, analyze the fault and dispatch the correct maintenance man to effect repairs. Maintenance alarm summaries will be available on demand as well as being printed out once each day. These maintenance alarms will allow work scheduling and maximum use to be made of maintenance personnel.

SYSTEM TYPES

Proprietary central computerized control systems are marketed by each of the major temperature control manufacturers. These systems all have common features and can accomplish a similar range of tasks. Each manufacturer, however, uses coding and computer languages which are unique to his system and cannot be decoded by any other system. Once a basic system has been selected and installed, all subsequent additions must be obtained from the original manufacturer. It is important, therefore, when selecting a system, to thoroughly investigate its expansion potential in relation to future predicted requirements.

Central computer control systems are usually modular in design and additional hardware and software can be added later if the necessary provisions have been made in the basic system.

Each manufacturer's system comprises standard "off the shelf" hardware, but the application is always tailored to the specific project. Basically, any system is composed of four major parts.

a. Interface panels which are located at strategic points throughout the building (usually equipment rooms). These panels form the focal point of all signals to and from a particular area.

b. Transmission system between the Central Console and all interface panels. This transmission system can be a simple single-core cable for digital transmission or multi-core cable for multiplex transmission.

c. Central control console and associated hardware located in a control room (usually close to boiler room and chief engineer's office).

The console, computer and associated hardware is the point at which the operator enters all instructions and retrieves all data. In addition, the computer generates routine instructions according to the program contents.

d. Software (program) generated by the manufacturer in conjunction with the prospective user. Software input via magnetic tapes, paper tapes, or cards, contains the basic operating instructions for the computer and is stored in the form of "bits" of information either in core memory or in bulk memory (cheaper than core for large quantities of data).

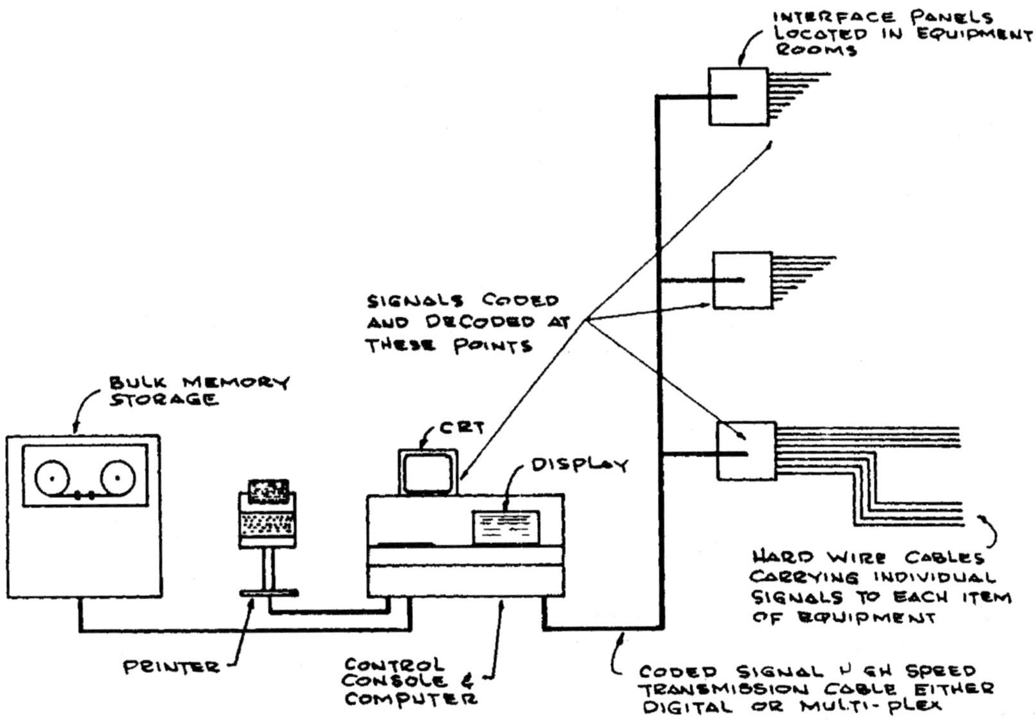

TYPICAL CENTRAL CONTROL SYSTEM

INTERFACE REQUIREMENTS

To obtain maximum benefit from a central console system it is necessary to measure, monitor and control many different items of equipment.

The central control system must be interfaced with the existing equipment to obtain this information in rational form, and to exert its control functions.

Figure 73 shows a table of desirable interface points and the type of function required for typical building systems. Interface of two types of signal is required; binary signals which comprise only two alternatives (on/off,open/close, etc.) and analog signals which measure a value against a particular scale (temperature, pressure, etc.).

Binary signals are used as instructions to start and stop equipment, open and close valves and dampers, etc. They are also used for retrieving data such as on/off, open/closed, etc.

Analog signals are used as instructions to raise or lower temperature set points, adjust damper positions, raise or lower pressure set points, etc. They are also used for retrieving data such as temperature, pressure, humidity, etc.

To select a central control system, it is first necessary to assemble a complete list of all desired interface points under the two categories of binary and analog and to arrange them in groups served by individual interface panels.

Depending on the type of existing controls, motor starters, contactors, etc., modifications and additions may be required to allow satisfactory interface. For instance, if a motor starter does not have spare auxiliary contacts, a relay must be added to the control circuit.

SIGNAL TRANSMISSION

To make effective use of the computer's capabilities, information between the computer console and the interface panels must be transmitted at high speed in a format easily handled by the computer. The most convenient method of transmission is digital where information is represented by pulses arranged serially and transmitted through a single core conductor. Digital cable is commonly coaxial, although some systems use a twisted pair of insulated wires. Voice intercom is carried over a separate screened cable.

Multiplex cable is multi-core, signals being multiplexed and transmitted in parallel. Typically the cable will comprise between 50 and 100 separate wires and have an overall diameter of one inch. It is more difficult to install this cable than coaxial, particularly in existing buildings where empty conduits and throughways are not available.

Some systems still use multiplex cable on the grounds that analog signals can be transmitted in unmodified form, whereas, they must be converted from analog to digital form with a small loss of accuracy at the interface panel for digital transmission. Conversely, analog signals are sensitive to interference from "spikes" and other spurious signals induced by adjacent building wiring, whereas, digital signals are not subject to interference.

FIGURE 73: INTERFACE POINTS

ITEM	START/STOP LEAD/LAG	READ OUT	RESET	ALARM 1	ALARM 2
Ventilation					
Outside air temperature (dry bulb)		X			
Outside air temperature (wet bulb)		X			
Air temperature on pre-heat coil (dry bulb)		X		X	
Air temperature on pre-heat coil (wet bulb)		X			
Air temperature on cooling coil (dry bulb)		X			
Air temperature off cooling coil (dry bulb)		X			
Air temperature off humidifier (dry bulb)		X			
Air temperature off humidifier (wet bulb)		X			
Air temperature off re-heat coil (dry bulb)		X			
Critical area temperature (dry bulb)		X		X	
Critical area humidity		X		X	
Return air temperature (dry bulb)		X			
Return air temperature (wet bulb)		X			
Fresh air/exhaust air/recirculation air dampers		X	X		
Air velocity beyond filters					X
Supply fan	X	X		X	X
Humidifier spray pump	X	X			X
Supply air flow switch		X		X	
Exhaust fan	X	X		X	X
Heating coil control set-point			X		
Cooling coil control set-point			X		
Humidifier control set-point			X		
Hot Water Heating					
Boiler fuel input	X	X	X		
Boiler supply temperature		X	X	X	
Boiler return temperature		X			
Boiler high limit temperature				X	
Hot water pumps	X	X		X	X
Steam					
Boiler fuel input	X	X	X		
Boiler steam pressure		X	X	X	
Boiler condensate feed temperature		X			
Boiler high steam pressure limit				X	
Boiler low water level limit				X	
Boiler feed pumps	X	X		X	X
Chilled Water					
Chiller power/fuel input	X	X	X		
Chiller supply temperature		X	X	X	
Chiller return temperature		X		X	

FIGURE 73: (continued)

ITEM	START/STOP LEAD/LAG	READ OUT	RESET	ALARM 1	ALARM 2
Chilled Water (continued)					
Condenser supply temperature		X	X		
Condenser return temperature		X	X		
Chilled water pumps	X	X		X	X
Chilled water flow rate		X			
Condenser water pump	X	X		X	X
Condenser water flow rate		X			
Cooling tower fan	X	X			X
Air cooled condenser fan	X	X			X
Domestic Hot Water					
Domestic hot water heater fuel/power input	X	X	X		
Domestic hot water storage temperature		X	X		
Domestic hot water high temperature limit				X	
Domestic hot water return temperature		X			
Domestic hot water circulating pumps	X	X			X
Cold Water					
Cold water pressure boost pumps	X	X		X	X
Cold water boosted pressure		X		X	
Other Services					
Control system air pressure		X		X	
Electrical					
Total building power input KVA & KW		X		X	
Primary isolation switch		X			
Load interrupter switch		X			
Transformer winding temperature		X		X	
Transformer secondary voltage fixture		X		X	
Emergency generator	X	X		X	X
Parking lot lights	X	X			
Selected lighting circuits	X	X			

NOTE: Alarm 1 denotes Critical Alarm
Alarm 2 denotes Maintenance Alarm

When selecting a central control system, the different characteristics of the transmission methods must be analyzed and a choice made based on the particular circumstances. Generally, digital transmission systems provide more options for later additions as multiplex systems have a finite limit to the number of points that can be connected.

CONTROL CONSOLE AND HARDWARE

The control console and hardware items such as printers, graphic display, CRT, and input/output keyboards are of comparable performance for any manufacturers' system. Individual options are available on request to suit the prospective users particular requirements. For instance, different levels of access into the computer from the keyboard are available and may comprise from three to five levels. Access would range from restriction to normal operation for the lowest level up to reprogramming for the highest level, and would be controlled by key-switch. It is vital to protect memory from accidental erasure.

The central processor or computer will be supplied with integral core memory. Capacities typically vary from 16 K words to 64 K words, each word containing 16 bits of information. Core memory is expensive compared with bulk memory, and if the core capacity will be exceeded, then the addition of a bulk memory unit should be considered but first make sure that the computer selected can be interfaced with external memory.

Bulk memory can be either disc or tape, both with random access. Disc memory is preferred for control applications and typically has a capacity of 192 K words.

SOFTWARE

Software or programs for common applications are available from the manufacturer. Each of the major controls manufacturers has a library of application programs which have been developed over a number of years. The cost of these programs is low compared to custom written software. When selecting a central control system, investigate the library of programs available as the extent and range of the library is a good indices of a particular manufacturer's commitment and ability in the computer control field.

Programs and their applications are:

Executive Program - This program is essential to the function of the computer itself as it controls the priority of signal processing, directs information in and out of memory, controls the operation of all associated hardware and encodes/decodes incoming and outgoing signals. An essential function is the orderly shut-down of the computer on power failure to ensure that all information is safely stored in permanent memory before all power is lost. Beware of systems that store information in volatile memory which evaporates on power failure.

Start-Stop Program - This is the simplest program and yet with modifications becomes the most effective energy saver, particularly when used in conjunction with other programs and routines.

In its most simple form the program is arranged to start and stop equipment according to pre-determined times allowing for weekends and holidays. The program can also open and close dampers and turn lights on and off. It can, in addition, reschedule temperature from occupied level to unoccupied level, as all of these functions are binary in nature.

Manual override can be provided with automatic reversion to program control at the end of an occupied or unoccupied period.

Interface with other programs allows the basic program to be overridden and equipment turned on and off to meet some other criteria.

Start sequences can be arranged in cascade to avoid simultaneous starting and the cumulative effect of inrush currents causing a high peak electrical demand.

Load Shedding Program - This program is used in conjunction with the start/stop program to override normal operation and turn off equipment as a pre-determined peak load value is approached. In its simplest form the program would shut down selected equipment according to the measurement of peak load. The program can be modified to include a rolling priority feature whereby equipment would be turned off on an assigned priority basis for an adjustable period of time, after which it would be returned to service and a lower priority piece of equipment turned off. The rolling priority would only be activated as a pre-determined peak load value is approached and equipment turned off would be limited to that necessary to maintain the load below a given level.

Interface with a profile routine would allow a continuous profile of electrical demand to be made and predicted load in the immediate future to be calculated. This predicted load would then initiate the rolling priority to alter the slope of the prediction curve. The current maximum peak load experienced in the previous eleven months would be held in memory and used as the criteria for load shedding. This value would be constantly modified based on monitored information.

Optimizing Programs - These programs tend to be more sophisticated and require interface with an arithmetic routine to allow optimum settings to be calculated. The program interfaces with the start/stop program, load shedding program, and reset program to modify operation of equipment. Optimizing programs can provide economizer cycle control of outdoor air according to enthalpy and loads. Chilled water and boiler water temperatures can be modified to provide optimum equipment operation according to load. Some optimizing programs are available from the manufacturer's library but others may have to be custom written.

Reset Programs - These programs reset the control points of control thermostats and pressure controllers, and interface with the load shedding and optimizing programs. In place of, or in addition to, turning off equipment for load shedding and energy conservation, loads can be trimmed, i.e., if a peak load is approaching, space conditions can be reset to reduce cooling or heating loads; outdoor air dampers can also be reset.

FIGURE 74: DIAGRAM SHOWING INTERFACE BETWEEN PROGRAMS

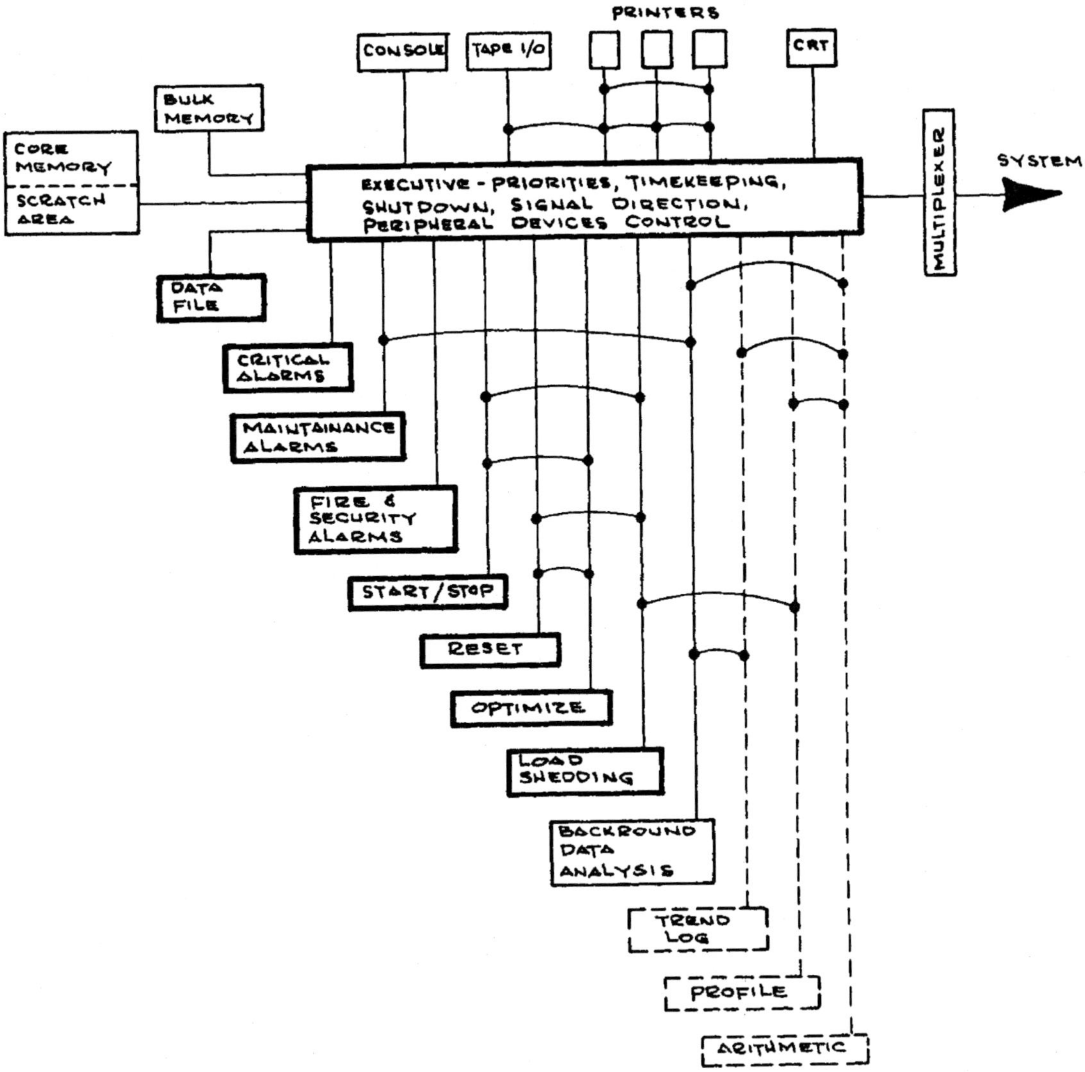

Alarm Programs - These programs report alarms generated either outside the computer (fire alarm pull boxes, boiler high pressure, etc.) or in computer software. Upper and lower limits can be set in software for values of any monitored point and arranged to report an alarm when exceeded.

Alarm programs can also be interfaced with the start/stop program to monitor hours run and report a maintenance alarm when a designated number of running hours is exceeded. Information on particular maintenance required can be stored in memory and be printed out when the alarm is reported.

Figure 74 shows typical relationships between programs and the interface between one and another. The program functions described are an indication only of what can be achieved with a central computer control system and are intended to stimulate investigation into all possibilities, particularly in reference to energy conservation.

Section 22

Alternative Energy Sources

ALTERNATIVE ENERGY SOURCES

BACKGROUND

Traditional sources of energy for existing buildings are (1) electricity, generated off-site by a utility company and distributed to the building, and (2) fossil fuels, i.e., oil, gas, coal, methane burned at the building site.

Alternative sources of energy include solar energy for heating, cooling and electric power generation, total energy systems, wind energy, fuel cells, geothermal energy, nuclear power, methane gas generated from liquid wastes, solid waste, tidal power, hydropower, ocean thermal differences, coal gasification, and hydrogen from the electrolysis of sea water.

Of these alternative sources, only solar energy for heating and cooling and total energy systems are included in this Section:

- -The technologies are well-known.
- -The expertise for design and installation is available.
- -Installation and operating costs can be estimated from available data.
- -Minimal modifications, only, to the building structure or site are required.
- -Each system is compatible with, and will be enhanced by, the other energy conservation measures outlined in ECM 1 and ECM 2.
- -Hardware is commercially available.

Solar energy and total energy systems can reduce the need for significant quantities of conventional energy used to serve the building's environmental systems. They reduce air pollution and operating costs, and save fossil fuels. They can be effectively used together in the same building.

The other alternative sources of energy are not included in this Section due to one or more of the following factors:

- -Hardware is not commercially available, nor are the future requirements for hardware sufficiently identified.
- -Electric power from alternative sources can be simply substituted in the future for conventionally generated electric power, and no provisions for building modifications are required now.
- -Expertise is not available to design such systems.

Engineering feasibility studies of alternative energy systems can be done only by qualified consultants who can select equipment for analysis, estimate costs, and predict the annual performance over a 20-year period. Because the initial investment of these systems is significantly higher than for conventional systems, special care is necessary to insure that the life cycle cost economics are accurately determined.

A. SOLAR ENERGY

Direct use of the sun's energy through windows, skylights, or glass doors is the most efficient way to use solar energy. Refer to ECM 1, Sections 4 and 9 "Heating and Ventilating" and "Lighting".

This Section outlines additional methods of utilizing solar energy with solar collectors for domestic water heating, space heating and cooling, pre-heating combustion air, tempering outdoor air, heating heavy oil, and regenerating desiccants which are used for dehumidification in air-conditioning applications.

1. Domestic Hot Water Heating

Solar water heaters have been used for more than 30 years throughout the world, especially in Israel, Australia, Japan, and Florida. Although many solar water heaters are still in operation, numerous others were abandoned when repairs were required because gas, oil, or electric heaters cost less to buy and energy was expensive. The high costs, now, of fossil fuels and electricity have again awakened an interest in solar water heaters(and other solar energy systems).

In office buildings and stores with 25 or more full-time employees, solar water heaters may be economically feasible. Religious buildings' demand for hot water is insufficient to warrant purchasing a solar water heater, unless they include extensive food preparation facilities which are used bi-weekly, or include school or office facilities which are used more than 35 hours/week. When solar energy is economically feasible for space heating, heating domestic hot water with the same system will be economical, as well.

Solar water heaters are commercially available in the United States. The components are offered separately, or together in a complete package. They include a flat plate collector, storage tank (existing storage tanks can be used), piping, controls, circulating pump, and, in climates where the collector is subjected to freezing weather, a heat exchanger with a secondary pump, piping circuit and anti-freeze. See the following diagram for a typical arrangement. Refer to the Section on solar heating and cooling for details of these components and for system design considerations.

For normal hot water use, approximately one square foot of collector per one gallon of hot water used per day is adequate. If kitchens or other processes require hot water at elevated temperatures, collector area required is from 25% to 50% more.

The solar collector can be mounted on the main roof, the roof of a building extension, or on-site.

It is, generally, not economically feasible to provide 100% of the domestic hot water requirements unless:

-The collector is also used for other services.

-Hot water temperature requirements do not exceed 90°F.

-An existing storage tank is oversized for the building requirements.

-The building is located in a temperate or hot climatic zone.

Solar Collector

Solar collectors convert direct and diffuse radiation from the sun to heat water. Solar collectors are manufactured with one or two coverplates, metallic absorber plate with an absorbing or selective surface, back insulation, side insulation and an enclosing frame. (See the following diagram.) Each manufacturer offers different features and anticipated performance for his product. The following guidelines should be considered when weighing alternative offerings:

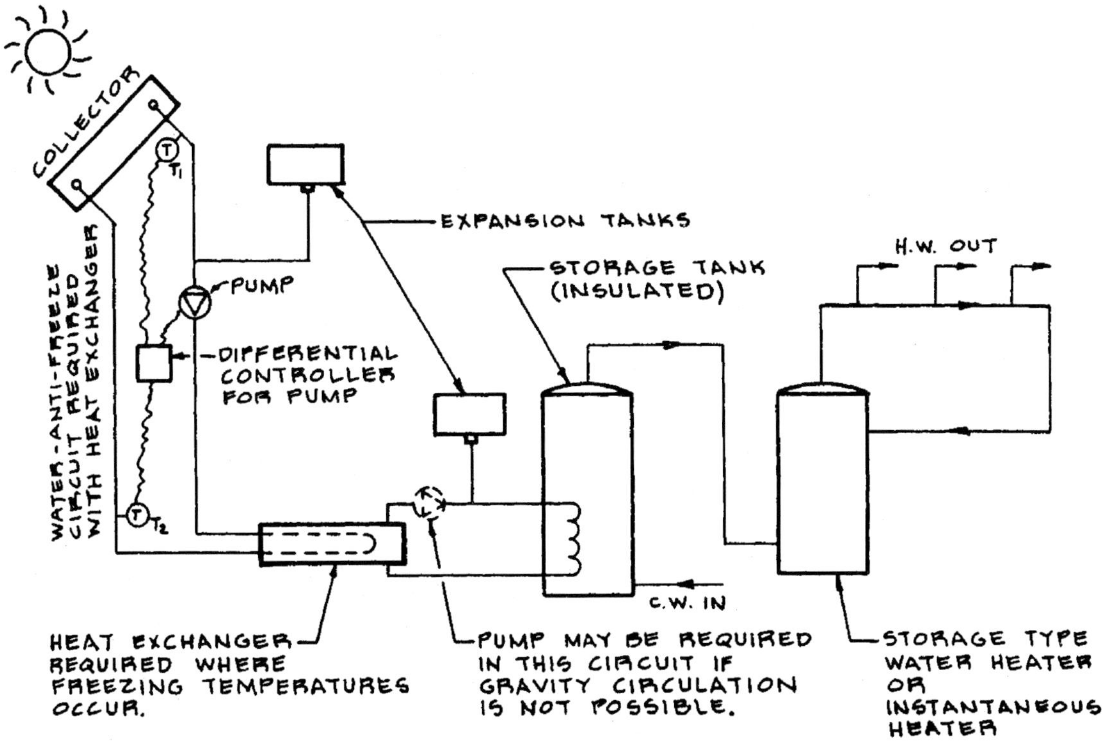

SCHEMATIC DIAGRAM OF SOLAR WATER HEATING SYSTEM

The thermal performance of the solar water heating system depends upon:

-Climatological and meterological conditions.

-The solar collector size, design, and construction.

-The orientation of the collector and tilt angle with the horizontal.

-The size of the storage tank.

-The temperature of the domestic hot water.

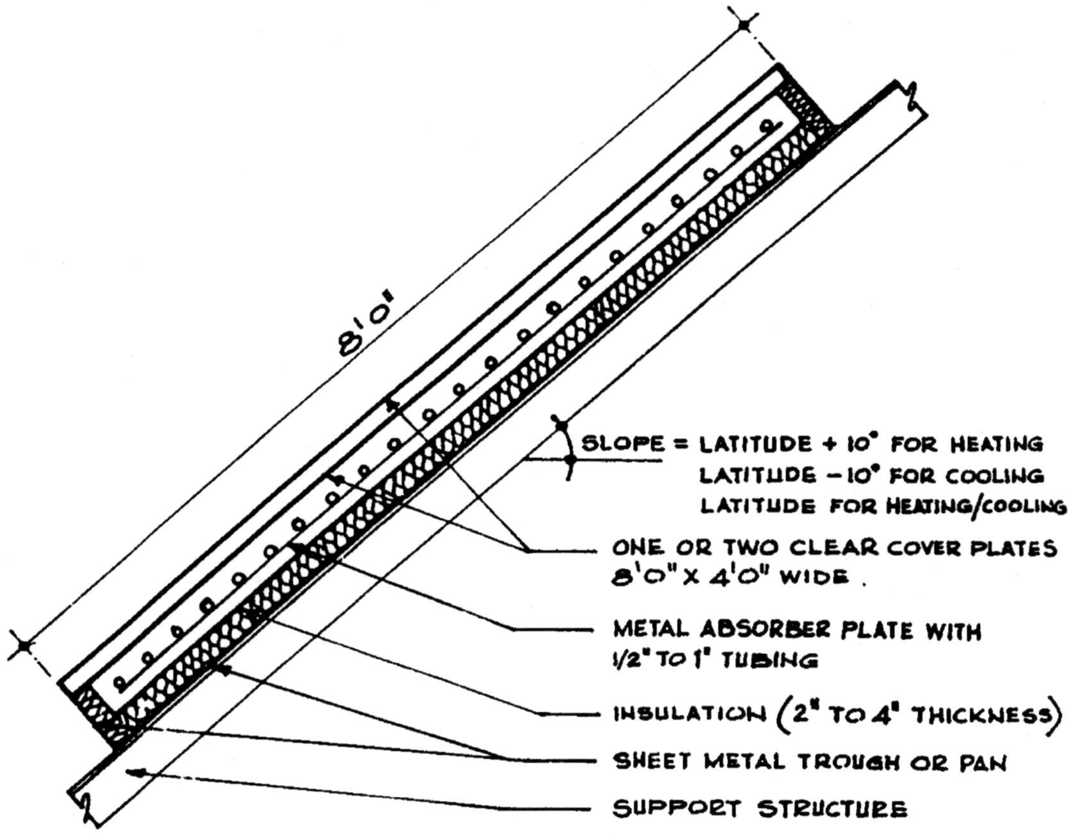

SECTION THROUGH SOLAR COLLECTOR

For Capacity:

-Two transparent coverplates are better than one.

-Copper absorber plates have better heat transfer than aluminum or steel of the same configuration.

-The greater the "R" value of the back and side insulation, the less the heat losses.

-Non-reflective glass is superior to highly reflective glass.

-Selective surfaces are more effective than flat black, or other colored absorptive coatings.

-Moisture-proof casings improve the performance by reducing fogging on the coverplates.

-Reflectors installed in front of collectors increase the concentration of solar energy on the collector plant and increase the thermal performance.

For Durability:

-Copper is the most resistant to corrosion and pitting.

-Moisture-proof casings reduce corrosion.

-Expansion joints reduce buckling and mechanical failure.

-Tempered glass resists breakage from the impact of hail or other objects. Wired glass resists breakage but also reduces transmission.

The best solution is the collector and system which provides the lowest life cycle costs per million Btu/s delivered to the storage tank or hot water taps.

-For available manufacturers of solar water heaters, consult Reference 72.

GENERAL GUIDELINES FOR SOLAR COLLECTOR INSTALLATION

CAUTION: These are general provisions only, and not design standards. Each installation must be examined individually.

-Provide adequate structural support to carry the dead weight (12 lbs./sq.ft.) of the collector plus wind loading.

-Orient the collector 10° west of due south if possible, but 20° either side of due south will not materially affect the performance. If the orientation must be farther to the east or west, plan on additional collector area.

- If the collector is used solely to supply hot water, a fixed tilt of latitude plus 10° will usually be optimum - variations of 10° up or down will not seriously affect yearly performance.

- Collector must not be shaded more than 10% of the time, or a larger collector area will be required.

- Provide approximately one square foot of collector and one gallon of hot water storage capacity per 16 gallons of hot water used per week. If kitchens or other processes require hot water at elevated temperature, provide one square foot of collector per 20 gallons of hot water used per year.

- The existing domestic hot water heating system, even though inefficient or undersized will usually be adequate to provide heat to supplement the solar collector system.

2. Space Heating and Cooling

Prior to 1972, only about twenty solar heated buildings were erected in the United States, and of those, only one was a commercial building; the rest were single family residences. Within the past three years, there has been a proliferation of solar energy activity in the United States resulting in many dozens of buildings now in design and/or under construction. While a majority of these are new buildings, the same design and construction techniques are applicable to retrofit existing buildings.

The National Science Foundation sponsored a program to retrofit four existing school buildings for solar heating. These installations are now providing valuable data on the performance of solar heating systems under varying climatic conditions. (See Reference 67 and 71.) Also, under a National Science Foundation grant, the installation of solar collectors for heating and cooling an existing school building in Atlanta, Georgia has been started.

The authors completed an analysis and preliminary design for NASA to retrofit an existing 53,000 sq.ft. office building in Edwards, California for solar heating and cooling. For the General Services Administration energy conservation demonstration building in Manchester, New Hampshire the authors designed a solar energy system after the building plans were completed, to provide heating, cooling, and domestic hot water for a portion of the building. Many more solar heating and a few cooling projects are described in Reference 72.

These installations have already provided designers with some guidelines, opportunities, and pitfalls to be avoided; the literature should be studied before conducting a feasibility study of solar energy for heating or cooling.

The hardware for solar space heating and cooling includes solar collectors, piping, controls and storage systems, which, in general, are similar in type to those used for heating domestic hot water, but with the following notable differences:

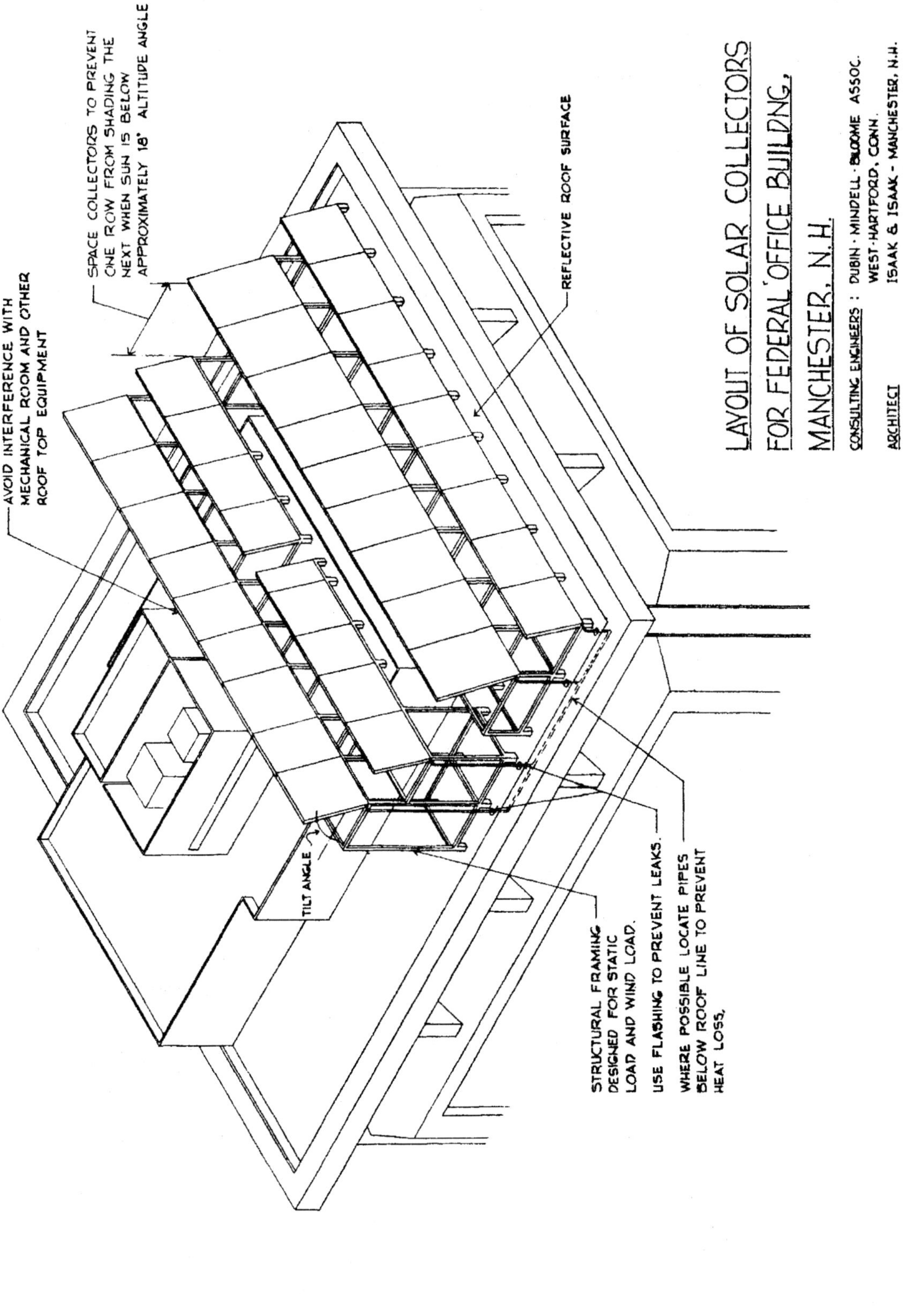
AVOID INTERFERENCE WITH MECHANICAL ROOM AND OTHER ROOF TOP EQUIPMENT
SPACE COLLECTORS TO PREVENT ONE ROW FROM SHADING THE NEXT WHEN SUN IS BELOW APPROXIMATELY 18° ALTITUDE ANGLE
REFLECTIVE ROOF SURFACE
TILT ANGLE
STRUCTURAL FRAMING DESIGNED FOR STATIC LOAD AND WIND LOAD.
USE FLASHING TO PREVENT LEAKS.
WHERE POSSIBLE LOCATE PIPES BELOW ROOF LINE TO PREVENT HEAT LOSS,
LAYOUT OF SOLAR COLLECTORS
FOR FEDERAL OFFICE BUILDNG,
MANCHESTER, N.H.
CONSULTING ENGINEERS : DUBIN · MINDELL · BLOOME ASSOC.
WEST · HARTFORD, CONN.
ARCHITECT ISAAK & ISAAK - MANCHESTER, N.H.

-The required collector area and storage volumes must be larger and are more costly.

-Collectors must produce hotter water temperatures for space heating. Temperatures up to 180°F are desirable, but lower temperatures can be used.

-The orientation of the collector is somewhat more critical.

-Absorber plates with selective surfaces rather than flat black coating will be more economically feasible for heating and cooling applications than for solar water heating alone.

-The interface with the existing heating and/or cooling system is critical.

-Storage systems must have a greater heat storage capacity per square foot of collector.

-Rocks or phase changing salts can be used for thermal storage instead of water in some cases.

-Collectors can be air instead of liquid heating types in selected applications.

-A full sized back-up heating system is required to supplement the solar system.

In most areas of the country for most buildings - even more so for existing buildings compared to new ones - the cost of a solar system to provide 100% of the heating and/or cooling system is prohibitive. However, in the existing heating system, burner/broiler, or electric heat can be left intact, or modified as necessary, to supplement the solar collectors.

Space Heating

a) Collectors: Flat plate collectors absorb both diffuse and direct radiation. Depending upon their construction, and meteorological and climate conditions each square foot of collector can collect from 50,000 to 150,000 Btu's per heating season. Reflectors can increase the capacity from 15% to 25% at relatively low installation costs. The efficiency of the collector is directly proportional to the collection temperature, so that the characteristics of the heating system, in the building which interfaces with the solar collector and storage units, has a major influence on the overall economic feasibility of the complete system. The optimum tilt for collectors, used solely for heating and domestic hot water, is latitude plus 15°. Any variations in tilt should be towards a steeper angle rather than a flatter one. Orientation should be 5°-10° west of due south, but not more than 10° variation either side of this azimuth. A larger variation will require more collector area for the same thermal performance.

With water type collectors, the higher the temperatures required, the greater are the benefits of absorber plates with selective surfaces. Absorptive surfaces, flat black, green, or even dark red, have an an absorption coefficient of 0.9 to 0.95, but have the same emissivity. A selective surface can have an absorption rate of approximately the same, but have an emission rate of about 10% to 15% of the radiation it receives. In many cases, only one layer of glass or plastic will be required when selective surfaces are used, when two might otherwise be required.

The total cost/benefits of the collector and the storage system and the heating system with which they interface must be analyzed for each of the major variations in the collector design and installation to optimize economic feasibility.

It is necessary to consider both thermal performance (heat collection) and the mechanical performance (structure and durability). The authors prepared the solar collectors for General Services Administration. The National Bureau of Standard has also prepared performance specifications. Both are available to the private sector. (See Reference 69 and 75.)

The size of the collector varies with use, climate, and the HVAC system with which it operates. The collector area required for heating ranges from 20% to 50% of the total building floor area to provide 20% to 70% of the annual heating and/or cooling energy requirements.

Collectors can be located on a building roof, on an adjacent building or garage or on the ground. If a building is being modified, then provisions made at the same time for a solar collector system can ease future installations. For example, the entire south wall can often be used as a collector, if it is designed with that conception in mind.

Air heating collectors are most suitable for space heating applications only (when cooling is not being considered) and for relatively small buildings. Except for building of less than 2,000 square feet in area, air collectors are not commercially available, but can be specially designed. They are particularly applicable for tempering make-up air or outdoor air for ventilation. In many cases, no storage systems will be required for tempering ventilation air when the load exceeds the collector area for five days per week. Most applications using air collectors will be at lower temperatures than water type collectors and selective surfaces with a well-designed air collector are not necessary.

b) Storage: The most practical storage system is hot water in metal or concrete tanks. From two to five gallons of storage are required per square foot of collector. The exact amount depends upon climatic conditions - the number of consecutive cloudy days, the type heating system, and the cost/benefits ratio of storage tank costs and collector performance vs. the operating costs of the supplementary heating system. (Cooling with solar energy may affect the storage tank size.)

Since a supplementary heating system is required in virtually all cases, the costs of extra storage volume (and space required) can be offset on a life cycle cost basis by operating the supplementary back-up system somewhat longer and reduce the volume of storage. Where the entry into a building is too small to deliver a tank, the tank can be fabricated by welding plates in the building or can be constructed of concrete. With hot air systems, rocks (about 100 lbs. per square foot of collector) can be used for heat storage in place of water. Latent heat storage using phase changing salts (salt hydrates, eutectic salts, and waxes and parafins) require less volume than water for equal thermal storage. The potential for the development of these storage types is very good, but at present they are not commercially available.

The tank and ducts, or piping from collector to the tank, should be insulated to limit the thermal losses, at any storage or collection temperature, to 5% of the system capacity per 24 hours. A minimum o of two storage tanks should be considered - three is better - to permit storage of water at various temperature levels, so that the water temperature is not degraded by water at lower temperatures from the collector when it cannot deliver water at higher tank temperatures for periods of time. With one tank only, collection is limited, and the economic feasibility reduced since the collectors cannot be used to their maximum potential.

Controls: The control system, at the least, must operate the circulating pump(s), divert warm water from one storage tank to another or direct to the load, operate fans (in air system) to blow air through the collectors or storage in response to climatic conditions and building load. There are many variations in controls which cannot be covered within the scope of this manual. Controls are not a major cost of the installation unless elaborate monitoring, data collection, and research is undertaken.

Anti-Freeze Protection: In areas where the outdoor temperature drops below 32°F. anti-freeze protection is required for liquid flow collectors. The following systems may be used:

- A secondary circuit with an anti-freeze solution, i.e. glycol, tri-ethylene glycol, propylene glycol, or light oil with a heat exchanger and a separate pump.

- Draining the collector when outdoor temperature drops below freezing and refilling when temperature rises above freezing.

- A closed circuit drainage system with nitrogen as a purging agent.

- Bleeding a small amount of heated storage water through the collector in locations where temperatures drop below 32°F for only brief periods of time, and the days are warm and clear.

Each system has advantages and disadvantages. If the storage water bleed system shows only a small loss of thermal energy, compared to useful heat collected on a seasonal basis (less than 5% loss), it is the least costly and least troublesome system.

The nitrogen system is still under development and more data is required before it can be recommended without qualification.

Draining and refilling involves problems with incomplete draining under some conditions and difficulties with venting when the system is refilled. Vents are also subject to freezing causing them to be inoperative.

The use of anti-freeze with a secondary circuit (See Diagram) introduces higher costs for the heat exchanger and secondary circuit, and corrosion is accelerated when anti-freeze solutions reach high temperatures in the summer. Traditionally, however, most current installations use anti-freeze with a secondary circuit and heat exchanger in cold climates.

INTERFACE BETWEEN SOLAR COLLECTOR AND EXISTING HEAT SYSTEM

The existing heating and cooling system influences the size and type of solar collector system. Existing radiation, fan coil, or induction units, sized to handle the average peak heating load with electricity, steam, or water above 180°F, limit the hours of useful collection unless the heating system is modified by one of the following methods:

- Replace some or all of the existing terminal units with oversized fan coil units which can operate with lower temperature water (Approx.120°F)
- Add an air handling unit with oversized fan and coils to supply part, or all, of the areas to be heated.
- Add radiant heating coils.
- Add new heating coils to existing air handling units or forced air furnaces.

Heating with radiant hot water panels, furnaces, or air handling units are the most compatible with solar energy heating systems, since they can either use the same coils or be fitted with new ones to accommodate the lower collector water temperatures.

Solar collectors combined with existing or new heat pumps can be very effective. Low temperature fluid delivered from the collector in cold and overcast weather serves as a heat source for a heat pump. The heat pump boosts the temperature to a higher level and the useful hours of solar energy collection are substantially increased. Solar assist heat pump installations should be considered especially in existing buildings that are already equipped with heat pumps or with electrically-driven chillers which can be converted to heat pump.

Cooling and Dehumidification: The common means for utilizing solar energy for cooling and dehumidification are with absorption chillers, or with desiccant systems for dehumidification.

With absorption systems, the heated fluid serves as the heat source in the absorption generator. Most existing units require 230° to 270°F for generation, and solar collectors on the market cannot provide those temperatures for any appreciable lengths of time. However, newer low temperature absorption units are now available in limited sizes to operate at generator temperatures as low as 180°F with only a slight loss of rated capacity.

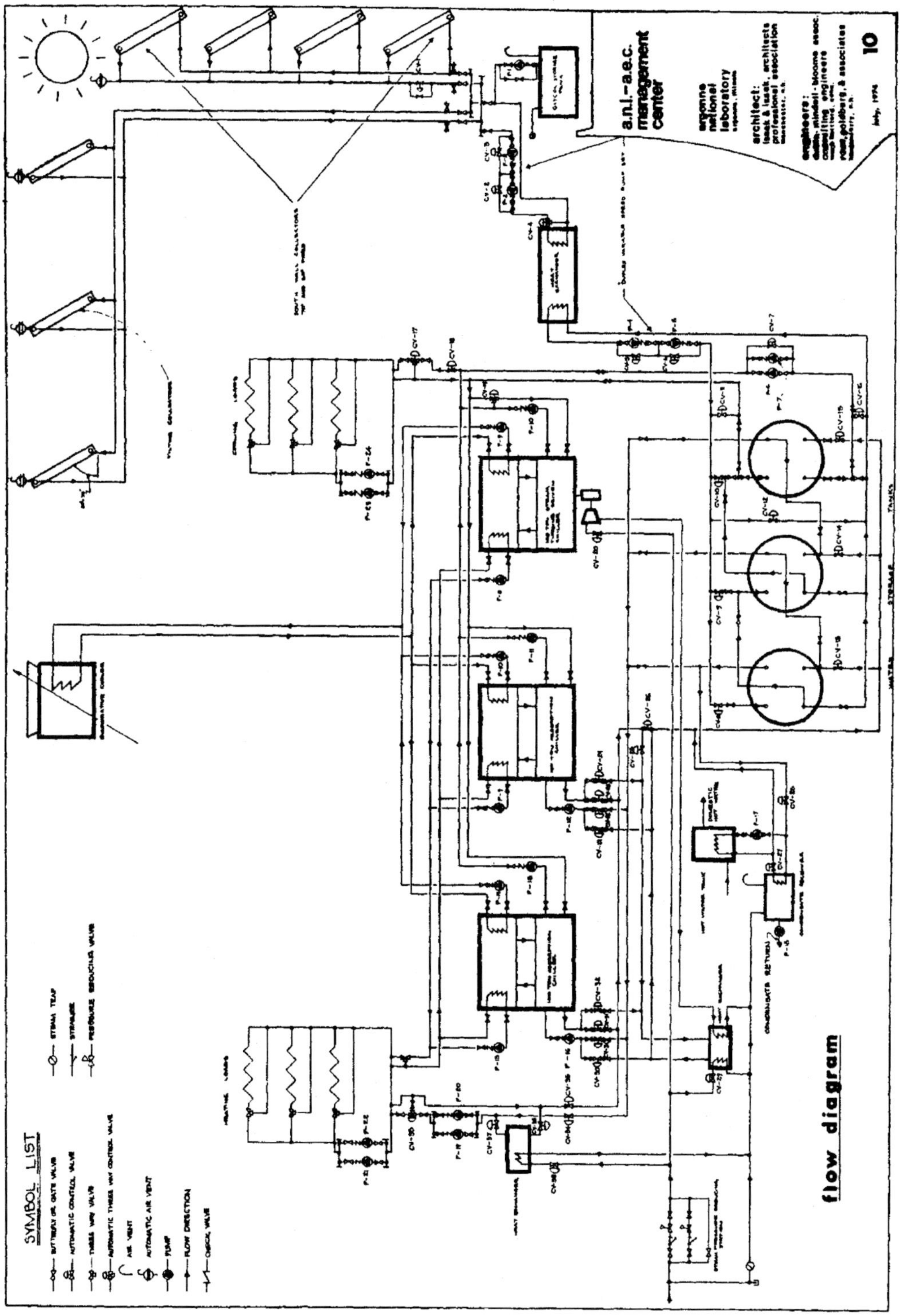

a.n.l.-a.e.c.
management
center
argonne
national
laboratory
10
SYMBOL LIST
flow diagram

Flat plate collectors, with selective surfaces, and supplementary reflectors, can operate at temperatures up to 200° in some climatic zones for a sufficient number of hours per cooling season to operate absorption units at high capacity. However, there are not many areas in the country where absorption units operating from solar collected heat alone can supply even 50% of the yearly cooling requirements of the building.

For cooling alone, the tilt of the collector should be equal to latitude less 10°. Where the collector is used for both heating and cooling, the heating efficiency suffers, if the angle is adjusted for cooling. A detailed analysis is required to determine optimum tilt angles of the collector for year-round use.

There will be very few stituations where solar cooling will be economic until the collector performances are improved and/or absorption units are produced which will operate at 160 to 180° generator temperatures. Both developments appear to be imminent within the next 5 or 6 years.

Regenerating Desiccants with Solar Heat

This system which has been used only in demonstration installations up to now appears to be cost effective. In areas where evaporative cooling is presently impractical because of high relative humidity, evaporative cooling combined with desiccant dehumidification can provide adequate comfort. Heat from the solar collector can supplement heat from other sources to regenerate the desiccant. Since regeneration temperature are lower than those required for absorption cooling, the collector is more cost effective than with absorption unit.

Estimated Costs

Costs vary widely with the size and type installation and the construction conditions encountered. Collectors cost from $5.00 to $12.00 per square foot; storage systems from $1.00 to $3.00 per gallon; piping and insulation from $3.00 per square foot of collector and upward; and controls and wiring approximately $2.00 per square foot of collector. Complete installations vary from $10.00 per square foot (unusual to fund one this low) to $30.00 per square foot. Costs and collector performance are continually changing; estimates should be based on preliminary design concepts and layouts, with bids from contractors taken early, before final design is started.

OTHER SYSTEMS TO UTILIZE SOLAR ENERGY

Tempering Heavy Oil

For large installation with #6 oil, flat plate solar collectors can be effectively used to preheat the oil before combustion. The oil storage tanks can be fitted with heating coils through which the heated fluid from the collector is circulated. No additional storage system is required.

Tempering Combustion Air

The efficiency of oil or gas combustion increases as the temperature of combustion air is increased. Solar air or water collector, without any additional storage facilities, can be used to temper the air for combustion. In the same manner, it is used to temper make up air for ventilation.

ELECTRIC POWER

The technology for using solar energy for the direct generation of electric power with photo-voltaic cells is well known. Solar cells are used in the outer space program to generate electricity and in remote areas of the world to generate small quantities of electricity, primarily for signal purposes. In Delaware, there is a demonstration house equipped with solar cells to supply electrical energy and most of the annual requirements for thermal energy. The hardware available for buildings is limited and too costly to consider now. The area required for solar collectors for heating and cooling will be ample to accommodate the future installation of solar cells for power generation.

Solar energy can also be used to generate high temperature hot water, or steam which can power a turbine or heat actuate engine for power generation, but these systems are not commercially available and the costs are not comparable with conventional generating methods.

B. TOTAL ENERGY SYSTEMS

BACKGROUND

An alternative to purchased utility power, is on-site electrical power generation using generators driven by combustion turbines or engines. For an on-site power plant to qualify as a total energy plant (TE), the heat energy that is normally lost when electricity is being generated must be recovered and put to useful work.

The lure of total energy lies in the recovered heat, since its equivalent would have to be purchased in the form of other fuels. The recovered heat can be used for heating domestic hot water, tempering outdoor ventilation air, cooling with absorption refrigeration, space heating or producing process steam.

Since all TE installations require added investment, the system can be economically advantageous only where the energy savings are sufficient to amortize (in a period short enough to be attractive to the investor) both the higher capital costs and the costs of service, maintenance and replacement parts. When electrical energy was plentiful and service reliable, TE's appeal was limited to economic considerations only. However, power plant and transmissions facility interruptions responsible for the increasing frequency of blackouts and brownouts--provide additional reasons for considering reliable decentralized TE systems.

TE plants consume 25 to 40% less raw source energy than a combination of a fossil-fueled central electrical generating plant and on-site boilers, which make TE a highly beneficial national energy conservation measure.

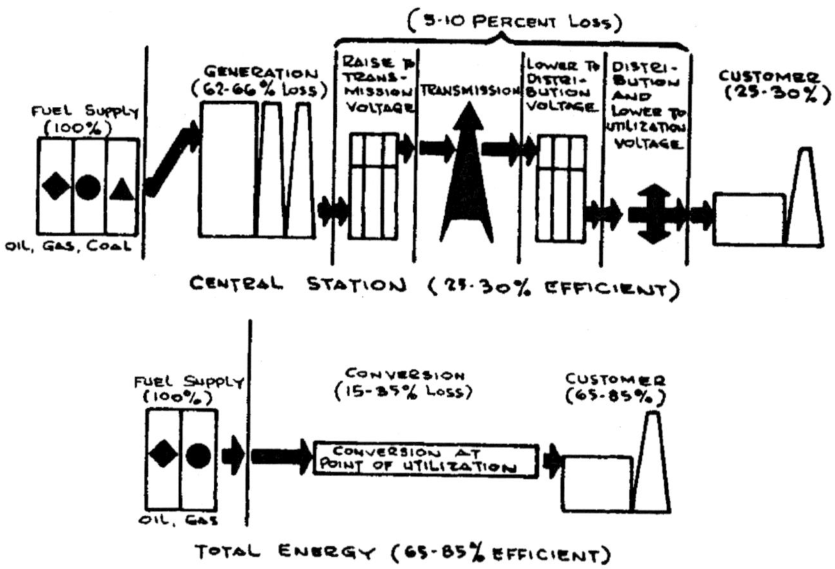

EFFICIENCY COMPARISON: TOTAL ENERGY VS. CENTRAL STATION

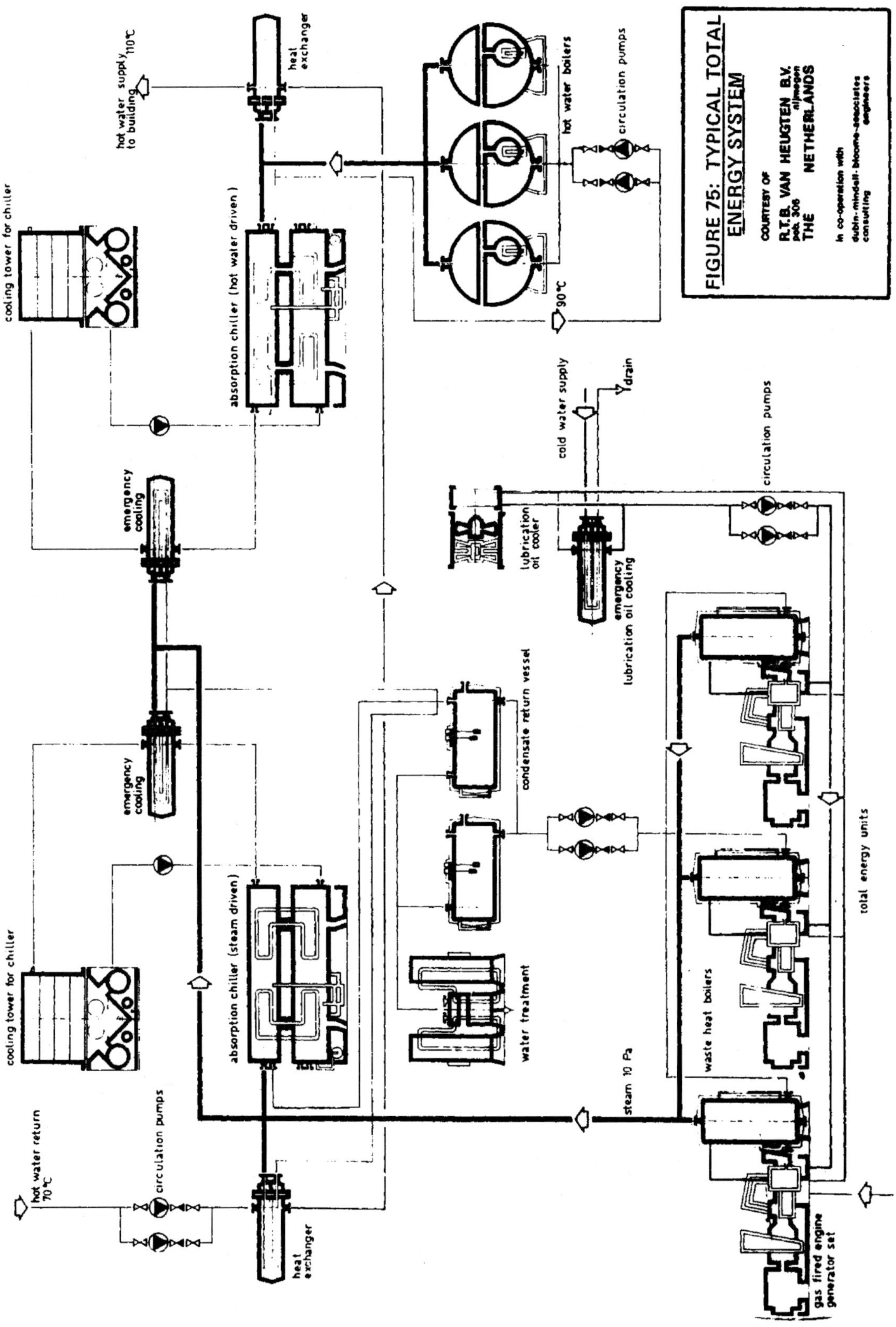

FIGURE 75: TYPICAL TOTAL ENERGY SYSTEM

COURTESY OF
R.T.B. VAN HEUGTEN B.V.
pob. 306 nijmegen
THE NETHERLANDS

in co-operation with
dubin-mindell-bloome-associates
consulting engineers

Most total energy systems to date have been installed in shopping centers, large department stores and hospitals, where long hours of operation provide an opportunity to utilize the equipment for many hours in the year and amortize the added investment through fuel savings. Existing retail stores and office buildings which are larger in area than 50,000 square feet, and heated and air conditioned, should now be considered for total energy systems.

Total, or selective energy systems using steam-driven turbines are beyond the scope of retrofitting in the context of this manual. Only systems employing engines or combustion turbines are included here-in. For a detailed flow diagram see Fig 75.

GUIDELINES FOR SELECTING BUILDINGS WHERE TE SYSTEMS MAY PROVE FEASIBLE

-Floor area greater than 50,000 square feet.

-Occupancy greater than 38 hours per week; additional hours will enhance the economic feasibility.

-Building includes both heating and air conditioning systems.

-Gas or oil rates low enough to compete with electric rates- (Fuel at 20¢/therm or lower vs purchased electricity at 3.2¢ kwh or higher).

-Skilled operating personnel available in-house, or by contract.

-Owners fiscal policy permits payback periods in excess of 8 years.

-Adequate space available to accommodate prime movers and generating equipment.

-Electrical and thermal loads are in appropriate balance. (Must have use for at least 50% of the waste heat).

-Where a number of buildings with different functions can be served together, the diversity of loads will enhance the economic feasibility.

If most or all of these guidelines are satisfied, then the second step is to conduct a feasibility study.'

PREPARING AN ENGINEERING AND ECONOMIC FEASIBILITY ANALYSIS

The following major factors must be considered.*

1. The load-profile for the building:

   The hourly requirements for waste heat on a 48 hour basis, for each season of the year and the simultaneous electrical load requirements (including minimum, average and peak demands), and consumption.

The load profile is influenced by the design and operation of the mechanical and electrical systems. It is especially important to reduce the peak electrical loads to enhance the economic viability of TE systems, since peak electrical loads determine the size and cost of the engines, turbines and generators.

*For an example of a detailed study refer to "Total Energy", a technical report from Educational Facilities Laboratories, by Dubin-Mindell-Bloome Associates, 1969.

The load-profile curves are essential to determine the amount of waste heat that can be recovered. This amount represents the energy and dollar savings of a TE plant.

2. Equipment and Fuel

(a) Prime Movers

Although the efficiency of reciprocating engines is considerably higher than that of gas turbines, if there is continuous use for waste heat the total efficiencies of both prime movers with heat recovery equipment are about the same. For a project that has a high demand for waste heat in relation to the electrical load, a turbine probably will be a more favorable selection than an engine. This would occur in areas with high heating and cooling loads. Where the demand for waste heat is lower in relation to the electrical loads, engines will be more economical.

Large slow-speed engines cost more initially, but require less maintenance and service, and use less fuel than smaller, high-speed engines. Both engine types must be evaluated against turbines, which are lighter and require less maintenance, but may cost more initially.

Engines and combustion turbines operate most efficiently when they are loaded to 60% or more of rated capacity. Since the electric loads vary for different conditions of occupancy, the most efficient total energy plant utilizes generating set modules so that each prime mover is always operating close to full rated capacity.

A series of prime mover-generator components or modules should be selected so that they match those loads which exist for the longest periods of time during the year. Usually these periods are the normal daytime hours of maximum loading and the unoccupied hours of minimum loading. The largest modules which will still permit operation at maximum efficiency during the greatest number of operating hours should be selected. Standby requirements dictate that if any module were to be out of service, there would be enough capacity of those remaining on-line to carry the maximum load. The liklihood of more than one module being out of service is not great, and therefore the minimum standby requirement would be equal to the largest module of the group of modules required to carry the maximum load.

(b) Heat Recovery

For gas turbines, the waste heat available is in the exhaust gases. Heat is recovered by means of a waste-heat boiler which resembles a conventional boiler. A duct, similar to a breeching, is connected directly from the exhaust of the turbine to the boiler. After passing through the boiler the gases are discharged to the atmosphere.

The over-all system efficiency for diesels is quite comparable to the gas engine. In most installations, 90 percent of the jacket water heat is recoverable and 60 to 65 percent of the exhaust gas heat is recoverable -- a total of about 45 percent of the input energy to the engine.

Gas turbines produce about twice as much recoverable heat as reciprocating engines of the same size. Generally, a turbine yields 7 to 13 pounds of steam per hour (15 psi steam) per kilowatt generated. An over-all system efficiency of slightly over 65 percent is possible in total energy plants, if all the recoverable heat is used.

(c) Storage Systems

Heat storage systems can be used where waste heat requirements do not coincide simultaneously with waste heat availability. Often the heat cannot be used immediately, but can be used at later periods when heat requirements exceed the heat available.

OTHER CONSIDERATIONS FOR TE PLANTS

Conventional electric power is supplied at 60 hertz. However, on-site power generation, particularly with high-speed turbines, provides the option of generating power at frequencies other than 60 hertz with little initial cost penalty. High frequency lighting may provide additional energy conservation benefits (Refer to Section 19 Lighting).

This page blank

Section 23

Cost Estimates and Economic Analysis

COST ESTIMATES AND ECONOMIC ANALYSIS

A. COST ESTIMATES

This Section includes methods of determining costs for many of the energy conservation opportunities discussed in the manual and an economic analysis model to be used in determining the financial feasibility of energy conservation investments. Detailed costs presented in Appendix B are based on single contracts. The cost per item would be less if it was part of a larger overall contract. Costs include labor, materials, overhead and profit. Special conditions should be noted.

Cost estimates for other items are presented throughout the manual as guides to indicate the order of magnitude of expense involved. When exact costs are required, detailed estimates should be obtained from local contractors.

Construction costs vary throughout the country due to:

a) wage levels

b) material costs, availability, and shipping

c) sales taxes

d) level of competition

e) climate and seasonal construction periods

f) union practices.

Costs included in Appendix B are based on Metropolitan New York wages and material costs, January 1975. In order to apply these costs to other areas of the country, cost indices for the following cities are presented, using New York City as a base of 100.

New York	100
Chicago	93
Los Angeles	93
Miami	89
Phoenix	88
Denver	82
Seattle	82
Atlanta	80
Houston	79

Adjust the New York costs to that of the nearest major city to achieve an adjusted cost prior to applying an escalation factor.

Continually increasing labor and materials costs will rapidly escalate the costs presented in the Appendix; therefore, it is avisable for the user to obtain updated local costs from his specific project from a local Consulting Engineer, Architect, Engineering Cost Consultant, a General Contractor. However, to approximate escalation use on of the professionally prepared construction cost indices: Boeckhs, F. W. Dodge Co., or articles on cost escalation in trade journals. By using these publications and the city cost indices, the user can determine a fairly accurate up-dated price.

The cost of fuel varies widely throughout the country and estimates should be based on the prevailing costs in the applicable area. Fuel suppliers can provide unit costs which can be converted to Btu's by using conversion factors in the "Energy Use Audit" in ECM 1.

The cost of electricity should be based on a rate schedule obtained from the local electric utility company. Where users are subject to peak demand charges, changes in peak demand resulting from energy conservation in addition to reduced consumption must be taken into account when analyzing future electrical costs.

B. ECONOMIC ANALYSIS

Appendix C presents an Energy Conservation Cost Analysis Model applicable to commercial building ownership. The model is designed to evaluate on a total life cycle cost basis, the economic and financial considerations of proposed energy conservation improvements to existing buildings, and to provide architects, engineers and building owners with a means of determining whether various energy conservation measures are within the owners economic capabilities.

Financial considerations should be an integral part of the total matrix of economic and non-economic factors considered in the evaluation of alternative energy conservation measures. The model presented in Appendix C provides a means of evaluating those variables associated with a proposed energy conservation measure which can be quantified in dollar terms. The model cannot evaluate non-quantifiable items such as patriotism, conformance with national policy, fuel availability, etc.

The cost model developed for this manual is designed to evaluate energy conservation measures which involve capital expenditures. The cost model can be used to evaluate conservation measures which do not involve capital expenditures or borrowing by simply leaving blank those portions of the model applicable to capital investments and debt service. In general, wherever the capital cost of the energy conservation measure divided by the yearly energy saved is three or less, payback will usually occur between six to ten years at a rate of return high enough to justify the investment. In these cases, it is suggested that analysis by cost model is not necessary.

1. PURPOSE OF ENERGY CONSERVATION INVESTMENT ANALYSIS MODELS

The purpose of the model is to provide a means of determining whether proposed energy conservation improvements for existing buildings are economically viable under present or reasonably foreseeable future circumstances and to allow building owners to objectively evaluate in economic terms alternative energy conservation improvements to existing buildings. The model is designed to evaluate energy conservation improvements in terms of their total life cycle cost savings. The ultimate output of the model is the discounted rate of return from an energy conservation investment. Those investments which provide the highest discounted rate of return should be given priority over those which produce a lesser return. Those energy improvements which produce a negative rate of return in terms of this model may have other overriding benefits which cannot be neglected. The model will also allow building owners to compare the rate of return which can be expected from investments in energy conservation improvements with the rate of return which can be obtained from alternative investments of their available capital resources.

2. DESCRIPTION OF THE MODEL

The nature of the entity owning a building materially affects the type of factors which must be considered in developing an investment analysis model. The Commercial Building Ownership Model is designed to be used for energy conservation improvements which are being made by an entity which is subject to Federal and/or State income taxation. It is applicable for use by most private individuals, for profit corporations, and real estate investment and development partnerships, joint ventures, syndicates and trusts.

If economic analysis is necessary for entities such as buildings owned by Federal, State and local governments, public agencies, and non-profit corporations, institutions and foundations, the Commercial Building Ownership Model can be utilized but must be modified to account for various applications of the tax laws.

In general, income tax, investment tax credit and depreciation are not taken into consideration. Property taxes, increases in rental income, and debt service are not applicable to government owned property but may apply to certain private non-profit organizations and institutions. In questionable cases, the advice of tax experts should be sought for specific interpretation of tax laws.

3. GENERAL OPERATION OF THE MODEL

This Section presents general narrative instructions regarding the operation of the energy conservation investment analysis model. Appendix C contains specific line item by line item instructions for operation of the model.

The purpose of the model is to isolate costs, revenues, and changes in operating expenses associated with an energy conservation improvement, and to evaluate these items separately and independently from the general operating income and expenses of a building. In using the model to evaluate the economic viability of a particular energy conservation measure, care must be taken to consider only those variables which are associated with the energy conservation measure. The design criteria for the development of the cost model was to make it as simple as possible consistent with the overall objective of providing a means of accurately evaluating the economic viability of an energy conservation investment. In this simplification process, some relatively minor items have been deleted. The Commercial Building Ownership Model has been designed for use by a "typical" taxpayer. There are a number of specific Federal income tax provisions which apply to certain subgroups of taxpayers, particularly high-income individuals owning substantial real estate holdings.

Such high-income owners of real estate should consult with a tax lawyer or Certified Public Accountant if they wish to refine the results of the model to fit their particular tax situation.

4. LIFE CYCLE COSTING

The evaluation of the economic viability of an energy conservation improvement should be based on total cost savings, discounted to reflect the time value of money, which are expected to accrue over the useful life of the improvement. The forms shown in Appendix C contain enough columns to evaluate improvements with useful lives of up to ten years. Many improvements will have useful lives greater than ten years, in some cases as long as the life of the building (40 years). The useful life assigned to an energy conservation improvement should be no greater than the remaining useful life of the building in which it is installed.

As a practical expedient, a building owner may choose an evaluation period shorter than an improvement's estimated useful life. Using a shortened useful life can greatly simplify the computations involved in computing the rate of return applicable to a particular energy conservation improvement. The rate of return computed using a truncated useful life of not less than ten years can be reasonably accurate given that as time increases, the present value of energy savings and the certainty with which such savings can be predicted both decrease. Assumptions of cost savings projected far into the future are inherently more risky than current assumptions, and the discounted present value of such far future savings is significantly less than more current savings.

5. MULTIPLE CONSERVATION MEASURES

The total energy savings resulting from multiple energy conservation improvements to a single building will generally be less than the cumulative energy savings resulting from each individual conservation measure. Projected cost savings from multiple energy conservation improvements may, therefore, be somewhat overstated unless the engineering data used to compute energy savings took into consideration the diminishing return effect of multiple energy conservation improvements to a single structure.

6. GENERAL PRICE LEVEL INFLATION

In preparing inputs for the cost model it is suggested that all costs be presented in current year dollars. This assumption assumes that increased expenses resulting from inflation will be offset by inflation generated revenue increases. In the case of religious and non-profit institutions, an inflation factor of 6% - 12% should be included.

7. MATERIALS NEEDED

In order to evaluate an energy conservation improvement using the cost model an analyst should have a desk top electronic calculator. Standard real estate investment analysis interest rate, discount, and mortgage tables are also required. A recommended source of such tables is the Realty Bluebook published by Professional Publishing Corporation, P.O. Box 4187 San Rafael, California, 94903 (telephone - 415/472-1964). Familiarity with income taxes as they apply to real estate would be helpful but not required for operation of the model.

This page blank

Section 24

Computer Programs for Energy Analysis

COMPUTER PROGRAMS FOR ENERGY ANALYSIS

BACKGROUND

Load calculations computed manually are based on a given set of steady state conditions (usually maximum and minimum expected conditions outdoors and occupied conditions indoors). While these calculations give a reasonable indication of the size of equipment required to meet maximum loads (invariably conservative) they give no indication of energy requirements to meet the myriad part load conditions that actually occur in an average year. A rough analysis of yearly energy requirements can be obtained by using heating degree days and cooling degree hours but this totally ignores the building and system reactions to constantly changing conditions.

Closer approximations of yearly energy requirements can be made by making a series of calculations for different outdoor conditions during both occupied and unoccupied times and correcting for the length of time these conditions are expected to exist. Accuracy of results improves with the number of different conditions selected and the number of separate calculations but this can develop into a long tedious and costly process if done by hand.

Computers are extremely good at making a series of repetitive calculations and programs have been developed independently by many different organizations to perform this task. Because these programs have been developed independently, each one has characteristics and functions which are unique. One program,for instance, may be very good at calculating heating and cooling loads but skimpy in its approach to analyzing HVAC system reaction to those loads while another may carefully analyze interreaction between HVAC systems but base its building load calculations on extrapolations of one typical day/month.

PROGRAM TYPES

Calculation programs fall into two categories: those used for solving design problems (Design programs) and those used for calculating energy usage (Energy programs). There is some overlap between categories as some of the energy programs have sub-routines which can be used independently for design problems.

Refer to Figure # 76 for Design Programs and Figure # 77 for Energy Programs that are currently available. These lists are assembled from data extracted from various sources and are presented only as an aid to the Manual user. They are not necessarily a complete list of all programs neither are they an indication of equal quality and performance. When selecting programs for a particular application, the prospective user must obtain detailed information from the programs authors and then make his own evaluation based on his requirements, in-house computer capabilities and projected costs. See Appendix D for Sources of Computer Programs.

Design Programs have been developed to allow rapid solution of particular design problems such as pressure drop in piping and duct work systems, illumination levels obtained from a given lighting fixture layout, etc. The value of these programs is that they allow many more alternative schemes to be considered in the initial design stages than would be possible with hand calculations and thus increase choices available to the designer.

Some programs allow the input of desired results and as output give the parameters of design. For example, some lighting programs will design a fixture layout based on inputs of room size, interreflectances and desired lighting profile.

Design programs for equipment selection must be used with discretion as some programs only recognize certain makes of equipment and will not necessarily allow selective comparisons to be made.

Energy Programs have been developed to calculate the yearly energy requirements for buildings and systems combined, for buildings only and for systems only.

Smaller programs or parts of larger programs are also available to calculate yearly energy requirements for discreet portions of a building or system considered in isolation.

Energy programs first calculate the building and/or system loads for a given set of conditions and then simulate the operation of the building and/or systems to determine the reaction to continuously changing conditions and calculate yearly loads.

Energy programs calculate loads in many different ways, some require "U" values, internal loads and equipment loads to be calculated by hand and the results used as input. Other more complex programs will not accept a "U" value input but require the physical characteristics of each building surface to be fully described in scientific units (i.e., Conductivity in Btu ft./sq. ft. hour degree F not the more commonly used Btu inch/sq. ft. hour degree F). Most programs have the facility for printing out the building and/or system loads before the energy analysis phase is started but some programs will not divulge the calculated "U" value for instance and the load print-outs must be carefully screened for input errors at this stage.

Once a valid load is developed, the energy analysis phase of the program can be initiated. This part of the program simulates the building and systems reaction to changing weather conditions, occupancy, internal loads, equipment loads, part load efficiencies, etc. The energy analysis is achieved by simulating the building and systems as mathematical models in the computer and recalculating loads on a time basis. Less complex programs may only calculate loads for two conditions on one day in each month and extrapolate for yearly energy loads while more complex programs calculate loads for each of the 8,760 hours in a year and integrate these for yearly energy loads. Calculations on an hourly basis allow the dynamic flow of energy to be simulated by accounting for the thermal lag of building materials and the effect of storage on loads. This effect can be important resulting in reduced loads and possible changes of operating techniques.

FIGURE 76: DESIGN PROGRAMS

TYPE	NAME	AUTHOR
Building Form		
Form Generation	Form Ge eration	University of Southern California
Shading Analysis	SHADOW	University of Texas
	SUNNY	University of Texas
Solar Gain Analysis	SUNSET	Dubin-Mindell-Bloome Associates
Building Exterior Envelope		
Glass Comparison	Glass Comparison	Libbey-Owens-Ford
Building Interior Planning		
Optimization	ARK-2	Perry, Dean & Stewart
	B.O.P.	Skidmore, Owings & Merrill
Lighting		
Conventional	Lighting II	APEC
	Lighting	Dalton, Dalton, Little & Newport
	Interior Lighting Analysis & Design	Giffels Associates, Inc.
ESI	Lighting Program	Isaac Goodbar
	Lighting Program	Illumination Computing Service
	Lighting Program	Ian Lewen
	Lumen II	Smith, Hinchman & Grylls
Daylighting	Daylighting	Libbey-Owens-Ford
Power		
Distribution Network	Electrical Feeder II	APEC
	Electrical Feeder Sizing	Dalton, Dalton, Little & Newport
	Three Phase Fault Analysis	Giffels Associates, Inc.
Demand Study	Electrical Demand Load Study	Giffels Associates, Inc.
HVAC		
Equipment Selection	HCC-III (Mini-Deck)	APEC
	Equipment Selection	Carrier Air Conditioning Co.
	Equipment Selection	Trane Co.
Duct Design	Duct Program	APEC
	Several	Dalton, Dalton, Little & Newport
	Several	Giffels Associates, Inc.
Air Handing Unit Design	Fan Static Calculations	Giffels Associates, Inc.
Domestic Water		
Piping Design	Piping Program	APEC
	Several	Dalton, Dalton, Little & Newport
	Several	Giffels Associates, In .
Vertical Transportation		
Elevator Design	Elevator Design	Dover Corporation
	Elevator Design	Otis Elevator
Operation & Maintenance		
Automated Control Systems		Honeywell, Inc.
		Johnson Control Service
		Powers Regulator Company
		Robertshaw Controls
Solar Energy Systems		
Solar Collector Models	Flat Plate Collector	Honeywell, Inc.
	Parabolic Trough Collector	Honeywell, Inc.
	Flat Plate Collector	Westinghouse

Of the available energy programs, some are biased towards building simulation and others are biased towards system simulation. When selecting a program, this bias should be recognized.

PROGRAM INPUTS

Each of the Design and Energy programs requires input of information ranging from size and shape of the building to complete weather data.

It is not the intent of this manual to detail all types of input but rather to caution the prospective user on certain aspects of input data. In general, great care should be taken in assembling input data as this sets the basis for all subsequent calculations. Computers have no discretionary powers and will carry out the literal meaning of input instructions. Check and double check all input data for ambiguities - this is cheaper than using expensive computer time to make new runs.

Energy programs require some form of weather data input. Less complex programs use weather data developed from ASHRAE tables or AFM 88-8. These sources deal with averaged weather conditions which reduces the precision of simulation but will still give satisfactory results where more expensive procedures are not justified. The limitation of use must, however, be carefully analyzed and understood by the prospective user.

More complex programs accept an input of NOAA weather data either in card form or on magnetic tape. This raw data is generally for hourly observation periods before 1965 and for three hourly observation periods from January 1, 1965 except for a few stations which have continued with hourly observation. One set of cards or a tape usually contains data for a number of consecutive years. Some programs have a sub-routine for processing this data and reducing it to a series of typical days ranging from one/month to one/week while others use hourly readings direct for one selected year.

When using weather data, make sure that the selected year is typical for the geographic location or use the data as a basis for obtaining a 10 year average of all pertinent readings.

When using weather tapes or cards, always remember that the data is for actual weather conditions for that year and may be far from typical for that geographic location.

PROGRAM OUTPUTS

The output of a computer program is a series of coded bits of information and is usually assembled into tabular form under "English Language" headings for easy handling. Outputs may be limited to yearly energy or may be increasingly detailed to show the energy load on a monthly or weekly or daily basis for each system and sub-system. Some programs will give only one form of output while others have the facility to give as much or as little information desired by the user and can also provide outputs at intermediate points in the

FIGURE 77: ENERGY PROGRAMS

Name	Author
	A. Commercial Programs
ECUBE	American Gas Association
HCC-111	APEC
Energy Analysis	Caudill Rowlett Scott
AXCESS	Electric Energy Association
Glass Comparison	Libbey-Owens-Ford
Energy Program	MEDSI
Energy Analysis	Meriwether & Associates
Building Cost Analysis	PPG Industries
TRACE	TRANE Company
Energy Program	Westinghouse Corp.
HACE	WTA Computer Services, Inc.
	B. Research Programs/Negotiable
CADS	UCLA
SIMSHAC	Colorado State University
FINAL	Dalton, Dalton, Little & Newport
HVAC Load	Giffels Associates, Inc.
Energy Program	Honeywell, Inc.
NBSLD (Honeywell)	Honeywell, Inc.
Energy Program	University of Michigan
NBSLD	National Bureau of Standards
B.E.A.P.	Pennsylvania State University
Post Office Program	
DEROB	University of Texas
TRANSYS	University of Wisconsin
	C. In-House Program/Proprietary
Energy Program	General Electric Company
Residential & Small Commercial	Honeywell, Inc.
Energy Program	IBM

program. Two cautions regarding output are required:

a) Do not demand more output than needed. A mass of data is intimidating and can also be misleading. Select carefully the output information required to analyze a problem.

b) Remember that the computer output is only the solution to a set of calculations not the answer to a problem. It is a good source of quantitative data but must be analyzed with skilled engineering judgement and applied with imagination to obtain the full benefit of any particular program.

PROGRAM SELECTION

The choice of computer program must be made on the basis of the prospective user's needs, his expertise and his in-house capabilities. The following questionaire will help establish these.

a) What degree of sophistication is required? If the problem is to evaluate various system designs then the appropriate design program will suffice and choice should be limited strictly to programs dealing with the particular type of system. If the problem is to evaluate energy use of alternative building shapes and construction then use an energy program biased towards buildings. Conversely, if the problem is to evaluate various system reactions in a given building then use an energy program biased towards systems.

b) How much money is available for computer calculations? The cost of computer time and data compiling varies widely from program to program and is not necessarily related to program complexity. Obtain estimates of run time from the program author.

c) What are the prospective user's in-house capabilities? Some programs such as NBSLD and AXCESS are only available for use on in-house computers while others such as RFM and APEC can be used either in-house or remote batch or time sharing. Still others are limited to use on the author's computer operated by the author's personnel. Obtain information on program application and availability from program authors.

FUTURE DEVELOPMENTS

The use of computers is a rapidly evolving science, programs are continuously updated as more experience is gained. New methods of presenting data are being developed.

At present, the available programs are rarely compatible with one another but efforts are being made to establish a common logic, algorithms and language so that the output of one program (say lighting calculations) may be used as input to another program (say energy analysis of building HVAC systems).

APPENDIX A
SELECTED CODES AND STANDARDS

COMPLYING WITH BUILDING REGULATIONS

Whenever Energy Conservation Opportunities are implemented, specific and careful consideration must be given to the various local or state codes, occupational regulations pertaining to health and safety and laws and ordinances or directives which regulate building construction, operation, maintenance, repair and use. The Energy Conservation Opportunities listed in this handbook have brief descriptions of those building regulations that may need to be considered. While this information is useful in alerting owners, managers and tenants to potentially applicable building regulations, the following information will be useful in understanding and determining specific building regulations applicable to any Energy Conservation Opportunity undertaken in any particular building.

BUILDING REGULATIONS

Building Regulations are diverse and complex and they are enforced by many regulatory agencies at different levels of township, city, county, state and federal government. Depending on the size of the jurisdiction, the delegation of authorities, administrative practices and technical expertise available, responsibility and enforcement may be well defined or regulations may be loosely applied. In some instances, building regulations may be vague, they may contradict each other and they may be outmoded.

Often, regulations adopt other regulations by reference, frequently with some modifications or exceptions. Some regulations have stipulated requirements that indirectly, adversely effect energy conservation, while other, more recent regulations specifically address energy conservation. The variety of jurisdictions, regulations, procedures and interpretations makes compliance complicated and without uniformity. Yet, with a general understanding of building regulation law and a general knowledge of how, when and by whom these regulations are applied, owners, managers and tenants can achieve compliance and energy conservation.

Building regulations are sets of legal requirements having to do with the physical structure of new and existing buildings. These regulations are adopted to accomplish and promote the public health, safety and general welfare. There is a recognized principle in the regulation of private property that every person ought to use his property so as not to injure that of his neighbor and that private interests must be subordinated to the general interests of the community. The authority to promulgate and enforce regulations lies with each state. In many instances, that authority has been delegated to other jurisdictions within a state - townships, cities, counties. The effectiveness of the control which building regulations have over the construction, occupancy, operation, maintenance, repair and alteration of buildings largely depends upon issuing permits and licenses, and upon inspections. Buildings can be subject to inspections at any time. Permits and licenses are often, but not always required.

RESPONSIBILITY

In general, with respect to the liability of building owners and tenants, for compliance with building regulations, the rule appears to be that ordinarily, the duty to comply rests with the one in possession and in control, or the one who undertakes the planning and construction. An owner cannot evade his duty to comply by leasing the building. A building owner or managing agent can insist a tenant make corrections to conform to building regulations, but the ultimate responsibility lies with the owner and his agents. Building regulations may provide that a licensed architect or engineer certify conformity with the regulations. Where defects of construction or departures from construction plans are in violation of building regulations because of the architect's or engineers lack of skill in preparing plans or negligence in supervision, the architect or engineer is liable and responsible. Where such violations are solely the fault of a construction contractor, that contractor is responsible.

ENERGY CONSERVATION

Although new energy conservation regulations, when enacted, may be retroactively applied to existing buildings, to date, the emphasis has been on new construction.

Many state governments have authorized energy conservation regulations but few have actually adopted them. California, Wisconsin, Washington and Colorado now have such regulations in effect, while Ohio, and North Carolina have adopted regulations to take effect by 1976. Oregon, Minnesota, Connecticut, New Jersey, Arizona, Idaho, Alabama, Nebraska and Kansas are in the process of adopting mandatory regulations, and Maryland, Georgia and New York have voluntary statewide codes in which they are incorporating energy conservation measures. Iowa, Indiana, Virginia, Tennessee, Illinois and West Virginia lack authority for such regulations, while Michigan, Florida, Massachusetts, New Mexico, Rhode Island and Montana have legislative authority, but are awaiting national standards before promulgating code requirements. These standards are being developed by the American Society of Heating, Refrigeration and Air Conditioning Engineers in "Design and Evaluation Criteria for Energy Conservation in New Buildings" (90-P).

Most energy conservation regulations adopted thus far apply to housing construction, although more general applicability is expected. Controls are expressed through design criteria, limiting heat loss or gain through minimal insulation values for materials or building components. Another method of budgeting amounts of energy allowed for various building types in BTU's per year has been adopted in Ohio.

ENFORCEMENT

Violations of building regulations may be determined by building regulation officials in regular, scheduled inspections of new construction, maintenance, repairs, alterations or remodelings. Violations may also be found in the routine inspection of existing buildings by building regulation officials. Buildings can be inspected at reasonable times and in a reasonable manner with or without prior knowledge of any violation by a building regulation official without the need of a search warrant.

Occupying, operating, maintaining, repairing or altering a building contrary to the provisions of the building regulations or without a license or permit when these are required, are violations of building regulations. Usually when a violation is found in an inspection, a first warning, sometimes in writing, is given. Generally, a second notice or followup inspection is made within a reasonable time to determine wyether the violation has been corrected. If it has not been remedied, a specified number of days in which to comply is usually designated. If proper action is not taken by the end of that period, sanctions may be used to enforce building regulations: fines, stop work orders, denial of occupancy or arrest. Since building regulation officials have the power and authority to interpret building regulations and inspect, in addition to the courts, many building regulations provide some type of appeal from the actions, orders or decisions of building regulation officials and for deviations from the strict ledger of the regulations. Generally, this is accomplished by appeal boards, arbitration boards or administrative officials.

SCOPE OF BUILDING REGULATIONS

Building regulations, in so far as they are concerned with the building itself, set up requirements for materials and methods of construction in a building code. The building code is usually concerned with structural strength and stability to provide reasonable safety to the community and to the occupants of a building as well as to the building itself. Other building regulations are concerned with other aspects of the public safety, health and welfare.

Similarly, fire prevention codes may be a part of, or separate from building codes. These regulations govern such equipment as sprinkler systems and standpipes for fire fighting, the means of safe exit, fire limits or fire zones related to land use, the combustibility of materials of construction, the area and height of buildings as this effects fire fighting and the number of building occupants.

Health and safety regulations such as OSHA Standards of the Occupational Safety and Health Administration (OSHA), U. S. Department of Labor or Federal air and water pollution regulations are a part of **all** building regulations. Frequently, there are other health regulations, either in the building code or as separate codes, regulating plumbing, sewage, drainage, light, ventilation, heating and air conditioning.

Planning regulations controlling the use of land and buildings; and zoning regulations governing types of land use, the types of activities permitted in buildings and the size, height, set back and land coverage of buildings are often separate codes. These codes frequently overlap each other and the building code.

Certain Public Utility regulations for water, sewer, electricity, gas and central steam plants are in effect building regulations. These may, in some instances, overlap other codes, but they are, in the main, separate codes.

Certain traffic regulations are also in effect building regulations. These are always separate from the building code.

Architectural Review Boards, Fine Arts Commissions, Property Owners or Tenant's Associations and Historic Preservation commissions all have independent regulations that complement and/or supplement any or all of the other applicable codes.

Finally, business and professional regulations require the licensing of architects, engineers, construction contractors, electricians and plumbers in most states.

APPLICATION OF BUILDING REGULATIONS

Building regulations are applicable to all privately owned real property. These regulations are not, however, applicable to real property owned by the Federal Government; but where private property is leased by the Federal Government for a term of years, building regulations do apply.

BUILDING REGULATIONS AND ENERGY CONSERVATION OPPORTUNITIES

The many building regulations apply to the various Energy Conservation Opportunities contained in this manual in a variety of ways. While this is complex, grouping Energy Conservation Opportunities into four (4) basic types provides a way to generally describe the requirements imposed by b uilding regulations. The basic types of Energy Conservation Opportunities and their building regulation considerations are briefly explained in the following sections.

OPERATING CHANGES

Operating changes are seldom subject to building codes, but they may be subject to health and safety standards, or occupational regulations defining working conditions. Two types of operating changes need to be considered:

> Operating changes which involve some reduction in comfort or environmental quality levels, and which, therefore, may be subject - building regulations, and
>
> routine measures with no building regulations implications.

The former type of operational changes may offer greater potential for energy conservation, but they may also offer conflicts with building regulations, particularly occupational health and safety standards. Included in these types of changes are temperature setbacks in winter, higher temperature during the cooling season, and uncontrolled humidity levels. Also included are reduced ventilation or exhausts which economize on fan power and heating/ cooling loads, but which like the other changes must conform to minimum building standards or OSHA regulations. Similarly, the reduction of light levels by removing lamps may violate certain quality level standards or egress requirements specified by fire safety regulations.

The latter includes such changes as rescheduling occupancy or cleaning hours to economize on fuel or power usage, simple operational changes such as turning off lights and electrical equipment and the systematic control of blinds or drapes. Also included here is the rearrangement of work stations, repainting with lighter colors to reduce lighting and as a result to reduce the cooling load while maintaining required comfort levels.

MAINTENANCE AND REPAIR

This category includes among other Energy Conserving Opportunities, items of general upkeep which pertain to the building shell, such as weatherstripping, recaulking, window cleanning and repair, and routine painting. Site maintenance includes tree trimming and snow removal, for example which can contribute to energy savings. These measures are seldom submect to regulation since comfort and service levels are usually not effected. When painting or replacing materials such as carpeting, Fire Code provisions governing flame spread, smoke developed and fuel contributed may apply.

Mechanical and electrical systems within a building also require regular maintenance and repair. These may be subject to operational regulations and inspections. Cleanning or blowdown of boilers and condensers increases operating efficiency, but can involve procedures governed by mechanical or safety regulations. Similarly, routine maintenance of lighting can include relamping with more efficient lamps, but the quality of distribution of light may be affected. Occupational standards such as OSHA can apply. The replacement of ballasts with a different type is subject to electrical code requirements. Repair work on either mechanical or electrical systems also can involve inspections, and those repairs may need to be made by licensed personnel.

BUILDING MODIFICATIONS

These include construction changes affecting the building shell, interior, mechanical and electrical systems. All these are subject to building regulations where there are health or safety implications.

Minor modifications to the building shell can include extensive painting and redecoration for improved lighting efficiency, the installation of blinds or shading devices and special glazing treatments to reduce heating or cooling loads. These and the removal of interior partitions or reducing their height to improve both lighting and HVAC operation have definite building regulation implications.

More major energy conservation construction opportunities include new insulation for exterior walls or roofs, reglazing with insulated or reflective glass, the change of entries to reduce infiltration and the lowering of ceiling heights to reduce loads and increase efficiency. These measures will be subject to building regulations since structural changes or new materials can be involved and fire assemblies of walls, floors, and ceilings or exitways may be affected. The modification of building facades with sunscreens, color or texture changes may be subject to covenent or zoning controls, architectural board or property owners association reviews, fine arts committee or historic preservation commission approvals in certain jurisdictions.

Modifications to mechanical or electrical systems including lighting must comply with building code regulations, occupational standards, health and safety codes, public utility regulations, fire prevention codes and boiler, refrigeration, electrical, plumbing and mechanical codes.

The HVAC system offers major potential for energy savings, through reduction of exhausts and fresh air intake, reclamation of heat from flue gasses or changes in piping and ductwork. These and other measures can reduce fan horsepower fuel consumption and improve overall operating efficiency, however, any change must satisfy mechanical code requirements, and health and safety standards for ventilation or comfort levels.

Any changes in power distribution and switching circuits must conform to electrical codes and be performed by licensed electricians. Alterations of lighting layout, fixture types or the height of suspended fixtures may be subject to building coees, as well as occupational health and safety regulations.

SITE MODIFICATIONS

This category of Energy Conservation Opportunities includes landscaping changes, ponding, fences, walls, berms and paving modifications which can reduce wind infiltration, control solar radiation and, thereby, reduce heating/cooling loads. Tree removal or new plant materials may be subject to covenants, fine arts commission, historic society or architectural review board approval along with other site plan changes. Fencing or extensive regrading must be cleared through local planning authorities charged with site plan approval and in some jurisdictions soil conservation, drainage, and erosion are strictly controlled. Any changes that reduce building setback or required buffers may be subject to zoning regulations, which also deal with sign control and outdoor lighting for walkways and parking lots. Changes that restrict visibility of roadways may be subject to traffic regulations. Site changes affecting fire lanes are subject to the requirements of the fire regulations.

APPLICABLE BUILDING REGULATIONS

The building regulations generally applicable to the four categories of Energy Conservation Opportunities are listed below:

Operating Changes and Maintenance and Repair

- OSHA Standards
- Building Code
- Fire Prevention Code
- Health and Safety Code
- Boiler, Refrigeration, Electrical, Plumbing, Mechanical and Elevator Codes
- Public Utility Regulations

Building Modifications

- o Planning Regulations
- o Zoning Regulations
- o Architectural Review Boards, Fine Arts Commissions, Property Owners or Tenants Associations and Historic Societies
- o Building Codes
- o Fire Prevention Code
- o OSHA Standards
- o Health and Safety Code
- o Boiler, Refrigeration, Electrical, Plumbing, Mechanical and Elevator Codes.

Site Modifications

- o Planning Regulations
- o Zoning Regulations
- o Traffic Regulations
- o Architectural Review Boards, Fine Arts Commissions, Property Owners or Tenants Associations and Historic Societies
- o Building Codes
- o Fire Prevention Code
- o Public Utility Regulations

These building regulations are represented by a variety of organizations - departments, authorities, commissions, bureaus, offices, sections, committees and companies - depending upon the structure of state, county, city and township government.

COMPLIANCE PROCEDURES

The compliance process has three basic, general steps. These are briefly explained in the following section.

1) Building Regulations/Government Agency Authorities

   Determine the building regulations that apply to the Energy Conservation Opportunities selected, and the governmental agencies having authority. Identify the permits, inspections and approvals required and their costs. Establish for each required permit the information necessary to obtain those permits, and the procedures for issuing them, making inspections and obtaining approvals and certifications.

2) Plans/Prices/Permits

   Prepare any drawings, specifications, instructions or written descriptions necessary to secure prices, carry out the work and obtain permits when these are required. Price the work, obtain the required permits and with the permit issuing agencies determine the schedule for inspections and approvals corresponding to the schedule for carrying out the work.

3) Construction/Inspection/Approval

Carry out the work, securing the necessary inspections and approvals at the times required. Upon completion, obtain final approvals, certifications of compliance and/or operating, use and occupancy permits when these are required.

BUILDING REGULATIONS/GOVERNMENT AGENCY AUTHORITIES

Applicable building regulations can be identified by visiting the township, city, county or state government agencies having jurisdiction. Staff of these agencies are public servants, and part of their duties include consulting with the public on the interpretation and applicability of building regulations. This consultation will be most valuable when specific Energy Conservation Opportunities are used as examples. Individual agencies must be consulted, since in most cases, one agency cannot speak for another.

For buildings located in rural communities where access to building regulation agencies, architects, engineers, and contractors is difficult, the Cooperative Extension Service of the U.S. Department of Agriculture may be of service. The bulletin (PA819) "This is Cooperative Extension" states that there are "more than 3,000."

The addresses of the Directors of State Extension Services can be obtained by writing to Extension Service, U.S. Department of Agriculture, Washington, D.C. 20250 for publication PA819.

PLANS/PRICES/PERMITS

The drawings, specifications, instructions or written descriptions necessary to secure prices and botain permits for carrying out Energy Conservation Opportunities can be provided for a fee by engineers and architects. This is a principal part of their traditional, professional services. Fees are usually related to the cost of new construction, remodeling or alteration, while fees for repairs and maintenance may be just a single service charge, depending upon the complexity of the work to be performed.

Whenever possible, decisions of building regulation officials should be confirmed in writing, preferably by the official.

BUILDING REGULATION INFORMATION

A basic understanding of building regulations can be gained from two sources:

- The Occupational Safety and Health Administration (OSHA), Construction Standards and General Information applicable nationally.
- The model codes used in part to develop, and often in part incorporated by reference in state, county, city and township building regulations.

CONSTRUCTION STANDARDS

OSHA Construction Standards and General Information is available through the Occupational Safety and Health Administration, 1726 M Street N.W., Suite 1010, Washington, D.C. 20006.

For specific information about compliance with OSHA Standards, contact the local offices of OSHA. These are listed under the U.S. Department of Labor heading in the telephone directory of more than 100 major cities.

MODEL CODES

There are four model building codes available in the United States. Copies of these codes can be secured as references by contacting the following organizations:

- The National Building Code is available from the American Insurance Association, 85 John Streer, New York, New York, 10038. The telephone number is AC212-433-4400. Branch offices are located at 120 South LaSalle Street, Chicago, Illinois, 60603; 550 California Street, 14th Floor, San Francisco, California, 94104 and 610 South Broadway, Los Angeles, California, 90054.
- The Uniform Building Code is available from the International Conference of Building Officials, 5360 South Workman Mill Road, Whittier, California, 90601. The telephone number is AC213-699-0541.
- The Southern Standard Building Code is available from the Southern Building Code Congress, 3617 8th Avenue, South, Birmingham, Alabama, 35222. The telephone number is AC205-252-8930.
- The BOCA-Basic Building Code is available from the Building Officials and Code Administrators International, Inc., 1313 East 60th Street, Chicago, Illinois, 60637. The telephone number is AC312-947-2580.

APPENDIX B
COST DATA

COSTS FOR SUPPLYING PLASTIC FILM TINT TO WINDOWS
STANDARD COOL QUALITY

1. Costs:

   (a) Include supply and installation complete.

   (b) are as of January 1975

	Per 10 windows		Per 200 windows		Per 400 windows	
Area of Window	12 S.F.	30 S.F.	12 S.F.	30 S.F.	12 S.F.	30 S.F.
Cost of Sq. Ft.	2.00	2.00	1.70	1.60	1.40	1.35
Cost per Window	24.00	60.00	20.00	48.00	17.00	41.00

COST STUDY FOR INSTALLATION OF VENETIAN BLINDS AND/OR VERTICAL LOUVER BLINDS TO WINDOWS IN EXISTING MASONRY BUILDING

1. Costs exclude:

   (a) Removal of any existing blinds or drapes.

   (b) Adapting or fitting blinds around existing window mounted air conditioners, louvers or fans, etc.

   (c) Any interior redecoration.

2. Costs are:

   (a) as of January 1975

   (b) for supply and installation by a specialist contractor dealing in venetian and louver blind work.

VENETIAN BLINDS

For Quantities of:	10 Windows		50-200 Windows		Over 400 Windows	
Approx. Window Area	12 S.F.	30 S.F.	12 S.F.	30 S.F.	12 S.F.	30 S.F.
2" Venetian Blind including all hardware, supplied and installed in existing building.	$9.75	21.72	8.68	19.34	8.33	18.36

P.V.C. VERTICAL LOUVER DRAPES

For Quantities of:	10 Windows		200 Windows		400 Windows	
Approx. Window Area	12 S.F.	30 S.F.	12 S.F	30 S.F.	12 S.F	30 S.F.
Vertical P.V.C. Louver Drapes including all hardware, supplied and installed in existing building.	$48.00	75.00	30.00	45.00	30.00	45.00

COST STUDY FOR IMPROVING GLAZING TO EXISTING WINDOWS

1. Costs exclude:

   Removal of (a) interior furniture and fixtures, including blinds, drapes, etc., and (b) interior fixed draft excluding louvers for existing openable sash.

   (c) Fitting glazing or any new frames around any existing window-mounted air conditioners, louver or fan outlets, etc., within border of window frame.

   (d) Structural changes.

   (e) Any interior redecoration.

2. Costs presume that:

   (a) Additional window frames where stated and included below are to be of custom made aluminum.

   (b) Limited access available to windows in occupied buildings.

   (c) That curtain wall buildings are air conditioned and therefore that windows are generally fixed non-operable type.

3. Costs:

   (a) Are as of January 1975.

   (b) Include General Conditions, Overhead and Profit element of 20%.

COST STUDY FOR IMPROVING GLAZING IN EXISTING WINDOWS

		$ COST PER WINDOW $					
		10 WINDOWS		50-200 WINDOWS		OVER 400 WINDOWS	
		Window 12 SF	Window 30 SF	Window 12 SF	Window 30 SF	Window 12 SF	Window 30 SF
1)	Remove existing metal frame and fixed glazing and replace with extruded alum.frame and adjustable trim with fixed double glazing. Work from building interior.	274	524	202	376	188	339
2)	Same as (1) above but with operating sash.	313	620	228	441	213	392
3)	Leave existing glazing intact and add new extruded alum.frame and adjustable trim with single fixed glazing. All work from building interior.	130	283	94	198	85	176
4)	Remove existing single glazing from curtain wall frame and replace with sealed double glazing including adapting glazing rebates in curtain wall to receive thicker glass.	211	386	204*	380*	194*	369*
5)	Remove existing single glazing from curtain wall frame and replace with one layer of reflective glass and one layer of clear glass including adapting rebates in curtain wall to receive new glazing.	257	473	250*	425*	240*	411*

* Includes the use of stage hoists on exterior face of building.

BLOCK WINDOWS WITH INSULATED PANELS (SHUTTERS) DURING UNOCCUPIED PERIODS

1. Costs exclude:

   (a) Any special conditions such as projecting interior window air conditioners.

   (b) Removal of drapes, blinds, etc.

2. Costs:

   (a) Include General Conditions, Overhead and Profit element of 20%.

   (b) Are as of January 1975.

3. Based on a minimum of 10 windows each size 3' x 4' costs are:

   (a) For custom made single piece shutters clipped in place — $ 65.00 per opening

   (b) For custom made hinged folding shutters — $103.00 per opening

COST STUDY FOR REMOVAL OF GLAZING

For all examples: -

1. Costs exclude:

   (a) Scaffolding costs.
   (b) Moving of internal furniture, etc.
   (c) Repainting walls internally around affected window openings.

2. Costs:

   (a) Include General Conditions, Overhead and Profit element of 20%.
   (b) Are as of January 1975.
   (c) Based on a quantity of 10 windows each size 3' x 4'.
   (d) Presume windows accessible at ground level.

3. Escalation:

   The materials included in this cost study have not been subject to the excessive increases in costs prevalent in materials with petro-chemical content. It is suggested that a rate of 8% per annum be used when updating costs included herein.

Example 1 - Masonry Structures

Costs exclude specifically unusual or difficult to match brick.

(a)	Removing aluminum windows and filling opening with masonry.	$170.00 per opening
(b)	Removing aluminum windows and filling opening with insulated sandwich panel and backup.	$198.00 per opening

Example 2 - Light Frame Construction

Costs exclude specifically repainting or restaining to achieve closer match, or repainting whole wall.

(a)	Removing window and filling opening with light construction.	$165.00 per opening
(b)	Removing window and replacing with sash for 50% of area and filling balance of opening with light construction.	$156.00 per opening

COST STUDY FOR WINDOW CLEANING

1. Costs presume that:

   (a) Window to be cleaned is between 20 and 30 sq. ft. in area (giving 40 to 60 sq. ft. of area to be cleaned).

   (b) Masonry type anchors and window cleaning belts are used for exterior work in lieu of ladders or staging.

   (c) A full day's work is available for one window cleaner at the same building or location.

   (d) Work would be undertaken by building maintenance type contractor.

2. Costs:

   (a) are as of January 1975

   (b) include General Conditions, Overhead and Profit element of 20%.

Cost per Window on basis of 50± windows = $ 1.85 per window.

INCREASE NATURAL ILLUMINATION BY ADDITIONAL WINDOWS, AND/OR SKYLIGHTS

1. Costs:

   (a) Are as of January 1975.

   (b) Include General Conditions, Overhead and Profit element of 20%.

2. Costs exclude removal of interior furniture and fixtures including drapes and blinds.

SKYLIGHTS

Provision of 48" x 48" acrylic double glazed domes based on a quantity of 6 to be installed and excluding (a) any major structural alterations (b) Electrical and Mechanical work and (c) redecoration internally. Cost $500.00 per unit.

IMPROVE BUILDING JOINTS

1. Costs presume that:

   (a) For example "0-10 windows" ladders can be used to undertake work.

   (b) For all other examples, stage hoists used on exterior face of building.

2. Costs:

   (a) Are as of January 1975.

   (b) Include General Conditions, Overhead and Profit element of 20%.

Cost Study for Raking Existing Caulking and Re-caulking Around Windows

	COSTS PER WINDOW					
	10 Windows		50-200 Windows		Over 400 Windows	
	Window 12 SF	Window 30 SF	Window 12 SF	Window 30 SF	Window 12 SF	Window 30 SF
Rake out old caulking and re-caulk around edges of windows.	$18	25	17	27	18	27

(Preformed closure costs not included in this study.)

Caulking

Rake out old caulking and recaulk $1.22 L.F.

Premoulded Seal

Assumed all work done in conjunction with caulking windows, etc., therefore additional scaffolding, etc., not required (or work can be accomplished by ladder to 30' height).

Rake out old caulking and recaulk using 1/2" x 1/2" premoulded polyfoam material. $2.03 L.F.

Ditto but using 1" x 1" premoulded polyfoam material. $2.80 L.F.

WEATHERSTRIP DOORS, WINDOWS AND OTHER OPENINGS TO PREVENT INFILTRATION

1. Costs exclude:

   (a) Removal of any blinds, drapes, etc.

   (b) Fitting weatherstripping around any existing windows, mounted air conditioners, fans, etc.

   (c) Any adjustment to frame or new threshold/sills which might be necessary.

   (d) Redecoration to door or window frames.

2. Costs presume that:

   (a) No field foreman would be required for up to 50 windows, but that foreman (non-working) would be required for over 50 windows.

   (b) This work would be undertaken by an outside contractor.

3. Costs:

   (a) Are as of January 1975.

   (b) Include General Conditions, Overhead and Profit element of 20%.

	Costs per Door	
	3' x 7' Door	Pair 6' x 7' Doors
Weatherstrip Door with Wood Frame –	$51.00	$ 74.00
Weatherstrip Door with Metal Frame with aluminum and rubber –	$96.00	$138.00

Costs per Window

	10 Windows		50-200 Windows		Over 400 Windows	
	Window 12 S.F.	Window 30 S.F.	Window 12 S.F.	Window 30 S.F.	Window 12 S.F.	Window 30 S.F.
Weatherstrip double hung window	$33.00	$47.00	$26.00	$39.00	$23.00	$35.00

COST STUDY FOR SURFACE MOUNTED DOOR CLOSERS TO SINGLE DOORS

1. Costs exclude:

   a) Any necessary redecoration to door and frame.

2. Costs presume that:

   a) Door is capable of receiving closer without any additional blocking, cutting, etc., being required.

   b) Work is undertaken by an outside General Contractor and in conjunction with other work on the building.

3. Costs:

   a) are as of January 1975.

   b) Include General Conditions, Overhead and Profit element of 20%.

   c) Would be cheaper by a third to one quarter if fixing was undertaken by janitor or building maintenance staff.

Cost of supplying and fixing door closers to single doors based on a quantity of about 10 doors = $ 91.00 per door

COSTS FOR INSTALLING REVOLVING DOORS

1. Cost exclude:

    a) any readjustment necessary to existing Electrical and Mechanical installation.
    b) additional swinging or hinged doors for use by handicapped, etc.

2. For installing the door in a new opening it is presumed that:

    a) the external wall is masonry.
    b) the street and ground floor levels are compatible.
    c) the new opening leads directly into an entrance hall area.

3. Costs:

    a) include General Conditions, Overhead and Profit element of 20%.
    b) are as of January 1975

4. Providing Revolving Door and installing in existing opening:

    Including removing existing aluminum tube section doorway and sidelights - finished opening 8' x 8' high. Installing new 7' diameter aluminum revolving door and trim opening.

    Cost per unit installed = $10,300.00

5. Providing Revolving Door and installing in new opening:

    Form new opening in external wall and install new 7' diameter aluminum revolving door.

    Cost per unit installed = $13,000.00

ADD VESTIBULE TO BUILDING ENTRY.

1. Costs exclude:

   a) Modification to existing doors.
   b) Heating and cooling of vestibule.

2. Costs:

   a) Include General Conditions, Overhead and Profit element of 20%.
   b) Are as of January 1975.
   c) Include for black anodized aluminum tube section gray tinted glass, concealed head door closers and light roof.

Cost per unit installed complete $3,105.00

EPOXY RESIN INSULATION
COATING ON EXTERIOR WALLS

1. Costs presume that:

   (a) Building to be treated is a 100' x 50' three-story office building 10'0" floor to floor.

   (b) An area of 10,000 S.F. will be treated.

   (c) Exterior wall construction is brickwork or blockwork.

   (d) Resin will be applied in two coats, total 4 mm thick.

   (e) Swing staging will be required for application of resin.

2. Costs:

   (a) are as of January 1975

   (b) include General Conditions, Overhead and Profit element of 20%.

---

4 mm Epoxy Resin on External Walls	-	42 ¢ per S.F.

ADD RIGID INSULATION TO OUTSIDE OF EXTERIOR WALLS

1. Costs presume that:-

   (a) building to be insulated is a 100' x 50' 3 story office building with 10' floor to floor height.

   (b) an area of 10,000 S.F. will be insulated.

   (c) no external trim, fixtures or paint required to be removed nor complications with windows, other openings and fascias, etc.

2. Costs:

   (a) are as of January 1975.

   (b) include general conditions, overhead and profit element of 20%.

3. Escalation: Cement costs are expected to increase again in 1975 and may be in the order of 5%.

   Metal lath and nail costs doubled in 1974 but as of the present increases of this magnitude are not expected for the coming year.

   Rigid insulation and stucco to outside of exterior walls.

   (a) using $1\frac{1}{2}$" rigid insulation $3.12 S.F.

   (b) using 2" rigid insulation $3.24 S.F.

ADD INSULATION TO CORE OF EXTERIOR WALLS

All as described and specified for solution using blown insulation (not batt type) of fiberglass, perlite and urea formaldehyde (urethane not costed due to its difficulties with meeting fire codes).

1. Costs Exclude:

   a) any special guarantees which may be called for.
   b) movement of furniture and occupant relocation.
   c) major patchwork.

2. Costs include:

   a) start up through clean up.
   b) reasonably good accessibility to wall space.
   c) for well planned operation, i.e. one-trip job with no delays and no premium time.
   d) general conditions overhead and profit element of 20%.
   e) Costs are as of January 1975.

3. Costs are based on a 100' x 50' plan three story building with 10' floor to floor heights.

4. Escalation: Urethane material costs in the past year are reported to have increased by as much as 15% and will be increasing still more in 1975.

   Urea-formaldehyde material costs in the past year are reported to have increased by 10% and will be increasing, by about 5% in 1975.

   Fiberglass and perlite report similar cost increases.

COSTS

a) Blown insulation (from interior of building)

   Fiberglass $1.45 S.F.
   Perlite $1.83 S.F.
   Urea formaldehyde from 75-80 ¢ S.F.

b) Blown insulation (from exterior of building)

   Fiberglass $1.24 S.F.
   Perlite $1.61 S.F.
   Urea formaldehyde from 65-75¢ S.F.

ADD INSULATION TO INSIDE OF EXTERIOR WALLS

1. Costs presume that:

    a) Building to be insulated is a 100' x 50' three story office building with 10' floor to floor height.

    b) An area of 10,000 S.F. to be insulated.

2. Costs Exclude:

    a) Resetting electrical receptacles and other E. & M. services.

    b) Framing out to door and window openings and setting door and window trim and cover base.

3. Costs:

    a) are as of January 1975.

    b) include General Conditions, Overhead and Profit element of 20%.

Costs per Square Foot

Insulate interior faces of exterior walls with 1-1/2" rigid insulation	$2.03 S.F.

APPLY SPRAY-ON INSULATION TO EXTERIOR OF ROOFS

1. Costs Exclude:

   a) any special guarantees or roof bonds which may be called for.
   b) extra masking,any unusual edge or flashing condition.
   c) any new flashings or expansion joints required.

2. Costs Include:

   a) start up through clean up.
   b) work on flat relatively unobstructed roof.
   c) General Conditions, Overhead and Profit element of 20%.

3. Costs are based on a roof of 10,000 S.F. and are as of January 1975.

a)	Cost using 2" polystyrene insulation and butyl roofing	$2.08 S.F.
b)	Cost using 2" urethane and elastomeric roofing	$2.15 S.F.
c)	Additional cost to be added to (a) or (b) above for spraying elastomeric roofing with aluminum spray (3 mils thick)	30¢ S.F.

CHANGE COLOR OF ROOF

(A) GENERALLY

1. Costs exclude:
   (a) any special guarantees, etc. which might be called for.

2. Costs include:
   (a) General Conditions, Overhead and Profit element of 20%.

   (b) Start up through clean up.

3. Costs based on roof area of 10,000 S.F. and are as of January 1975.

4. Escalation:

   Increases in the past year have been very heavy ranging from 50% to 150% for asphaltic and coal tar pitch products.

   For the coming year increases are difficult to forecast owing to pending legislation. However, it is believed that further increases may be expected whilst other construction trades show signs of tapering off.

(B) COSTS

1. Painting of sloping metal roofs

   In particular costs:

   (i) exclude scaffolding around outside of building.
   (ii) presume roof can support weight of workers and existing painting is non asphaltic.
   (iii) presume building is metal prefabricated type with low eave heights.
   (iv) presume roof ladders can be used for working purposes with rope and safety belt security,
   (v) presume light hand cleaning and 2 coat built up spray-on aluminum coating used for finishing.

   (a) Costs using 2 coat built up spray-on aluminum coating as finishing — 43¢ S.F.

   (b) Cost using single coat of non-fibrated aluminum paint — 44¢ S.F.

   (c) Cost using single coating of black asphaltic cut-back or emulsion — 45¢ S.F.

APPLY RIGID INSULATION TO EXTERIOR OF FLAT ROOF.

1. Costs exclude:

   (a) unusual roof conditions, catwalks and walkways.

2. Costs include;

   (a) start up through clean up.

   (b) work on flat relatively unobstructed roof.

   (c) General Conditions, Overhead and Profit element of 20%.

3. Costs are based on a roof of 10,000 S.F. at a height above ground level of 4 to 8 stories.

   For tall buildings costs would not increase a great deal providing easy access was available to a suitable elevator.

4. Escalation.

   Costs for asphaltic and tar products increased significantly in 1974. Signs of stability however are beginning to appear in the market and increases for 1975 may be marginal for these materials.

   Costs using 4 ply tar and gravel 20 year bonded roof and the following insulation materials.

   (a) 2" rigid urethane = $2.61 S.F.

   (b) 2" fiberglass = $2.16.S.F.

   (c) 2" fiberboard = $2.35 S.F.

   (d) 2" cellular glass = $2.38 S.F.

   (e) 2" perlite = $2.35 S.F.

APPLY RIGID INSULATION TO EXTERIOR OF SLOPING ROOF

1. Costs exclude:-

   a) any special guarantees, etc. which might be called for.

2. Costs

   a) are as of January 1975
   b) include General Conditions, Overhead and Profit element of 20%.
   c) include for roofing a 10,000 S.F. pitched surface of a 4-story building.
   d) include for scaffolding up to eave. level on all faces and for a materials hoist.

3. Escalation

Asphalt shingle material costs increased by approximately 20% in 1974. Forecasts for 1975 indicate that increases will not be as severe.

Fiberglass shingle materials increased by 6% in 1974. Increases for 1975 are expected not to exceed the 1974 figure.

Labor costs will undoubtedly continue their annual increase of from 5 to 10%.

All materials in this particular cost study are currently available.

Costs

a) Using 265# asphalt shingles and 2" insulation $1.79 S.F.

b) Alternative using 260# fiberglass shingles and 2" insulation $1.70 S.F.

APPLY SPRAY ON INSULATION TO UNDERSIDE OF ROOF

1. Costs exclude:

   (a) any special guarantees, etc. which might be called for.
   (b) moving furniture if required.
   (c) protection of sensitive equipment such as computers.

2. Costs include:

   (a) for all work to be done at night on double-time wages
   (b) removal of existing 2'x4' ceiling and suspension system and replacement with new.
   (c) cleaning surface and masking ducts, etc. prior to application of spray.
   (d) final clean up.
   (e) General Conditions, Overhead and Profit element of 20%.

3. Costs are based on a ceiling of 10,000 S.F. as might be encountered in a large supermarket.

4. Escalation:

   It is reported that costs for insulation material increased by as much as 20% in 1974. Increases will undoubtedly occur in 1975 but generally show signs of tapering off. Labor costs will undoubtedly continue to increase at rates varying between approximately 5 and 10% per annum.

(a)	Cost using 1" mineral fibre spray	$2.65 S.F.
(b)	Cost using 1" urethane plastic spray	$3.14 S.F.

APPLY RIGID OR BATT INSULATION TO UNDERSIDE OF ROOF

Costs exclude:

a) any special guarantees, etc. which might be called for.
b) moving furniture if required.
c) protection of sensitive equipment such as computers.

Costs include:

a) for all work to be done at night on double-time wages.
b) removal of existing 2'x4' ceiling and suspension system and replacement with new.
c) cleaning surfaces and masking ducts, etc. prior to installation of insulation materials.
d) final clean up.
e) General Conditions, Overhead and Profit element Of 20%.

Costs are based on an area of 10,000 S.F.

Escalation:

It is reported that costs for insulation materials increased by as much as 20% in 1974. Increases will undoubtedly occur in 1975 but generally show signs of tapering off. Labor costs will undoubtedly continue their annual upward trend at rates varying between approximately 5 and 10% per annum.

Alternative (A) using rigid insulation board.

a)	2" styrofoam	=	\$2.78 S.F.
b)	2" fiberboard	=	\$3.12 S.F.
c)	2" fiberglass	=	\$2.95 S.F.
d)	2" expanded glass	=	\$2.98 S.F.
e)	2" perlite	=	\$3.12 S.F.

Alternative (B) using batt insulation.

a)	3" fiberglass	=	\$2.24 S.F.
b)	3" mineral fiber	=	\$2.22 S.F.

PROTECT INSULATION WITH VAPOR BARRIER

1. Costs exclude:

   a) removal of existing ceiling and replacement with new.
   b) replacement of any deteriorated insulation.

2. Costs include:

   a) work to underside of flat relatively unobstructed roof.
   b) start up through clean up.
   c) General Conditions, Overhead and Profit element of 20%.

3. Costs are based on an area of 10,000 S.F. and are as of January 1975.

Application of sheet vapor barrier by stapling including lapping and gluing	14¢ S.F.

COST STUDY FOR REPAINTING EXISTING WALLS AND CEILINGS

1. Costs exclude:

   a) Removal of interior furniture and fixtures, including drapes and blinds.

2. Costs presume that:

   a) Work would be completed in sections.

   b) Paint would be roller applied.

   c) Walls are in good condition and do not need excessive scraping or filling of cracks, etc.

   d) No field foreman would be required for areas of approximately 10,000 S.F., but that foreman (non-painting) would be required for greater areas.

3. Costs:

   a) are as of January 1975.

   b) include General Conditions, Overhead and Profit element of 20%

Wash, touch up and apply two coats gloss paint on smooth finish plaster:

Areas of approximately:

10,000 S.F.	20,000 S.F.	50,000 S.F.	100,000 S.F.
27¢ S.F.	24¢ S.F.	23¢ S.F.	23¢ S.F.

Clean down and apply two coats flat finish on concrete block walls:

Areas of approximately:

10,000 S.F.	20,000 S.F.	50,000 S.F.	100,000 S.F.
34¢ S.F.	32¢ S.F.	31¢ S.F.	29¢ S.F.

COST STUDY FOR REPAINTING EXISTING WALLS AND CEILINGS (ONE COAT)

1. Costs exclude:

   a) Removal of interior furniture and fixtures, including drapes and blinds.

2. Costs presume that:

   a) Work would be completed in sections.

   b) Paint would be roller applied.

   c) Walls are in good condition and do not need excessive scraping or filling of cracks, etc.

   d) No field foreman would be required for areas of approximately 10,000 S.F., but that foreman (non-painting) would be required for greater areas.

3. Costs:

   a) are as of January 1975.

   b) include General Conditions, Overhead and Profit element of 20%.

Wash, touch up and apply one coat gloss paint on smooth finish plaster:

Areas of approximately:

10,000 S.F.	20,000 S.F.	50,000 S.F.	100,000 S.F.
20¢ S.F.	17¢ S.F.	16¢ S.F.	16¢ S.F.

Clean down and apply one coat flat finish on concrete block walls:

Areas of approximately:

10,000 S.F.	20,000 S.F.	50,000 S.F.	100,000 S.F.
24¢ S.F.	23¢ S.F.	20¢ S.F.	20¢ S.F.

CHANGE COLOR AND/OR TEXTURE OF INTERIOR WALLS

1. Costs exclude:

   a) Removal of interior furniture and fixtures.

2. Costs presume that:

   a) Work would be completed in sections.
   b) Paint would be roller applied.
   c) No field foreman would be required for areas of approximately 10,000 S.F. but that foreman would be required for greater areas.

3. Costs
   a) Are as of January 1975.
   b) Include General Conditions, Overhead and Profit element of 20%.

1. Apply 3/8" gypsum board on furring strips taped and spackled and painted with two coats of Latex paint.

   Areas of approximately:-

10,000 S.F.	20,000 S.F.	50,000 S.F.	100,000 S.F.
88¢ S.F.	86¢ S.F.	86¢ S.F.	85¢ S.F.

2. Apply 3/8" gypsum board on furring strips taped and spackled and painted with one coat of Latex paint.

   Areas of approximately:-

10,000 S.F.	20,000 S.F.	50,000 S.F.	100,000 S.F.
81¢ S.F.	80¢ S.F.	79¢ S.F.	79¢ S.F.

COST STUDIES IN CONNECTION WITH PLUMBING WORK

A. Costs presume that:

(a) Work would be undertaken by outside plumbing sub-contractor.

B. Costs:

(a) are as of January 1975.

(b) include General Conditions, Overhead and Profit element of 20%

(c) The examples given cannot be taken in isolation and may be much higher if only a small volume or amount of work is undertaken.

---

1. Replacing screw-down type faucet with spring operated, self-closing type faucet, ½" iron pipe size including all accessories.

   Cost per Faucet - $51.00

2. Insulating hot or cold water tanks with 3# density glass fiber with foil scrim kraft facing, finished with pre-sized glass cloth jacket.

Material Thickness	Cost per S.F. of surface area
1"	$ 2.60
1½"	2.70
2"	2.95
3"	3.60

3. Insulating hot and cold water pipes with pre-sized glass fiber material with a standard jacket.

Pipe Sizes	Price per L.F. Installed Insulation Thickness: 1"	
	1"	1½"
	$	$
1/2"	1.35	2.05
3/4"	1.40	2.10
1"	1.45	2.15
1¼"	1.50	2.20
1½"	1.55	2.25
2"	1.60	2.35
2½"	1.65	2.45
3"	1.70	2.50
4"	2.00	3.05
5"	2.25	3.15
6"	2.55	3.30
8"	3.15	4.20
10"	3.85	5.00
12"	4.50	5.60
14"	5.20	6.45
16"	6.00	7.20
18"	6.70	7.60
20"	8.25	8.50
24"	9.00	9.70

For costing purposes, add to total lineal feet of piping three (3) lineal feet for each fitting or pair of flanges to be insulated.

COST STUDIES IN CONNECTION WITH HEATING, VENTILATING AND AIR CONDITIONING

1. Saving energy by continued efficient operation of equipment through proper and effective service contracts.

Presuming an Inspection and Labor type service contract is let for servicing simple air conditioning and fan coil units on an annual fee basis to cover cost of labor (for inspection, maintenance, breakdown repairs) and material (such as filters, oil, grease, etc.), then suggested costs might be:

A/C Units

System Tonnage	2.5	5.	7.5	10	15	20	25	30	40
Price	$168	202	253	308	410	495	595	685	865

Fan Coil Units

System Tonnage	1/2	3/4	1	1½	2
Price	$25	35	45	70	95

2. Addition of a 7-day cycle time switch clock to existing control system and fixing within 10' of equipment.

Cost per installation, including labor, material, overhead and profit = $80.00*

3. Replacing thermostat with electric clock thermostat in same position, standard voltage 115 volts 60 cycle with day and night control.

Cost per installation, including labor, material, overhead and profit = $132.00*

4. Provision of a control system for a Heating, Ventilating and Air Conditioning installation (i.e. to control heating, cooling, operation of equipment, humidity, sequence of operations, etc.)

Costs per installation, including labor, material, overhead and profit:

1-12 tons	$ 240.00 per ton
over 12 tons	88.00 per ton

These costs are approximate only and do not include for removal of existing inefficient control system.

*The cost depends very much on local conditions. If this is the only electrical work being undertaken, the cost could be greater by two or three times.

COST STUDIES IN CONNECTION WITH ELECTRICAL WORK

1. Costs exclude:

   a) Removal of furniture and other floor mounted fixtures to gain access to fixtures overhead.

   b) Staging for lighting above 10' level.

2. Costs presume that:

   a) Lighting fixtures are easily accessible and that the changes or cleaning costs below are in contiguous areas.

   b) Labor for changing bulbs or cleaning fixtures will be building maintenance staff and the items are estimated accordingly and exclude any element of overhead and profit.

A. Replacing existing incandescent light bulbs with similar bulbs but of lower wattage.

Approx. number of bulbs	10	50	100	200
Price	$ 11.00	41.00	80.00	158.00

B. Removing one 40 watt 48" long fluorescent tube from existing 4 or 2 tube fixture and placing in storage.

Approx. number of fixtures	10	50	100	200
Price	$ 9.00	32.00	63.00	123.00

C. Cleaning ceiling-mounted 48" 4-tube fluorescent fixture or incandescent fixture of similar size.

Approx. number of Fixtures	10	50	100	200
Price	$ 14.00	58.00	115.00	224.00

COST STUDY FOR LOWERING CEILING HEIGHT AND REPLACING EXISTING LIGHT FIXTURES WITH TASK ORIENTED FIXTURES

1. Costs presume that:

   (a) Location of offices is in upper stories of multi-story building.

   (b) An area of 30' x 50' is to be renovated, comprised of a series of small rooms. (If one room only is to be renovated, costs would be appreciably higher.)

   (c) 40 light fixtures would be replaced with 25 new four-30 watt rapid start cool white 3'0" x 3'0" recessed fixtures.

   (d) New suspended ceiling of 2' x 4' mineral fiber board would be installed.

2. Costs:

   (a) Are as of January 1975.

   (b) Include General Conditions, Overhead and Profit element of 20%.

TOTAL COST (1,500 S.F.)	COST PER S.F.	COST PER FIXTURE
$ 6,400.00	$ 4.27	$ 256.00

APPENDIX C
ECONOMIC MODEL & EXAMPLE

A. COMMERCIAL BUILDING OWNERSHIP MODEL

General Sequence

The Commercial Building Ownership Energy Conservation Investment Analysis Model consists of five basic tables. Table A is a summary table which presents in tabular form the discounted rate of return and other data on a proposed energy conservation measure. It contains multiple columns to facilitate comparison of alternative conservation measures. Table B is used to compute after-tax cash flow from an energy conservation improvement. Table C is used to convert this stream of yearly after-tax cash flow into a discounted rate of return. Tables D and E are supporting tables in which much of the input data to Table B is computed. In completing these tables, the following order of preparation should be followed:

1. Complete all applicable items in Table E and post the results computed to Tables B and D as indicated. Non-applicable items should be left blank.

2. As applicable, compute line items 2, 3, 4, and 5 on Table B.

3. Compute net operating profit (Table B-line 7). This involves adding up the items previously computed and posted to Table B. Enter the net operating profit by year computed in Table B in Column (a) of Table D.

4. Complete Table D and enter the projected income tax computed in Table D on line 11 of Table B.

5. Complete the computations on Table B. Post after-tax cash flow (Table B-line 13) to Table C. Complete the computations in Table C. This will yield the discounted rate of return on the investment, which is the end result of the cost model.

6. Post the discounted rate of return (Table C) and Payback Period and Maximum Cash Required (Table B-line 14) to Table A.

Specific Instructions

Energy Savings (Table B-line 1 and Table E-1). In projecting energy savings, it is necessary to project the estimated future cost of energy and the amount of energy expected to be saved. The engineering data presented elsewhere in this manual will allow a building owner to determine the projected BTU savings per square foot from alternative energy conservation measures. These BTU savings per square foot can be translated into gross BTUs saved per year by multiplying BTUs saved by the gross square footage of the building. Gross BTUs saved can be converted into a dollar cost by multiplying by the projected cost per BTU for the particular source of energy being utilized. If the projected cost per BTU for a particular source of energy is expected to increase at a greater rate than general inflation, the amount of this increase in excess of general price level inflation should be factored into the projected energy cost savings.

Increase in Rental Income (Table B-Line 6 and Table E-2). Enter on this line the projected net change, if any, in rental income resulting from an energy conservation improvement. It should include any rent loss during the construction or installation period of an energy conservation improvement, and any projected decrease in the amount of rentable space as a result of the installation of such an improvement. Line 6 should also include any projected change in rental income due to the nature of an energy conservation measure, e.g., decreases resulting from elimination of windows, increases resulting from improved interior aesthetics, modernized building skin, etc. If appropriate,a factor should be added for the increased marketability (rentability) resulting from a building's ability to operate more effectively than competitors under conditions of mandated energy restrictions or curtailments and/or during periods of unavailability of or shortages of fuel. Many factors listed above will necessitate value judgements by persons familiar with the rental market in the area where a building is located. If the net change in rental income is negative, it should be so indicated on Table B by the use of parentheses.

Net Investment (Table B-Line 8 and Table E-3). The net investment for purposes of the private ownership cost model is the amount of cash which is expended at the time the improvement is made. It is the purchase price of the improvement less the principal amount of any note, loan, or mortgage used to finance the improvement. Salvage proceeds, if any, from the sale of replaced assets should also be deducted from the purchase price in determining the net investment.

Debt Service (Table B-Line 9 and Table E-4). Completion of this line item is only applicable if a portion of the cost of an energy conservation improvement is financed by a loan, note, mortgage, or some other form of indebtedness. When no loan is involved the loss of income from other potential investments should be weighed against the return from investment in energy conservation. The annual debt service on the amount borrowed is the sum of principal and interest payments paid during a given calendar year. In order to compute the annual debt service for a loan, it is necessary to determine the amount of the loan, the payback period in years, the payback provisions, and the annual interest rate. With these, the annual debt service can be computed using standard mortgage amortization tables. Alternatively, the amount of annual payments broken down by interest and principal can be obtained from the bank,loan company, or other institution where financing is expected to be obtained.

Basis of New Asset (Table E-5). The basis of a new asset is the sum of the purchase price of the asset plus the book value (original cost less accumulated depreciation) of the old asset replaced by the new asset less any trade-in allowance received on the old asset. Income tax laws do not allow the recognition of a gain or loss resulting from disposal of an old asset when the old asset is traded in on a new asset. Any gain or loss resulting from such a trade-in is simply added to the basis of the new asset being purchased. If an old asset being replaced by a new asset is sold rather than traded in, any gain or loss resulting from the sale of the asset is recognized at the time of the sale.

Depreciation (Table E-6). The accounting definition of depreciation is the loss of usefulness, expired utility, or dimunition of service yield from a fixed asset caused by wear and tear from use, misuse or obsolescence. Both accounting and tax theory assume that depreciable assets have a limited useful life and therefore the cost of such assets should be written off (depreciated) over that useful life. Depreciation represents the theoretical reduction in value of property improvements by reason of physical deterioration or economic obsolescence. In practice, properly maintained improved real estate reduces in value at a far slower rate than allowed by income tax laws and accounting theory. This does not affect an investor's right to write off (depreciate) an investment over the shortest period allowed by tax statutes.

The useful life of an improvement to a building is defined for tax purposes by specific provisions in Federal tax laws. If a taxpayer uses the Asset Depreciation Range System (ADR), he must use a useful life for the energy conservation improvements which falls within the specific ranges specified by the ADR provisions of Federal tax statutes. If a taxpayer does not use the ADR system, the "facts and circumstances" surrounding each particular investment should be used to determine its depreciable life. That is, the asset should be depreciated over its estimated economic/physical useful life. In determining this useful life, the taxpayer's past experience is very important. The useful life of building improvements should not be any longer than the remaining useful life of the building itself. If improvements are made by a lessee, such improvements should be amortized over a useful life no longer than the remaining life of the lease.

Generally speaking, in order to maximize the after-tax benefits of an energy conservation improvement, a building owner should depreciate such an improvement as fast as possible. Therefore, he should use the shortest useful life allowed by the ADR system, if he uses the ADR system to depreciate his assets, or the shortest useful life under the "facts and circumstances" method if he does not use the ADR system. If a taxpayer uses the ADR system for part of his capital acquisitions during a given year, he must use the ADR system for all of his capital acquisitions made during that year. A taxpayer cannot use the ADR system for determining the useful life of some of his assets unless he uses it for all of them.

Several different rates of depreciation are available for investments in energy conservation improvements. Straight line depreciation assumes that the diminution in value of a fixed asset occurs evenly over its useful life. Using straight line depreciation, a building improvement with a useful life of forty years would be depreciated by 2.5 percent (100 ÷ 40) of its original cost for forty years. There are several methods of accelerated depreciation such as declining balance and sum of the years digit method which provide for a greater amount of depreciation at the beginning of an asset's useful life and a lesser amount during the last years of its useful life. The theory behind accelerated depreciation is that an asset loses value more quickly during its first years than during its last years. Accelerated depreciation changes the timing of depreciation taken on a piece of property. It does not change the total amount of depreciation which can be taken on a piece of property. The total amount of depreciation which can be taken on any property can be no more than its basis (see previous section). Declining balance depreciation is a very common form used for real estate investments. Under declining balance depreciation, the straight line depreciation rate (100 divided by the useful life of an asset) is multiplied by a specific percentage -- 125 percent, 150 percent, or 200 percent -- to obtain an accelerated rate. In the example cited previously, a building improvement with a forty year useful life would have a straight line depreciation rate of 2.5 percent, a 125 percent declining balance rate of 3.125 percent, a 150 percent declining balance rate of 3.75 percent, and a 200 percent declining balance rate of 5.00 percent. These accelerated rates are applied to the undepreciated balance in each year to obtain the depreciation expense for that year. This results in a steadily declining depreciation expense over the entire life of the asset. Generally speaking, the maximum depreciation rates available for energy conservation improvements would be as follows:

1. Improvements to real property such as office buildings and retail facilities -- 150 percent declining balance.

It should be noted that under certain circumstances a portion of such accelerated depreciation may be subject to certain special taxation when the property is sold. Furthermore, there are special taxes on accelerated depreciation deductions for certain high-income individuals.

Gain on Disposal of Retired Asset (Table E-7). The gain or loss on the sale of an asset which is replaced by an energy conservation improvement is computed by subtracting the original cost of the retired asset from the sum of the accumulated depreciation taken on such a retired asset plus any salvage proceeds received from the disposal of the retired asset.

Investment Tax Credit (Table B-Line 12 and Table E-8). The investment tax credit is only applicable to personal property used in a trade or business. A building's structure and improvements are considered real property and as such are not eligible for an investment tax credit. As a general rule, building components which can be picked up and moved, e.g., window air conditioning units, bookcases, etc., are considered personal property and are eligible for the investment tax credit. Structural portions of a building such as central heating and air conditioning units, roofs, windows, etc., are real property and are not eligible for the investment tax credit. Energy conservation improvements which are structural portions of the building are not eligible for the investment tax credit. For such structural components, the investment tax credit line should be left blank on Table B.

An investment tax credit is received in the year in which qualifying property is placed in service. The amount of credit depends upon the useful life of the qualifying property. If the useful life of the property is less than three years, no credit is available. If the useful life of the property is at least three years but less than five years, the amount of credit is seven percent times one-third of the purchase price of the property. If the useful life of the property is at least five years but less than seven years, the amount of credit is seven percent times two-thirds of the cost of the property. If the useful life of the property is greater than seven years, the amount of the credit is seven percent of the purchase price. The amount of investment credit may not exceed a taxpayer's tax liability. If tax liability exceeds $25,000, the amount of credit is limited to $25,000 plus 50 percent of the tax liability in excess of that amount. It should be noted that the investment credit has been changed several times since it was originally introduced in 1962. Because it is regarded by the Federal government as a means of manipulating the economy, it is reasonable to assume that the provisions of the present investment tax credit will be changed as economic conditions change.

Labor and Material Savings (Table B-Line 2 and 3). The amount to be entered in these line items should be the projected net change in labor and material costs expected to result from the installation of a particular energy conservation improvement. In some instances, no savings will be made in these categories. If there are no projected labor or material savings, Table B-Lines 2 and 3 should be left blank. In other circumstances, labor and material costs may actually increase as a result of energy conservation improvements. If this is the case, negative amounts, designated by parentheses, should be entered as appropriate.

Property Taxes (Table B-Line 4). This line should be used to enter projected changes in local property taxes expected to result from installation of major energy conservation improvements in a building. As a practical matter, the majority of energy conservation improvements will not result in property being reassessed and therefore there will be no increase in property taxes as a result of such energy conservation improvements. In those circumstances where an energy conservation improvement will increase the assessed value of a building, property tax applicable to the improvement can be estimated by multiplying the estimated increase in the value of the building by the local property tax assessment ratio times the local property tax rate. For example, a $100,000 energy conservation addition to a building in a jurisdiction which assesses property at 50 percent of fair market value and has a tax rate of 4 percent of assessed valuation would result in an annual increase in property taxes of $2,000 ($100 000 X .50 X .04).

Reduction in Other Operating Expenses (Table B-Line 5). For certain energy conservation improvements, this may be an important line item. This is where such miscellaneous and hard to quantify cost savings such as productivity changes resulting from installation of energy conservation measures and projected increased marginal income resulting from the extension of the useful life of a building as a result of an energy conservation improvement should be entered. This line should also be used to enter other miscellaneous cost savings or increases applicable to a particular energy conservation improvement which are not covered elsewhere in Table B. In regards to productivity increases, general instructions for their computation cannot be provided since they are contingent upon facts and circumstances of individual buildings and specific energy conservation improvements.

Net Operating Profit (Table B-Line 7). Net operating profit resulting from an energy conservation improvement is a summation of energy savings and the reduction in other operating expenses plus any change in rental income. It is the sum of lines 1 through 6 on Table B.

Income Tax (Table B-Line 11 and Table D-Column 6). Taxable income is computed by subtracting depreciation and interest expense from the sum of net operating profit and any gain on the diposal of retired assets. Income tax is computed by multiplying taxable income times an individual's or corporation's tax rate. This income tax rate varies among individuals and organizations. The income tax rate used should be a composite rate of Federal, state, and local income taxes which would apply to a marginal change in the taxpayer's taxable income. Negative income tax figures can occur and are based on the assumption that the property owner has other taxable income to which the initial tax losses resulting from the energy conservation improvement can be applied.

After-Tax Cash Flow (Table B-Line 13). The amount of after-tax cash flow is the key factor in determining the economic viability of an energy conservation improvement for a taxable entity. It represents the amount of cash which will be generated by an energy conservation improvement after payment of applicable income taxes. It is computed by subtracting income tax (Table B-Line 11) from pre-tax cash flow (Table B-Line 10). Investment tax credit, if any, is added to pre-tax cash flow in determining after-tax cash flow.

Computation of Rate of Return (Table C). The present value of a dollar received currently is greater than the present value of a dollar to be received at some time in the future. A dollar to be received at some date far into the future has virtually no present value. The ultimate economic evaluation of an energy conservation investment is the after-tax cash flow expected to be generated by the investment discounted to reflect the time value of money. A common method of evaluating the time value of returns received from real estate investments and capital improvement investments is to determine the discount rate which when applied to the projected stream of after-tax cash flow from the investment yields a present value of zero. This method is used in the Commerical Building Ownership Energy Conservation Cost Model. If an energy conservation investment yields a cash flow which when discounted at the rate of ten percent has zero present value, the discounted rate of return is ten percent. Such a ten percent rate of return is equivalent to a fixed income investment which yields ten percent per year for the same time period for which the energy conservation improvement was evaluated and at the end of that period returns the amount of principal invested. In evaluating alternative energy conservation improvements, the improvement with the highest after-tax discounted rate of return should be given first priority. If the discounted rate of return available from an energy conservation improvement is less than the projected rate of return from alternative investments which an individual or a corporation can make, the energy conservation improvement cannot be justified solely on economic grounds.

In order to compute the discounted rate of return for a particular cash flow, the discount rate for a particular interest rate should be applied to that cash flow. These discount rates are available in standard real estate investment analysis tables. The rate which produces a zero present value is arrived at by a trial and error procedure, which with practice can be done quickly. In the example presented in this section (see Table C), a 12 percent discount factor was applied to the projected after-tax cash flow from a hypothetical energy conservation improvement and yielded a present value of $253. When the discount factor was increased to 13 percent, the present value computed to negative $912. Through interpolation, it can be said that the discounted rate of return for the investment presented in the example is approximately 12.2 percent.

Computation of Payback Period (Table A). The payback period is the number of years required to recoup the out-of-pocket cash investment in an energy conservation investment. It is the number of years required before the cumulative after-tax cash flow (Table B-Line 14) from an energy conservation investment becomes a positive amount.

Computation of Maximum Cash Required (Table A). The maximum cash required is the amount of cash, after financing and income taxes, needed to purchase and install an energy conservation improvement. It is the largest negative amount appearing in line 14 of Table B. A building owner cannot install an energy conservation investment, regardless of its life cycle rate of return, unless he has cash available for investment equal to or greater than the maximum cash required. The cost model presented in this chapter shows the net amount of cash required each year. This net amount is the sum of positive and negative cash flows. Within a given year, timing differences in the receipt of these positive and negative streams of cash may result in interim cash requirements which are greater than net cash required for the year as a whole.

ECONOMIC ANALYSIS EXAMPLE - COMMERCIAL BUILDING OWNERSHIP

The intent of the following example is to examine a specific energy conservation improvement, installation of double pane windows and its effects on building systems.

Building Description:

Location: Cleveland, Ohio

Type: Office

Latitude: 41°N

Occupancy: 40 hours/week, 450 occupants

Size/Area: 90,000 square feet

150 x 100 = 15,000 sq. feet per floor

6 stories

Window/Area: 3500 sq. ft. each exposure

Lighting/Power: 3.5 watts/sq. ft.

West exposure shaded by adjacent building

Heating season degree days: 6000

Annual mean daily Langleys: 325

Annual dry bulb degree hours above 78°F DB: 2800

Annual wet bulb degree hours above 66°F WB: 2000

Winter design conditions:

Outdoor: 7°F

Indoor: 68°F - occupied hours

58°F - unoccupied hours

Summer design conditions:

Outdoor: 86°F DB, 74°F WB

Indoor: 75°F DB

Peak heating load: 2,731,000 Btu/hr.

Peak cooling load: 391 tons

Present windows - single pane: U = 1.1

New windows - double pane: U = 0.65

It is assumed that the installation of new windows will also reduce the rate of infiltration since all caulking/sealers will be replaced.

Present infiltration rate: 1 cfm per foot of crack

New infiltration rate: 0.75 cfm per foot of crack

Total crack length: 5832 feet

Installed Equipment:

Boilers:

Two 50 HP firetube hot water boilers - rated capacity 1,674,000 Btu/hr. output each

Fuel - No. 2 oil, 140,000 Btu/gal.

Efficiency at rated capacity: 80%

Seasonal efficiency: 65%

Water temperature from boiler: 240°F

Return temperature to boiler: 210°F

Hot water pump: 189 gpm @ 50 ft. head, 1750 rpm

Chillers:

Two 200-ton electric driven centrifugal chillers

Coefficient of Performance at rated capacity: 4.0

Seasonal Coefficient of Performance: 2.5

Water temperature leaving chiller: 44°F

Water temperature entering chiller: 54°F

Chilled water pump: 940 gpm @ 90 ft. head, 1750 rpm

Systems:

It is assumed that each floor is provided with its own air handling system with centrifugal fan providing 22,200 cfm at 3 inch static pressure, 1000 rpm.

We will first determine the magnitude of the reduction of heating and cooling loads and then reflect those reductions with modifications to the installed heating and cooling systems.

Effect on Heating Load:

The effect of the installation of double pane windows on the heating load will be twofold. First and foremost, there will be a reduction of conduction losses. Second, the load due to infiltration will be reduced.

Conduction/Solar

Entering Figure #34 at 6,000 degree days and intersecting with the 325 Langley and north exposure single glazing line, the present heat loss is 122 x 10^3 Btu/sq.ft. per year. Since the west exposure is shaded, the same figure is used for heat loss through the west windows. The effect of solar radiation is included regardless of whether the exposure is shaded or unshaded to account for diffuse radiation.

Figure 34 is used similarly to determine total heat loss through all four exposures for single and double glazing and is summarized in the following table:

EXPOSURE	AREA SQ. FT.	SINGLE GLAZING YEARLY HEAT LOSS Btux10^3/sq. ft.	SINGLE GLAZING TOTAL HEAT LOSS Btux10^6	DOUBLE GLAZING YEARLY HEAT LOSS Btux10^3/sq. ft.	DOUBLE GLAZING TOTAL HEAT LOSS Btux10^6	SAVINGS BTUX10^6
NORTH	3500	122	427	65	228	199
SOUTH	3500	85	298	45	158	140
EAST	3500	100	350	55	193	157
WEST	3500	122	427	65	228	199
				TOTAL SAVINGS		695 x 10^6 Btu/Yr.

Infiltration:

Present infiltration rate: 1 cfm/ft. x 5832 ft. = 5832 cfm

New infiltration rate: 0.75 x 5832 = 4374 cfm

Reduction in infiltration = 5832 - 4374 = 1458 cfm

Savings due to reduction in infiltration are determined from Figure 17 using 68°F for 40 hour/week and 58°F for 128 hour/week and 6000 degree days.

Results are:

Occupied hours:	(40×10^6) (1458/1000) =	58.3×10^6
Unoccupied hours:	(100×10^6) (1458/1000) =	145.8×10^6
		204.1×10^6 Btu/year

Total reduction in heat load due to installation of double pane windows:

Conduction/solar savings	695×10^6
Infiltration savings	204.1×10^6
Total Savings	899.1×10^6 Btu/year

Total gallons of oil saved is 9880 gallons. The new peak heating load after installation of double pane windows is calculated to be 2251×10^3 Btu/hr.

Effect on Cooling Load

The effect of the installation of double pane windows on the cooling load will be to reduce solar and conduction gains and infiltration rate. It is assumed that the building is cooled during occupied times only.

Conduction Gains:

Entering Figure 51 at 2800 degree hours and intersecting with 40 occupied hours per week the difference in heat gain for single and double pane windows is 800 Btu/sq. ft./year. For 14,000 sq. ft. of glazing, the yearly savings is 11.2×10^6 Btu/year.

Solar Gains:

Entering Figure 50 at 2800 degree hours and intersecting with 325 Langleys, yearly solar heat gains are determined and tabulated below:

EXPOSURE	AREA SQ. FT.	SINGLE GLAZING YEARLY HEAT GAIN $Btux10^3$/sq. ft.	SINGLE GLAZING TOTAL HEAT GAIN $Btux10^6$	DOUBLE GLAZING YEARLY HEAT GAIN $Btux10^3$/sq. ft.	DOUBLE GLAZING TOTAL HEAT GAIN $Btux10^6$	SAVINGS $BTUX10^6$
NORTH	3500	40	140	35	122.5	17.5
SOUTH	3500	112	392	93	325.5	66.5
EAST	3500	114	399	107	374.5	24.5
WEST	3500	114	399	107	374.5	24.5
					TOTAL SAVINGS	$133x10^6$ Btu/yr.

Infiltration:

Reduction in infiltration = 1458 cfm

Savings due to reduction in infiltration are determined from Figure 21 using 50% RH for 40 hours/week operation.

Entering with 2000 degree hours, the energy used is (6×10^6) (1458/1000) = 8.75×10^6 Btu/year.

Total reduction in cooling load due to installation of double pane windows:

Conduction savings: 11.2×10^6

Solar savings: 133×10^6

Infiltration savings: 8.75×10^6

Total Savings 152.95×10^6

The new peak cooling load is calculated at 376 tons.

We are now able to make other modifications to the heating and cooling systems to take advantage of the reductions in peak loads.

Heating System Modifications:

Reduce the water flow rate.

Peak load reduction enables us to reduce the water flow rate from 189 gpm to 150 gpm. From the pump laws, the new operating point is determined to be 150 gpm at 31.5 ft. head and 1389 rpm. Entering Figure 61 for the initial and new operating points gives a savings in pumping energy of 25 x 10^6 Btu/year for 4000 hours of operation. This is equivalent to an electrical energy saving of 7325 KWH.

Reduce the air volume.

As a result of the load reduction and from the fan laws, it is determined that the air volume may be reduced from 22,000 cfm to 18,138 cfm at 2 inches pressure and 817 rpm. Savings due to the reduction in fan volume will be determined later as further modifications are likely.

Cooling System Modifications:

Reduce the water flow rate.

The chilled water pump flow rate can be reduced from 940 gpm to 903 gpm to compensate for the load reduction. The pump laws show that the new operating point is 903 gpm at 83 ft. head and 1683 rpm. From Figure 62, the savings in pumping energy is 10 x 10^6 Btu/year for 1000 hours of operation which is equivalent to 2930 KWH.

Reduce the air volume.

The reduction in cooling load will allow the air volume to be reduced to 21,000 cfm on each floor. However, we have previously determined that the new heating load requires an air volume of 18,138 cfm. It is desirable to match the air volumes for both heating and cooling loads as closely as possible, therefore, it would be advisable to reduce the cooling load further.

Remove unnecessary lighting.

In order to effect a further reduction in cooling load and to balance the air volumes for the heating and cooling seasons, lighting is reduced by an average of 1 watt/sq. ft. giving an additional reduction of the cooling load of 307 x 10^3 Btu/hr. The air volume required to meet the new cooling load is 18,950 cfm. The total reduction in cooling load, hence the total savings in electrical energy is equal to the sum of the cooling load reduction due to the installation of double pane windows (152.95 x 10^6 Btu/yr.) plus the the reduction due to the removal of lighting (307 x 10^3 Btu/yr). Applying the seasonal COP of 2.5 and converting to KWH, the yearly savings is 17,962 KWH.

Reducing the lighting load has the effect of increasing the peak heating load since it was calculated for occupied periods. An air volume of 20,510 cfm is required to meet the revised peak, however, by increasing the hot air temperature differential by 2°F, we can meet the peak load with the same air volume required for the cooling load.

Savings due to reduction in air volume.

At the initial conditions of 22,200 cfm at 3 inch pressure, electrical energy input is 15 KW per hour of operation or 60,000 KWH for 4000 hours. From the fan laws, the new operating point is determined to be 18,950 cfm at 2.2 inches pressure and 854 rpm. Figure 24 may be used to determine energy consumption for the new operating conditions - 100 x 10^6 Btu or 29,300 KWH. The savings in electrical input is 30,700 KWH for each fan or a total of 184,200 KWH for all six fans.

Savings during the cooling season are based on 1000 hours of operation and are 4750 KWH per fan or 28,500 KWH total.

SUMMARY OF SAVINGS

ITEM	ENERGY SAVED GALLONS/OIL OR KWH/YEAR	IMPLEMENTATION COST
1. Heating load reduction	9,880 gal. or 1,383.2x10^6Btu	$175,000
2. Cooling load reduction	17,962 KWH	Included in Items #1 & 5
3. Hot water pump modification	7,325 KWH	$ 100
4. Chilled water pump modification	2,930 KWH	$ 100
5. Reduction in lighting	290,304 KWH	$ 10,000
6. Reduction in air volume	212,700 KWH	$ 1,000
TOTALS	531,221 KWH or 1813.1x10^6 Btu	$186,200
	3196.3 x 10^6 Btu	

Economic Analysis

As mentioned in Section 23, Economic Analysis, the effects of some energy conservation improvements are not cumulative, but rather have a diminishing returns effect as the heating/cooling loads are progressively reduced. However, in the example under consideration, the only item which has a positive/negative effect is the reduction in lighting levels which reduces the cooling load but increases the heating load. This effect was taken into account when the fan volumes were modified.

The Commercial Building Ownership Model will be utilized as follows to determine the economic feasibility of the proposed energy conservation investment.

____
(Building)

Table A. ENERGY CONSERVATION INVESTMENT ANALYSIS SUMMARY COMMERCIAL BUILDING OWNERSHIP

	Type of Investment			
1. After-Tax Rate of Return (Table C)	12.2%			
2. Payback Period in Years (Table B-Line 14)	7			
3. Maximum Cash Required (Table B-Line 14)	$26,200			

Note: Cost saving resulting from multiple energy conservation investments in a single building will be less than the cumulative cost savings from each investment, unless the engineering data used to compute energy savings took into consideration the diminishing returns effect of multiple energy conservation improvements to a single structure.

TABLE B: ENERGY CONSERVATION INVESTMENT COMPUTATION OF CASH FLOW—COMMERCIAL BUILDING OWNERSHIP

____
(Building)

____
(Type of Investment)

Line Number	Reduction (Increases) in Operating Expenses	Year											
		0	1	2	3	4	5	6	7	8	9	10	Total
1	Energy Savings (Table E-1)		$19,492	$20,660	$21,892	$23,216	$24,607	$26,080	$27,658	$29,308	$31,072	$32,927	$256,912
2	Labor Savings												
3	Materials Savings												
4	Property Taxes												
5	Other												
6	Net Increase (Decrease) in Rental Income (Table E-2)												
7	Net Operating Profit (Line 1 + 2 + 3 + 4 + 5 + 6)		19,492	20,660	21,892	23,216	24,607	26,080	27,658	29,308	31,072	32,927	256,912
8	Net Investment (Table E-3)	$26,200											26,200
9	Debt Service (Table E-4)		13,274	13,274	13,274	13,274	13,274	13,274	13,274	13,274	13,274	13,274	132,740
10	Pre-Tax Cash Flow (Line 7-8-9)	(26,200)	6,218	7,386	8,618	9,942	11,333	12,806	14,384	16,034	17,798	19,653	97,972
11	Income Tax (Table D-Column F)		2,716	3,557	4,460	5,444	6,499	7,639	8,880	10,211	11,660	13,223	74,289
12	Investment Tax Credit (Table E-8)												
13	After-Tax Cash Flow (Line 10 - 11 + 12)	(26,200)	3,502	3,829	4,158	4,498	4,834	5,167	5,504	5,823	6,138	6,430	23,683
14	Cumulative After-Tax Cash Flow (Cumulative Total of Line 13)	(26,200)	(22,698)	(18,869)	(14,711)	(10,213)	(5,379)	(212)	5,292	11,115	17,253	23,683	

____
(Building)

____
(Type of Investment)

Table C. ENERGY CONSERVATION INVESTMENT
COMPUTATION OF AFTER-TAX RATE OF RETURN
COMMERCIAL BUILDING OWNERSHIP

Year	After Tax Cash Flow (Table B, Line 13)	12 % Discount Factor	Present Value	13 % Discount Factor	Present Value
0	$(26,200)	1.000	$(26,000)	1.000	$(26,000)
1	3,502	0.892	3,124	0.884	3,096
2	3,829	0.797	3,052	0.783	2,998
3	4,158	0.711	2,956	0.693	2,881
4	4,498	0.635	2,856	0.613	2,757
5	4,834	0.567	2,741	0.542	2,620
6	5,167	0.506	2,615	0.480	2,480
7	5,504	0.452	2,488	0.425	2,339
8	5,823	0.403	2,347	0.376	2,189
9	6,138	0.360	2,210	0.332	2,038
10	6,430	0.321	2,064	0.294	1,890
Total			253		(912)

Table D. ENERGY CONSERVATION INVESTMENT COMPUTATION OF INCOME TAX

COMMERCIAL BUILDING OWNERSHIP

__________ (Building)

__________ (Type of Investment)

Year	(a) Net Operating Profit (Table B-Line 7)	(b) Depreciation (Table E-6)	(c) Interest (Table E-4)	(d) Gain (Loss) on Disposal of Retired Asset (Table E-7)	(e) Taxable Income (Column a-b-c+d)	(f) Income Tax @ % (Table B-Line 11)
0						
1	$19,492	$5,060	$9,000		$ 5,432	$ 2,716
2	20,660	5,060	8,487		7,113	3,557
3	21,892	5,060	7,913		8,919	4,460
4	23,216	5,060	7,269		10,887	5,444
5	24,607	5,060	6,549		12,998	6,499
6	26,080	5,060	5,742		15,278	7,639
7	27,658	5,060	4,838		17,760	8,880
8	29,308	5,060	3,826		20,422	10,211
9	31,072	5,060	2,692		23,320	11,660
10	33,9[illegible]7	5,060	1,422		26,445	13,223

(Building)

(Type of Investment)

Table E. ENERGY CONSERVATION INVESTMENT DATA SHEET COMMERCIAL BUILDING OWNERSHIP

1. Projected Energy Savings (Table B-Line 1)
(ELECTRICAL)

Year	(a) BTU Savings Per Year (000,000)	(b) Projected Cost Per Million BTU	(c) Energy Savings (Column a X b)
0			
1	1813.06	$ 8.79	$15,937
2	1813.06	9.32	16,898
3	1813.06	9.87	17,895
4	1813.06	10.47	18,983
5	1813.06	11.10	20,125
6	1813.06	11.76	21,322
7	1813.06	12.47	22,609
8	1813.06	13.22	23,969
9	1813.06	14.01	25,401
10	1813.06	14.85	26,924

2. Increase in Rental Income (Table B-Line 6)

Year	(a) Rent Loss During Construction	(b) Rent Loss Due to Decrease In Rentable Space	(c) Rent Increase (Decrease) in Remaining Space	(d) Net Increase (Decrease) in Rental Income (Column c-a-b)
0				
1				
2				
3				
4				
5				
6				
7				
8				
9				
10				

____
(Building)

____
(Type of Investment)

Table E. **ENERGY CONSERVATION INVESTMENT DATA SHEET (continued) COMMERCIAL BUILDING OWNERSHIP**

1. Projected Energy Savings (Table B-Line 1)
(FUEL OIL)

Year	(a) BTU Savings Per Year (000,000)	(b) Projected Cost Per Million BTU	(c) Energy Savings (Column a X b)
0			
1	1383.2	$2.57	$3,555
2	1383.2	2.72	3,762
3	1383.2	2.89	3,997
4	1383.2	3.06	4,233
5	1383.2	3.24	4,482
6	1383.2	3.44	4,758
7	1383.2	3.65	5,049
8	1383.2	3.86	5,339
9	1383.2	4.10	5,671
10	1383.2	4.34	6,003

2. Increase in Rental Income (Table B-Line 6)

Year	(a) Rent Loss During Construction	(b) Rent Loss Due to Decrease In Rentable Space	(c) Rent Increase (Decrease) in Remaining Space	(d) Net Increase (Decrease) in Rental Income (Column c-a-b)
0				
1				
2				
3				
4				
5				
6				
7				
8				
9				
10				

____
(Building)

____
(Type of Investment)

Table E. ENERGY CONSERVATION INVESTMENT DATA SHEET (continued) COMMERCIAL BUILDING OWNERSHIP

3. Financing

Purchase price	$ 101,200
Less:	
Salvage proceeds received from sale of old asset	--
Amount financed	75,000
Net Investment (Table B-Line 8)	$ 26,200

4. Debt Service

Amortization of Amount Financed

Payback period in years	10
Annual interest rate	12 %

DEBT SERVICE SCHEDULE

Year	Interest (Table D Column c)	Principal	Total Debt Service (Table B-Line 9)	Remaining Principal Balance
0				75,000
1	$9,000	$ 4,274	$13,274	70,726
2	8,487	4,787	13,274	65,939
3	7,913	5,361	13,274	60,578
4	7,269	6,005	13,274	54,573
5	6,549	6,725	13,274	47,848
6	5,742	7,532	13,274	40,316
7	4,838	8,436	13,274	31,880
8	3,826	9,448	13,274	22,432
9	2,692	10,582	13,274	11,850
10	1,422	11,850	13,274	0

(Building)

(Type of Investment)

Table E. ENERGY CONSERVATION INVESTMENT DATA SHEET (continued) COMMERCIAL BUILDING OWNERSHIP

Basis of New Asset

Purchase price $ 101,200

Plus: Book value (original cost less accumulated depreciation) of old asset replaced by new asset

Less: Trade-in received on old asset replaced by new asset

Basis of new asset (Use this figure for completing the depreciation schedule shown in Table E-6.) $ 101,200

Depreciation

Useful life in years (not longer than remaining life of building or lease) 20

Depreciation method used Straight Line

DEPRECIATION SCHEDULE

Year	Depreciation Expense (Table D-Column b)	Undepreciated Balance
0		$101,200
1	$5,060	96,140
2	5,060	91,080
3	5,060	86,020
4	5,060	80,960
5	5,060	75,900
6	5,060	70,840
7	5,060	65,780
8	5,060	60,720
9	5,060	55,660
10	5,060	50,600

____
(Building)

____
(Type of Investment)

Table E. ENERGY CONSERVATION INVESTMENT DATA SHEET (continued) COMMERCIAL BUILDING OWNERSHIP

7. Gain (Loss) on Disposal of Retired Asset

Salvage proceeds received from disposal of retired asset	$ 0
Less: Original cost of retired asset	-
Plus: Accumulated depreciation on retired asset	-
Gain (loss) on disposal of retired asset (Table D-Column d)	$ 0

8. Investment Tax Credit

Eligible: Yes ____ No X

Useful life in years ____________

Amount of Credit: (Table B-Line 12)

Useful life less than 3 years	-0-
Useful life at least 3 years but less than 5 years (purchase price X .07 X .333)	
Useful life at least 5 years but less than 7 years (purchase price X .07 X .667)	
Useful life greater than 7 years (purchase price X .07)	

APPENDIX D

SOURCES FOR COMPUTER PROGRAMS TO CALCULATE LOADS & ENERGY USE

American Gas Association
1515 Wilson Boulevard
Arlington, Virginia 22209
PROGRAM: E. CUBE

Automated Procedures for Engineering Consultants, Inc.
Grant Deneau Tower
Suite M-15
Dayton, Ohio 45402
PROGRAM: HCC-III
Lighting II
Piping
Duct

Charles D. Beach, P. C.
1515 Mariposa Avenue
Boulder, Colorado 80302
PROGRAM: Development Work on Solar Simulations for Westinghouse

University of California, Los Angeles
School of Architecture and Urban Planning
Los Angeles, California 90024
PROGRAM: CADS: Computer-Aided Design Systems

Caudill Rowlett Scott
1111 West Loop South
Houston, Texas
PROGRAM: Energy Analysis

Colorado State University
University Computer Center
Fort Collins, Colorado 80523
PROGRAM: Simulation Model for Solar Heated and Cooled Buildings (SIMSHAC)

Consultant Computation Bureau
594 Howard Street
San Francisco, California 94105
PROGRAM: Post Office Program

Dalton, Dalton, Little & Newport
3650 Warrensville Center Road
Cleveland, Ohio 44122
PROGRAM: See Program Lists

Dover Corporation
Elevator Division
P.O. Box 2177
Memphis, Tennessee 38101
PROGRAM: Elevator Design

Dubin-Mindell-Bloome Associates
42 West 39th St.
New York, N.Y. 10018
PROGRAM: Sunset

Electric Energy Association
90 Park Avenue
New York, New York 10016
PROGRAM: Access

Environmental Research Lab., Inc.
Thunderbird Industrial Air Park
7410 E. Helm Drive
Scottsdale, Arizona 85260
PROGRAM: Lighting

University of Florida
Solar Energy Department
237 Mechanical Engineering Building
Gainesville, Florida 32601
PROGRAM: Energy Program

General American Transportation Corporation
7449 N. Natches
Niles, Illinois 60648
PROGRAM: Post Office Program

General Electric Company
Lincoln Tower
P.O. Box 1661
Louisville, Kentucky 40210
PROGRAM: Energy Program

Giffels Associates, Inc.
Marquette Building
243 W. Congress
Detroit, Michigan 48226
PROGRAM: See Program Lists

Isaac Goodbar
Edison Price Inc.
409 East 60th St.
New York, N.Y. 10022
PROGRAM: Lighting Design

Hitman Associates
9190 Red Branch Road
Columbia, Maryland 21043
PROGRAM: Post Office Program

Honeywell, Inc.
1500 West Dundee Road
Arlington Heights, Illinois 60004
PROGRAM: Energy Program

Honeywell, Inc.
2600 Ridgeway Parkway
Minneapolis, Minnesota 55413
PROGRAM: NBSLD (Honeywell Version)
HELHAK-1; Parabolic Trough Collector
Flat Plate Collector Program

Honeywell, Inc.
Residential Division
2600 Ridgeway Parkway
Minneapolis, Minnesota 55413
PROGRAM: Residential and Small Commercial Control Simulation

IBM
Real Estate and Construction Division
1000 Westchester
White Plains, New York 10614
PROGRAM: Energy Analysis

Illumination Computing Service
P. O. Box 162
Arcadia, California 91006
PROGRAM: Lighting Analysis (ESI)

Johnson Service Company
507 E. Michigan Street
Milwaukee, Wisconsin 53201
PROGRAM: Control Systems

Libbey-Owens-Ford
811 Madison
Toledo, Ohio 43695
PROGRAM: Glazing Comparison
Daylighting Program

Mechanical Engineering Data Services
30 Finler Drive
Maryland Heights, Missouri 63043
PROGRAM: MEDSI Energy Program

Ross F. Meriwether & Associates, Inc.
Northwood Executive Building
1600 Northeast Loop 410
San Antonio, Texas 78209
PROGRAM: Library of Energy Programs

University of Michigan
College of Architecture & Urban Planning
Ann Arbor, Michigan 48104
PROGRAM: Energy Program

NASA
Urban Systems Project Office
Code EZ
Johnson Space Center
Houston, Texas 77058
PROGRAM: Modified Post Office Program

National Bureau of Standards
Institute for Applied Technology
Washington, D. C. 20234
PROGRAM: NBSLD

Otis Elevator Company
260 Eleventh Avenue
New York, New York 10001
PROGRAM: Elevator Design

Pennsylvania State University
Department of Architectural Engineering
University Park, Pennsylvania 16802
PROGRAM: B.E.A.P.: Building Energy Analysis

Perry, Dean & Stewart
955 Park Square
Boston, Massachusetts 02116
PROGRAM: ARK-2

PPG Industries
One Gateway Center
Pittsburgh, Pennsylvania 15222
PROGRAM: Building Cost Analysis

Post Office Program
SEE:
Consultants Computation Bureau
General American Transportation Corporation
Hitman Associates
National Aeronautics and Space Administration

Powers Regulator Co.
3400 Oakton Street
Skokie, Illinois 60076
PROGRAM: Control Systems

Robertshaw Control Company
1800 Glenside Drive
Richmond, Virginia 23226
PROGRAM: Control System

Skidmore, Owings and Merrill
30 West Monroe Street
Chicago, Illinois 60603
PROGRAM: Building Optimization Program (BOP)

Smith, Hinchman & Grylls Associates
455 West Fort Street
Detroit, Michigan 48226
PROGRAM: Lumen II

University of Southern California
Department of Architecture
University Park
Los Angeles, California 90007
PROGRAM: Form Generation

University of Texas
Numerical Simulation Laboratory
School of Architecture
Austin, Texas 78712
PROGRAM: DEROB
SUNNY
DYNAWALB

TRANE Company
3600 Pommel Creek Road
LaCrosse, Wisconsin 54601
PROGRAM: T.R.A.C.E.; Equipment Selection

Westinghouse Corporation
Building Environmental Systems
Westinghouse Building
Gateway Center
Pittsburgh, Pennsylvania 15222
PROGRAM: Energy Program

Westinghouse Electric Company
Box 1693
Baltimore, Maryland 21203
Mail Stop: 999
PROGRAM: Flat Plate Collector Performance
Solar System Simulation

University of Wisconsin
Solar Energy Lab
Engineering Research Building
1500 Johnson Drive
Madison, Wisconsin 53706
PROGRAM: TRNSYS

WTA Computer Services, Inc.
2357 59th Street
St. Louis, Missouri 63110
PROGRAM: H.A.C.E.: Heating & Air
Conditioning Energy Analysis
Program

APPENDIX E
ENERGY CONSERVATION EXAMPLE

ENERGY SAVINGS FOR TYPICAL OFFICE BUILDING

A. ASSUMPTIONS:

OFFICE BUILDING

Location	-	Chicago, Illinois
Gross area per floor	=	10,000 sq.ft. (100' x 100')
Height	=	
Floor-to-floor height	=	10 Stories (100,000 sq.ft. total area)
Window/wall area ratio	=	33%

CONDITIONS BEFORE OPERATIONAL CHANGES

Windows	U = 1.1 Clear single glazing
Shading by curtains:	
Cooling season	SC = .67
Heating season	SC = 1.0
Wall	U = 0.3 (Excluding glazed areas)
Roof	U = 0.2
Outdoor Ventilation Air	= 19,800 cfm @ 30 cfm/person during occupied periods only.
Average Infiltration	= 1/2 air change/hr.
Occupancy	= 45 hrs/week
Occupants	= 660 people

Interior Heat Gains:	
Lights	= 4.0 W/sq.ft.
Office Equipment	= 0.5 "
Fans	= 1.0 "
Total	= 5.5 W/sq.ft.

Boiler Efficiency	= 60%
Chiller COP	= 3.0 (Average)
Indoor Temperature	= 75° Heating Season (No setback)
	= 75° Cooling Season @ 50% RH (No cooling during unoccupied periods)

Domestic Hot Water Flow Rate	= 2 gpm 2 140°F
Lighting Level	= 80 fc
Ventilation Rate	= 150,000 cfm

B. SUMMARY OF ENERGY USED IN CHICAGO OFFICE BUILDING

1. Before Conservation Measures Implemented

a. Site Energy

Heating (Oil)	106,00	Btu/sq.ft./yr.
Cooling (Electric)	12,800	"
Lighting "	31,400	"
Power "	13,000	"
Domestic Hot Water (Oil)	4,400	"
Total	168,000	Btu/sq/ft./yr.

b. Raw Source Energy

Heating	106,400	Btu/sq.ft./yr.
Cooling*	42,600	"
Lighting*	104,600	"
Power*	43,300	"
Domestic Hot Water	4,400	"
Total	301,300	Btu/sq.ft./yr.

2. After Selected Conservation Measures Implemented

(Measures include No. 12, 21, 27, 31 and 34 of Part C)

a. Site Energy

Heating	28,800	Btu/sq.ft./yr.
Cooling	5,900	"
Lighting	12,400	"
Power	9,700	"
Domestic Hot Water	2,100	"
Total	58,900	Btu/sq.ft./yr.

b. Raw Source Energy

Heating	28,800	Btu/sq.ft./yr.
Cooling*	19,600	"
Lighting*	41,300	"
Power*	32,300	"
Domestic Hot Water	2,100	"
Total	124,100	Btu/sq./ft.

* The figures for raw source energy reflect the true national energy cost for electricity. The equivalent Btu's have been modified to account for the efficiency of conversion from thermal energy to electrical energy at the generating station. Generating efficiency has been taken as 30%.

C. ANALYSIS OF ENERGY SAVINGS FOR CHICAGO OFFICE BUILDING

HEATING (76,000 gal. #2 oil/yr.)	ANNUAL ENERGY SAVINGS	SITE ENERGY SAVINGS Btu/sq/ft/yr	% SAVINGS OF SITE ENERGY	ESTIMATED INITIAL COST
1. 12° Night Setback	27,000 gal.	38,000	23%	$ 7,500
2. Caulk Windows	10,000 "	13,800	8%	11,880
3. Increase Boiler Efficiency From 60% to 66%	6,500 "	9,200	5%	500
4. Reduce Outside Air During Occupied Hours	1,800 "	2,500	1%	- ---
5. 7° Daytime Setback	1,000 "	1,400	1%	- ---
6. (1 + 5)	28,000 "	39,200	23%	7,500
7. (1 + 2 + 3 + 4 + 5)	39,000 "	54,600	32%	19,880
8. Add Night Barriers U=0.1	29,000 "	41,000	24%	71,280
9. Add Storm Windows	26,000 "	36,400	22%	54,780
10. (8 + 3)	33,300 "	46,600	28%	71,780
11. (9 + 3)	30,500 "	42,700	25%	55,280
12. (1 + 2 + 3 + 4 + 5 + 8)	55,400 "	77,500	46%	91,160
COOLING (375,000 KWHR/yr.)				
13. Reduce Outside Air During Occupied Hrs in Cooling Season	11,000 KWHR	400	1%	- ---
14. Raise Room Temperatures and Humidity to 78°F, 55% RH	18,400 "	600	---	- ---
15. Raise Chilled Water Temperature 5°	18,750 "	600	---	- ---
16. Clean Chiller Condenser	37,500 "	1,300	1%	1,200
17. (13 + 14 + 15 + 16)	76,000 "	2,600	2%	1,200
18. Reflective Coating(SC-.25) Applied to Glass on S,E.& W. exposures	32,000 "	1,100	1%	16,000
19. Economizer Cycle	103,000 "	3,500	2%	8,700
20. Reduce Lighting to 3.0 W/sq.ft.	33,000 "	1,100	1%	- ---
21. (13 + 14 + 15 16 + 18 + 19 + 20)	201,000 "	6,800	4%	25,900

ENERGY SAVINGS FOR TYPICAL OFFICE BUILDING APPENDIX

	ANNUAL ENERGY SAVINGS	SITE ENERGY SAVINGS Btu/sq/ft/yr	% SAVINGS OF SITE ENERGY	ESTIMATED INITIAL COST
LIGHTING (922,000 KWHR/yr)				
22. Remove 25% of Bulbs	225,000 KWHR	7,700	5%	$ 1,600
23. Shut off Perimeter Lights when Daylight is Available	165,000 "	5,600	8%	6,000
24. Replace 40W Bulbs with 35 W	112,000 "	3,800	2%	2,000
25. Shut Off Lights for 1 hour/day	100,000 "	3,400	2%	1,500
26. Install New System @ 2W/sq.ft.	450,000 "	15,300	9%	200,000
27. (23 + 25 + 26)	560,000 "	19,100	11%	207,500
POWER (380,000 KWHR/yr)				
28. Modify duct system to reduce static pressure	54,000 "	1,800	1%	11,250
29. Shut Off Chilled Water Pump Condenser Water Pump and Cooling Tower Fan When Chiller Is Off	39,000 "	1,300	--	- ---
30. Shut Off Radiation Pumps	5,000 "	200	--	- ---
31. (28 + 29 + 30)	98,000	3,300	2%	11,250
DOMESTIC HOT WATER (3,100 gal.oil/yr.				
32. Lower Water Temperature to 100°F	750 gal.	1,000	--	- ---
33. Reduce Flow Rate to 1/2 gpm	750 "	1,000	--	1,500
34. (32 + 33)	1,500 "	2,000	1%	1,500
TOTALS				
35. (12 + 21 + 27 + 31 + 34)	56,900 gal. 859,000 KWHR	109,100	65%	$337,318

D. EFFECTS OF SELECTED MODIFICATIONS

The effects of various operational and structural modifications were computed as savings in:

1) Amount of fuel consumed on-site annually

2) Per cent of each segment of energy consumed

3) Per cent of total energy consumed on-site annually.

While comparing the absolute energy savings for various modifications, it should be borne in mind that the figures in this analysis are valid only for buildings in Chicago, or a similar climate. For example, the small effect that cooling modifications have in this example on overall energy consumption for this building is due to the fact that cooling itself is a relatively small segment of the overall energy consumption here. This is entirely a function of climate and internal loads. The same building in a warmer climate with the same cooling modifications would have a higher savings of primary energy and show a higher percentage savings of the total, while the relative percentage savings among cooling items might remain the same. For this reason, the percentage reduction in each segment of energy provides a useful measure for evaluating the effectiveness of each option in any climate.

These examples are, of course, not inclusive, but merely illustrate the relative effectiveness of certain measures.

The savings shown for each option reflect only the savings for that segment of energy under which the item is listed. When an item, such as lighting, affects savings in more than one segment, it is listed under each one. Since savings for individual items are not directly additive, savings for selected combinations of options are included to show sample optimal savings resulting from the interaction between several modifications.

GLOSSARY

Absorption Chiller: A refrigeration machine using heat as the power input to generate chilled water.

Absorption Coefficient: The fraction of the total radiant energy incident on a surface that is absorbed by the surface.

Absorptivity: The physical characteristic of a substance describing its ability to absorb radiation.

Ambient: Surrounding (i.e. ambient temperature is the temperature in the surrounding space).

Activated Carbon: A form of carbon capable of absorbing odors and vapors.

Air Changes: Expression of ventilation rate in terms of room or building volume. Usually air changes/hour.

Ballast: A device used in a starting circuit for fluorescent and other types of lamps.

Blow Down: The discharge of water from a boiler or cooling tower sump that contains a high proportion of total dissolved solids.

British Thermal Unit (Btu): A heat unit equal to the amount of heat required to raise one pound of water one degree Fahrenheit.

Building Envelope: All external surfaces which are subject to climatic impact; for example, walls, windows, roof, floor, etc.

Building Load: Heating load is the rate of heat loss from the building at steady state conditions when the indoor and outdoor temperatures are at their selected design levels (design criteria). The heating load always includes infiltration and may include ventilation loss and heat gain credits for lights and people.

Cooling load is the rate of heat gain to the building at a steady state condition when indoor and outdoor temperatures are at their selected design levels, solar gain is at its maximum for the building configuration and orientation, and heat gains due to infiltration, ventilation, lights, and people are present.

Cavity Ratio: Number indicating room cavity proportions, which is calculated using length, width, and height.

Centrifugal Chiller: A refrigeration machine using mechanical energy input to drive a centrifugal compressor to generate chilled water.

Centrifugal Fan: Device for propelling air by centrifugal action. Forward curved fans have blades which are sloped forward relative to direction of rotation. Backward curved fans have blades which are sloped backward relative to direction of rotation. Backward curved fans are generally more efficient at high pressures than forward curved fans.

Coefficient of Performance: Ratio of the tons of refrigeration produced to energy required to operate equipment.

Coefficient of Utilization: Ratio of lumens on a work plane to lumens emitted by the lamps.

Cold Deck: A cold air chamber forming part of a ventilating unit.

Condensate: Water obtained by changing the state of water vapor (i.e. steam or moisture in air) from a gas to a liquid usually by cooling.

Condenser: A heat exchanger which removes latent heat from a vapor changing it to its liquid state. (In a refrigeration chiller the components which reject heat.)

Conductance, Thermal: A measure of the thermal conducting properties of a composite structure such as wall, roof, etc., including the insulating boundary layers of air.

Conductivity, Thermal: A measure of the thermal conducting properties of a single material expressed in units of Btu per inch thickness (sq. ft.) (hour) (degree F temperature difference).

Cooling Tower: Device that cools water directly by evaporation.

Damper: A device used to vary the volume of air passing through an air outlet, inlet, or duct.

Degree Day: The difference between the median temperature of any day and 65°F when the median temperature is less than 65°F.

Degree Hour: The difference between the median temperature for any hour and a selected datum.

Demand Factor: The ratio of the maximum demand of a system, or part of a system, to the total connected load of a system or part of the system under consideration.

Desiccant: A substance possessing the ability to absorb moisture.

Direct Expansion: Generic term used to describe refrigeration systems where the cooling effect is obtained directly from the refrigerant (e.g. refrigerant is evaporated directly in a cooling coil in the air stream).

Disability Glare: Spurious light from any source, which impairs a viewer's ability to discern a given object.

Double Bundle Condenser: Condenser (usually in refrigeration machine) that contains two separate tube bundles allowing the option of either rejecting heat to the cooling tower or to another building system requiring heat input.

Dry Bulb Temperature: The measure of the sensible temperature of air.

Economizer Cycle: A method of operating a ventilation system to reduce refrigeration load. Whenever the outdoor air conditions are more favorable (lower heat content) than return air conditions, outdoor air quantity is increased.

Efficacy of Fixtures: Ratio of usable light to energy input for a lighting fixture or system (lumens/watt).

Energy Requirement: The total yearly energy used by a building to maintain the selected inside design conditions under the dynamic impact of a typical year's climate. It includes raw fossil fuel consumed in the building and all electricity used for lighting and power. Efficiencies of utilization are applied, and all energy is expressed in the common unit of Btu's.

Enthalpy: For the purpose of air conditioning enthalpy is the total heat content of air above a datum usually in units of Btu/lb. It is the sum of sensible and latent heat and ignores internal energy changes due to pressure change.

Equivalent Sphere Illumination: That illumination which would fall upon a task covered by an imaginary transparent hemisphere which passes light of the same intensity through each unit area.

Evaporator: A heat exchanger which adds latent heat to a liquid changing it to a gaseous state. (In a refrigeration system it is the component which absorbs heat.)

Foot-Candle: Energy of light at a distance of one foot from a standard (sperm oil) candle.

Heat Gain: As applied to HVAC calculations, it is that amount of heat gained by a space from all sources, including people, lights, machines, sunshine, etc. The total heat gain represents the amount of heat that must be removed from a space to maintain indoor comfort conditions.

Heat Loss: The sum cooling effect of the building structure when the outdoor temperature is lower than the desired indoor temperature. It represents the amount of heat that must be provided to a space to maintain indoor comfort conditions.

Heat, Latent: The quantity of heat required to effect a change in state.

Heat, Sensible: Heat that results in a temperature change but no change in state.

Heat, Specific: Ratio of the amount of heat required to raise a unit mass of material one degree to that required to raise a unit mass of water one degree.

Heat Pump: A refrigeration machine possessing the capability of reversing the flow so that its output can be either heating or cooling. When used for heating, it extracts heat from a low temperature source and increases the temperature to the point where it can be used.

Heat Transmission Coefficient: Any one of a number of coefficients used in the calculation of heat transmission by conduction, convection, and radiation, through various materials and structures.

Hot Deck: A hot air chamber forming part of a ventilating unit.

Humidity, Relative: A measurement indicating moisture content of air.

Infiltration: The process by which outdoor air leaks into a building by natural forces through cracks around doors and windows, etc. (usually undesirable).

Insolation: The amount of solar radiation on a given plane. Expressed in Langleys or Btu/ft.^{2}

Langley: Measurement of radiation intensity. One Langley = 3.68 Btu/ft.^{2}

Life Cycle Cost: The cost of the equipment over its entire life including operating and maintenance costs.

Load Leveling: Deferment of certain loads to limit electrical power demand to a predetermined level.

Load Profile: Time distribution of building heating, cooling, and electrical load.

Lumen: Unit of luminous flux.

Luminaire: Light fixture designed to produce a specific effect.

Make-Up: Water supplied to a system to replace that lost by blow down, leakage, evaporation, etc.

Manchester, New Hampshire Project: A demonstration building commissioned by GSA (Isaak and Isaak, Architects) and developed by Dubin-Mindell-Bloome Associates to incorporate energy conserving architectural features and mechanical and electrical systems.

Modular: System arrangement whereby the demand for energy (heating, cooling) is met by a series of units sized to meet a portion of the load.

Orifice Plate: Device inserted in a pipe or duct which causes a pressure drop across it. Depending on orifice size it can be used to restrict flow or form part of a measuring device.

Orsat Apparatus: A device for measuring the combustion components of boiler or furnace flue gases.

Piggyback Operation: Arrangement of chilled water generation equipment whereby exhaust steam from a steam turbine driven centrifugal chiller is used as the heat source for an absorption chiller.

Power Factor: Relationship between KVA and KW. When the power factor is unity, KVA equals KW.

R-Value: The resistance to heat flow expressed in units of sq. ft. hour degree F/Btu.

Raw Source Energy: The quantity of energy input at a generating station required to produce electrical energy including all thermal and power conversion losses.

Roof Spray: A system that reduces heat gain through a roof by cooling the outside surface with a water spray.

Seasonal Efficiency: Ratio of useful output to energy input for a piece of equipment over an entire heating and cooling season. It can be derived by integrating part load efficiencies against time.

Software: Term used in relation to computers normally describes computer programs and other intangibles.

Sol-Air Temperature: The theoretical air temperature that would give a heat flow rate through a building surface equal in magnitude to that obtained by the addition of conduction and radiation effects.

Ton of Refrigeration: A means of expressing cooling capacity—1 ton = 12,000 Btu/hour cooling.

'U' Value: A coefficient expressing the thermal conductance of a composite structure in Btu per square foot hour degree F temperature difference.

Veiling Reflection: Reflection of light from a task, or work surface, into the viewer's eyes.

Vapor Barrier: A moisture impervious layer designed to prevent moisture migration.

Wet Bulb Temperature: The lowest temperature attainable by evaporating water in the air without the addition or subtraction of energy.

ABBREVIATIONS

ASHRAE: American Society of Heating, Refrigeration and Air-Conditioning Engineers
CFM: Cubic ft. per minute
COP: Coefficient of performance
CU: Coefficient of utilization
DB: Dry bulb temperature
DX: Direct expansion
ESI: Equivalent sphere illumination
IES: Illuminating Engineers Society
HID: High intensity discharge (lamps)
HZ: Hertz
HVAC: Heating, ventilating and air conditioning
KVA: Kilovoltampere
O.A.: Outside air
P.F.: Power factor
TD: Temperature difference
TE: Total energy (system)
WB: Wet bulb temperature

BIBLIOGRAPHY AND SELECTED REFERENCES

ECM1 and 2

1. The Ad Hoc Committee on Energy Efficiency in Large Buildings. Report to the Interdepartmental Fuel and Committee of the State of New York. Albany: State of New York, March 7, 1973.

2. American Institute of Architects Research Corporation. Energy Conservation in Building Design. Washington, D.C.: American Institute of Architects, May 1, 1974.

3. American Institute of Plant Engineers. Better Ideas for Conserving Energy. Cincinnati, Ohio, August 1, 1974.

4. American Society of Heating, Refrigerating, and Air-Conditioning Engineers. Applications Volume. New York, 1974.

5. American Society of Heating, Refrigerating, and Air-Conditioning Engineers. Design and Evaluation Criteria for Energy Conservation in New Buildings. (Proposed Standard 90-P). New York, 1974.

6. American Society of Heating, Refrigerating, and Air-Conditioning Engineers. Handbook of Fundamentals. New York, 1972.

7. American Society of Heating, Refrigerating, and Air-Conditioning Engineers. Equipment Volume. New York, 1972.

8. American Society of Heating, Refrigerating, and Air-Conditioning Engineers. Systems Volume. New York, 1973.

9. Blackwell, H. R. "Application procedures for evaluation of veiling reflections in terms of ESI", Journal of IES, April 1973.

10. Brown, J. L. and C. Mueller. "Brightness discrimination and brightness Contrast", in Vision and Visual Perception. New York: John Wiley and Sons.

11. Boynton, R. M. and D. E. Boss. "The effect of background luminance upon visual search performance". University of Rochester. Report to the Illuminating Engineering Research Institute.

12. Citizens' Advisory Committee on Environmental Quality. Citizen Action Guide to Energy Conservation. Washington, D. C.: Government Printing Office.

13. Commercial Refrigeration Manufacturer's Association. Retail Food Store Energy Conservation. Washington, D. C.

14. Daylighting Committee of the Illuminating Engineering Society. "Recommended Practice of Daylighting". New York: IES, 1962.

15. Departments of the Air Force, Army, and Navy. Engineering Weather Data. Washington, D. C.: Government Printing Office, 1967.

16. Dorsey, Robert T., "An Extension of the Energy Crisis: More Light on Light-

ing." American Institute of Architects Journal, June, 1972.

17. Dubin, Fred S., "Check-off List on Energy Conservation with Regard to Power." Electrical Consultant, 89:4.

18. Dubin, Fred S., "Energy Conservation in Building Operations." Address to Regional Directors of the Building's Management Division of the General Services Administration, April 3, 1973.

19. Dubin, Fred S., "Energy Conservation Needs New Architecture and Engineering." Public Power, March/April, 1972.

20. Dubin, Fred S., "Energy Conservation through Building Design and a Wiser Use of Electricity." Address to the Annual Conference of the American Public Power Association: San Francisco, California, June 26, 1972.

21. Dubin, Fred S. "If you Want to Save Energy." American Institute of Architects Journal, December, 1972.

22. Dubin, Fred S., and Margot Villecco. "Energy for Architects." Architecture Plus, July, 1973.

23. Dubin-Mindell-Bloome Associates, P.C. A Study of Air Conditioning, Heating, and Ventilating Design for Veterans Administration Hospital Kitchens and Laundries. Washington, D.C. Research Staff, Office of Construction, Veterans Administration, February, 1970.

24. Dubin-Mindell-Bloome Associates, P.C. A Study of Design Criteria and Systems for Air Conditioning Existing Veterans Administration Hospitals, Washington, D.C.: Research Staff, Office of Construction, Veteran's Administration, March, 1971.

25. Dubin-Mindell-Bloome Associates, P.C. Energy Conservation and Office Building Design: Research, Analysis, and Recommendations: Office Building, Manchester, New Hampshire. May 15, 1973.

26. Dubin-Mindell-Bloome Associates, P.C. Energy Conservation Design Guidelines for Office Buildings. Washington, D.C.: General Services Administration, Public Buildings Service, January 1974.

27. Dubin-Mindell-Bloome Associates. Report to Connecticut General Insurance Corporation on Energy Conservation Opportunities. December, 1973.

28. "Energy System Limits Waste." Design Engineering Journal, February, 1974.

29. Fanger, P. O. Thermal Comfort: Analysis and Applications in Environmental Engineering. New York: McGraw-Hill, 1972.

30. Federal Energy Administration. Tips for Energy Savers. Washington, D.C.: Government Printing Office.

31. Federal Energy Office. Project Independence Background Paper. Washington, D. C., February 11-12, 1974.

32. Gatts, Robert R., Robert G. Massey and John C. Robertson (U. S. Department of Commerce, National Bureau of Standards). Energy Conservation, Program Guide for Industry and Commerce. Washington, D.C.: Government Printing Office, 1974.

33. Givoni, B., Man and Architecture. London: Elsevier Publishing Company, 1969.

34. Griffith, J. W., "Analysis of Reflected Glare and Visual Effect from Windows". Presented to the National Technical Conference of the Illuminating Engineering Society. Detroit, Michigan, September 8-13, 1963.

35. "GSA Sets Energy Saving Steps." Design Engineering Journal. May 1974.

36. Handbook of Air-Conditioning, Heating, and Ventilating (Clifford Strock and Richard L. Koral, editors). New York: Industrial Press, 1965.

37. "Heat Recovery Cuts Energy Use". Design Engineering Journal. February 1974.

38. Heating, Piping and Air Conditioning Seminar on "Energy Conservation for Existing Buildings", Proceedings of New York, December 1974.

39. Henderson, R. L., J. F. McNelis, and H. G. Williams. "A survey and analysis of important visual tasks in offices". The Annual Conference of the Illuminating Engineering Society, 1974.

40. IES Lighting Handbook. New York: Illuminating Engineering Society, 1972.

41. "Innovative Lighting Fixture Aids." Actual Specifying Engineer, April, 1972.

42. Jennings, Burgess H., Necati Ozisik, and Lester F. Schutrum. "New Solar Research Data on Windows and Draperies." American Society of Heating Refrigerating and Air Conditioning Engineers Journal reprint.

43. Kaufman, John E. "Energy Utilization: Optimizing the Uses of Energy for Lighting." Lighting Design & Application. October, 1973.

44. "Life Support Systems for a Dying Planet". Progressive Architecture. October 1971.

45. Lighting Survey Committee. "How to Make a Lighting Survey." Report to the Illuminating Engineering Society, November 1963.

46. McClure, Charles J. R., "Optimizing Building Energy Use." American Society of Heating, Refrigerating and Air-Conditioning Engineers Journal. September 1971.

47. Murphy, E. E. and J. G. Dorsett. "Demand Controllers Disconnect Some Loads During Peak Periods." Transmission & Distribution. July 1971.

48. Northeast Utilities Company. Guidelines for Implementing Northeast Utilities Systems' Energy Management Program. Hartford, Connecticut, 1973.

49. Office of Emergency Preparedness. The Potential for Energy Conservation. Washington, D. C.: Government Printing Office, October 1972.

50. Olivieri, Joe B. "Heat Recovery Systems - Part I." Air Conditioning, Heating & Refrigeration News, January 1, 1972.

51. Olivieri, Joe B. "Heat Recovery Systems - Part II." Air Conditioning, Heating & Refrigeration News, January 17, 1972.

52. Phipps, H. Harry. "Energy Systems Analysis - Why and How." Building Systems Design. May 1972.

53. Plant Engineering. "Getting the most from your electrical power system: selected articles". Barrington, Illinois: Plant Engineering Library.

54. Price, Seymour. Air Conditioning for Building Engineers and Managers. New York: Industrial Press, 1970.

55. "The Role of Refrigeration in Energy Conservation and Environmental Protection." Seminar of the Air-Conditioning and Refrigeration Institute: New Orleans, Louisians, January 26, 1972.

56. Ross & Baruzzini, Inc. Energy Conservation Applied to Office Lighting. Washington, D.C.: Federal Energy Administration, April 1, 1975. Contract No. 14-01-0001-1845.

57. Ross & Baruzzini, Inc. Lighting Systems Study. Washington, D.C.: General Services Administration, Public Building Service, March 1974.

58. Saunders, J. E. "The role of the level and diversity of horizontal illumination in an appraisal of a simple office task". Lighting Research and Technology, 1969:1.

59. Shlaer, S. "The relation between visual acuity and illumination". Journal of General Physiology, 1937:21.

60. Stewart, John L. "Report on TC 4.5-Fenestration." American Society of Heating Refrigerating and Air-Conditioning Engineers Journal, November 1972.

61. Syska & Hennessy, Engineers and Tishman Research Corporation. A Study of the Effects of Air Changes and Outdoor Air on Interior Environment, Energy Conservation, and Construction and Operating Costs: Phase II. Washington, D.C.: General Services Administration, December 1973.

62. "Three Hundred Hints to Save Energy." Congressional Record. October 30, 1973.

63. U. S. Department of Commerce, National Bureau of Standards. Technical Options for Energy Conservation in Buildings. Washington, D. C.: Government Printing Office, July 1973.

64. U. S. Department of Commerce, National Oceanic and Atmospheric Administra-

tion, National Weather Service. Operations of the National Weather Service. Silver Springs, Maryland: October, 1970.

SOLAR ENERGY

65. American Society of Heating, Refrigerating, and Air-Conditioning Engineers. "Low Temperature Applications of Solar Energy." New York: ASHRAE, 1967.

66. Barber, E.M. and D. Watson. "Criteria for the Preliminary design of solar heated buildings". Guilford, Conn.: Sunworks, Inc., 1974.

67. Colorado State University and University of Wisconsin. Design and Construction of a Residential Solar Heating and Cooling System. Washington,= D.C.: National Science Foundation, Report No. NSR/RANN/SE/GE-40451/PR/74/2.

68. Daniels, Farrington. Direct Use of the Sun's Energy. New York: Ballantine, 1974.

69. Dubin-Mindell-Bloome Associates. "Specifications of solar collectors for the Federal Office Building, Manchester, New Hampshire". Boston: General Services Administration.

70. Duffie, J. A. and W. A. Beckman. Solar Energy Thermal Processes. New York: John D. Wiley and Sons, 1974.

71. General Electric Company. Final Report of the Solar Heating Experiment on the Grover Cleveland School, Boston, Mass. Washington, D. C.: National Science Foundation, Report No. NSF/RANN/74/064.

72. Report of the Solar Heating and Cooling Committee of BRAB, National Academy of Engineers. Washington, D. C.: National Science Foundation, 1974.

73. Shurcliffe, W. A. "Solar heating buildings, a brief survey", Solar Energy Digest (P.O. Box 17776, San Diego, California 92117).

74. "Solar Energy Industry Report". (1001 Connecticut Ave., N.W., Washington, D. C. 20036).

75. U. S. Department of Commerce, National Bureau of Standards. Performance Criteria for the Solar Heating/Cooling of Residences. Washington, D. C.: Government Printing Office.

76. Yellot, John I. "Utilization of sun and sky radiation for heating and cooling of buildings". ASHRAE Journal, December 1973.

TOTAL ENERGY SYSTEM REFERENCES

77. Ahner, D. J. "Environmental Performance". Based on a paper presented at the General Electric Gas Turbine State of the Art Engineering Seminar, June, 1971.

78, Barrangon, Maurice. "Conservation of Resources, How Total Energy Saves Fuel". GATE Information Digest. March, 1971

79. Becker, Herbert P. "The Concept of Total Energy--Lecture No. 1". Fall Lecture Series. ASHRAE at the United Engineering Center, October, 1967.

80. Bjerklie, John W., P.E. "Small Gas Turbines for Total Energy Systems." Actual Specifying Engineer. August, 1971.

81. Dubin, Fred S., P.E. Editorial, "Life Support Systems for a Dying Planet"; Available Now: Systems that Save Energy"; and "Can Building Codes Help Protect the Environment?" Progressive Architecture. October, 1971. "The New Architecture and Engineering," presented at The Workshop on Total Energy Conservation in Public Buildings. State University of New York, Albany. January, 1972.

82. "TR-2---Total Energy for Schools and Colleges." A technical report prepared for the Educational Facilities Laboratories of the Ford Foundation. "Total Energy Systems for Mass Housing." Paper delivered at the Urban Technology Conference 2. July, 1972. San Francisco.

83. Dubin-Mindell-Bloome Associates. "Total Energy." A technical report from Education Facilities Laboratories. May, 1970.

84. Federal Register. "Standards of Performance for New Stationary Sources." Environmental Protection Agency. December, 1971. Washington, D. C. Vol. 36, No. 247, Part 11.

85. Henderson, John O. "Turbines or Engines for Total Energy?" Caterpillar Tractor Company Gas Age. May, 1966.

86. Holdren, John and Herrera, Philip. <u>Energy</u>, Sierra Club, Copyright 1971.

87. Huber, Ernest G., P. E. "Total Energy Concept Applied to Small Diesel Engines and Gas Turbines." Air Force Civil Engineer. February. 1970. File No. 3207.

88. Kennedy, J. J. "Watch the Basic Details when installing Gas Prime Movers." Plant Engineering. February, 1970. File No. 3202.

89. "Total Energy Evaluated." Energy International. February, 1972.

90. Williams, J. R. "Operational Experience with Total Energy Plants in the U. S. A."

INDEX